AF324263

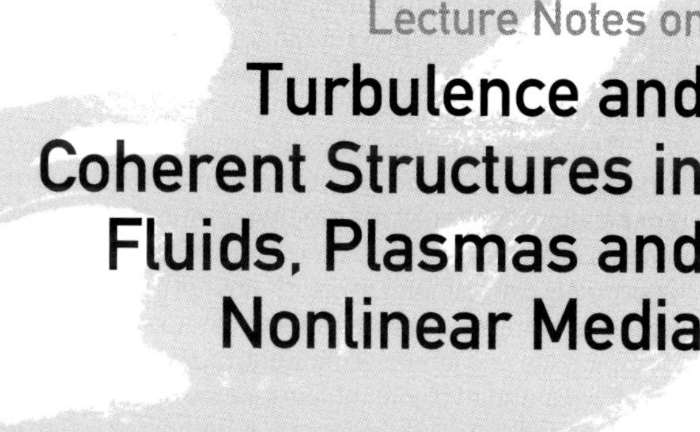

Turbulence and Coherent Structures in Fluids, Plasmas and Nonlinear Media

WORLD SCIENTIFIC LECTURE NOTES IN COMPLEX SYSTEMS

Editor-in-Chief: A.S. Mikhailov, *Fritz Haber Institute, Berlin, Germany*

H. Cerdeira, *ICTP, Triest, Italy*
B. Huberman, *Hewlett-Packard, Palo Alto, USA*
K. Kaneko, *University of Tokyo, Japan*
Ph. Maini, *Oxford University, UK*

AIMS AND SCOPE

The aim of this new interdisciplinary series is to promote the exchange of information between scientists working in different fields, who are involved in the study of complex systems, and to foster education and training of young scientists entering this rapidly developing research area.

The scope of the series is broad and will include: Statistical physics of large nonequilibrium systems; problems of nonlinear pattern formation in chemistry; complex organization of intracellular processes and biochemical networks of a living cell; various aspects of cell-to-cell communication; behaviour of bacterial colonies; neural networks; functioning and organization of animal populations and large ecological systems; modeling complex social phenomena; applications of statistical mechanics to studies of economics and financial markets; multi-agent robotics and collective intelligence; the emergence and evolution of large-scale communication networks; general mathematical studies of complex cooperative behaviour in large systems.

Published

World Scientific Lecture Notes
in Complex Systems – Vol. 4

editors

Michael Shats

Horst Punzmann

Australian National University, Australia

Lecture Notes on
Turbulence and Coherent Structures in Fluids, Plasmas and Nonlinear Media

NEW JERSEY • LONDON • SINGAPORE • BEIJING • SHANGHAI • HONG KONG • TAIPEI • CHENNAI

Published by

World Scientific Publishing Co. Pte. Ltd.

5 Toh Tuck Link, Singapore 596224

USA office: 27 Warren Street, Suite 401-402, Hackensack, NJ 07601

UK office: 57 Shelton Street, Covent Garden, London WC2H 9HE

British Library Cataloguing-in-Publication Data
A catalogue record for this book is available from the British Library.

Lecture Notes on
TURBULENCE AND COHERENT STRUCTURES IN FLUIDS, PLASMAS
AND NONLINEAR MEDIA

ISBN 981-256-698-8

Printed by Mainland Press Pte Ltd

Preface

The problem of turbulence and coherent structures is of key importance in many fields of science and engineering. It is an area which is vigorously researched across a diverse range of disciplines such as theoretical physics, oceanography, atmospheric science, magnetically confined plasma, nonlinear optics, etc. Modern studies in turbulence and coherent structures are based on a variety of theoretical concepts, numerical simulation techniques and experimental methods, which cannot be reviewed effectively by a single expert. The main goal of these lecture notes is to introduce state-of-the-art turbulence research in a variety of approaches (theoretical, numerical simulations and experiments) and applications (fluids, plasmas, geophysics, nonlinear optical media) by several experts.

This book is based on the lectures delivered at the 19th Canberra International Physics Summer School held at the Australian National University in Canberra (Australia) from 16-20 January 2006. The Summer School was sponsored by the Australian Research Council's Complex Open Systems Research Network (COSNet).

The lecturers aimed at (1) giving a smooth introduction to a subject to readers who are not familiar with the field, while (2) reviewing the most recent advances in the area. This collection of lectures will provide a useful review for both postgraduate students and researchers new to the advancements in this field, as well as specialists seeking to expand their knowledge across different areas of turbulence research.

The material covered in this book includes introductions to the theory of developed turbulence (G. Falkovich) and statistical and renormalization methods (D. McComb). The role of turbulence in ocean energy balance is addressed in a review by H. Dijkstra. A comprehensive introduction to the complex area of the theory of turbulence in plasma (J. Krommes) is complemented by a review of experimental methods in plasma turbulence (M. Shats and H. Xia). An introduction to the main ideas and modern capabilities of numerical simulation of turbulence is given by J. Jimenez.

Experimental methods in fluid turbulence studies are illustrated in the lectures by J. Soria describing the particle image velocimetry. Finally, the relatively new field of the physics of vortex flows in optical fields is reviewed by A. Desyatnikov.

The Summer School in Canberra was accompanied by a workshop on the same topic. The Workshop Proceedings (editors J. Denier and J. Frederiksen) will also be published by World Scientific under the same title as these Lecture Notes ("Turbulence and Coherent Structures in Fluids, Plasmas and Nonlinear Media"). References in this book to the Workshop papers are given as "I. Jones, Workshop Proceedings".

Michael Shats
Convenor of the 19th Canberra International Physics Summer School
Canberra, August 2006

Contents

Chapter 1

Introduction to Developed Turbulence

Gregory Falkovich

Weizmann Institute of Science, Rehovot 76100 Israel
E-mail: gregory.falkovich@weizmann.ac.il
Homepage: http://www.weizmann.ac.il/home/fnfal/

This is a short course on developed turbulence, weak and strong. The main emphasis is on fundamental properties like universality and symmetries. Two main notions are explained: i) fluxes of dynamical integrals of motion, ii) statistical integrals of motion.

Contents

1.1. Introduction

Turbulence is a state of a physical system with many interacting degrees of freedom deviated far from equilibrium. This state is irregular both in time and in space. Turbulence can be maintained by some external influence or it can be decaying turbulence on the way of relaxation to equilibrium. As the term suggests, it first appeared in fluid mechanics and was later generalized for far-from-equilibrium states in solids and plasma. For example, obstacle of size L placed into fluid moving with velocity V provides for a turbulent wake if the Reynolds number is large: $Re = VL/\nu \gg 1$. Here ν is the kinematic viscosity. At large Re, flow perturbations produced at the scale L have their viscous dissipation small compared to the nonlinear effects.

Nonlinearity produces motions of smaller and smaller scales until viscous dissipation stops this at a scale much smaller than L so that there is a wide (so-called inertial) interval of scales where viscosity is negligible and nonlinearity dominates. Another example is the system of waves excited on a fluid surface by wind or moving bodies and in plasma and solids by external electromagnetic fields. The state of such system is called wave turbulence when the wavelength of the waves excited strongly differs from the wavelength of the waves that effectively dissipate. Nonlinear interaction excites waves in the interval of wavelengths (called transparency window, or inertial interval) between the injection and dissipation scales.

The ensuing complicated and irregular dynamics calls for a statistical description based on averaging either over regions of space or intervals of time. Here we focus on a single-time statistics of steady turbulence that is on the spatial structure of fluctuations. Because of the conceptual simplicity of the inertial range, it is natural to ask if our expectation of universality— that is, freedom from the details of external forcing and internal friction—is true at the level of a physical law. Another facet of the universality problem concerns features that are common to different turbulent systems. This quest for universality is motivated by the hope of being able to distinguish general principles that govern far-from-equilibrium systems, similar in scope to the variational principles that govern thermal equilibrium.

Constraints on dynamics are imposed by conservation laws, and therefore conserved quantities must play an essential role in turbulence. The conservation laws are broken by pumping and dissipation, but both factors do not act in the inertial interval. For example, in the incompressible turbulence, the kinetic energy is pumped by external forcing and is dissipated by viscosity. According to the idea suggested by Richardson in 1921, the kinetic energy flows throughout the inertial interval of scales in a cascade-like process. The cascade idea explains the basic macroscopic manifestation of turbulence: the rate of dissipation of the dynamical integral of motion has a finite limit when the dissipation coefficient tends to zero. For example, the mean rate of the viscous energy dissipation does not depend on viscosity at large Reynolds numbers. That means that a symmetry of the inviscid equation (here, time-reversal invariance) is broken by the presence of the viscous term, even though the latter might have been expected to become negligible in the limit $Re \to \infty$.

The cascade idea fixes only the mean flux of the respective integral of motion demanding it to be constant across the inertial interval of scales. We shall see that flux constancy determines the system completely only for weakly nonlinear system (where the statistics is close to Gaussian). To describe an entire turbulence statistics of strongly interacting systems, one has to solve problems on a case-by-case basis with most cases still unsolved. Particularly difficult (and interesting) are the cases with broken scale invariance where knowledge of flux does not allow one to predict even the order of magnitude of high moments. We describe the new concept of statistical integrals of motion which allows for the description of system with broken scale invariance.

1.2. Weak wave turbulence

From a theoretical point of view, the simplest case is the turbulence of weakly interacting waves. Examples include waves on the water surface, waves in plasma with and without magnetic field, spin waves in magnetics. We assume spatial homogeneity and denote a_k the amplitude of the wave with the wavevector $\mathbf{k}$. When the amplitude is small, it satisfies the linear equation

$$\partial a_k/\partial t = -i\omega_k a_k + f_k(t) - \gamma_k a_k \; . \tag{1.1}$$

Here the dispersion law ω_k describes wave propagation, γ_k is the linear damping rate and f_k describes pumping. For the linear system, a_k is different from zero only in the regions of $\mathbf{k}$-space where f_k is nonzero. To describe wave turbulence which involves wavenumbers outside the pumping region, one must account for the interaction between different waves. Considering for a moment wave system as closed (that is, without external pumping and dissipation) one can describe it as a Hamiltonian system using wave amplitudes as normal canonical variables (see, for instance, the monograph[1]). At small amplitudes, the Hamiltonian can be written as an expansion over a_k, where the second-order term describes non-interacting waves and high-order terms determine the interaction:

$$H = \int \omega_k |a_k|^2 \, d\mathbf{k} \tag{1.2}$$

$$+ \int \left(V_{123} a_1 a_2^* a_3^* + c.c. \right) \delta(\mathbf{k}_1 - \mathbf{k}_2 - \mathbf{k}_3) \, d\mathbf{k}_1 d\mathbf{k}_2 d\mathbf{k}_3 + \mathrm{O}(a^4) \; .$$

Here $V_{123} = V(\mathbf{k}_1, \mathbf{k}_2, \mathbf{k}_3)$ is the interaction vertex and c.c. means complex conjugation. In the Hamiltonian expansion, we presume every subsequent term smaller than the previous one, in particular, $\xi_k = |V_{kkk} a_k| k^d / \omega_k \ll 1$. Wave turbulence that satisfies this condition is called weak turbulence. Here d is the space dimensionality which can be 1, 2 or 3.

The dynamic equation which accounts for pumping, damping, wave propagation and interaction has thus the following form:

$$\partial a_k/\partial t = -i\delta H/\delta a_k^* + f_k(t) - \gamma_k a_k \; . \tag{1.3}$$

It is likely that the statistics of the weak turbulence at $k \gg k_f$ is close to Gaussian for wide classes of pumping statistics (that has not been shown rigorously though). This is definitely the case for the random force with the statistics not very much different from Gaussian. We consider here and below a pumping by a Gaussian random force statistically isotropic and homogeneous in space, and white in time:

$$\langle f_k(t) f_{k'}^*(t') \rangle = F(k)\delta(\mathbf{k} + \mathbf{k}')\delta(t - t') . \tag{1.4}$$

Angular brackets mean spatial average. We assume $\gamma_k \ll \omega_k$ (for waves to be well defined) and that $F(k)$ is nonzero only around some k_f.

Since the dynamic equation (1.3) contains a quadratic nonlinearity then the statistical description in terms of moments encounters the closure problem: the time derivative of the second moment is expressed via the third one, the time derivative of the third moment ix expressed via the fourth one etc. Fortunately, weak turbulence in the inertial interval is expected to have the statistics close to Gaussian so one can express the fourth moment as the product of two second ones. As a result one gets a closed equation for the single-time pair correlation function[1] $\langle a_k a_{k'} \rangle = n_k \delta(\mathbf{k} + \mathbf{k}')$

$$\frac{\partial n_k}{\partial t} = F_k - \gamma_k n_k + I_k^{(3)},$$

$$I_k^{(3)} = \int \left(U_{k12} - U_{1k2} - U_{2k1} \right) d\mathbf{k}_1 d\mathbf{k}_2, \tag{1.5}$$

$$U_{123} = \pi \left[n_2 n_3 - n_1(n_2 + n_3) \right] |V_{123}|^2 \delta(\mathbf{k}_1 - \mathbf{k}_2 - \mathbf{k}_3)\delta(\omega_1 - \omega_2 - \omega_3) .$$

It is called kinetic equation for waves. The collision integral $I_k^{(3)}$ results from the cubic terms in the Hamiltonian i.e. from the quadratic terms in the equations for amplitudes. It can be *interpreted* as describing three-wave interactions: the first term in the integral (1.5) corresponds to a decay of a given wave while the second and third ones correspond to a confluence with other wave.

One can estimate from (1.5) the inverse time of nonlinear interaction at a given k as $|V(k, k, k)|^2 n(k) k^d / \omega(k)$. We define k_d as the wavenumber where this inverse time is comparable to $\gamma(k)$ and assume that nonlinearity dominates over dissipation at $k \ll k_d$. As has been noted, wave turbulence appears when there is a wide (inertial) interval of scales where both pumping and damping are negligible, which requires $k_d \gg k_f$, the condition analogous to $Re \gg 1$. This is schematically shown in Fig. 1.

The presence of frequency delta-function in $I_k^{(3)}$ means that in the first order of perturbation theory in wave interaction we account only for resonant processes which conserve the quadratic part of the energy $E = \int \omega_k n_k \, d\mathbf{k} = \int E_k dk$. For the cascade picture to be valid, the collision integral has to converge in the inertial interval which means that energy exchange is small between motions of vastly different scales, the property called interaction locality in k-space. Consider now a statistical steady state established under the action of pumping and dissipation. Let us multiply (1.5) by ω_k and integrate it over either interior or exterior of the ball with radius k. Taking $k_f \ll k \ll k_d$, one sees that the energy flux through any spherical surface (Ω is a solid angle), is constant in the inertial interval and is equal to the energy production/dissipation rate ϵ:

$$P_k = \int_0^k k^{d-1} dk \int d\Omega\, \omega_k I_k^{(3)} = \int \omega_k F_k\, d\mathbf{k} = \int \gamma_k E_k\, dk = \epsilon \ . \tag{1.6}$$

This (integral) equation determines n_k. Let us assume now that the

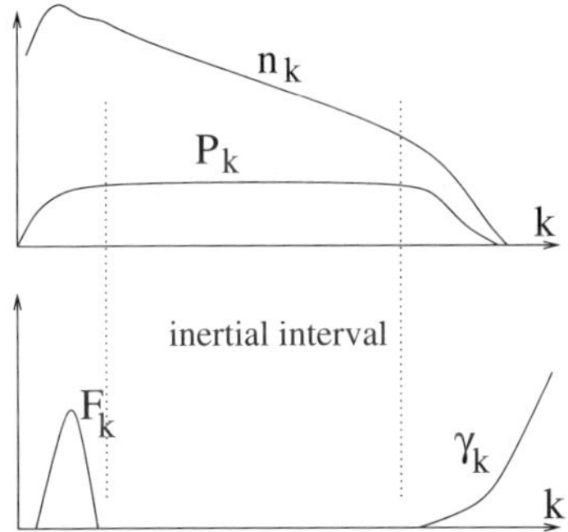

Fig. 1.1.   A schematic picture of the cascade.

medium (characterized by the Hamiltonian coefficients) can be considered isotropic at the scales in the inertial interval. In addition, for scales much larger or much smaller than a typical scale (like Debye radius in plasma or the depth of the water) the medium is usually scale invariant: $\omega(k) = ck^\alpha$ and $|V(\mathbf{k}, \mathbf{k}_1, \mathbf{k}_2)|^2 = V_0^2 k^{2m} \chi(\mathbf{k}_1/k, \mathbf{k}_2/k)$ with $\chi \simeq 1$. Remind that we presumed statistically isotropic force. In this case, the pair correlation function that describes a steady cascade is also isotropic and scale invariant:

$$n_k \simeq \epsilon^{1/2} V_0^{-1} k^{-m-d} \ . \tag{1.7}$$

One can show that (1.7), called Zakharov spectrum, turns $I_k^{(3)}$ into zero.[1]

If the dispersion relation $\omega(k)$ does not allow for the resonance condition $\omega(k_1) + \omega(k_2) = \omega(|\mathbf{k}_1 + \mathbf{k}_2|)$ then the three-wave collision integral is zero and one has to account for four-wave scattering which is always resonant. In other words, whatever the $\omega(k)$ relationship is, one can always find four wavevectors that satisfy $\omega(k_1) + \omega(k_2) = \omega(k_3) + \omega(k_4)$ and $\mathbf{k}_1 + \mathbf{k}_2 = \mathbf{k}_3 + \mathbf{k}_4$.

The collision integral that describes scattering,

$$I_k^{(4)} = \frac{\pi}{2} \int |T_{k123}|^2 \big[n_2 n_3 (n_1 + n_k) - n_1 n_k (n_2 + n_3)\big] \delta(\mathbf{k} + \mathbf{k}_1 - \mathbf{k}_2 - \mathbf{k}_3)$$

$$\times \delta(\omega_k + \omega_1 - \omega_2 - \omega_2) \, d\mathbf{k}_1 d\mathbf{k}_2 d\mathbf{k}_3 , \tag{1.8}$$

conserves the energy and the wave action $N = \int n_k \, d\mathbf{k}$ (the number of waves). Pumping generally provides for an input of both E and N. If there are two inertial intervals (at $k \gg k_f$ and $k \ll k_f$), then there should be two cascades. Indeed, if $\omega(k)$ grows with k, then the absorbtion of a finite amount of E at $k_d \to \infty$ corresponds to an absorption of an infinitely small N. It is thus clear that the flux of N has to go in the opposite direction, that is, to large scales. A so-called inverse cascade with the constant flux of N can thus be realized at $k \ll k_f$. A sink at small k can be provided by a wall friction in the container or by long waves leaving the turbulent region in open spaces (like in sea storms). The collision integral $I_k^{(3)}$ involves products of two n_k, so that the flux constancy requires $E_k \propto \epsilon^{1/2}$, while for the four-wave case $I_k^{(4)} \propto n^3$ gives $E_k \propto \epsilon^{1/3}$. In many cases (when there is a complete self-similarity) this knowledge is sufficient to obtain the scaling of E_k from a dimensional reasoning without actually calculating V and T. For example, short waves on a deep water are characterized by the surface tension σ and density ρ, so the dispersion relation must be $\omega_k \sim \sqrt{\sigma k^3/\rho}$, which allows for the three-wave resonance, and thus $E_k \sim \epsilon^{1/2}(\rho\sigma)^{1/4}k^{-7/4}$. For long waves on a deep water, the surface-restoring force is dominated by gravity so that the gravity acceleration g replaces σ as a defining parameter and $\omega_k \sim \sqrt{gk}$. Such dispersion law does not allow for the three-wave resonance so that the dominant interaction is four-wave scattering which permits two cascades. The direct energy cascade corresponds to $E_k \sim \epsilon^{1/3}\rho^{2/3}g^{1/2}k^{-5/2}$. The inverse cascade carries the flux of N which we denote Q, it has the dimensionality $[Q] = [\epsilon]/[\omega_k]$ and corresponds to $E_k \sim Q^{1/3}\rho^{2/3}g^{2/3}k^{-7/3}$.

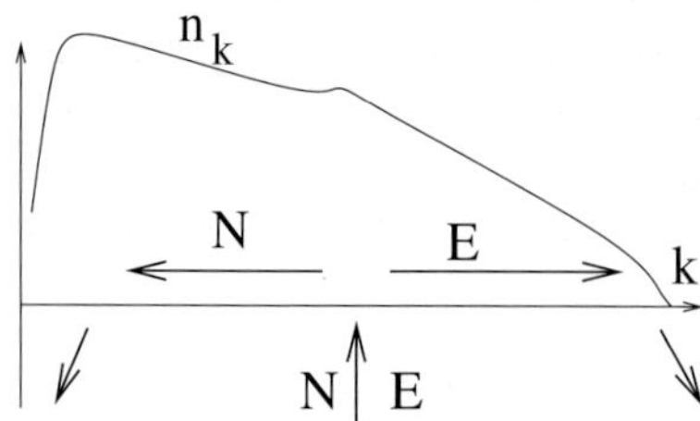

Fig. 1.2. Two cascades under four-wave interaction.

Since the statistics of weak turbulence is near Gaussian, it is completely determined by the pair correlation function, which is in turn determined by the respective flux. We thus conclude that weak turbulence is universal in the inertial interval.

1.3. Strong wave turbulence

One cannot treat wave turbulence as a set of weakly interacting waves when the wave amplitudes are big enough (so that $\xi_k \geq 1$) and also in a particular case of linear (acoustic) dispersion relation $\omega(k) = ck$ for arbitrarily small amplitudes. Indeed, there is no dispersion of wave velocity for acoustic waves so that waves moving in the same direction interact strongly and produce shock waves when viscosity is small. Formally, there is a singularity due to coinciding arguments of delta-functions in (1.5) (and in the higher terms of perturbation expansion for $\partial n_k/\partial t$), which is thus invalid at however small amplitudes. Still, some features of the statistics of acoustic turbulence can be understood even without closed description. We discuss this in a one-dimensional case which pertains, for instance, to sound propagating in long pipes. Since weak shocks are stable with respect to transverse perturbations,[2] quasi one-dimensional perturbations may propagate in 2D and 3D as well. In the frame of reference moving with the sound velocity, the weakly compressible 1d flows ($u \ll c$) are described by the Burgers equation[2,3]

$$u_t + uu_x - \nu u_{xx} = 0 \; . \tag{1.9}$$

Burgers equation has a propagating shock-wave solution $u = 2v\{1 + \exp[v(x - vt)/\nu]\}^{-1}$ with the energy dissipation rate $\nu \int u_x^2 \, dx$ independent of ν. The shock width ν/v is a dissipative scale and we shall consider acoustic turbulence produced by a pumping correlated on much larger scales (for example, pumping a pipe from one end by frequencies much less than cv/ν). After some time, it will develop shocks at random positions. Here we consider the single-time statistics of the Galilean invariant velocity difference $\delta u(x,t) = u(x,t) - u(0,t)$. The moments of δu are called *structure functions* $S_n(x,t) = \langle [u(x,t) - u(0,t)]^n \rangle$. Quadratic nonlinearity makes the time derivative of the second moment to be expressed via the third one:

$$\frac{\partial S_2}{\partial t} = -\frac{\partial S_3}{3\partial x} - 4\epsilon + \nu \frac{\partial^2 S_2}{\partial x^2} \; . \tag{1.10}$$

Here $\epsilon = \nu\langle u_x^2 \rangle$ is the mean energy dissipation rate. Equation (1.10) describes both a free decay (then ϵ depends on t) and the case of a permanently acting pumping which generates turbulence statistically steady at scales less than the pumping length. In the first case, $\partial S_2/\partial t \simeq S_2 u/L \ll \epsilon \simeq u^3/L$

(where L is a typical distance between shocks) while in the second case $\partial S_2/\partial t = 0$ so that $S_3 = 12\epsilon x + \nu \partial S_2/\partial x$.

Consider now a limit $\nu \to 0$ at fixed x (and t for decaying turbulence). Shock dissipation provides for a finite limit of ϵ at $\nu \to 0$, then

$$S_3 = -12\epsilon x \ . \tag{1.11}$$

This formula is a direct analog of (1.6). Indeed, the Fourier transform of (1.10) describes the energy density $E_k = \langle |u_k|^2 \rangle/2$ which satisfies the equation $(\partial_t - \nu k^2)E_k = -\partial P_k/\partial k$ where the k-space flux is

$$P_k = \int_0^k dk' \int_{-\infty}^{\infty} dx S_3(x)k' \sin(k'x)/24 \ .$$

It is thus the flux constancy that fixes $S_3(x)$ which is universal, i.e., it is determined solely by ϵ and depends neither on the initial statistics of decay nor on the pumping for steady turbulence. On the contrary, other structure functions $S_n(x)$ are not given by $(\epsilon x)^{n/3}$. Indeed, the scaling of the structure functions can be readily understood for any dilute set of shocks (when shocks do not cluster in space) which seems to be the case both for smooth initial conditions and a large-scale pumping in Burgers turbulence. In this case, $S_n(x) \sim C_n|x|^n + C'_n|x|$, where the first term comes from the regular (smooth) parts of the velocity (the right x-interval in Fig. 1.3.), while the second term comes from $O(x)$ probability to have a shock in the interval x. The scaling exponents, $\xi_n = d \ln S_n/d \ln x$, thus behave as follows: $\xi_n = n$ for $n \leq 1$ and $\xi_n = 1$ for $n > 1$. That means that the probability density function (PDF) of the velocity difference in the inertial interval $P(\delta u, x)$ is not scale-invariant, i.e., the function of the re-scaled velocity difference $\delta u/x^a$ cannot be made scale-independent for any a. As one goes to smaller scales, the low-order moments decrease faster than the high-order ones, that means that the smaller the scale, the more probable are large fluctuations. In other words, the level of fluctuations increases with the resolution. When the scaling exponents ξ_n do not lie on a straight line, this is called an anomalous scaling since it is related again to the symmetry (scale invariance) of the PDF broken by pumping and not restored even when $x/L \to 0$. As an alternative to the description in terms of structures (shocks), one can relate the anomalous scaling in Burgers turbulence to additional integrals of motion. Indeed, the integrals $E_n = \int u^{2n} dx/2$ are all conserved by the inviscid Burgers equation. Any shock dissipates the finite amount of E_n in the limit of $\nu \to 0$, so that similarly to (1.11), one denotes $\langle \dot{E}_n \rangle = \epsilon_n$ and obtains $S_{2n+1} = -4(2n+1)\epsilon_n x/(2n-1)$ for integer n.

Note that $S_2(x) \propto |x|$ corresponds to $E(k) \propto k^{-2}$, since every shock gives $u_k \propto 1/k$ at $k \ll v/\nu$, such that the energy spectrum is determined by

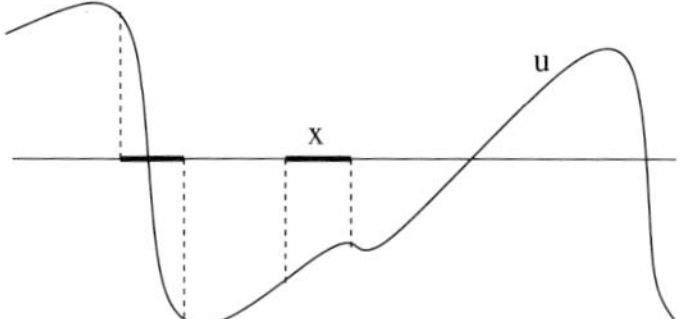

Fig. 1.3. Typical velocity profile in Burgers turbulence

the type of structures (shocks) rather than by energy flux constancy. Similar ideas were suggested for other types of strong wave turbulence assuming
them to be dominated by different structures. Weak wave turbulence, being a set of weakly interacting plane waves, can be studied uniformly for
different systems.[1] On the contrary, when the nonlinearity is strong, different structures appear. Broadly, one distinguishes conservative structures
(like solitons and vortices) from dissipative structures which usually appear
as a result of finite-time singularity of the non-dissipative equations (like
shocks, light self-focussing, or wave collapse). For example, nonlinear wave
packets are described by nonlinear Schrödinger equation,

$$i\Psi_t + \nabla^2\Psi + T|\Psi|^2\Psi = 0 \ . \tag{1.12}$$

Weak wave turbulence is determined by $|T|^2$ and is the same both for $T < 0$
(wave repulsion) and $T > 0$ (wave attraction). At high levels of nonlinearity, different signs of T correspond to dramatically different physics: at
$T < 0$ one has a stable condensate, solitons and vortices, while at $T > 0$
instabilities dominate and wave collapse is possible at $d = 2, 3$. No analytic
theory is yet available for strong turbulence described by (8.1).

Since the parameter of nonlinearity $\xi(k)$ generally depends on k, then
there may exist weakly turbulent cascade up to some k_*, such that $\xi(k_*) \sim 1$
and strong turbulence beyond this wavenumber. In this case weak and
strong turbulence can coexist in the same system. Presuming that some
mechanism (for instance, wave breaking) prevents the development of wave
amplitudes corresponding to $\xi_k \gg 1$, one may hypothesize that some cases
of strong turbulence correspond to the balance between dispersion and nonlinearity local in k-space so that $\xi(k)$ =const throughout its domain in k-
space. That would correspond to the spectrum $E_k \sim \omega_k^3 k^{-d}/|V_{kkk}|^2$ which
is ultimately universal, or independent even of the flux (only the boundary
k_* depends on the flux). For gravity waves, this gives $E_k = \rho g k^{-3}$, the
same spectrum one obtains presuming wave profile to have cusps (another
type of dissipative structure leading to whitecaps in stormy sea[4]). It is
unclear if such flux-independent spectra are realized.

1.4.   Incompressible turbulence

Incompressible fluid flow is described by the Navier-Stokes equation

$$\partial_t \mathbf{v}(\mathbf{r},t) + \mathbf{v}(\mathbf{r},t) \cdot \nabla \mathbf{v}(\mathbf{r},t) - \nu \nabla^2 \mathbf{v}(\mathbf{r},t) = -\nabla p(\mathbf{r},t) \,, \quad \mathrm{div}\, \mathbf{v} = 0 \,. \quad (1.13)$$

We are again interested in the structure functions $S_n(\mathbf{r},t) = \langle [(\mathbf{v}(\mathbf{r},t) - \mathbf{v}(0,t)) \cdot \mathbf{r}/r]^n \rangle$ and consider distance r smaller than the force correlation scale for a steady case and smaller than the size of turbulent region for a decay case. We treat first the three-dimensional case. Similar to (1.10), one can derive[2] the Karman-Howarth relation between S_2 and S_3:

$$\frac{\partial S_2}{\partial t} = -\frac{1}{3r^4}\frac{\partial}{\partial r}(r^4 S_3) + \frac{4\epsilon}{3} + \frac{2\nu}{r^4}\frac{\partial}{\partial r}\left(r^4 \frac{\partial S_2}{\partial r}\right) \,. \quad (1.14)$$

Here $\epsilon = \nu \langle (\nabla \mathbf{v})^2 \rangle$ is the mean energy dissipation rate. Neglecting time derivative (which is zero in a steady state and small comparing to ϵ for decaying turbulence) one can multiply (1.14) by r^4 and integrate: $S_3(r) = -4\epsilon r/5 + 6\nu dS_2(r)/dr$. Kolmogorov considered the limit $\nu \to 0$ for fixed r and *assumed* nonzero limit for ϵ which gives the so-called 4/5 law[2,5,6] :

$$S_3 = -\frac{4}{5}\epsilon r \,. \quad (1.15)$$

This relation is a direct analog of (1.6) and (1.11). It also means that the kinetic energy has a constant flux in the inertial interval of scales (the viscous scale η is defined by $\nu S_2(\eta) \simeq \epsilon \eta^2$). On the first sight, it might appear from (1.13) that the dissipation rate of turbulent energy would vanish as $\nu \to 0$ (or as $Re \to \infty$), but an important feature of turbulence is that the average rate of the energy dissipation per unit mass, $\langle \epsilon \rangle$, remains finite in this limit: no matter how small the viscosity, or how high the Reynolds number, or how extensive the scale-range participating in the energy cascade, the energy flux remains equal to that injected at the stirring scale. Historically, this is the first example of what is now called "anomaly" in theoretical physics: a symmetry of the equation (here, time-reversal invariance) remains broken even as the symmetry-breaking factor (viscosity) becomes vanishingly small.[7] If one screens a movie of steady turbulence backwards, we can tell that something is indeed wrong!

The law (1.15) implies that the third-order moment is universal, i.e. it does not depend on the details of the turbulence production but is determined solely by the mean energy dissipation rate. The rest of the structure functions have never been derived. Kolmogorov[6] (and also Heisenberg, von Weizsacker and Onsager) presumed the pair correlation function to be determined only by ϵ and r which would give $S_2(r) \sim (\epsilon r)^{2/3}$ and the energy spectrum $E_k \sim \epsilon^{2/3} k^{-5/3}$. Experiments suggest that $\zeta_n = d\ln S_n/d\ln r$ lie on a smooth concave curve sketched in Fig. 1.4. . While ζ_2 is close to

2/3 it has to be a bit larger because experiments show that the slope at zero $d\zeta_n/dn$ is larger than $1/3$ while $\zeta(3) = 1$ in agreement with (1.15). Like in Burgers, the PDF of velocity differences in the inertial interval is not scale invariant in the 3D incompressible turbulence. So far, nobody was able to find an explicit relation between the anomalous scaling for 3D Navier-Stokes turbulence and either structures or additional integrals of motion.

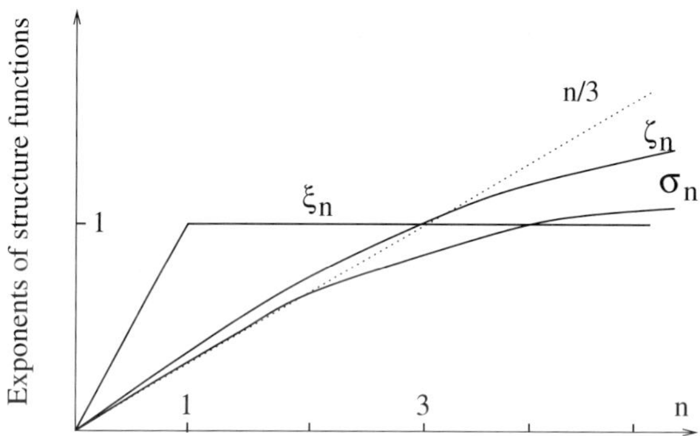

Fig. 1.4. The scaling exponents of the structure functions ξ_n for Burgers, ζ_n for 3D Navier-Stokes and σ_n for the passive scalar. The dotted straight line is $n/3$.

While not exact, Kolomogorov's approximation $S_2(\eta) \simeq (\epsilon\eta)^{2/3}$ can be used to estimate the viscous scale: $\eta \simeq LRe^{-3/4}$. The number of degrees of freedom involved into 3D incompressible turbulence can thus be roughly estimated as $N \sim (L/\eta)^3 \sim Re^{9/4}$. That means, in particular, that detailed computer simulations of water or oil pipe flows ($Re \sim 10^4 \div 10^7$) or turbulent cloud ($Re \sim 10^6 \div 10^9$) are out of question for a foreseeable future. To calculate correctly at least the large-scale part of the flow, it is desirable to have some theoretical model to parameterize the small-scale motions. The main obstacle here is the lack of qualitative understanding and quantitative description of how turbulence statistics changes with the scale. This breakdown of the scale invariance in the inertial range is another example of anomaly (effect of pumping scale does not disappear even at the limit $r/L \to 0$). Such an anomalous (or multifractal) scaling, is arguably an important feature of turbulence, which sets it apart from the usual critical phenomena: one needs to work out the behavior of moments of each order independently without succumbing to dimensional analysis. Anomalous scaling in turbulence is such that $\zeta_{2n} < n\zeta_2$ so that S_{2n}/S_2^n for $n > 2$ increases as $r \to 0$. The relative growth of high moments means that strong fluctuations become more probable as the scales become smaller. Its practical importance is that it limits our ability to produce realistic models for small-scale turbulence.

2D Turbulence. Large-scale motions in shallow fluid can be approximately considered two-dimensional. When the velocities of such motions are much smaller than the velocities of the surface waves and the velocity of sound, such flows can be considered incompressible. Their description is important for understanding atmospheric and oceanic turbulence at the scales larger than the atmosphere height and the ocean depth. Vorticity $\omega = curl\,\mathbf{v}(= \nabla \times v)$ is a scalar in a two-dimensional flow. It is advected by the velocity field and dissipated by viscosity. Taking *curl* of the Navier-Stokes equation one gets

$$d\omega/dt = \partial_t\omega + (\mathbf{v} \cdot \nabla)\omega = \nu\nabla^2\omega \ . \tag{1.16}$$

Two-dimensional incompressible inviscid flow just transports vorticity from place to place and thus conserves spatial averages of any function of vorticity, $\Omega_n \equiv \int \omega^n d\mathbf{r}$. In particular, we now have the second quadratic inviscid invariant (in addition to energy) which is called *enstrophy*: $\Omega_2 = \int \omega^2 \, d\mathbf{r}$. Since the spectral density of the energy is $|\mathbf{v}_k|^2/2$, while that of the enstrophy is $|\mathbf{k} \times \mathbf{v}_k|^2$, then (similarly to the cascades of E and N in wave turbulence under four-wave interaction) one expects that the direct cascade (towards large k) is that of enstrophy, while the inverse cascade is that of energy, as was suggested by Kraichnan.[8] What about other Ω_n? The intuition developed so far might suggest that the infinity of dynamical conservation laws must bring about anomalous scaling. Turbulence never fails to defy natural expectations as we shall see. Again, for the inverse energy cascade, there is no consistent theory except for the flux relation that can be derived similarly to (1.15):

$$S_3(r) = 4\epsilon r/3 \ . \tag{1.17}$$

The inverse cascade is observed in the atmosphere (at the scales $30 \div 500$ km) and in the laboratory experiments.[9] Experimental data suggest that there is no anomalous scaling that is $S_n \propto r^{n/3}$. In particular, $S_2 \propto r^{2/3}$ which corresponds to $E_k \propto k^{-5/3}$. It is ironic that probably the most widely known statement on turbulence, the 5/3 spectrum suggested by Kolmogorov for 3D, is not correct in this case (even though the true scaling is close) while it is probably exact in the Kraichnan's inverse 2D cascade. Qualitatively, it is likely that the absence of anomalous scaling in the inverse cascade is associated with the growth of the typical turnover time (estimated, say, as $r/\sqrt{S_2}$) with the scale. As the inverse cascade proceeds, the fluctuations have enough time to get smoothed out, opposite to the direct cascade in 3D, where the turnover time decreases in the direction of the cascade.

Passive Scalar Turbulence. Before discussing the direct (enstrophy) cascade, we describe a similar, yet somewhat simpler problem of passive scalar turbulence which allows one to introduce the necessary notions of Lagrangian description of the fluid flow. Consider a scalar quantity $\theta(\mathbf{r}, t)$, which is subject to molecular diffusion and advection by the fluid flow, but has no back influence on the velocity (i.e. passive):

$$\partial_t \theta + (\mathbf{v} \cdot \nabla)\theta = \kappa \nabla^2 \theta + \varphi \; . \tag{1.18}$$

Here κ is molecular diffusivity. The examples of passive scalar are smoke in the air, salinity in the water and temperature when one can neglect thermal convection. Without pumping, dissipation and diffusion, ω and θ are advected in the same way in the same 2D flow: they are both Lagrangian invariants satisfying $d\omega/dt = d\theta/dt = 0$. However, note that vorticity is related to velocity, while the passive scalar is not. If the source φ produces fluctuations of θ on some scale L, then the inhomogeneous velocity field stretches, contracts and folds the field θ producing progressively smaller and smaller scales: this is the mechanism of the scalar cascade. If the rms velocity gradient is Λ, then molecular diffusion is substantial at the scales less than the diffusion scale $r_d = \sqrt{\kappa/\Lambda}$. For scalar turbulence, the ratio $Pe = L/r_d$, called Peclet number, plays the role of the Reynolds number. When $Pe \gg 1$, there is an inertial interval with a constant flux of θ^2:

$$\langle (\mathbf{v}_1 \cdot \nabla_1 + \mathbf{v}_2 \cdot \nabla_2)\theta_1 \theta_2 \rangle = 2P \; , \tag{1.19}$$

where $P = \kappa \langle (\nabla \theta)^2 \rangle$ and subscripts denote the spatial points. In considering the passive scalar problem, the velocity statistics is presumed to be given. Still, the correlation function (1.19) mixes $\mathbf{v}$ and θ and does not generally allow one to make a statement on any correlation function of θ. The proper way to describe the correlation functions of the scalar at the scales much larger than the diffusion scale is to employ the Lagrangian description that is to follow fluid trajectories.[10] Indeed, if we neglect diffusion, then the equation (1.18) can be solved along the characteristics $\mathbf{R}(t)$ which are called Lagrangian trajectories and satisfy $d\mathbf{R}/dt = \mathbf{v}(\mathbf{R}, t)$. Presuming zero initial conditions for θ at $t \to -\infty$ we write

$$\theta\Big(\mathbf{R}(t), t\Big) = \int_{-\infty}^{t} \varphi\Big(\mathbf{R}(t'), t'\Big) \, dt' \; . \tag{1.20}$$

In that way, the correlation functions of the scalar $F_n = \langle \theta(\mathbf{r}_1, t) \ldots \theta(\mathbf{r}_n, t) \rangle$ can be obtained by integrating the correlation functions of the pumping along the trajectories that satisfy the final conditions $\mathbf{R}_i(t) = \mathbf{r}_i$. We consider a pumping which is Gaussian, statistically homogeneous and isotropic in space and white in time:

$$\langle \varphi(\mathbf{r}_1, t_1)\varphi(\mathbf{r}_2, t_2) \rangle = \Phi(|\mathbf{r}_1 - \mathbf{r}_2|)\delta(t_1 - t_2),$$

where the function Φ is constant at $r \ll L$ and goes to zero at $r \gg L$. The pumping provides for symmetry $\theta \to -\theta$ which makes only even correlation functions F_{2n} nonzero. The pair correlation function is as follows:

$$F_2(r,t) = \int_{-\infty}^{t} \Phi\Big(R_{12}(t')\Big)\, dt' \; . \tag{1.21}$$

Here $R_{12}(t') = |\mathbf{R}_1(t') - \mathbf{R}_2(t')|$ is the distance between two trajectories and $R_{12}(t) = r$. The function Φ essentially restricts the integration to the time interval when the distance $R_{12}(t') \leq L$. Simply speaking, the stationary pair correlation function of a tracer is $\Phi(0)$ (which is twice the injection rate of θ^2) times the average time $T_2(r,L)$ that two fluid particles spent within the correlation scale of the pumping. The larger r, the less time it takes for the particles to separate from r to L, and the less is $F_2(r)$. Of course, $T_{12}(r,L)$ depends on the properties of the velocity field. A general theory is available only when the velocity field is spatially smooth at the scale of the scalar pumping L. This so-called Batchelor regime happens, in particular, when the scalar cascade occurs at the scales less than the viscous scale of fluid turbulence.[10–12] This requires the Schmidt number ν/κ (called Prandtl number when θ is temperature) to be large, which is the case for very viscous liquids. In this case, one can approximate the velocity difference $\mathbf{v}(\mathbf{R}_1,t) - \mathbf{v}(\mathbf{R}_2,t) \approx \hat{\sigma}(t)\mathbf{R}_{12}(t)$ by the Lagrangian strain matrix $\sigma_{ij}(t) = \nabla_j v_i$. In this regime, the distance obeys the linear differential equation

$$\dot{\mathbf{R}}_{12}(t) = \hat{\sigma}(t)\mathbf{R}_{12}(t) \; . \tag{1.22}$$

The theory of such equations is well-developed and is related to what is called Lagrangian chaos since fluid trajectories separate exponentially as typical for systems with dynamical chaos (see, e.g.[10,13]): at t much larger than the correlation time of the random process $\hat{\sigma}(t)$, all moments of R_{12} grow exponentially with time and $\langle \ln[R_{12}(t)/R_{12}(0)] \rangle = \lambda t$, where λ is called a senior Lyapunov exponent of the flow (note, that for the description of the scalar we need the flow taken backwards in time which is different from that taken forward because turbulence is irreversible). Dimensionally, $\lambda = \Lambda f(Re)$, where the limit of the function f at $Re \to \infty$ is unknown. We thus obtain:

$$F_2(r) = \Phi(0)\lambda^{-1}\ln(L/r) = 2P\lambda^{-1}\ln(L/r) \; . \tag{1.23}$$

In a similar way, one shows that for $n \ll \ln(L/r)$ all F_n are expressed via F_2 and the structure functions $S_{2n} = \langle [\theta(\mathbf{r},t) - \theta(0,t)]^{2n} \rangle \simeq (P/\lambda)^n \ln^n(r/r_d)$ for $n \ll \ln(r/r_d)$. This can be generalized for an arbitrary statistics of pumping as long as it is finite-correlated in time.[10]

2D Enstrophy cascade. Now, one can use the analogy between passive scalar and vorticity in 2D.[8,14] For the enstrophy cascade, one derives the flux relation analogous to (1.19):

$$\langle (\mathbf{v}_1 \cdot \nabla_1 + \mathbf{v}_2 \cdot \nabla_2)\omega_1\omega_2 \rangle = 2D \, , \qquad (1.24)$$

where $D = \langle \nu(\nabla\omega)^2 \rangle$. The flux relation along with $\omega = curl\,\mathbf{v}$ suggests the scaling $\delta v(r) \propto r$, so that velocity is close to spatially smooth (of course, it cannot be perfectly smooth to provide for a nonzero vorticity dissipation in the inviscid limit, but the possible singularitites are indeed no stronger than logarithmic). This makes the vorticity cascade similar to the Batchelor regime of the passive scalar cascade with a notable change in that the rate of stretching λ acting on a given scale is not a constant but is logarithmically growing when the scale decreases. Since λ scales as vorticity, the law of renormalization can be established from dimensional reasoning and one gets $\langle \omega(\mathbf{r},t)\omega(0,t) \rangle \sim [D\ln(L/r)]^{2/3}$ which corresponds to the energy spectrum $E_k \propto D^{2/3}k^{-3}\ln^{-1/3}(kL)$. High-order correlation functions of vorticity are also logarithmic, for instance, $\langle \omega^n(\mathbf{r},t)\omega^n(0,t) \rangle \sim [D\ln(L/r)]^{2n/3}$. Note that both passive scalar in the Batchelor regime and vorticity cascade in 2D are universal, which is determined by the single flux (P and D respectively) despite the existence of the high-order conserved quantities. Experimental data and numerical simulations support these conclusions.[9,10]

1.5. Zero modes and anomalous scaling

Let us now return to the Lagrangian description and discuss it when velocity is not spatially smooth, for example, that of the energy cascades in the inertial interval. One can assume that it is Lagrangian statistics which is determined by the energy flux when the distances between fluid trajectories are in the inertial interval. That assumption leads, in particular, to the Richardson law for the asymptotic growth of the interparticle distance:

$$\langle R_{12}^2(t) \rangle \sim \epsilon t^3 \, , \qquad (1.25)$$

first established from atmospheric observations (in 1926) and later confirmed experimentally for energy cascades both in 3D and in 2D. There is no consistent theoretical derivation of (1.25) and it is unclear whether it is exact (likely to be in 2D) or just approximate (possible in 3D). Semi-heuristic argument usually presented in textbooks is based on the mean-field estimate: $\dot{\mathbf{R}}_{12} = \delta\mathbf{v}(\mathbf{R}_{12},t) \sim (\epsilon R_{12})^{1/3}$ which upon integration gives: $R_{12}^{2/3}(t) - R_{12}^{2/3}(0) \sim \epsilon^{1/3}t$. For the passive scalar it gives, by virtue of (1.21), $F_2(r) \sim \Phi(0)\epsilon^{-1/3}[L^{2/3} - r^{2/3}]$ as suggested by Oboukhov and Corrsin.[15,16] The structure function is then $S_2(r) \sim \Phi(0)\epsilon^{-1/3}r^{2/3}$. Experiments measuring the scaling exponents $\sigma_n = d\ln S_n(r)/d\ln r$ generally give σ_2 close to 2/3 but higher exponents deviate from the straight line even stronger than the exponents of the velocity in 3D. Moreover, the scalar exponents σ_n are anomalous even when advecting velocity has a normal scaling like in 2D energy cascade.

To describe multi-point correlation functions or high-order structure functions one needs to study multi-particle statistics. Here an important question is what memory of the initial configuration remains when final distances far exceed initial ones. To answer this question one must analyze the conservation laws of turbulent diffusion. We now describe a general concept of conservation laws which, while conserved only on the average, still determine the statistical properties of strongly fluctuating systems. In a random system, it is always possible to find some fluctuating quantities who's ensemble averages do not change. We now ask a more subtle question: is it possible to find quantities that are expected to change on the dimensional grounds but they stay constant.[7,10] Let us characterize n fluid particles in a random flow by inter-particle distances R_{ij} (between particles i and j) as in Figure 1.5. . Consider homogeneous functions f of inter-particle distances with a nonzero degree ζ, i.e. $f(\lambda R_{ij}) = \lambda^\zeta f(R_{ij})$. When all the distances grow on average, say according to $< R_{ij}^2 > \propto t^a$, then one expects that a generic function grows as $f \propto t^{a\zeta/2}$. How to build (specific) functions that are conserved on average, and which ζ-s they have? As the particles move in a random flow, the n-particle cloud grows in size while fluctuations in the shape of the cloud decrease in magnitude. Therefore, one may look for suitable functions of size and shape that are conserved because the growth of distances is compensated by the decrease of shape fluctuations.

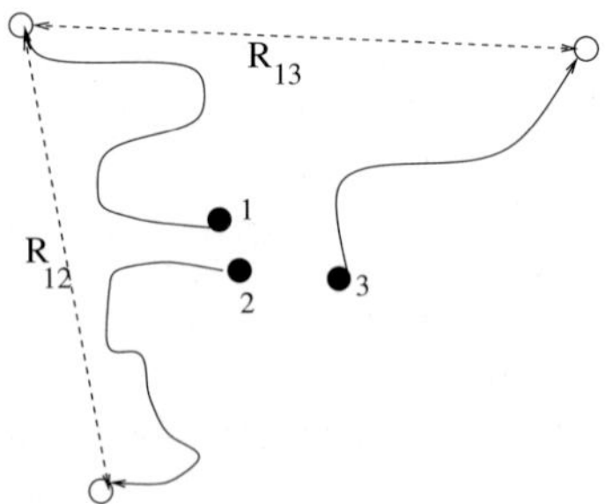

Fig. 1.5.	Three fluid particles in a flow

For the simplest case of Brownian random walk, inter-particle distances grow by the diffusion law: $\langle R_{ij}^2(t)\rangle = R_{ij}^2(0) + \kappa t$, $\langle R_{ij}^4(t)\rangle = R_{ij}^4(0) + 2(d+2)\left[R_{ij}^2(0)\kappa t + \kappa^2 t^2\right]/d$, etc. Here d is the space dimensionality. Two particles are characterized by a single distance. Any positive power of this distance grows on the average. For three particles, one can build conserved quantities by taking the differences where all powers of t cancel out: $f_2 = \langle R_{12}^2 - R_{13}^2 \rangle$, $f_4 = \langle 2(d+2)R_{12}^2 R_{13}^2 - d(R_{12}^4 + R_{13}^4) \rangle$, etc. These polynomials are called harmonics since they are zero modes of the Laplacian in the 2-dimensional space of $\mathbf{R}_{12}$, $\mathbf{R}_{13}$. One can write the Laplacian as $\Delta = R^{1-2d}\partial_R R^{2d-1}\partial_R + \Delta_\theta$, where $R^2 = R_{12}^2 + R_{13}^2$ and Δ_θ is the angular Laplacian on $2d - 1$-dimensional unit sphere. Introducing the angle, $\theta = \arcsin(R_{12}/R)$, which characterizes the shape of the triangle, we see that the conservation of both $f_2 = \langle R^2 \cos 2\theta \rangle$ and $f_4 = \langle R^4[(d+1)\cos^2 2\theta - 1] \rangle$ can be also described as due to cancellation between the growth of the radial part (as powers of t) and the decay of the angular part (as inverse powers of t). For n particles, the polynomial that involves all distances is proportional to R^{2n} (i.e. $\zeta_n = n$) and the respective shape fluctuations decay as t^{-n}.

The scaling exponents of the zero modes are thus determined by the laws that govern decrease of shape fluctuations. The zero modes, which are conserved statistically, exist for turbulent macroscopic diffusion as well. However, there is a major difference since the velocities of different particles are correlated in turbulence. Those mutual correlations make shape fluctuations decaying slower than t^{-n} so that the exponents of the zero modes, ζ_n, grow with n slower than linearly. This is very much like the total energy of the cloud of attracting particles does not grow linearly with the number of particles. Indeed, power-law correlations of the velocity field lead to super-diffusive behavior of inter-particle separations: the farther particles are, the faster they tend to move away from each other, as in Richardson's law of diffusion. That is the system behaves as if there was an attraction between particles that weakens with the distance, though, of course, there is no physical interaction among particles (but only mutual correlations because they are inside the correlation radius of the velocity field). Let us stress that while zero modes of multi-particle evolution exist for all velocity fields—from those that are smooth to those that are extremely rough as in Brownian motion—only those non-smooth velocity fields with power-law correlations provide for an anomalous scaling. Zero modes were discovered in[18–20] and then described in.[21–23]

The existence of multi-particle conservation laws indicates the presence of a long-time memory and is a reflection of the coupling among the particles due to the simple fact that they are all in the same velocity field.

We shall now ask: How does the existence of these statistical conservation laws (called martingales in the probability theory) lead to anomalous scaling of fields advected by turbulence? According to (1.20), the correlation functions of θ are proportional to the times spent by the particles within the correlation scales of the pumping. The structure functions of θ are differences of correlation functions with different initial particle configurations as, for instance, $S_3(r_{12}) \equiv \langle [\theta(\mathbf{r}_1) - \theta(\mathbf{r}_2)]^3 \rangle = 3\langle \theta^2(\mathbf{r}_1)\theta(\mathbf{r}_2) - \theta(\mathbf{r}_1)\theta^2(\mathbf{r}_2) \rangle$. In calculating S_3, we are thus comparing two histories: the first one with two particles initially close to the position $\mathbf{r}_1$ and one particle at $\mathbf{r}_2$, and the second one with one particle at $\mathbf{r}_1$ and two particles at $\mathbf{r}_2$— see Fig 1.6.. That is, S_3 is proportional to the time during which one can distinguish

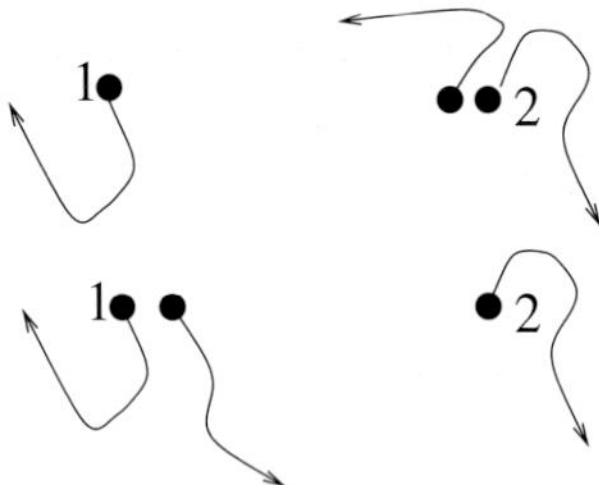

Fig. 1.6.   Two configurations (upper and lower) determining the third structure function

one history from another, or to the time needed for an elongated triangle to relax to the equilateral shape. That time decreases as r_{12} grows: the further away the particles, the faster they loose correlations.

Quantitative details can be worked out for the white-in-time velocity[17] (profound insight of Kraichnan was that it is spatial rather than temporal non-smoothness of the velocity that is crucial for an anomalous scaling):

$$\langle v^i(\mathbf{r}, t)v^j(0, 0) \rangle = \delta(t)\Big[D_0\delta_{ij} - d_{ij}(\mathbf{r}) \Big] ,$$

$$d_{ij} = D_1\, r^{2-\gamma}\Big[(d + 1 - \gamma)\, \delta^{ij} + (\gamma - 2)r^i r^j r^{-2} \Big] . \qquad (1.26)$$

Here the exponent $\gamma \in [0, 2]$ is a measure of the velocity nonsmoothness with $\gamma = 0$ corresponding to a smooth velocity, while $\gamma = 2$ corresponds to a velocity which is very rough in space (distributional). Richardson-Kolmogorov scaling of the energy cascade corresponds to $\gamma = 2/3$. Lagrangian flow is a Markov random process for the Kraichnan ensemble (1.26). Every fluid particle undergoes a Brownian random walk with the so-called eddy diffusivity D_0. The PDF $P(r,t)$ for two particles to be separated by r after time t satisfies the diffusion equation (see e.g.[10])

$$\partial_t P = L_2 P , \quad L_2 = d_{ij}(\mathbf{r})\nabla^i \nabla^j = D_1(d - 1)r^{1-d}\partial_r r^{d+1-\gamma}\partial_r , \qquad (1.27)$$

with the scale-dependent diffusivity $D_1(d-1)r^{2-\gamma}$. The asymptotic solution of (1.27) is $P(r,t) = r^{d-1}t^{d/\gamma}\exp\left(-\mathrm{const}\, r^\gamma/t\right)$ (lognormal for $\gamma = 0$). For $\gamma = 2/3$, it reproduces, in particular, the Richardson law. Multiparticle probability distributions also satisfy diffusion equations in the Kraichnan model as well as all the correlation functions of θ. Multiplying (1.18) by $\theta_2 \ldots \theta_{2n}$ and averaging over the Gaussian statistics of $\mathbf{v}$ and φ one derives

$$\partial_t F_{2n} = L_{2n} F_{2n} + \sum_{l,m} F_{2n-2}\Phi(\mathbf{r}_{lm})\,, \qquad L_{2n} = \sum d_{ij}(\mathbf{r}_{lm})\nabla_l^i \nabla_m^j\,. \quad (1.28)$$

This equation enables one, in principle, to derive inductively all steady-state F_{2n} starting from F_2. The equation $\partial_t F_2(r,t) = L_2 F_2(r,t) + \Phi(r)$ has a steady solution $F_2(r) = 2[\Phi(0)/\gamma d(d-1)D_1][dL^\gamma/(d-\gamma) - r^\gamma]$, which has the Corrsin-Oboukhov form for $\gamma = 2/3$. Further, F_4 contains the so-called forced solution having the normal scaling 2γ but also, remarkably, a zero mode Z_4 of the operator L_4: $L_4 Z_4 = 0$. Such zero modes necessarily appear (to satisfy the boundary conditions at $r \simeq L$) for all $n > 1$ and the scaling exponents of Z_{2n} are generally different from $n\gamma$ that is anomalous. In calculating the scalar structure functions, all terms cancel out except a single zero mode (called irreducible because it involves all distances between $2n$ points). Analytical and numerical calculations of Z_n and their scaling exponents σ_n give[10] σ_n lying on a convex curve (see Fig. 1.4.) which saturates[23] to a constant at large n. Such saturation[24] is a signature that most singular structures in a scalar field are shocks like in Burgers turbulence, the value σ_n at $n \to \infty$ is the fractal codimension of fronts in space.

The existence of statistical conserved quantities breaks the scale invariance of scalar statistics in the inertial interval and explains why scalar turbulence knows about pumping "more" than just the value of the flux. Note that both symmetries, one broken by pumping (scale invariance) and another by damping (time reversibility) are not restored even when $r/L \to 0$ and $r_d/r \to 0$.

For the vector field (like velocity or magnetic field in magnetohydrodynamics) the Lagrangian statistical integrals of motion may involve both the coordinate of the fluid particle and the vector it carries. Such integrals of motion were built explicitly and related to the anomalous scaling for the passively advected magnetic field in the Kraichnan ensemble of velocties.[10] Doing that for velocity that satisfies the Navier-Stokes equation remains a task for the future.

This course has popular[7] and technical[10] versions. I am grateful to my co-authors, K. Gawędzki, M. Vergassola and K. Sreenivasan, for teaching me many things. This work was supported by the Israel Science Foundation.

Bibliography

1. V. Zakharov, V. L'vov and G. Falkovich, *Kolmogorov Spectra of Turbulence* (Springer-Verlag, Berlin 1992).
2. L. Landau, and E. Lifshits, *Fluid Mechanics* (Pergamon Press, Oxford, 1987).

3. W. E, K. Khanin, A. Mazel and Ya.G. Sinai, *Phys. Rev. Lett.* **78**, 1904 (1997); *Annals of Math.* **151**, 877 (2000); U. Frisch and J. Bec, in *Les Houches 2000: New Trends in Turbulence*, ed. M. Lesieur, Springer EDP-Sciences (2001).

4. O. Phillips, *The Dynamics of the Upper Ocean* (Univ. Press, Cambridge, 1977).

5. U. Frisch, *Turbulence*(Univ. Press, Cambridge, 1995).

6. A. N. Kolmogorov, 1941, *C. R. Acad. Sci. URSS* **30**, 301.

7. G. Falkovich and K. Sreenivasan, *Physics Today*, April 2006.

8. R. H. Kraichnan, *Phys. Fluids* **10**, 1417 (1967).

9. P. Tabeling, *Phys. Rep.* **362**, 1, (2002).

10. G. Falkovich, K. Gawędzki and M. Vergassola, *Rev. Mod. Phys.*, **73**, 913 (2001).

11. G. K. Batchelor, *J. Fluid Mech.* **5**, 113 (1959).

12. R. H. Kraichnan, *J. Fluid Mech.* **64**, 737 (1974).

13. T. Antonsen and E. Ott, *Phys. Rev. A* **44**, 851 (1991).

14. G. Falkovich and V. Lebedev, *Phys. Rev. E* **50**, 3883 (1994).

15. S. Corrsin, *J. Appl. Phys.* **22**, 469 (1951).

16. A. M. Obukhov, *Izv. Akad. Nauk SSSR, Geogr. Geofiz.* **13**, 58 (1949).

17. R. H. Kraichnan, *Phys. Fluids* **11**, 945-963 (1968).

18. M. Chertkov et al, *Phys. Rev. E* **52**, 4924 (1995).

19. K. Gawędzki and A. Kupiainen, *Phys. Rev. Lett.* **75**, 3834 (1995).

20. B. Shraiman and E. Siggia, *C.R. Acad. Sci.*, **321**, 279 (1995).

21. M. Chertkov and G. Falkovich, *Phys. Rev Lett.* **76**, 2706 (1996).

22. D. Bernard, K. Gawędzki and A. Kupiainen, *Phys. Rev. E* **54**, 2564 (1996).

23. E. Balkovsky and V. Lebedev, *Phys. Rev. E* **58**, 5776 (1998).

24. A. Celani et al, *Phys. Fluids* **17**, 287 (2001).

Chapter 2

Renormalization and Statistical Methods

David McComb

*School of Physics, University of Edinburgh,
James Clerk Maxwell Building, The King's Buildings,
Mayfield Road, Edinburgh EH9 3JZ, Scotland, UK.
E-mail: w.d.mccomb@ed.ac.uk*

This chapter is based on the three lectures given as *Renormalization for beginners* and is divided into three main sections, each corresponding to one of the lectures. After a brief introduction in the first section, we give an overview of renormalization in physics, concluding with a short discussion of its potential for application to turbulence. The second main section then discusses renormalized perturbation theories and two-point turbulence closures. In the interests of completeness, this is supplemented by material from the Workshop lecture *Two-point closures revisited*. Lastly we briefly discuss renormalization group as used in particle theory and the theory of critical phenomena, and then concentrate on its application to macroscopic fluid turbulence.

Contents

                     David McComb

2.1. Introduction

We begin by considering the meaning of the term 'renormalization'. The term originated in quantum field theory in the late 1930s, and referred to what was purely a method of removing divergences. Since then it has been variously regarded as *ad hoc* or 'mysterious' although physical interpretations can be given in terms of 'bare' and 'dressed' particles, where 'bare' particles are not observable.

The actual technique originated earlier in macroscopic physics as a method of taking collective effects into account. It is also very important in statistical and condensed matter physics, where physical motivation (and interpretation) is predominant.

Early examples of renormalization include the following:

(1) Effective mass of a body moving in a fluid (1830).
(2) Eddy viscosity in turbulence: collective effect of eddies renormalizes the fluid viscosity: Boussinesq (1877).
(3) Weiss theory of magnetism: collective action of molecular magnets leads to an effective field (1907).
(4) Debye-Hückel theory of electrolytes: potential of one electron is screened by a cloud of electrons around it (1923).

Discussions of renormalization can be found in the books by Brown (1993)[1] and McComb (2004).[2]

2.2. Overview of renormalization in physics with application to turbulence

In this section we define the many-body problem and introduce the concepts of renormalization and quasi-particles. Perturbation theory is discussed in terms of the λ-expansion, along with re-expansion in a control parameter, such as density or temperature. We discuss mean-field theory, self-similarity and scale invariance. The Renormalization Group (RG) transformation is introduced and the physical significance and nature of fixed points discussed.

2.2.1. *The basic programme of statistical physics*

This can be stated algorithmically as follows:

(1) Obtain the probability distribution of the microscopic arrangements of the system.
(2) Obtain the normalization of this distribution.
(3) The inverse of the normalization is the partition function Z.
(4) Use the bridge equation to obtain the free energy F.
(5) Use partial differentiation to obtain thermodynamic quantities of interest from F.

We shall now expand on the details of this process for non-interacting, and then interacting systems. A fuller treatment of this material, and additional references, may be found in the book by McComb.[2]

2.2.1.1. *Non-interacting N-body systems*

Consider N particles, each of mass m in a box. The system has energy eigenstates E_i for integer i. At equilibrium the probability distribution takes the form,

$$P(E_i) = Z^{-1} e^{-E_i/kT} \equiv Z^{-1} e^{-\beta E_i},$$

where k is the Boltzmann constant, T is absolute temperature and Z is the partition function, which is given by

$$Z = \sum_i e^{-E_i/kT} \equiv \sum_i e^{-\beta E_i}.$$

As individual particles are non-interacting, we can write:

$$Z = (Z_1)^N,$$

where Z_1 is single-particle partition function. Hence the bridge equation can be written as:

$$F = -kT \ln Z = -NkT \ln Z_1.$$

That is, Z factorises. As we shall see, this is no longer the case when there are interactions between particles.

2.2.1.2. *Interacting N-body system*

Imagine that we now switch on a potential between pairs of particles: the total energy of the system becomes

$$H = \sum_i H_i + \sum_{i,j} H_{ij}$$

where H is the energy function or system Hamiltonian, $H_i = p_i^2/2m$ is the kinetic energy of ith particle, $H_{ij} = V_{ij}(q_i - q_j)$ is the interaction potential, with q_i, q_j being the positions of the ith and j^{th} particles respectively.

Now, due to the coupling term, the partition sum no longer factorises. In general the problem of working out the partition function is insoluble: this is known as the **the many-body problem**. Specific *ad hoc* solutions may exist for certain specific problems but in general we have to resort to approximate methods. However, first we introduce two important concepts, **quasi-particles** and **renormalization**.

2.2.1.3. *Quasi-particles*

Let us replace the interaction term in the Hamiltonian by the **average** effect of all other particles on one particle. That is, we approximate the Hamiltonian by

$$H \simeq \sum_i H_i',$$

where H_i' is the **effective** Hamiltonian of the ith quasi-particle. Each of these effective Hamiltonians has a portion of the interaction energy added to their single-particle form. In this way we can replace the interacting system by a system of non-interacting quasi-particles. Then we can use elementary statistical mechanics to calculate Z.

2.2.1.4. *Renormalization*

In order to describe the process of forming a quasi-particle, we borrow the term **renormalization** from quantum field theory.

A renormalization process is one where we make the replacement:

$$\text{'bare' quantity} + \text{interactions} \rightarrow \text{'dressed' quantity}.$$

As specific examples we have:

(1) An electron in a cloud of electrons suffers an effective charge renormalization due to the collective effect of other electrons.
(2) Collective vibrations of coupled oscillations: replace these by a set of non-interacting renormalised oscillators with renormalized natural frequencies which depend on the energy.

2.2.2. *Theoretical approaches*

There are two broad theoretical approaches. These are:

(1) **Perturbation expansion**: this is the only truly general method. Known generically as the λ-expansion. It never works! (Or, at least, not straightaway.) Usually necessary to cure divergent integrals and also to re-expand in some other control parameter, such as density or temperature.
(2) **Mean-field theory**: this involves both mean-field and self-consistent steps. For best results, it relies on a variational principle. Gives quite good but rarely perfect results.

In practice these two approaches are not mutually exclusive and may be combined. However we shall treat them separately in turn and then indicate some applications where they are combined.

2.2.3. *Perturbation theory*

2.2.3.1. *The λ-expansion*

The classical partition function for an interacting gas is:

$$Z = \frac{1}{N!h^{3N}} \int d\mathbf{p}_1 \dots \int d\mathbf{p}_N \int d\mathbf{q}_1 \dots \int dq_N e^{-E(\mathbf{q},\mathbf{p})/kT}.$$

We can factor out the integration with respect to momenta by writing

$$E(\mathbf{q},\mathbf{p}) = \sum_{i=1}^{N} \frac{p_i^2}{2m} + \Phi(\mathbf{q}),$$

where $\Phi(\mathbf{q})$ is some interaction potential. Consider the important general case of pair potentials:

$$\Phi(\mathbf{q}) = \sum_{i<j=1}^{N} \phi(|\mathbf{q}_i - \mathbf{q}_j|) \equiv \sum_{i<j=1}^{N} \phi_{ij}.$$

2.2.3.2. *The configuration integral*

Factorize the partition function: $Z = Z_0 Q$, where Z_0 is the perfect gas partition function and is given by:

$$Z_0 = \frac{V^N}{N!} \left(\frac{2\pi mkT}{k^2} \right) \frac{3N}{2},$$

while Q is configuration integral:

$$Q = \frac{1}{V^N} \int d\mathbf{q}_1 \dots \int d\mathbf{q}_N e^{-\Phi(\mathbf{q})/kT}.$$

The nature of the problem depends on choice of potential. Examples: Hard-sphere; Coulomb; Lennard-Jones (intermolecular).

Expand out the exponential, introducing a book-keeping parameter ($\lambda = 1$):

$$Q = V^{-N} \int d\mathbf{q}_1 \dots \int d\mathbf{q}_N \sum_{s=0}^{\infty} \left(-\frac{\lambda}{kT} \right)^s \frac{1}{s!} \left(\sum_{i<j=1}^{N} \phi_{ij} \right)^s.$$

Problems encountered with this expansion depend on choice of ϕ_{ij}. For instance:

(1) Problems with divergent integrals: need regularization.
(2) Number of terms increases rapidly with s: combinatorial divergence
(3) Lack of small parameter: series is divergent.

For a dense gas expect an equation of state

$$PV = NkT[1 + A_1(T)n + A_2(T)n^2 + \dots]$$

where n is number density and $A_1, A_2 \dots$ are virial coefficients. Introduce **Mayer functions**

$$f_{ij} = e^{-\phi_{ij}/kT} - 1.$$

Expansion for configuration integral becomes

$$Q = \frac{1}{V^N} \int d\mathbf{q}_1 \dots \int d\mathbf{q}_N \left[1 + \sum_{i<j} f_{ij} + \sum_{i<j} \sum_{k<l} f_{ij} f_{kl} + \dots \right].$$

Terms in the expansion correspond to molecular clusters: lead to cluster integrals.

Introduce the free energy using the bridge equation: obtain F as a power series in n, the number density. The general procedure is very complicated and involves partial summations of various classes of terms. For small n, we can derive the Van der Waal's equation for an imperfect gas:

$$\left(p + \frac{a}{V^2} \right) (V - b) = NkT.$$

Here a/V^2 represents mutual attraction between molecules and b is the excluded volume due to finite size of molecules. This is best demonstrated using diagrams: a line segment stands for a Mayer function connecting two particles.

Two–particle cluster

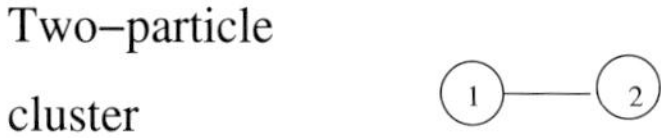

Three–particle clusters

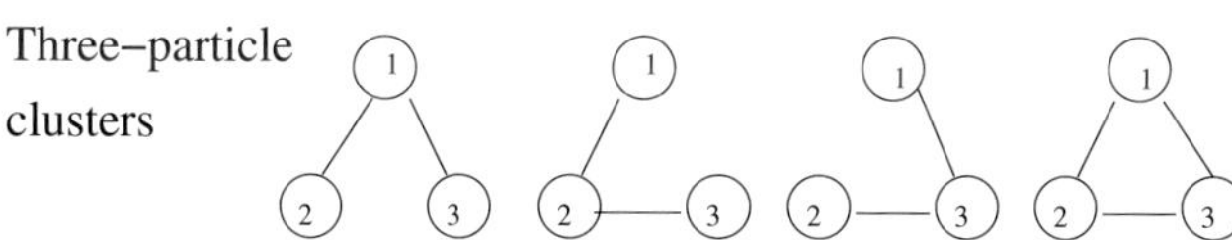

Fig. 2.1.   Mayer diagrams for two-particle and three-particle clusters of molecules.

Four–particle clusters

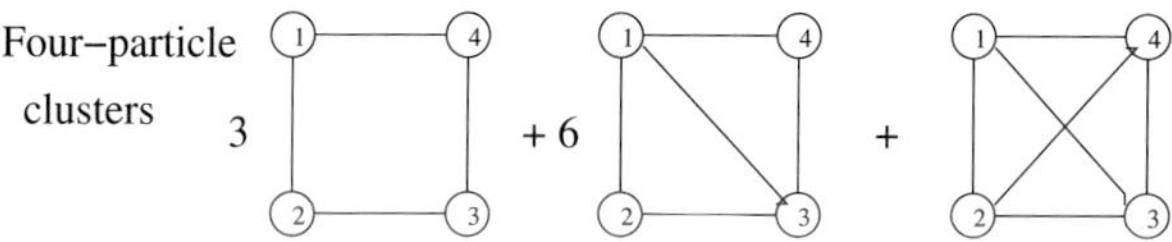

Fig. 2.2.   Mayer diagrams for four-particle clusters.

2.2.3.3.   *High-temperature expansions*

Starting point is the partition function:

$$Z = tr\ \exp[-\beta H],$$

where 'tr' stands for 'sum over states' and as usual $\beta \equiv 1/kT$. Expand out the exponential as a power series in $\beta \equiv 1/kT$. Thus:

$$\exp[-\beta H] = 1 - \beta H + \beta^2 H^2/2 + \ldots$$

For $T \gg T_c$, only a few terms needed. For $T \to T_c$, need lots of terms. The interesting thing is that it can still be used to find both critical temperature and critical exponents. The ratio test for convergence can be used to locate a simple pole which corresponds to the critical point.

2.2.3.4.   *Conclusions about perturbation expansions*

Re-expansion in powers of a control parameter such as density or temperature (and even fractional dimension) is usually necessary. Special techniques such as locating singularities or using Padé approximants may be helpful. Partial summation of restricted classes of terms may be possible. Identify classes of terms by the topology of equivalent diagrams: see Debye-Hückel revisited.

2.2.4.   *Mean-field theories*

Some well known ones are as follows:

(1) Debye-Hückel theory of the screened potential.
(2) Weiss theory of magnetism.
(3) Van der Waal's equation provides a mean-field theory for the gas-liquid phase transition.
(4) Bogoliubov variational principle and mean field theory of the Ising model.

2.2.4.1. *Debye-Hückel theory*

Consider an electron gas or an electrolyte as N electrons free to move in a uniform background of positive charge. The bare potential at r due to one electron at $r = 0$ is the Coulomb form:

$$\phi_0(r) = \frac{e}{r}.$$

The dressed potential at r due to one electron at $r = 0$ is

$$\phi(r) = \frac{e}{r} \exp[-r/l_D],$$

where l_D is Debye length:

$$l_D = \left[\frac{4\pi e^2 n_\infty}{kT} \right]^{-1/2},$$

and n_∞ is the uniform charge density.

(1) **Mean-field assumption.** There exists an effective potential $\phi(r)$ such that the probability of finding another electron at distance r from the first is

$$p(r) = \frac{e^{-e\phi(r)/kT}}{Z}.$$

(2) **Self-consistent assumption.** Solve Poisson's equation self-consistently for ϕ:

$$\nabla^2 \phi = 4\pi e n_\infty \left[e^{-e\phi/kT} - 1 \right].$$

Note that the dressed, or screened potential can be interpreted as charge renormalization:

$$e \to e \exp\left[-r/l_D\right].$$

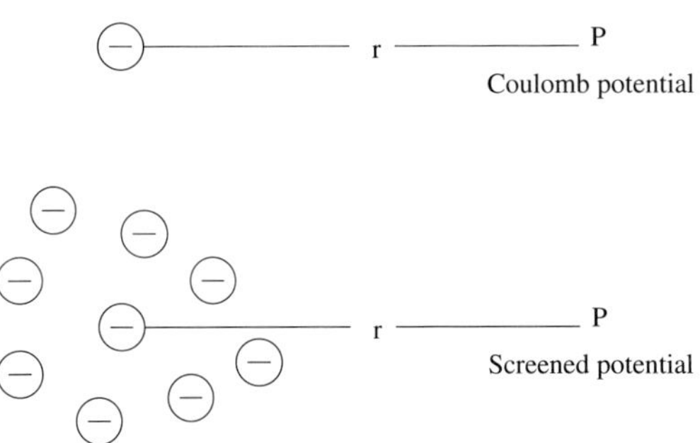

Fig. 2.3.   The effect of a cloud of electrons is to screen the potential.

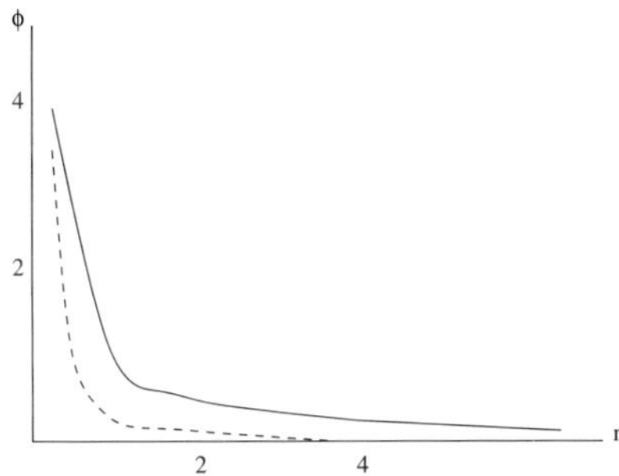

Fig. 2.4.   Screened potential (dotted line) and Coulomb potential (solid line).

2.2.4.2.   *Debye-Hückel revisited*

The problem of the electron gas can also be tackled using the virial expansion method. It is found that the Debye-Hückel result for the screened potential can be recovered by summing the simply-connected or ring diagrams in the Mayer series. The ring diagrams are shown up to sixth order in Fig. 2.5.. This is an example of the perturbation and mean field theories being combined. Another example which we do not discuss here is the use of the Bogoliubov variational principle to extend the mean-field theory of the Ising model: see, for example, the book by McComb.[2]

2.2.4.3.   *Weiss theory of ferromagnetism*

A magnet consists of a lattice of N spins, in an external field B. When spins align, there is a net magnetization M. If this happens when $B = 0$, then there is spontaneous symmetry breaking.

                           David McComb

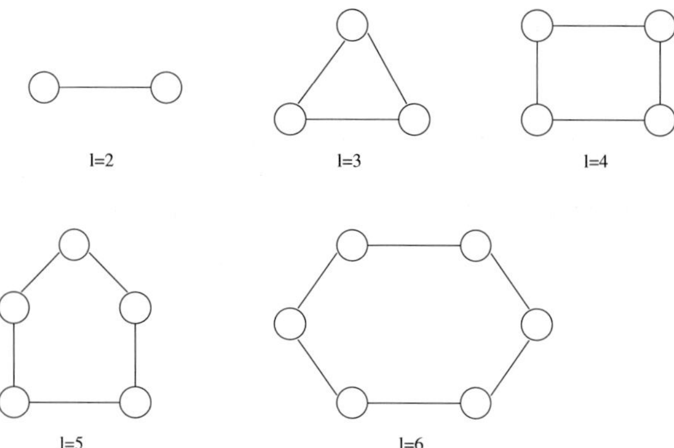

Fig. 2.5.   Debye-Hückel revisited: contribution from irreducible cluster diagrams to sixth order.

(1) **Mean-field assumption**: any one spin experiences an effective field B_E made up of external field B and molecular field B' due to all other spins.

$$B_E = B + B'.$$

(2) **self-consistent assumption**: $B' \propto M$.

Leads to

$$\frac{M}{M_\infty} = \tanh\left[\frac{Jz}{kT} \cdot \frac{M}{M_\infty}\right],$$

where M_∞ is saturation magnetization, J is exchange interaction, and z is the lattice coordination number or number of nearest neighbours.

This transcendental equation can be solved graphically for the critical temperature: $T_c = Jz/k$.

2.2.5. *Problems with many scales: the renormalization group*

Magnetization arises as fluctuations in which spins become aligned over a distance $\xi(T)$: the correlation length. At high temperatures $T \gg T_c$, thermal disorder ensures there is no correlation. At low temperatures $T \to 0$, spins line up, ultimately over the length of the lattice L. At intermediate temperatures

$$a \le \xi \le L.$$

At $T \to T_c$, fluctuations exist on all scales in between a and L. All these scales are equally important.

In contrast, consider problems with few scales, such as:

(1) Non-equilibrium dilute gas. Here we have two scales. The duration of molecular collision (short) and the time between collisions (long). The ratio of the two scales gives a perturbation parameter.
(2) Laminar flow of a viscous liquid along a circular pipe. Here the pipe of radius R provides a characteristic scale and the maximum velocity at $r = 0$, $U(0)$ provides another. This leads to a **universal** form

$$\frac{U(r)}{U(0)} = \left(1 - \frac{r^2}{R^2}\right),$$

which holds for all pipes and all fluids provided the Reynolds number is small.

2.2.5.1. *How do we tackle problems of many scales?*

Hope to find:

(1) geometrical similarity;
(2) self-similarity;
(3) scale invariance.

These will lead to:

(1) mathematical homogeneity;
(2) fractal structure;
(3) fixed points in the theory of magnetism.

These lead on to:

(1) Static scaling hypothesis (Widom 1965): 'Gibbs free energy $G(\theta_c, B)$ is a generalized homogeneous function'.
(2) Renormalization Group (Wilson 1971); RG derives the scaling form of the free energy.

Both of these theories lead to relationships between critical exponents.

2.2.5.2. *Real-space renormalization group (RG)*

Each RG transformation has two steps:

(1) 'Coarse-grain' our microscopic description.
(2) Re-scale lengths to try to restore the original picture.

If the original picture is restored, we call this a fixed point. As an example, consider blocking the lattice for the case of a square lattice with $d = 2$

Replace the 4 spin values in one block by an effective spin value at the centre of the block. Then the new lattice has to have its lattice constant reduced by a factor $b = 2$ in each direction to restore the original lattice size.

The second lattice has a factor of $b^d = b^2 = 4$ fewer lattice sites than the original one. This procedure is illustrated in Fig. 2.6..

If the 'spin field' is scale invariant, the new lattice may be similar to the old one in its properties. This is identified mathematically if a sequence of transformations reaches a fixed point. The transformations are actually on the Hamiltonian:

$$H_{n+1} = R_b H_n,$$

where R_b stands for the RG transformation (or RGT) and b is the spatial rescaling factor. A fixed point is identified when:

$$H_{n+1} = R_b H_n = H_n = H_N = H^*.$$

Either a capital letter for the subscript or an asterisk superscript is used to indicate a fixed-point value.

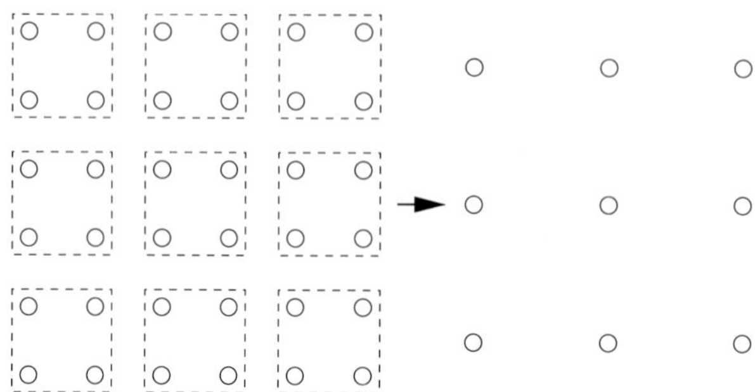

Fig. 2.6.   Blocking the lattice.

2.2.5.3.   *Discrete dynamical systems, recursion relations and fixed points.*

As an example: borrow £$X(0)$ at 1% interest per month, repay at £20 per month. After one month you owe:

$$X(1) = X(0) + 0.01X(0) - 20 = 1.01X(0) - 20.$$

After $n + 1$ months:

$$X(n + 1) = 1.01X(n) - 20.$$

Consider three different initial amounts:

$$X(0) = \pounds1000, \quad X(1) = \pounds990, \quad X(2) = \pounds979.90\ldots$$
$$X(0) = \pounds3000, \quad X(1) = \pounds3010, \quad X(2) = \pounds3020.10\ldots$$
$$X(0) = \pounds2000, \quad X(1) = \pounds2000, \quad X(3) = \pounds2000\ldots.$$

Evidently $X(0) = \pounds2000$ is a fixed point of the system.

2.2.5.4. *A system with two fixed points*

Consider the dynamical system,

$$X(n+1) = [X(n) + 4]X(n) + 2.$$

It is easily shown that it has two fixed points $c = -1$ and $c = -2$. Hence for $X(0) = -1, X(n) = -1$; or $X(0) = -2, X(n) = -2$, for all n.

Consider $X(0) = -1.01, -0.99, -2.4$ and sketch the solutions. From the graphs in Fig. 2.7.:

- $c = -1$ is a repelling fixed point (unstable equilibrium).
- $c = -2$ is an attractive fixed point (stable equilibrium).
- $X(0) = -0.99$ is not within the "basin of attraction" of $c = -2$.

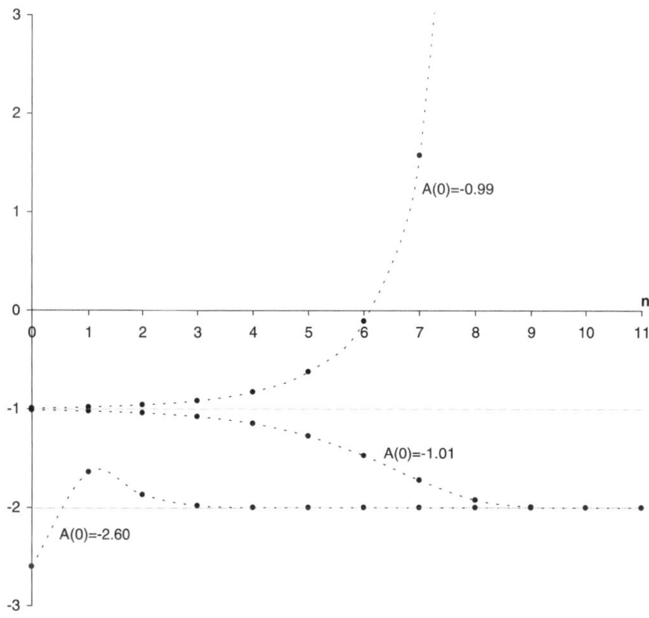

Fig. 2.7.   Behaviour of a system with two fixed points.

2.2.5.5. *Fixed points for a magnet*

We refer here to Figs. 8-10.

- Case 1. $T = 0$. All spins aligned. RGT must give same result however much we 'coarse grain'.

- Case 2. $T \to \infty$. All spins randomly oriented: hence RGT must give same results at every stage.
- Case 3. $T \geq T_c$. Spins aligned over a correlation length $\xi(T)$.

For finite ξ, a finite number of RGTs will hide ordering effects. For $\xi(T_c) \to \infty$, no finite number of RGTs can hide the correlation.

As a result, we can identify:

$$\text{fixed point} \equiv \text{critical point},$$

if $\xi(T_c) \to \infty$.

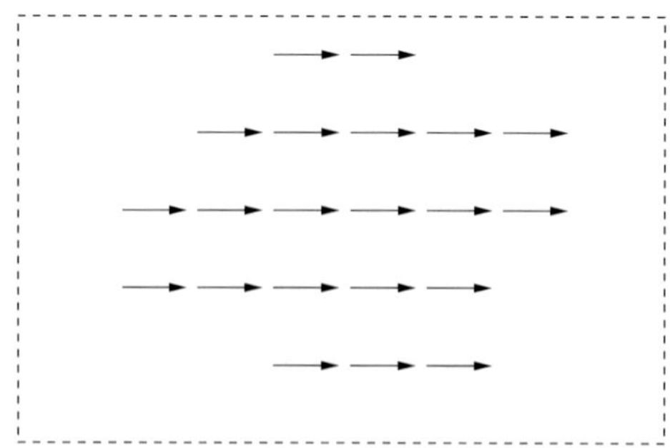

Fig. 2.8. At low temperatures the spins on a lattice line up in one direction.

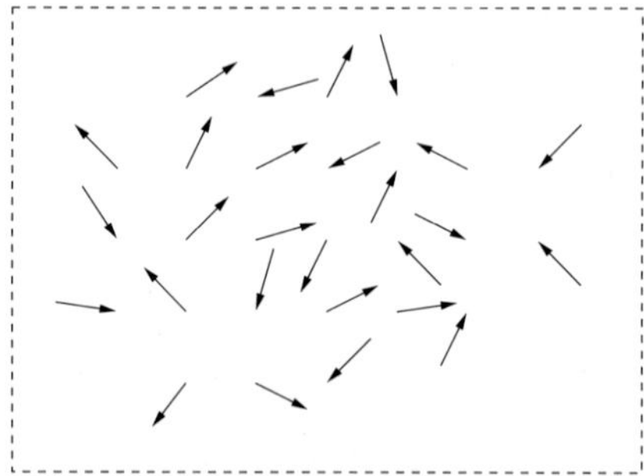

Fig. 2.9. At high temperatures the spins on a lattice are oriented at random.

2.2.6. *Application of renormalization methods to turbulence*

In wavenumber (k) space, the Navier-Stokes equations are equivalent to a quantum field theory with the Reynolds number as the coupling constant.

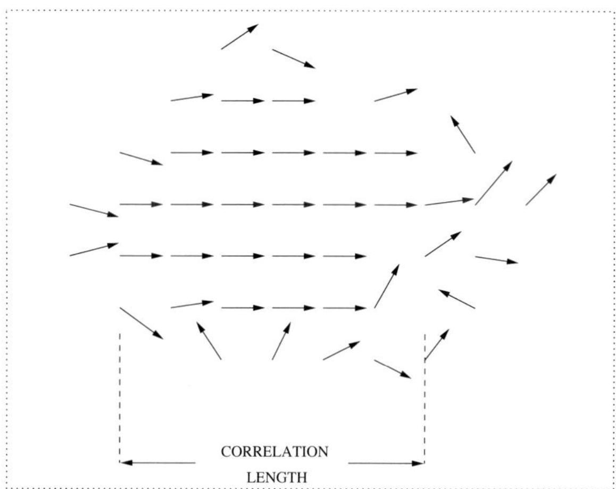

Fig. 2.10.   Correlations develop at just above critical temperature.

The molecular viscosity ν_0 may be renormalized by the collective effects of turbulent eddies to an effective form $\nu_T(k)$. The viscous response function $R_0(k; t - t') = \exp[-\nu_0 k^2(t - t')]$ may be renormalized to an effective form $R(k; t - t')$. Note that the latter is not necessarily of exponential form. Renormalization may lead to other effects such as eddy noise or stochastic backscatter.

Perturbation theory requires a soluble zero-order model. This is normally obtained by setting the nonlinear term equal to zero and introducing random stirring forces to generate random fluid motion. The stirring forces $\mathbf{f}(\mathbf{k}, \mathbf{t})$ are taken to have a Gaussian distribution. Their covariance is chosen to be of the form:

$$\langle \mathbf{f}(\mathbf{k}, \mathbf{t}) \cdot \mathbf{f}(\mathbf{k}', \mathbf{t}') \rangle = 2(2\pi)^3 D(k)\delta(\mathbf{k} + \mathbf{k}')\delta(t - t').$$

The stirring forces do work on the fluid at a rate given by

$$\varepsilon_W = \int d^3k\, D(k).$$

In general, the coupling constant (the Reynolds number) is large. Hence a successful renormalized perturbation theory (RPT) relies on the partial summation of certain classes of terms. However, if we apply RG to turbulence, then the coupling is determined by a local (in wavenumber) Reynolds number $\mathcal{R}_0(k)$. The local coupling is related to the energy spectrum $E(k)$ by

$$\mathcal{R}_0(k) = [E(k_0)]^{1/2}/\nu_0 k^{1/2}.$$

As Fig. 11 shows, this allows one to do perturbation theory as $k \to 0$ and $k \to \infty$. The former case has been much studied as it allows one to take over the results of the theory of critical phenomena. But it is remote from real turbulence.

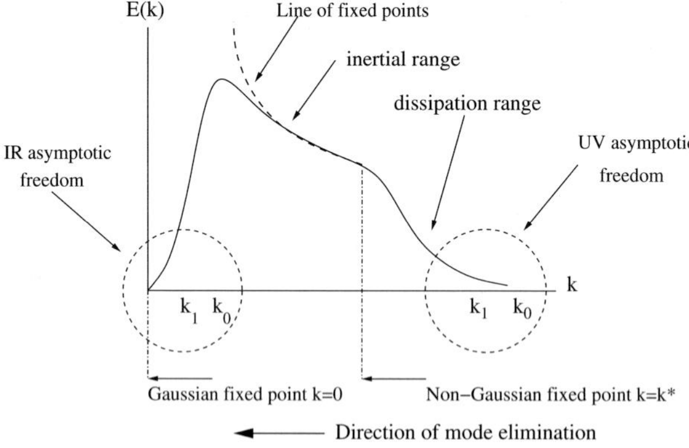

Fig. 2.11.   The turbulence energy spectrum indicating the application of RG methods.

2.3.   Renormalized perturbation theories and two-point turbulence closures

In this Section we begin with a short history of the closure problem, and then state the basic equations in wavenumber space (or k-space). This is followed by an outline of the quasi-normality theory, as the best known *ad hoc* method. We turn then to perturbation theory, as the only general method, and state the second-order covariance equations as an approximation to the required statistical closure. A comparison between these and the quasi-normality equations is used to motivate the introduction of renormalized perturbation theory or RPT; and at that point we list the main pioneering RPTs, outline the method (Kraichnan-Wyld perturbation theory) and state the resulting second-order covariance equations in terms of an unknown renormalized response function.

At this lowest non-trivial order, the turbulence closure problem can be seen as one of finding the renormalized response function. We then outline the derivation of the DIA equation for the response function, showing that it is a natural outcome of the Kraichnan-Wyld formulation. In Section 3.6.7 we state the resulting DIA response equation in detail, and anticipate later discussions by also stating the LET response equation in order to facilitate comparisons.

In Section 3.6.8 we discuss the failure of first-generation RPTs, as indicated by their incompatibility with the Kolmogorov '-5/3' spectrum (K41[3,4]), and in Section 3.6.8 we discuss the re-interpretation of this failure which led to the LET theory, which is compatible with K41. In Section 3.8 some numerical results are presented for the free decay of isotropic turbulence and also for stationary forced isotropic turbulence, as predicted by both DIA and LET theories. The thorny problem of unresolved and contentious issues is discussed in Section 3.8: as will be seen, some of these affect the way in which the turbulence community perceives RPTs and others are wider, affecting the whole of turbulence theory.

The section closes with a derivation of the LET response equation, making use of the latest developments in this subject, and also gives the extension to new single-time and Markovianized forms of the theory.

2.3.1. *A brief history of closures*

The turbulence closure problem was formulated for turbulent shear flows by Reynolds in the 1890s (see McComb (1990)[5] and references therein): the equation for the mean velocity $\bar{U}$ contains the unknown covariance of two fluctuating velocities $\langle uu \rangle$: the Reynolds stress.

It was later formulated for isotropic turbulence by Taylor in the 1930s (see Batchelor (1971)[6] and references therein): the equation for the correlation of two velocities $\langle uu \rangle$ contains the unknown correlation of three velocities $\langle uuu \rangle$. The equation for $\langle uuu \rangle$ contains the unknown $\langle uuuu \rangle$, and so on ...

The problem is seen as an open hierarchy of statistical equations (one-point for Reynolds, two-point for Taylor) which requires some *ansatz* for closure.

Eddy-viscosity, mixing-length and n-equation models are all effectively 'statistical closure approximations' for the single-point problem. The Heisenberg eddy viscosity and the quasi-normality theory are effectively closure approximations for the two-point problem. Quasi-normality in 1954 (see Leslie (1973)[7] or McComb (1990),[5] McComb (1995)[8]) was the first *formal* treatment of the closure problem: we solve the next equation in the hierarchy for $\langle uuu \rangle$ by factorizing $\langle uuuu \rangle$ in terms of $\langle uu \rangle \times \langle uu \rangle$.

Quasi-normality failed when computed numerically in the 1960s, as it predicted negative spectra: it is not physically realizable. Eulerian renormalized perturbation theories (DIA 1959);[9] (EFP 1964);[10] (SCF 1965)[11]) were physically realizable: but not compatible with K41.

Lagrangian-history theories (Kraichnan (1965),[12] (Kraichnan and Herring (1978),[13] Kaneda (1981),[14] Kida and Goto (1997)[15]) are claimed to be random Galilean invariant: the implication is that they are compatible with K41. The LET theory (McComb (1974)[16]) is compatible with K41: it is purely Eulerian.

2.3.2. *Basic equations in k-space*

The velocity field $\mathbf{u}(\mathbf{x}, t)$ can be expressed in terms of its Fourier transform $\mathbf{u}(\mathbf{k}, t)$, thus:

$$\mathbf{u}(\mathbf{x}, t) \equiv u_\alpha(\mathbf{x}, t) = \int d^3k \, u_\alpha(\mathbf{k}, t) \exp(i\mathbf{k} \cdot \mathbf{x}). \tag{2.1}$$

The Fourier transform pair is completed by

$$u_\alpha(\mathbf{k}, t) = \left(\frac{1}{2\pi}\right)^3 \int d^3x \, u_\alpha(\mathbf{x}, t) \exp(-i\mathbf{k} \cdot \mathbf{x}). \tag{2.2}$$

The covariance of velocities for homogeneous and isotropic turbulence takes the form:

$$\langle u_\alpha(\mathbf{k}, t) u_\beta(\mathbf{k}', t') \rangle = \delta(\mathbf{k} + \mathbf{k}') C_{\alpha\beta}(\mathbf{k}; t, t') = \delta(\mathbf{k} + \mathbf{k}') P_{\alpha\beta}(\mathbf{k}) C(k; t, t'), \tag{2.3}$$

where the projector $P_{\alpha\beta}(\mathbf{k})$ is expressed in terms of the Kronecker delta as

$$P_{\alpha\beta}(\mathbf{k}) = \delta_{\alpha\beta} - \frac{k_\alpha k_\beta}{|\mathbf{k}|^2}. \tag{2.4}$$

The continuity equation is

$$k_\alpha u_\alpha(\mathbf{k}, t) = 0, \tag{2.5}$$

and the solenoidal Navier-Stokes equation (NSE) takes the form

$$\left(\frac{\partial}{\partial t} + \nu_0 k^2\right) u_\alpha(\mathbf{k}, t) = M_{\alpha\beta\gamma}(\mathbf{k}) \int \mathrm{d}^3 j \, u_\beta(\mathbf{j}, t) u_\gamma(\mathbf{k} - \mathbf{j}, t). \tag{2.6}$$

The inertial transfer operator $M_{\alpha\beta\gamma}(\mathbf{k})$ is symmetric under the interchange of the indices β and γ and is given by

$$M_{\alpha\beta\gamma}(\mathbf{k}) = (2i)^{-1}[k_\beta P_{\alpha\gamma}(\mathbf{k}) + k_\gamma P_{\alpha\beta}(\mathbf{k})]. \tag{2.7}$$

It is sometimes convenient to define the linear operator $\mathcal{L}_0(k)$ as:

$$\mathcal{L}_0(k) = \left(\frac{\partial}{\partial t} + \nu_0 k^2\right). \tag{2.8}$$

If we wish to study the initial value problem posed by the free decay of turbulence then the turbulence ensemble is specified by the initial conditions. It is usual to do this by choosing the arbitrary velocity field at the initial time to be random with a Gaussian distribution. If we wish instead to study stationary turbulence, this requires the addition of a stirring force $f_\alpha(\mathbf{k}, t)$ to the Navier-Stokes equation. In this case we choose the stirring forces to be random with a Gaussian distribution and complete their specification by taking their autocorrelation to be:

$$\langle f_\alpha(\mathbf{k}, t) f_\beta(\mathbf{k}', t') \rangle = 2(2\pi)^3 P_{\alpha\beta}(\mathbf{k}) D(k) \delta(\mathbf{k} + \mathbf{k}') \delta(t - t'), \tag{2.9}$$

where $D(k)$ is the arbitrarily chosen force spectrum.

2.3.2.1. *Equation for the velocity covariance*

The velocity-field covariance $C_{\alpha\beta}(\mathbf{k}; t, t')$ is defined by equation (2.3). To obtain an equation for this, we multiply each term in (2.6) by $u_\sigma(-\mathbf{k}, t')$ and then average each term. The resulting equation is

$$\mathcal{L}_0(k) P_{\alpha\sigma}(\mathbf{k}) C(k; t, t') = M_{\alpha\beta\gamma}(\mathbf{k}) \int \mathrm{d}^3 j\, C_{\beta\gamma\sigma}(\mathbf{j}, \mathbf{k} - \mathbf{j}, -\mathbf{k}; t, t'). \quad (2.10)$$

Here $C_{\alpha\beta\gamma}(\mathbf{k}, \mathbf{j}, -\mathbf{k} - \mathbf{j}; t, t')$ stands for the three-velocity correlation. The aim of a closure approximation is to express the three-velocity correlation in terms of the pair correlation or covariance $C(k; t, t')$.

2.3.2.2. *Equation for the energy spectrum*

The energy spectrum $E(k, t)$ is related to the spectral density $C(k, t)$ by

$$E(k, t) = 4\pi k^2 C(k, t). \quad (2.11)$$

To obtain an equation for this, we first multiply each term in (2.6) by $u_\sigma(-\mathbf{k}, t)$. Then we form a second equation from (2.6) for $u_\sigma(-\mathbf{k}, t)$, multiply this by $u_\alpha(\mathbf{k}, t)$, add the two resulting equations together and average the final expression. The resulting equation is

$$\left(\frac{\partial}{\partial t} + 2\nu_0 k^2 \right) P_{\alpha\sigma}(\mathbf{k}) C(k, t) = M_{\alpha\beta\gamma}(\mathbf{k}) \int \mathrm{d}^3 j\, C_{\beta\gamma\sigma}(\mathbf{j}, \mathbf{k} - \mathbf{j}, -\mathbf{k}; t)$$

$$- M_{\sigma\beta\gamma}(\mathbf{k} \int \mathrm{d}^3 j\, C_{\beta\gamma\alpha}(\mathbf{j}, -\mathbf{k} - \mathbf{j}, \mathbf{k}; t). \quad (2.12)$$

Here $C_{\alpha\beta\gamma}(\mathbf{k}, \mathbf{j})$ stands for the three-velocity correlation.

We then set $\sigma = \alpha$, sum over α (noting that $Tr\, P_{\alpha\beta} = 2$) and multiply each term in (2.12) by $2\pi k^2$, to obtain:

$$\left(\frac{\partial}{\partial t} + 2\nu_0 k^2 \right) E(k, t) = T(k, t). \quad (2.13)$$

The energy transfer spectrum $T(k, t)$ is given by

$$T(k, t) = 2\pi k^2 M_{\alpha\beta\gamma}(\mathbf{k}) \int \mathrm{d}^3 j\, \{ C_{\beta\gamma\alpha}(\mathbf{k} - \mathbf{j}, -\mathbf{k}, t) - C_{\beta\gamma\alpha}(-\mathbf{k} - \mathbf{j}, \mathbf{k}, t) \}.$$

$$(2.14)$$

This is an example of the statistical closure problem: an equation for a second-order moment contains an unknown third-order moment. And so on ...

2.3.2.3. *Dissipation rate ε in k-space*

By definition, $\varepsilon = -dE/dt$ for freely decaying turbulence. Integrating over wavenumber, the energy balance becomes:

$$\frac{dE}{dt} = -\varepsilon = -\int_0^\infty 2\nu_0 k^2 E(k,t)dk. \tag{2.15}$$

This is because the inertial transfer term vanishes when integrated over all k. The region in k-space where the dissipation mainly occurs is characterised by the Kolmogorov dissipation wavenumber: $k_d = (\varepsilon/\nu_0)^{1/4}$.

2.3.2.4. *Inertial transfer in k-space*

Write $T(k,t)$ as

$$T(k,t) = \int_0^\infty S(k,j)\,dj, \tag{2.16}$$

where S depends on the triple moment. It can be shown that S is antisymmetric under the interchange $k \rightleftharpoons j$:

$$S(k,j;t) = -S(j,k;t). \tag{2.17}$$

Hence

$$\int_0^\infty T(k,t)dk = \int_0^\infty dk \int_0^\infty dj\, S(k,j;t) = 0, \tag{2.18}$$

is an exact symmetry which expresses conservation of energy.

For stationarity we must add an input spectrum $W(k) = 4\pi k^2 D(k)$, where $D(k)$ is defined by equation (2.9). Then $dE(k,t)/dt = 0$, and the energy balance becomes:

$$T(k) + W(k) - 2\nu_0 k^2 E(k) = 0. \tag{2.19}$$

At sufficiently high Reynolds numbers, assume there is a wavenumber κ such that input effects are below it and dissipation effects above it. That is, for a well-posed problem:

$$\int_0^\kappa W(k)dk \simeq \varepsilon \simeq -\int_\kappa^\infty 2\nu_0 k^2 E(k)dk. \tag{2.20}$$

We can obtain low-k and high-k balance equations by first integrating from zero up to κ and then from infinity down to κ. First,

$$\int_0^\kappa dk \int_\kappa^\infty dj\, S(k,j) + \int_0^\kappa W(k)dk = 0. \tag{2.21}$$

i.e. energy supplied directly by the input term to modes with $k \leq \kappa$ is transferred by the nonlinearity to modes with $j \geq \kappa$. Thus $T(k)$ behaves like a dissipation and absorbs energy. Second,

$$\int_\kappa^\infty dk \int_0^\kappa dj\, S(k,j) - \int_\kappa^\infty 2\nu_0 k^2 E(k)dk = 0. \tag{2.22}$$

i.e. nonlinearity transfers energy from modes with $j \leq \kappa$ to modes with $k \geq \kappa$, where it is dissipated into heat. In this range of wavenumbers $T(k)$ behaves like a source and emits energy which is then dissipated by viscosity.

2.3.3. *Quasi-normality hypothesis*

The covariance equation on the time diagonal takes the form:

$$\left(\frac{\partial}{\partial t} + 2\nu_0 k^2\right) C\left(k,t\right) = \int d^3 j L\left(\mathbf{k},\mathbf{j}\right)$$

$$\times \int_0^t ds R_0\left(k;t,s\right) R_0\left(j;t,s\right) R_0\left(\left|\mathbf{k}-\mathbf{j}\right|;t,s\right)$$

$$\times 2\left[C\left(j,s\right) C\left(\left|\mathbf{k}-\mathbf{j}\right|,s\right) - C\left(k,s\right) C\left(\left|\mathbf{k}-\mathbf{j}\right|,s\right)\right].(2.23)$$

The viscous response function is given by:

$$R_0(k;t,s) = \exp[-\nu_0 k^2(t-s)], \tag{2.24}$$

and the coefficient $L(\mathbf{k},\mathbf{j})$ is defined as:

$$L(\mathbf{k},\mathbf{j}) = -2M_{\alpha\beta\gamma}(\mathbf{k})M_{\beta\alpha\delta}(\mathbf{j})P_{\gamma\delta}(\mathbf{k}-\mathbf{j}). \tag{2.25}$$

The coefficient $L(\mathbf{k},\mathbf{j})$ is discussed later in Section 2.3.6.3.

2.3.4. *Perturbation theory*

Add a stirring force $f_\alpha(\mathbf{k},t)$ and a book-keeping parameter λ to the NSE:

$$\mathcal{L}_0(k)u_\alpha(\mathbf{k},t) = f_\alpha(\mathbf{k},t) + \lambda M_{\alpha\beta\gamma}(\mathbf{k})\int d^3 j\, u_\beta(\mathbf{j},t)u_\gamma(\mathbf{k}-\mathbf{j},t). \tag{2.26}$$

$\lambda = 0$ (linear system) or $\lambda = 1$ (nonlinear system): thus λ is also a control parameter. We make the perturbation expansion

$$u_\alpha(\mathbf{k},t) = u_\alpha^{(0)}(\mathbf{k},t) + \lambda u_\alpha^{(1)}(\mathbf{k},t) + \lambda^2 u_\alpha^{(2)}(\mathbf{k},t)\dots. \tag{2.27}$$

We take $u^{(0)}$ to be Gaussian, through our choice of the random stirring force f and calculate the higher-order coefficients $u^{(1)}$, $u^{(2)}$, … in terms of it. Then we can obtain an equation for the exact covariance C in terms of an expansion in the bare covariance C_0 and the viscous response function R_0.

2.3.4.1. *Second-order covariance equations*

The second-order equation for the velocity covariance is:

$$\left[\frac{\partial}{\partial t} + \nu_0 k^2\right] C(k;t,t')$$

$$= \int d^3 j L(\mathbf{k},\mathbf{j})\left[\int_0^{t'} ds R_0(k;t',s)C_0(j;t,s)C_0(|\mathbf{k}-\mathbf{j}|;t,s)\right.$$

$$\left. - \int_0^t ds R_0(j;t,s)C_0(k;s,t')C_0(|\mathbf{k}-\mathbf{j}|;t,s)\right] + \mathcal{O}\left(\lambda^3\right). \tag{2.28}$$

And on the time diagonal:

$$\left(\frac{\partial}{\partial t} + 2\nu_0 k^2\right) C\left(k, t\right)$$

$$= 2 \int d^3 j L\left(\mathbf{k}, \mathbf{j}\right) \int_0^t ds R_0\left(k; t, s\right) R_0\left(j; t, s\right) R_0\left(\left|\mathbf{k} - \mathbf{j}\right|; t, s\right) \times$$

$$\left[C_0\left(j, s\right) C_0\left(\left|\mathbf{k} - \mathbf{j}\right|, s\right) - C_0\left(k, s\right) C_0\left(\left|\mathbf{k} - \mathbf{j}\right|, s\right)\right] + \mathcal{O}\left(\lambda^3\right). \quad (2.29)$$

2.3.5. *Quasi-normality versus perturbation theory*

Compare equation (2.23) from QN (right hand side in terms of R_0 and C) with (2.29) from perturbation theory (right hand side in terms of R_0 and C_0): i.e. QN looks already partially renormalized. In fact the QN equation is an approximation (i.e. quasi-Gaussian) whereas equation (2.29) is an exact second-order truncation of the perturbation series and is fully Gaussian in nature. QN can be renormalized by the *ad hoc* replacement of ν_0 in the viscous response by an effective turbulence viscosity (introducing an adjustable constant in the process): this leads on to the EDQNM family of closures. A renormalization programme consists of making the consistent replacements $R_0 \to R$ and $C_0 \to C$ on the right hand side equations (2.28) and (2.29). Additionally we require some principle to determine R.

2.3.6. *Renormalised perturbation theory (RPT): the general idea*

There is no additional control parameter (e.g. density or temperature) so the perturbation series has to be renormalised by partial summation. The primitive perturbation series for the actual convariance C and response R can be written as coupled equations of the form:

(1) $\mathcal{L}_{0k} C_k =$ infinite series involving C^0 and R^0.
(2) $\mathcal{L}_{0k} R_k =$ infinite series involving C^0 and R^0.

Renormalisation means making the replacements

$$C_k^0 \to C_k \text{ and } R_k^0 \to R_k,$$

on the right hand side. In practice some other step is also needed.

The replacement of C_k^0 and R_k^0 by C_k and R_k can be justified (to some extent) on topological grounds using diagrams. It can also be partially justified by reversion of power series. It means replacing the mythical C_k^0 by the observable C_k. It also means replacing the observable R_k^0 by the mythical R_k. In this respect it differs from quantum field theory where bare Green functions are replaced by renormalised observable Green functions.

2.3.6.1. *Pioneering RPTs*

We now list the pioneering renormalized perturbation theories, as follows:

(1) Kraichnan (1959): direct interaction approximation (DIA). Successful in both qualitative and quantitative terms but does not give the K41 '-5/3' spectrum at large Reynolds numbers.[9]
(2) Wyld (1961): formal analysis of the Navier-Stokes perturbation expansion in terms of diagrams with renormalisation arising from purely topological considerstions. Showed DIA response equation emerged naturally at lowest nontrivial order.[17]
(3) Edwards (1964): derived Liouville equation for the p.d.f. of turbulent velocities. Approximation by Fokker-Planck equation determined renormalised response. Strong point is that is an expansion about Gaussian.[10] Weak point is its restriction to single-time forms. Can be related to DIA.
(4) Herring (1965): self-consistent field theory. A more abstract version of the Edwards theory. A two-time generalisation (1966) is very similar to DIA.[11,18]
(5) Lee (1965): corrected a double-counting problem in the Wyld analysis. This problem need not have arisen in the first place (McComb 1990).[5,19]

2.3.6.2. *Kraichnan-Wyld perturbation theory*

Add a stirring force $f_\alpha(\mathbf{k}, t)$ and a book-keeping parameter to the NSE, as given by (2.26), and repeated here for convenience:

$$\mathcal{L}_0(k)u_\alpha(\mathbf{k}, t) = f_\alpha(\mathbf{k}, t) + \lambda M_{\alpha\beta\gamma}(\mathbf{k}) \int \mathrm{d}^3 j \, u_\beta(\mathbf{j}, t)u_\gamma(\mathbf{k} - \mathbf{j}, t).$$

As before, $\lambda = 0$ (linear system) or $\lambda = 1$ (nonlinear system): thus λ is also a control parameter. If we scaled variables in a suitable way, we could replace λ by a Reynolds number. Hence the perturbation expansion

$$u_\alpha(\mathbf{k}, t) = u_\alpha^{(0)}(\mathbf{k}, t) + \lambda u_\alpha^{(1)}(\mathbf{k}, t) + \lambda^2 u_\alpha^{(2)}(\mathbf{k}, t) \ldots.$$

Introduce a simplified notation: the NSE becomes

$$\mathcal{L}_{0k}u_k = f_k + \lambda M_{kjl}u_j u_l.$$

Write the inverse of the operator on the left hand side as:

$$\mathcal{L}_{0k}^{-1} \equiv R_k^{(0)} \equiv \text{ the viscous response function.}$$

Zero-order solution is:

$$u_k^{(0)} = R_k^{(0)} f_k.$$

For convenience, re-write the NSE as:

$$u_k = u_k^{(0)} + \lambda R_k^{(0)} M_{kjl} u_j u_l.$$

Substitute the perturbation series into this, multiply out, and equate terms at each order of λ.

The coefficients in the perturbation series are given by:

$$\text{order } \lambda^0 : \ u_k^{(0)} = R_k^{(0)} f_k;$$

$$\text{order } \lambda^1 : \ u_k^{(1)} = R_k^{(0)} M_{kjl} u_j^{(0)} u_l^{(0)};$$

$$\text{order } \lambda^2 : \ u_k^{(2)} = 2 R_k^{(0)} M_{kjl} u_j^{(0)} R_l^{(0)} M_{lpq} u_p^{(0)} u_q^{(0)},$$

and so on. The exact covariance is then given by:

$$C_k = \langle u_k u_{-k} \rangle = \langle u_k^{(0)} u_{-k}^{(0)} \rangle + \langle u_k^{(0)} u_{-k}^{(2)} \rangle$$

$$+ \langle u_k^{(1)} u_{-k}^{(1)} \rangle + \langle u_k^{(2)} u_{-k}^{(0)} \rangle$$

$$+ \mathcal{O}\left(\lambda^4\right).$$

Substituting for the coefficients $u_k^{(1)}, u_k^{(2)}, \ldots$

$$C_k = C_k^{(0)} + 2 R_k^{(0)} M_{kjl} M_{lpq} R_l^{(0)} \langle u_k^{(0)} u_j^{(0)} u_p^{(0)} u_q^{(0)} \rangle$$

$$+ R_k^{(0)} M_{kjl} M_{-kpq} R^{(0)} \langle u_j^{(0)} u_l^{(0)} u_p^{(0)} u_q^{(0)} \rangle$$

$$+ 2 R_k^{(0)} M_{kjl} M_{lpq} R_l^{(0)} \langle u_k^{(0)} u_j^{(0)} u_p^{(0)} u_q^{(0)} \rangle + \mathcal{O}\left(\lambda^4\right).$$

Factorise the moments, using Gaussian statistics:

$$\langle u_k^{(0)} u_j^{(0)} u_p^{(0)} u_q^{(0)} \rangle = \delta_{kj} \delta_{pq} P_k P_p C_k^{(0)} C_p^{(0)}$$

$$+ \delta_{kp} \delta_{qj} P_k P_j C_k^{(0)} C_j^{(0)} + \delta_{kq} \delta_{jp} P_k P_q C_k^{(0)} C_q^{(0)}.$$

We can combine all the M's and P's into a simple coefficient $L(k, j)$: details are given in the next section.

2.3.6.3. *The L coefficients in turbulence theory*

The coefficient $L(\mathbf{k},\mathbf{j})$ is defined as:

$$L(\mathbf{k},\mathbf{j}) = -2M_{\alpha\beta\gamma}(\mathbf{k})M_{\beta\alpha\delta}(\mathbf{j})P_{\gamma\delta}(\mathbf{k}-\mathbf{j}). \tag{2.30}$$

The coefficient $L(\mathbf{k},\mathbf{j})$ can be evaluated as:

$$L(\mathbf{k},\mathbf{j}) = -\frac{\left[\mu\left(k^2+j^2\right)-kj\left(1+2\mu^2\right)\right]\left(1-\mu^2\right)kj}{k^2+j^2-2kj\mu} \tag{2.31}$$

where μ is the cosine of the angle between the vectors $\mathbf{k}$ and $\mathbf{j}$. Alternatively, the coefficient $L(\mathbf{k},\mathbf{k}-\mathbf{j})$ is defined as:

$$L(\mathbf{k},\mathbf{k}-\mathbf{j}) = -2M_{\alpha\beta\gamma}(\mathbf{k})M_{\beta\alpha\delta}(\mathbf{k}-\mathbf{j})P_{\delta\gamma}(\mathbf{j}) \tag{2.32}$$

The coefficient $L(\mathbf{k},\mathbf{k}-\mathbf{j})$ can be evaluated as:

$$L(\mathbf{k},\mathbf{k}-\mathbf{j}) = \frac{(k^4-2k^3j\mu+kj^3\mu)(1-\mu^2)}{k^2+j^2-2kj\mu} \tag{2.33}$$

2.3.6.4. *The Kraichnan-Wyld covariance equations*

The second-order equation for the velocity covariance is:

$$\left[\frac{\partial}{\partial t}+\nu_0 k^2\right]C(k;t,t')$$

$$= \int d^3 j L(\mathbf{k},\mathbf{j})\left[\int_0^{t'} ds R(k;t',s)C(j;t,s)C(|\mathbf{k}-\mathbf{j}|;t,s)\right.$$

$$\left. - \int_0^t ds R(j;t,s)C(k;s,t')C(|\mathbf{k}-\mathbf{j}|;t,s)\right]; \tag{2.34}$$

and on the time diagonal:

$$\left(\frac{\partial}{\partial t}+2\nu_0 k^2\right)C(k,t)$$

$$= 2\int d^3 j L(\mathbf{k},\mathbf{j})\int_0^t ds R(k;t,s)R(j;t,s)R(|\mathbf{k}-\mathbf{j}|;t,s)\times$$

$$[C(j,s)C(|\mathbf{k}-\mathbf{j}|,s)-C(k,s)C(|\mathbf{k}-\mathbf{j}|,s)]. \tag{2.35}$$

These equations are an exact second-order truncation of renormalized perturbation theory. Their derivation can rely on either the topology of Feynmann-type diagrams or reversion of power series. They satisfy all the required symmetries and in particular the closure conserves energy, displaying the correct antisymmetric behaviour of the transfer spectrum. With an appropriate choice of renormalized response function R they can predict the free decay of isotropic turbulence, without invoking arbitrary constants, in agreement with experiment and numerical simulation. The turbulence problem in this context becomes one of finding a principle to determine the renormalised response function. Such a principle leads to a renormalised perturbation theory or RPT.

2.3.6.5. *Renormalised response functions and RPTs*

All RPTs rely on some use of linear response theory which is in some sense renormalized. All RPTs can be interpreted as mean-field theories. The DIA response function corresponded to the summation of certain classes of terms to all orders in perturbation theory (Wyld 1961). The EFP (Edwards 1964) and SCF (Herring 1965) theories were single-time self-consistent theories which are cognate to DIA. A later two-time version version of SCF (Herring 1966) was very similar to DIA. All these theories have an incorrect interpretation of energy conservation in terms of their response function. The LET theory (McComb 1974) used a local (in wavenumber) energy balance to determine the response function.

2.3.6.6. *Derivation of DIA*

We now have our equation for C_k as:

$$C_k = C_k^{(0)} + R_k^{(0)} L(k,j) R_l^{(0)} C_j^{(0)} C_k^{(0)}$$

$$+ R_k^{(0)} L(k,j) \left[R_k^{(0)} C_j^{(0)} C_l^{(0)} - R_j^{(0)} C_l^{(0)} C_k^{(0)} \right] + \mathcal{O}\left(\lambda^4\right).$$

Let us remind ourselves what we are trying to do. Multiply the NSE through by u_{-k} and average:

$$\mathcal{L}_{0k} C_k = \langle f_k u_{-k} \rangle + M_{kjl} \langle u_j u_l u_{-k} \rangle.$$

The first term on the rhs is the input term. This can be worked out exactly and gives us a renormalised R_k.

The second term on the rhs is the energy transfer due to the nonlinearity. It can only be treated approximately. We deal with the input term first: recall $u_k^{(0)} = R_k^{(0)} f_k$ and $R_k^{(0)} R_k^{(0)} = R_k^{(0)}$. Re-write the expansion for C_k as:

$$C_k = R_k^{(0)} \langle f_k R_k^{(0)} f_{-k} \rangle + R_k^{(0)} L(k,j) R_l^{(0)} C_j^{(0)} C_k^{(0)} + \mathcal{O}\left(\lambda^4\right)$$

$$+ R_k^{(0)} L(k,j) \left[R_k^{(0)} C_j^{(0)} C_l^{(0)} - R_j^{(0)} C_l^{(0)} C_k^{(0)} \right] + \mathcal{O}\left(\lambda^4\right).$$

Express $C_k^{(0)}$ in terms of stirring forces in the first nonlinear term on the rhs.

$$C_k = R_k^{(0)} \left\langle f_k \left\{ R_k^{(0)} + R_k^{(0)} L(k,j) R_l^{(0)} C_j^{(0)} + \mathcal{O}\left(\lambda^4\right) \right\} f_{-k} \right\rangle$$

$$+ R_k^{(0)} L(k,j) \left[R_k^{(0)} C_j^{(0)} C_l^{(0)} - R_j^{(0)} C_l^{(0)} C_k^{(0)} \right] + \mathcal{O}\left(\lambda^4\right).$$

Identify the quantity in curly brackets as the exact response function (to order λ^4). Hence, recalling that $\langle f_k f_{-k} \rangle = W_k$:

$$C_k = R_k^{(0)} R_k W_k + R_k^{(0)} L(k,j) \left[R_k^{(0)} C_j^{(0)} C_l^{(0)} - R_j^{(0)} C_l^{(0)} C_k^{(0)} \right] + \mathcal{O}\left(\lambda^4\right).$$

This equivalence leads to

$$R_k = R_k^{(0)} + R_k^{(0)} L(k,j) R_l^{(0)} C_j^{(0)} + \mathcal{O}\left(\lambda^4\right).$$

Now take three steps with the C_k and R_k equations:

(1) Operate from the left with $\mathcal{L}_{0k}$.
(2) Replace $C_k^{(0)} \to C_k$ and $R_k^{(0)} \to R_k$ on rhs.
(3) Drop terms of order λ^4 and higher.

For the covariance we get

$$\mathcal{L}_{0k} C_k = W_k + L(k,j) \left[R_k C_j C_l - R_j C_l C_k \right]$$

For the response function we get

$$\mathcal{L}_{0k} R_k = \delta(t - t') + L(k,j) R_l C_j.$$

In the next section we write this equation with full notation restored.

2.3.6.7. *The DIA and LET response equations*

The DIA response equation is:

$$\left[\frac{\partial}{\partial t} + \nu_0 k^2 \right] R(k; t, t')$$
$$+ \int d^3 j\, L(\mathbf{k}, \mathbf{j}) \int_{t'}^{t} dt''\, R(k; t'', t') R(j; t, t'') C(|\mathbf{k} - \mathbf{j}|; t, t'')$$
$$= \delta(t - t'). \tag{2.36}$$

For convenience in making comparisons, we state the LET response equation at this stage, thus:

$$\theta\left(t - t'\right) \left(\frac{\partial}{\partial t} + \nu_0 k^2 \right) \theta\left(t - t'\right) R\left(k; t, t'\right) - \theta(t - t') R(k; t, t') \delta\left(t - t'\right)$$
$$+ \int d^3 j\, L\left(\mathbf{k}, \mathbf{j}\right) \theta\left(t - t'\right) \int_{t'}^{t} ds\, R\left(j; t, s\right) R\left(k; s, t'\right) \theta\left(t - s\right) C\left(|\mathbf{k} - \mathbf{j}|; t, s\right)$$
$$= \int d^3 j\, L\left(\mathbf{k}, \mathbf{j}\right) \theta\left(t - t'\right) \int_{0}^{t'} ds \frac{\theta\left(t - s\right) C\left(|\mathbf{k} - \mathbf{j}|; t, s\right)}{C\left(k; t', t'\right)} \times$$
$$\times \left\{ R\left(k; t', s\right) \theta\left(t - s\right) C\left(j; t, s\right) - R\left(j; t, s\right) \theta\left(t' - s\right) C\left(k; t', s\right) \right\}. \tag{2.37}$$

Apart from the addition of the second term on the left hand side, this is the LET response equation as it appears in the literature. The addition of this extra term as a consequence of time-ordering, fixes the problem of the singularity in the time-derivative of the response equation (2.37) which occurs when one takes $t = t'$.

More importantly, if we compare (2.37) with the DIA response equation (2.36), the additional terms on the right hand side of (2.37) cancel the infra-red divergence and ensure compatibility with the Kolmogorov K41 spectrum.

The fluctuation-dissipation relation can be used directly to calculate the LET theory instead of the above response equation. With this simplification it is much easier to calculate than DIA. We shall return to this later.

2.3.7. *Assessment of the pioneering RPTs*

DIA is both quantitatively and qualitatively very good at predicting the free decay of isotropic turbulence. However, it is not consistent with $K41$. We turn to the simpler Edwards theory to understand the difficulty. If we take the two-time dependences as $R_k \sim e^{-\omega_k(t-t')}$ and $C_k \sim q_k e^{-\omega_k(t-t')}$ then for stationary turbulence the equations become:

The Edwards equation for the spectral density q_k is:

$$\int d^3j \frac{L(k,j)q_{k-j}(q_k - q_j)}{\omega_k + \omega_j + \omega_{k-j}} = W(k) - 2\nu_0 k^2 q_k,$$

while the Edwards equation for the response ω_k is:

$$\omega_k = \frac{1}{2}\int d^3j \frac{L(k,j)q_{k-j}}{\omega_k + \omega_j + \omega_{k-j}}.$$

Re-arrange the energy (density) balance equation as:

$$W(k) + \int d^3j \frac{L(k,j)q_{k-j}q_j}{\omega_k + \omega_j + \omega_{k-j}} - \left(2\nu_0 k^2 + 2\omega_k\right) q_k = 0.$$

With the obvious interpretation:

$$\omega_k = \nu_T(k)k^2.$$

This leads on to the idea of the infinite Reynolds number limit and the infra-red divergence. Edwards took the limit $\nu_0 \to 0$ such that $\varepsilon = \text{constant}$. Then $K41$ applies for all k such that $0 \le k \le \infty$. That is: $q_k \sim \varepsilon^{2/3}k^{-11/3}$ and $\omega_k \sim \varepsilon^{1/3}k^{2/3}$ for all k. The energy balance becomes:

$$\int d^3j \frac{L(k,j)q_{k-j}(q_k - q_j)}{\omega_k + \omega_j + \omega_{k-j}} = \varepsilon\delta(k) - \varepsilon\delta(k - \infty).$$

The integral in the energy balance is well-behaved due to cancellations. The integral for the response function is divergent due to $q_{k-j} \to \infty$ as $k - j \to 0$.

2.3.8. *The local energy transfer (LET) theory*

It was pointed out (McComb 1974) that the above interpretation of the energy transfer spectrum in terms of ω_k is wrong. The *entire* transfer spectrum, not just some part of it, behaves like a sink for small k and like a source for large k. A re-interpretation led to

$$\omega_k = \frac{1}{2} \int d^3 j \, \frac{L(k,j)q_{k-j}(q_k - q_j)}{(\omega_k + \omega_j + \omega_{k-j})q_k}.$$

This is well-behaved and leads on to the two-time LET theory.

2.3.9. *Numerical computation of RPTs*

The first numerical investigation of an RPT was in 1964 by Kraichnan,[20] who calculated the free decay of DIA from several arbitrary initial spectra, and for initial values of the (microscale-based) Taylor-Reynolds number up to $R_\lambda \sim 35$. Later, Herring and Kraichnan investigated the DIA, EFP and SCF closures, along with the Test-field Model (TFM);[21] and later still the Lagrangian-history closures, ALHDI and SBALHDI.[22]

The free decay of the LET theory has been investigated,[23–25] using the same numerical methods and initial spectra as Kraichnan. Comparisons were made with results from DIA, TFM and the Lagrangian-history theories, along with representative experimental results. More recently Mc-Comb and Quinn[26] have obtained results for both DIA and LET with forcing, in order to study stationary turbulence.

All of this work shows that RPT closures perform quite well, in that they give good qualitative and quantitative agreement with the results of both experiments and direct numerical simulations. In particular, the DIA and LET closures are very similar under most circumstances.

These investigations have all been for three-dimensional turbulence but broadly similar conclusions have been drawn for two-dimensions by Frederiksen and co-workers,[27–30] in an extensive programme of calculations with application to atmospheric turbulence.

In order to illustrate the behaviour of RPT closures, we now present some of the calculations by my student (now Dr.) Anthony Quinn.[31] As well as considering forced turbulence, we also repeated calculations of decaying turbulence where the new feature is a comparison with a direct numerical simulation which had the same initial spectrum. Noting that the 'initial spectrum' for the DNS is the result of an ensemble average, its equivalence to the LET/DIA initial spectrum can only be an approximation and in Fig.2.12. this is represented by error bars. We note that as time evolves, the spectrum decreases in amplitude, but also spreads out as energy is transferred to higher wavenumbers by the nonlinear term; as indicated in Fig.2.13. by the development of the transfer spectrum. Note that at $t = 0$ the transfer spectrum is zero at all wavenumbers, corresponding to the Gaussian initial conditions. In Fig.2.14. we see that the energy decreases monotonically whereas the dissipation increases initially and then declines. Both forms of behaviour are, as one would expect and one notes, in reasonable agreement between the closures and the DNS. Fig.2.15. shows similar behaviour for the microscale but the results for skewness reinforce the view that it is the most sensitive index of differences between theories. However, one should not read too much into this one result. As we shall see when making comparisons for forced turbulence, the DNS results cannot be taken as representing the experimental position (itself subject to much uncertainty). Also, earlier calculations show LET can give values of the evolved skewness ranging from $0.4 - 0.5$, depending on choice of initial spectrum.[23]

Lastly, as the investigation of forced turbulence is rather preliminary in nature, we just show two reasonably encouraging results. In Fig.2.16., we plot the compensated energy spectrum at $R_\lambda = 232$ for LET and DIA, and compare with both the DNS result and the *ad hoc* result due to Pao (see[5] for a discussion) which is adjusted to give good agreement with experiment. We see that the closures agree with each other, but that neither they nor the Pao spectrum agree with the DNS at the lower numbers. To some extent, this reflects the fact that the role of the forcing is crucial at low wavenumbers. Evidently many more studies are needed in this area (as in most area of turbulence!) and also somewhat higher Reynolds numbers.

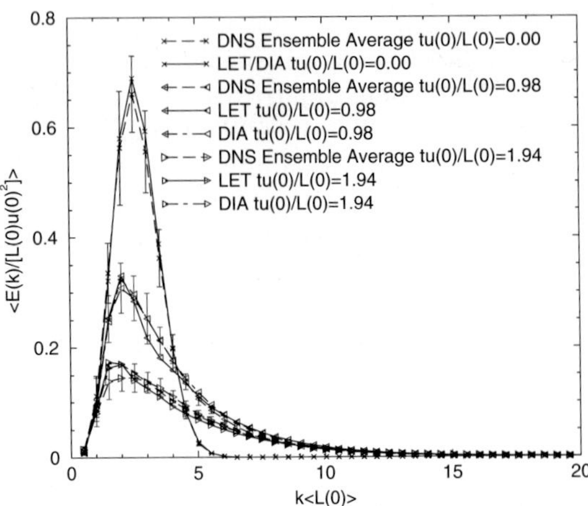

Fig. 2.12.   Free decay of the energy spectrum: DIA and LET compared with DNS at initial Taylor-Reynolds number R_λ=95.

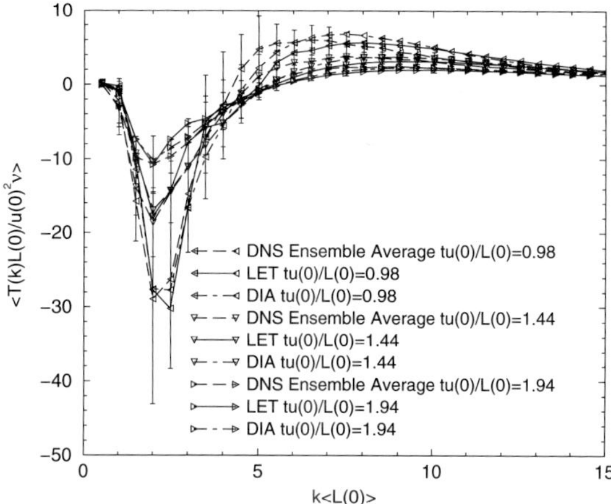

Fig. 2.13. The transfer spectrum for DIA and LET compared with DNS at initial Taylor-Reynolds number $R_\lambda=95$.

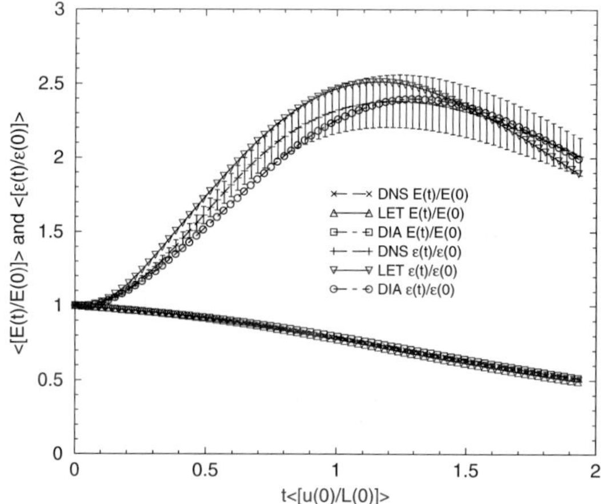

Fig. 2.14. Variation of energy and dissipation rate during free decay at initial Taylor-Reynolds number $R_\lambda=95$.

2.3.10. *Perceptions of RPTs*

They come from theoretical physics and are seen by the turbulence community (who are mostly fluid dynamicists) as being 'alien'. Articles in

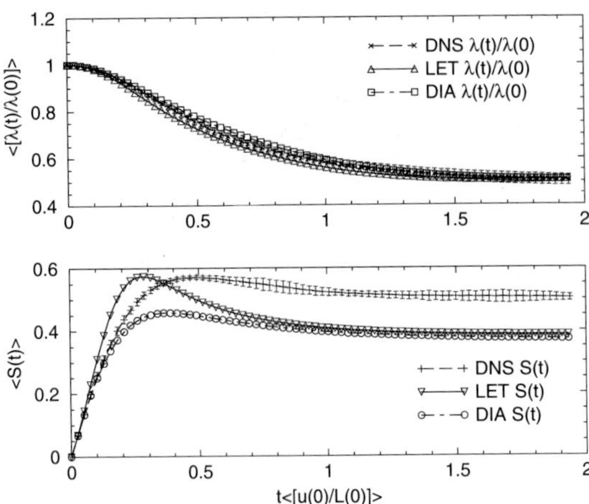

Fig. 2.15. Variation of skewness and microscale during free decay at initial Taylor-Reynolds number $R_\lambda=95$.

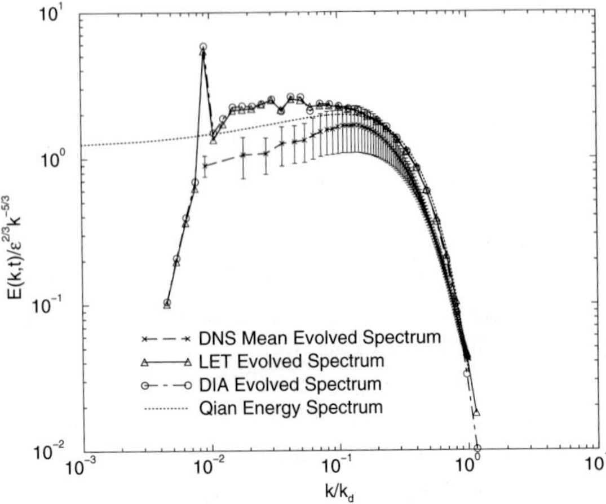

Fig. 2.16. Compensated energy spectra for forced turbulence at an evolved Taylor-Reynolds number of $R_\lambda = 232$.

journals which contain numerous Feynman-type diagrams and phrases like 'one-loop renormalisation' and 'Ward-Takahashi identities' merely bemuse most turbulence researchers. The community notes that such theories are never applied to 'real' problems in real space. In the absence of understanding, fluid dynamicists look to the theoretical physicists for some consensus. Instead they see disagreement and wild claims that they suspect cannot be

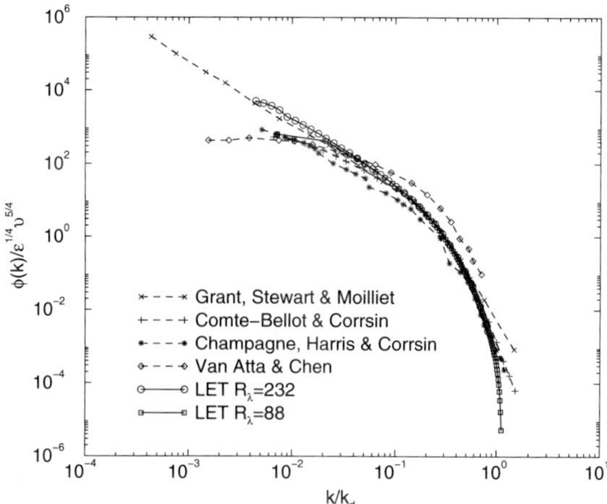

Fig. 2.17.   One-dimensional spectra for forced LET compared with experiment at an evolved Taylor-Reynolds number of $R_\lambda = 232$.

true. Accordingly they write off the whole field as being 'mired in controversy'.

2.3.10.1.   *What are the issues?*

The main unresolved issues can be summarized as follows:

(1) Disagreement on the causes of the failure of DIA and EFP to give $k^{-5/3}$. 'Lack of convective invariance' *versus* 'lack of scale invariance'.
(2) The need for *ad hoc* corrections to the Wyld formalism.
(3) Conflict between Wyld (diagram) and MSR (path-integral)[32] formalisms on vertex renormalisation.
(4) MSR extends ideas from canonical Hamiltonian systems in thermal equilibrium to macroscopic fluid motion. Is this valid?
(5) Lagrangian *versus* Eulerian formulations.
(6) Galilean invariance (GI): does it suppress vertex renormalisation?
(7) *IR* and *UV* divergences: do they exist?

2.3.10.2.   *Wider issues*

General disagreement on Galilean invariance and $K41$ make the turbulence picture even more confused. Thus:

A Galilean invariance (GI) is widely invoked, often as a 'low-speed' version of Lorentz invariance, and used to justify many things. We shall argue that it is a trivial symmetry in most cases, being 'satisfied' by the constant mean velocity: see comments on Issue 6.

B Correctness of Kolmogorov $k^{-5/3}$ spectrum is challenged by two groups:

(1) Those who believe that intermittency of the local dissipation rate invalidates the argument.

(2) Those who draw analogies with critical phenomena and see '$-5/3$' as the 'canonical dimension' corresponding to mean-field theory and wish to use RG to establish the 'anomalous dimension'.

2.3.10.3.   *Issue 1*

To a considerable extent, the 'disagreement' between Kraichnan and Edwards is due to the fact that they work with two-time and single-time theories respectively. Kraichnan argues that $C(\mathbf{x}, \mathbf{x}'; t, t') = \langle \mathbf{u}(\mathbf{x}, t)\mathbf{u}(\mathbf{x}'t')\rangle$ cannot be GI whereas the single-time form $C(\mathbf{x}, \mathbf{x}'; t, t)$ is GI. Accordingly, a closure in terms of an expansion in time-convolutions of two-time covariances (see equation (29)) cannot be correct on the time diagonal. The left hand side is invariant, the right hand side may not be. Edwards considers a time-independent situation and the infinite-Reynolds number limit at constant dissipation rate ε. Under these circumstances the $k^{-5/3}$ spectrum applies for all k such that $0 \leq k \leq \infty$. He finds that this leads to an infinite value of the response integral as $k \to 0$. This is interpreted as lack of scale invariance. Further study is needed to understand the relationship between these two approaches and recent work by McComb and Kiyani on the connection between DIA and EFP may assist in this.[33]

2.3.10.4.   *Issues 2, 3 and 4*

Lee[19] made *ad hoc* corrections to eliminate a double-counting problem with the Wyld formalism. However, a more correct initial procedure would have made this correction unnecessary.[5] It is not a real issue. MSR predicts some additonal vertex corrections which Wyld does not mention. Kraichnan (in unpublished work) agrees with MSR. (But see next point.) The MSR formalism is a *synthetic* formalism. In effect it has been set up to reproduce DIA and this it duly does. Some elements in it (e.g. Legendre transformation) need more critical examination.

2.3.10.5.   *Issue 5. Lagrangian versus Eulerian*

Kraichnan introduced Lagrangian-history coordinates.[12] He introduced a generalized velocity:

$\mathbf{u}(\mathbf{x}, t|s) \equiv$ velocity at time s of a particle which was at $\mathbf{x}$ at time t.

$t =$ labelling time (Eulerian).

$s =$ measuring time (Lagrangian).

On this basis DIA was reworked as LHDI. This achieved GI but the resulting equations were too complicated so it was abridged to ALHDI, but did not perform as well as hoped. Later Kraichnan and Herring[13] introduced strain-based ALHDI. This gave much better numerical predictions and is known as SBALHDI.

Kraichnan ends up working with $C(k; t|s)$ rather than $C(k; t, t')$ and claims that this form is compatible with random GI. Kaneda[14] produced a version of Kraichnan's LHDI formalism by working with measuring-time derivatives rather than labelling-time derivatives. Kaneda's theory is in terms of $C(k; t, t')$ and appears to be very similar (possibly indentical) to the purely Eulerian SCF theory. Kida[15] has, in effect, rederived Kaneda's Lagrangian equations.

2.3.10.6. *Issue 6. Galilean invariance and vertex renormalisation*

In recent years Galilean invariance has been invoked to constrain the nature of perturbation expansions of the NSE, and related equations in soft condensed matter. Arguing by analogy with Lorentz invariance in quantum field theory, Ward identities have been derived using GI. These imply that there is no vertex renormalisation which in turn leads to nontrivial relationships between critical exponents. This view has been challenged by McComb[34] who asserts that GI is satisfied trivially by the mean velocity. Recently Berera and Hochberg[35] have argued that there is an exception to this for the particular case $k = 0$.

2.3.10.7. *Issue 7. IR and UV divergences: do they exist?*

In quantum field theory there are well known divergences in the primitive (unrenormalised) perturbation theory. In recent decades, many theorists have claimed that such divergences exist in the perturbation expansion and then they claim to find ways of dealing with them. It is curious that none of the pioneers (Kraichnan, Wyld, Edwards, Herring, Lee, Balescu and Senatorski ...) noticed or commented on them! The perturbation expansion is in terms of R_0 and C_0, and as C_0 is not an observable we are unable to say whether or not there are divergences. The exception is where we calculate $C_0 = R_0 D(k)$, where $D(k)$ is the covariance of the stirring forces, and choose a power law for $D(k)$. In other words, we only get divergences if we ourselves put them into the problem.

2.3.10.8. *The wider issue of the K41 '-5/3' power law*

The moments/spectra of the velocity field are not *solutions* of the NSE. Either they are obtained by averaging operations on the actual solution of the NSE (the velocity field); or they are connected together by the open moment hierarchy of the NSE and are indeterminate.

The one exception to this general rule is that the second- and third-order moments are rigorously connected by conservation of energy. Scale-invariance leads to a *de facto* closure of the moment hierarchy. This enforces K41 and is in accord with experiment/DNS, although higher-order moments probably depart from K41 behaviour.

References to 'intermittency corrections' beg the question! If corrections to higher order moments exist, they may not be due to intermittency. References to 'intermittent dissipation rate' ignore the irrelevance of the dissipation rate to the K41 arguments. The Kolmogorov energy spectrum is determined by the *inertial transfer rate*. For stationary flows the two quantities are numerically equal. *Fine-scale* (or, better) *internal* intermittency is the fact that the turbulence cascade in any one realization is not space filling. This behaviour is part of the dynamics of turbulence and is true for virtually *all* length scales. Questions one might ask are: does it have consequences for energy conservation? For scale invariance? Bear in mind that both these properties are tested using averaged quantities.

Analogies with the theory of critical phenomena should be drawn with care. Dimensional analysis in equilibrium problems is a relatively weak tool which relies on the introduction of densities, and relates only to length. In turbulence, energy conservation associated with a flux through the modes is a controlling symmetry which has no analogue in equilibrium critical phenomena. There is no justification for calling K41 a *mean-field theory*. The '-5/3' law was derived by two different methods and neither is a mean-field theory. On the contrary, EFP, DIA and SCF *are* mean-field theories, yet are not compatible with K41. In K41 the exponent is determined by dimensional analysis, confirmed by a *de facto* closure of the Karman-Howarth equation. As we shall see the prefactor can be determined by renormalization group analysis.

2.3.11. *New developments in LET*

This work was done in conjunction with my student (now Dr) Khurom Kiyani.[36] A new symmetrized time-ordered covariance is introduced. This eliminates problems encountered in using exponential time dependences. It also reconciles conflicting requirements of time-reversal symmetry (covariance) and causality (response function) in the fluctuation dissipation relation. An improved derivation of the LET response equation has been given in terms of closing the Kraichnan-Wyld perturbation series at second-order by means of a local (in wavenumber) energy balance.

By specialising to a particular initial condition, the response equation is reduced to a fluctuation-dissipation relation. The instantaneous propagator (velocity-field response function) is trivially shown to be transitive with respect to intermediate time. The mean-field propagator ('covariance' response function) is non-trivially shown to be transitive with respect to intermediate times. This is a new result. Relationships have been obtained linking single-time covariances.

Modified two-time LET equations have been obtained. These eliminate minor problems which arose on the time diagonal when $t = t'$. A partial propagator representation has been introduced allowing the two-time LET equations to be represented solely in terms of single-time covariances. Single-time and Markovianized versions of LET have been derived. A time-independent form of the LET equations has been obtained and shown to be well-behaved in the limit of infinite Reynolds numbers.[33,37]

2.3.11.1. *Problems with time-ordering*

Isotropy also implies time-reversal symmetry, which requires that

$$C\left(k; t, t'\right) = C\left(k; t', t\right).$$ (2.38)

The renormalized response is not an observable but must nevertheless satisfy the causality condition

$$R(k; t, t') = 0 \qquad \text{for} \qquad t' > t.$$ (2.39)

It would be helpful if we could assume exponential forms for the covariance and renormalized response function, thus:

$$C(k; t - t') = C(k)e^{-\omega(k)|t-t'|}; \qquad R(k; t - t') = e^{-\omega(k)(t-t')}.$$ (2.40)

However, there is a basic problem with these forms in that the time-reversal symmetry of (2.38) is in practice not satisfied, and that differentiating the steady-state covariance with respect to difference time leads to a non-zero result at the origin, where $t = t'$.

2.3.11.2. *Derivation of mean propagator equation*

By using an integrating factor and integrating the covariance equation over time we can write it as

$$C_{\alpha\sigma}\left(\mathbf{k};t,t'\right) = R^{(0)}_{\alpha\varepsilon}\left(\mathbf{k};t,s\right)C_{\epsilon\sigma}\left(\mathbf{k};s,t'\right) +$$

$$+ \left[\lambda\int_{s}^{t}dt''\,R^{(0)}_{\alpha\epsilon}\left(\mathbf{k};t,t''\right)M_{\epsilon\beta\gamma}\left(\mathbf{k}\right)\times\right.$$

$$\left.\times\int d^3j\,\langle u_\beta\left(\mathbf{j},t''\right)u_\gamma\left(\mathbf{k}-\mathbf{j},t''\right)u_\sigma\left(-\mathbf{k},t'\right)\rangle\right], \qquad (2.41)$$

where s is some initial time and the integrating factor is

$$R^{(0)}_{\alpha\epsilon}\left(\mathbf{k};t,t''\right) = \begin{cases} P_{\alpha\epsilon}\left(\mathbf{k}\right)e^{-\nu_0 k^2\left(t-t''\right)} & t \geq t'', \\ 0 & t < t''. \end{cases} \qquad (2.42)$$

From the primitive perturbation series (27), we have

$$C_{\alpha\sigma}\left(\mathbf{k};t,t'\right) = C^{(0)}_{\alpha\sigma}\left(\mathbf{k};t,t'\right) + \lambda^2 C^{(2)}_{\alpha\sigma}\left(\mathbf{k};t,t'\right)\cdots. \qquad (2.43)$$

When (2.43) is substituted in (2.41) we can see that $R^{(0)}_{\alpha\epsilon}\left(\mathbf{k};t,s\right)$ acts as a zero-order response for the zero-order covariance, thus:

$$C^{(0)}_{\alpha\sigma}\left(\mathbf{k};t,t'\right) = \theta\left(t-s\right)R^{(0)}_{\alpha\epsilon}\left(\mathbf{k};t,s\right)C^{(0)}_{\epsilon\sigma}\left(\mathbf{k};s,t'\right). \qquad (2.44)$$

This is an exact result. We call this the zero-order or bare result. Rearranging (2.41) to prompt the next step,

$$C_{\alpha\sigma}\left(\mathbf{k};t,t'\right) = \left[R^{(0)}_{\alpha\varepsilon}\left(\mathbf{k};t,s\right)\right.$$

$$+ \frac{1}{C_{\epsilon\sigma}\left(\mathbf{k};s,t'\right)}\lambda\int_{s}^{t}dt''\,R^{(0)}_{\alpha\epsilon}\left(\mathbf{k};t,t''\right)M_{\epsilon\beta\gamma}\left(\mathbf{k}\right)\times$$

$$\left.\times\int d^3j\,\langle u_\beta\left(\mathbf{j},t''\right)u_\gamma\left(\mathbf{k}-\mathbf{j},t''\right)u_\sigma\left(-\mathbf{k},t'\right)\rangle\right]\times$$

$$\times C_{\epsilon\sigma}\left(\mathbf{k};s,t'\right). \qquad (2.45)$$

We postulate that we may write this in its renormalized form as

$$C_{\alpha\sigma}\left(\mathbf{k};t,t'\right) = \theta\left(t-s\right)R_{\alpha\epsilon}\left(\mathbf{k};t,s\right)C_{\epsilon\sigma}\left(\mathbf{k};s,t'\right). \qquad (2.46)$$

2.3.11.3. *The fluctuation-dissipation relationship*

Or, in its *isotropic* version as

$$C\left(k;t,t'\right) = \theta\left(t-s\right)R\left(k;t,s\right)C\left(k;s,t'\right), \qquad (2.47)$$

where the $\theta(t-s)$ incorporates the causality condition.

We have effectively replaced the zero-order equation (2.44) by its renormalized version using the replacements $C^{(0)} \to$ and $R^{(0)} \to R$. Choosing the time-ordering $t > t'$ say, is merely a matter of applying the Heaviside unit function to both sides:

$$\theta(t - t')C\left(k;t,t'\right) = \theta(t - t')\theta\left(t - s\right)R\left(k;t,s\right)C\left(k;s,t'\right). \qquad (2.48)$$

If we now set $s = t'$ in (2.48), which amounts to a choice of the initial condition, we get

$$\theta\left(t - t'\right)C\left(k;t,t'\right) = \theta\left(t - t'\right)R\left(k;t,t'\right)C\left(k;t',t'\right). \qquad (2.49)$$

This result takes the form of a fluctuation-dissipation relationship (or FDR).

2.3.11.4. *Symmetrized time-ordered covariance*

We now introduce a representation of the covariance which preserves the symmetry under interchange of time arguments as

$$C\left(k;t,t'\right) = \theta\left(t - t'\right)C\left(k;t,t'\right) + \theta\left(t' - t\right)C\left(k;t,t'\right)$$
$$-\delta_{t,t'}C\left(k;t,t'\right). \qquad (2.50)$$

Equation (2.50) may be written in the form of a time-ordered fluctuation-dissipation relation by using (2.49) to construct it:

$$C\left(k;t,t'\right) = \theta\left(t - t'\right)R\left(k;t,t'\right)C\left(k;t',t'\right)$$
$$+\theta\left(t' - t\right)R\left(k;t',t\right)C\left(k;t,t\right)$$
$$-\delta_{t,t'}C\left(k;t,t'\right). \qquad (2.51)$$

Turning now to the problem of the exponential forms as given by (2.40) we find that this time-ordered representation (2.50) has the required property that

$$\lim_{t \to t'} \frac{\partial}{\partial t}C(k;t,t') = 0. \qquad (2.52)$$

2.3.11.5. *Group-closure properties of the LET*

For the velocity field, we have the instantaneous propagator $\hat{R}_{\alpha\sigma}(\mathbf{k};t,s)$ defined by

$$u_\alpha(\mathbf{k},t) = \hat{R}_{\alpha\sigma}(\mathbf{k};t,s)u_\sigma(\mathbf{k},s). \qquad (2.53)$$

It is easily shown that this is transitive with respect to intermediate times:

$$\hat{R}_{\alpha\rho}(\mathbf{k};t,t') = \hat{R}_{\alpha\sigma}(\mathbf{k};t,s)\hat{R}_{\sigma\rho}(\mathbf{k};s,t'). \qquad (2.54)$$

For the mean-field propagator, $R_{\alpha\sigma}(\mathbf{k};t,s) = \langle\hat{R}_{\alpha\sigma}(\mathbf{k};t,s)\rangle$, we can show that the renormalized response is also transitive with respect to intermediate times.

$$R\left(k;t,t'\right) = R\left(k;t,s\right)R\left(k;s,t'\right), \qquad (2.55)$$

where we have specialized to the isotropic case. We can also write linked single-time covariances as

$$C(k;t,t) = \theta(t - s)R(k;t,s)R(k;t,s)C(k;s,s). \qquad (2.56)$$

2.3.12.   *Single-time LET equations*

With the *ansatz* of local energy transfer to determine the response, along with an assumption of an exponential relationship between the response function and the eddy damping, as given by equation (2.40), we find

$$\left(\frac{\partial}{\partial t} + 2\nu_0 k^2\right) C(k;t) = 2 \int d^3 j L(\mathbf{k},\mathbf{j}) D(k,j,|\mathbf{k}-\mathbf{j}|;t)$$
$$\times C(|\mathbf{k}-\mathbf{j}|;t)[C(j;t) - C(k;t)]$$
$$= -2\omega(k;t)C(k;t). \tag{2.57}$$

With eddy damping given by

$$\omega(k;t) = -\int d^3 j L(\mathbf{k},\mathbf{j}) D(k,j,|\mathbf{k}-\mathbf{j}|;t)$$
$$\times \frac{C(|\mathbf{k}-\mathbf{j}|;t)}{C(k;t)}[C(j;t) - Ck;t]. \tag{2.58}$$

The triple-mode damping function satisfies:

$$\frac{\partial D(k,j,|\mathbf{k}-\mathbf{j}|;t)}{\partial t} = 1 - \left[\left(\nu_0 k^2 + \nu_0 j^2 + \nu_0 |\mathbf{k}-\mathbf{j}|^2\right) + \right.$$
$$+ \omega(k;t) + \omega(j;t)\omega(|\mathbf{k}-\mathbf{j}|;t)\big]$$
$$\times D(k,j,|\mathbf{k}-\mathbf{j}|;t). \tag{2.59}$$

The initial conditions can be taken as:

$$C(k;t=0) = \frac{E(k;t=0)}{4\pi k^2}, \qquad D(k,j,|\mathbf{k}-\mathbf{j}|;t=0) = 0, \tag{2.60}$$

where $E(k;t=0)$ is some arbitrarily chosen initial energy spectrum. This is similar to the test-field model, but has an extra term in the equation for the eddy damping. The extra term cancels infra-red divergences and this means that (unlike the test-field model) it does not require an additional hypothesis and adjustable constant to be compatible with the Kolmogorov distribution.

This theory still has to be tried out on the standard test problems of isotropic turbulence. However, the two-time LET theory performs well on such problems and we may hope that the single-time and Markovianized forms will perform adequately while offering computational advantages. The theory has potential for generalization to form realizable Markovian closures to include the effect of waves.[38] The theory also has potential for application to inhomogeneous and shear flow turbulence.

2.4. Renormalization group (RG) applied to macroscopic fluid turbulence

We re-introduce RG, first as the field-theoretic version and then as the wavenumber-space version of Wilson theory. The importance of a conditional average is emphasised, along with its nontrivial nature in deterministic chaos. We then dicuss how the RG transformation may be implemented for the NSE equations, firstly by Gaussian perturbation theory at low wavenumbers then by the Two-field theory of turbulence, and lastly by non-Gaussian perturbation theory at high-wavenumbers. Note that a general, but elementary, discussion of these topics can be found in the book by McComb.[2]

2.4.1. *Three flavours of RG*

These are as follows:

A. Field-theoretic (Stuckelberg and Petermann (1953); Gell-Mann and Low (1954)): later Callan (1970) and Symanzik (1970)).
B. Theory of critical phenomena: RG applied to a continuum formulation in wavenumber space (Wilson 1972).
C. Extension of RG to macroscopic fluid motion as described by the NSE: Gaussian perturbation theory (Forster, Nelson and Stephen (1976)) and the Two-field theory of turbulence (McComb and Watt (1990)).

In this section we shall discuss A and B briefly before concentrating on C.

2.4.1.1. *A. Field-theoretic RG*

Consider a free field $\phi_0(\mathbf{k})$ with an associated bare particle of mass m_0 and coupling constant λ_0. Under an RGT with spatial rescaling parameter b the field becomes $\phi(\mathbf{k})$ with an associated particle of mass m and with coupling λ. Form invariance of the Hamiltonian requires:

$$m_0^2 \to m^2 = \mathcal{Z}_\phi m_0^2; \quad \lambda_0 \to \lambda = \mathcal{Z}_\phi^2 \lambda_0; \quad \phi_0 \to \phi \equiv \mathcal{Z}_\phi^{1/2} \phi_0,$$

where $\mathcal{Z}_\phi$ is the renormalisation constant. The n-particle 'bare' propagator is

$$G_0^{(n)}(\mathbf{k}_1, \mathbf{k}_2, \dots \mathbf{k}_n) = \langle \phi_0(\mathbf{k}_1)\phi_0(\mathbf{k}_2) \dots \phi_0(\mathbf{k}_n) \rangle.$$

The n-particle 'exact' propagator is

$$G^{(n)}(\mathbf{k}_1, \mathbf{k}_2, \dots \mathbf{k}_n) = \langle \phi(\mathbf{k}_1)\phi(\mathbf{k}_2) \dots \phi(\mathbf{k}_n) \rangle.$$

If we use the same functional measure for the averages, then $G^{(n)}$ transforms as $|\phi|^n$ and so:

$$G_0^{(n)}(\mathbf{k}_1, \ldots \mathbf{k}_n; \lambda_0, m_0, \varepsilon) = \mathcal{Z}_\phi^{n/2} G^{(n)}(\mathbf{k}_1 \ldots \mathbf{k}_n; \lambda, m, \varepsilon).$$

Note the occurrence of the parameter $\varepsilon = d - 4$, where d is the lattice/space dimension. Renormalisation invariance arises from the fact that the 'bare' parameters are constants. Hence, differentiating both sides of the RGT with respect to b:

$$b\frac{d}{db}G_0^{(n)} = 0 = b\frac{d}{db}\left(\mathcal{Z}_\phi^{n/2} G^{(n)}\right).$$

Use of the chain rule then leads on to the Callan-Symanzik differential equations.

Can we apply these ideas to real fluid turbulence? Let's note the following points:

- Renormalisation invariance occurs in the far dissipation range.
- The relevant fixed point is the 'top' of the inertial range.
- Differentiation implies 'closeness'.
- Doesn't look hopeful.
- Yet a book[39] has been written about the many attempts to do it!

2.4.1.2. *B. Theory of critical phenomena*

In the continuum limit, the lattice constant a leads to a UV cutoff $\Lambda = \pi/a$. The RG transformation becomes:

Average over modes with $\Lambda/b \leq k \leq \Lambda$, while leaving modes with $k \leq \Lambda/b$ unaffected. *This requires a conditional average.* Rescale the system in real space:

$$x \to x' = x/b; \qquad \int d^d x \to \int d^d x' = b^{-d} \int d^d x.$$

Rescale the system in wavenumber space:

$$k \to k' = bk; \qquad \int d^d k \to \int d^d k' = b^d \int d^d k.$$

Split up the Hamiltonian: $H_{GL} = H_0 + H_I$. H_0 is soluble while H_I contains the interactions. Solve for Z_0, the Gaussian partition functional. Implement RGT perturbatively

$$H'_{GL} = R_b H_{GL} = R_b H_0 + \langle R_B H_I \rangle_0 + \ldots .$$

Where $\langle \ldots \rangle_0$ denotes an average against Gaussian statistics. This procedure depends on our ability to evaluate the Gaussian functional integral.

The intractable conditional average is expanded in a series of Gaussian averages.

Results from the linearised analysis of real-space RG are used to identify the non-trivial ('critical') fixed-point. Both the exact and the zero-order (Gaussian) theories have the same fixed point for the $d > 4$. The corresponding critical exponents are those for mean-field theory of the Ising model. The exact theory calculates corrections to mean-field theory for $d < 4$. This is done as a perturbation expansion in $\varepsilon = d - 4$, treated as small. Set $\varepsilon = 1$ to get results for $d = 3$!

2.4.1.3.   *C. Extension of RG to fluid motion*

Consider real fluid turbulence at high Reynolds number. The velocity field (in k-space) $\mathbf{u}(\mathbf{k}, t)$ is defined on $0 \leq k \leq k_{max}$, with k_{max} chosen to capture (almost all) the dissipation. The inertial range of wavenumbers $k_b \leq k \leq k_t$ characterised by scaling behaviour with $E(k) \sim k^{-5/3}$. The 'internal' scale is set by the Kolmogorov dissipation wavenumber $k_d^0 = (\varepsilon/\nu_0^3)^{1/4}$. In practice, $k^* = k_t$ will be the nontrivial fixed point, where $k_t \sim 0.2k_d$. Trivial fixed points are at $k = 0$ ('high-temperature' fixed point) and $k \to k_{max}$ ('low-temperature' fixed point). In practice $k_{max} \sim 1.5k_d$.

2.4.2.   *RG Algorithm for turbulence*

Filter the velocity field $\mathbf{u}(\mathbf{k}, t)$ into $\mathbf{u}^-(\mathbf{k}, t)$ on $0 \leq k \leq k_1$ and $\mathbf{u}^+(\mathbf{k}, t)$ on $k_1 \leq k \leq k_0$. The RG algorithm consists of two steps:

(1) Solve the NSE on $k_1 \leq k \leq k_0$. Substitute that solution for the mean effect of the high-k modes into the NSE on $0 \leq k \leq k_1$. This results in an increment to the viscosity: $\nu_0 \to \nu_1 = \nu_0 + \delta\nu_0$.
(2) Rescale the basic variables so that the NSE on $0 \leq k \leq k_1$ looks like the original NSE on $0 \leq k \leq k_0$.

These steps are repeated for $k_2 < k_1$, $k_3 < k_2$, and so on; until a fixed point is reached and this defines the renormalised viscosity.

The basic problem lies in carrying out Step 1, which requires a nontrivial conditional average. Forster *et al*[40] evaded this problem by choosing k_0 to be small enough to exclude the energy cascade. This ensures that the stirring forces have a dominant role, with coefficients in a perturbation series evaluated as Gaussian averages. Here the fixed point corresponds to some form of universal behaviour as $k \to 0$. In contrast, McComb, Roberts and Watt (1992)[41] formulated an ensemble-based conditional average for chaotic systems. This allowed the elimination of bands of wavenumbers in the dissipation range with a fixed point corresponding to the top of the inertial range.

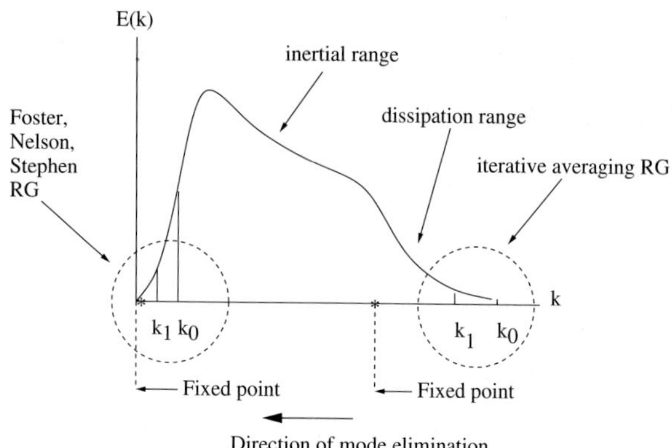

Fig. 2.18.   RG applied to low and high wavenumber regions.

2.4.3.   *Turbulence mode elimination: the basic problem*

Consider the k-space velocity field u_k on $0 \leq k \leq k_0$. Choose the maximum wavenumber k_0 such that

$$\int_0^{k_0} 2\nu_0 k^2 E(k) dk \simeq \int_0^{\infty} 2\nu_0 k^2 E(k) dk = \varepsilon. \qquad (2.61)$$

Filter the velocity field at $k = k_1 < k_0$,

$$\begin{aligned} u_k &= u_k^- \quad \text{for} \quad 0 \leq k \leq k_1, \\ &= u_k^+ \quad \text{for} \quad k_1 \leq k \leq k_0. \end{aligned} \qquad (2.62)$$

Decompose the NSE into u_k^- and u_k^+ modes:

$$L_{0k} u_k^- = M_k^- \{ u_j^- u_{k-j}^- + 2u_j^- u_{k-j}^+ + u_j^+ u_{k-j}^+ \}; \qquad (2.63)$$

$$L_{0k} u_k^+ = M_k^+ \{ u_j^- u_{k-j}^- + 2u_j^- u_{k-j}^+ + u_j^+ u_{k-j}^+ \}. \qquad (2.64)$$

2.4.3.1.   *Turbulence mode elimination: the general solution*

Objective: eliminate u^+ from the right hand side of the filtered NSE as given by (2.63). In principle this means solving equation (2.64) for u^+. Note that if we retain u^+ on the right hand side of (2.64), then this is straightforward, at least numerically. However, if we wish to eliminate u^+ totally, then a solution purely in terms of u^- involves an infinite iteration and integrals over time histories of the u^-. Take M^+ as expansion parameter and write the solution of (2.64) to order $(M^+)^2$:

$$u_k^+ = R_{0k} M_k^+ u_j^- u_l^- + 2 R_{0k} R_{0l} M_k^+ M_l^+ u_j^- u_p^- u_{l-p}^- + \mathcal{O}\left((M_k^+)^3\right). \quad (2.65)$$

We can further write this as:

$$u_k^+ = A\left[u_k^-, t\right] u_k^- \quad (2.66)$$

where A is a functional, standing for an infinite series, and time and wavenumber integrals are involved.

2.4.3.2. *Inadequacy of the filtered ensemble average*

Some researchers use the filtered ensemble average to eliminate modes in the first step of the RG algorithm. We can show that this is not an acceptable substitute for a proper conditional average. Define a filtered ensemble average over the high-k modes, such that:

$$\langle u^+ \rangle^+ = 0. \quad (2.67)$$

However, from (2.63), we have

$$u^- \equiv u^- [u^-, u^+], \quad (2.68)$$

and so

$$\langle u^- \rangle^+ \neq u^-. \quad (2.69)$$

Applying the filtered ensemble average to (2.63) we obtain:

$$L_{0k} \langle u_k^- \rangle^+ = M_k^- \{ \langle u_j^- u_{k-j}^- \rangle^+ + \langle 2 u_j^- u_{k-j}^+ \rangle^+ + \langle u_j^+ u_{k-j}^+ \rangle^+ \} \quad (2.70)$$

If we follow those who assume $\langle u^- \rangle^+ = u^-$, then the incorrectness of this follows by *reductio ad absurdum*. Equation (2.70) becomes:

$$L_{0k} u_k^- = M_k^- \{ u_j^- u_{k-j}^- + 2 u_j^- \langle u_{k-j}^+ \rangle^+ + \langle u_j^+ u_{k-j}^+ \rangle^+ \}. \quad (2.71)$$

Now consider the right hand side. The second term vanishes by (2.67). The third term vanishes by homogeneity: $M_k \delta(k) Q^+ (k - j) = 0$.

Hence the use of a filtered average instead of a conditional average leads to the absurd result that there is no turbulence problem, as from (2.71) the equation for explicit scales would be:

$$L_{0k} u_k^- = M_k^- u_j^- u_{k-j}^-. \quad (2.72)$$

2.4.4.   *Gaussian perturbation theory in the limit $k \to 0$*

The pioneering papers were by Forster *et al.*[40,42] Start from the NSE with a random stirring force f_k and decompose into f_k^- and f_k^+ on the same intervals as u_k^- and u_k^+. Re-write (2.63) and (2.64) introducing both a stirring force f_k, and a book-keeping parameter λ_0 (where $\lambda_0 = 1$) in front of the nonlinear term, as:

$$L_{0k}u_k^- = f_k^- + \lambda_0 M_k^- \{u_j^- u_{k-j}^- + 2u_j^- u_{k-j}^+ + u_j^+ u_{k-j}^+\}; \tag{2.73}$$

$$L_{0k}u_k^+ = f_k^+ + \lambda_0 M_k^+ \{u_j^- u_{k-j}^- + 2u_j^- u_{k-j}^+ + u_j^+ u_{k-j}^+\}. \tag{2.74}$$

The stirring force is chosen to have a Gaussian distribution such that

$$\langle f_k \rangle = 0; \qquad \langle f_k f_j \rangle = 2D_k \delta_{k+j}. \tag{2.75}$$

Choose D_k to be a power law

$$D_k = D_0 k^{-y}. \tag{2.76}$$

Further, under filtered ensemble average, we have:

$$\langle f_k^+ \rangle^+ = 0; \qquad \langle f_k^- \rangle^+ = f_k^-. \tag{2.77}$$

This form of average is appropriate in the present case because the distribution of f_k is the 'product of distributions' for the individual wavenumbers 'k'. Eliminate u^+ from (2.63) by means of the perturbation series

$$u_k^+ = u_k^{+(0)} + \lambda_0 u_k^{+(1)} + \lambda_0^2 u_k^{+(2)} + \ldots. \tag{2.78}$$

Substituting expansion (2.78) into (2.63) for u^- gives, to $\mathcal{O}\left(\lambda_0^2\right)$,

$$L_{0k}u_k^- = f_k^- + \lambda_0 M_k^- \left\{u_j^- u_{k-j}^- + 2u_j^- u_{k-j}^{+(0)} + u_j^{+(0)} u_{k-j}^{+(0)}\right\}. \tag{2.79}$$

Equating coefficients of powers of λ_0 gives the hierarchy of equations

$$u_k^{+(0)} = R_{0k}f_k^+; \tag{2.80}$$

$$u_k^{+(1)} = R_{0k}M_k^+ \left\{u_j^- u_{k-j}^- + 2u_j^- u_{k-j}^{+(0)} + u_j^{+(0)} u_{k-j}^{+(0)}\right\}; \tag{2.81}$$

and so on.

Now substitute for u^+ in (2.73) (keeping terms ordered as 'explicit', 'cross' and 'Reynolds'):

$$L_0 u_k^- = f_k^- + \lambda_0 M_k^- u_j^- u_{k-j}^- + 2M_k^- \{\overbrace{\lambda_0 u_j^- u_{kj}^{(0)}}^{A} + \lambda_0^2 u_j^- u_{k-j}^{(1)}\}$$

$$+ M_k^- \{\overbrace{\lambda_0 u_j^{(0)} u_{k-j}^{(0)}}^{B} + 2\lambda_0^2 u_j^{(0)} u_{k-j}^{(1)}\}. \tag{2.82}$$

If we average, term A vanishes due to (2.67) and B due to homogeneity. Thus (2.82) becomes:

$$L_{0k}u_k^- = f_k^- + \lambda_0 M_k^- u_j^- u_{k-j}^-$$
$$+ 2M_k^- \lambda_0^2 u_j^- \langle u_{k-j}^{(1)} \rangle + M_k^- 2\lambda_0^2 \langle u_j^{(0)} u_{k-j}^{(1)} \rangle. \tag{2.83}$$

With some re-arrangement:

$$L_{0k}u_k^- = f_k^- + \lambda_0 M_k^- u_j^- u_{k-j}^-$$
$$+ 2M_k^- R_{0k-j} M_{k-j}^+ \lambda_0^2 u_j^- u_p^- u_{k-j-p}^- - \delta\nu_0(k)k^2 u_k^-, \tag{2.84}$$

where

$$\delta\nu_0(k) = 4\lambda_0^2 k^{-2} M_k^- R_{0k-j} M_{k-j}^+ Q_j^{(0)}$$
$$= 8\lambda_0^2 k^{-2} M_k^- R_{0k-j} M_{k-j}^+ |R_{0j}|^2 D_k, \tag{2.85}$$

the second form following from equation (2.75), which gives an additional factor of 2.

Taking the limit $k \to 0$, the triple nonlinearity in u^- is an irrelevant variable and so may be neglected. This leaves the final equation

$$\left(L_0 + \delta\nu_0 k^2\right) u_k^- = f_k^- + \lambda M_k^- u_j^- u_{k-j}^-. \tag{2.86}$$

The viscosity increment $\delta\nu_0$ is given by its value at $k = 0$,

$$\delta\nu_0 = \frac{\lambda^2 D_0 A(d) S_d [\exp(\epsilon l) - 1]}{\nu_0^2 \Lambda^\epsilon (2\pi)^d \epsilon}, \tag{2.87}$$

with

$$\epsilon = 4 + y - d, \qquad A(d) = \frac{d^2 - d - \epsilon}{2d(d+2)}, \qquad S_d = \frac{2\pi^{d/2}}{\Gamma(d/2)}, \tag{2.88}$$

where D_0 represents the effect of the force spectrum and l is used in defining the cutoff between the u^+ and u^- modes, $\Lambda \exp(-l)$, where Λ is our upper cutoff wavenumber. Due to the requirement $k \to 0$, this theory doesn't describe actual turbulence, but is instead a theory of the long-wavelength properties of stirred hydrodynamics.

2.4.5. *The two-field theory of turbulence*

This was due to McComb and Watt,[43,44] and was based on a formal conditional average[41] and on *iterative averaging*.[45] We start from the NSE and decompose into u^+ and u^- modes, as in equations (2.63) and (2.64). Now k_1 $(< k_0)$ is defined by

$$k_1 = hk_0 = (1 - \eta)k_0,$$

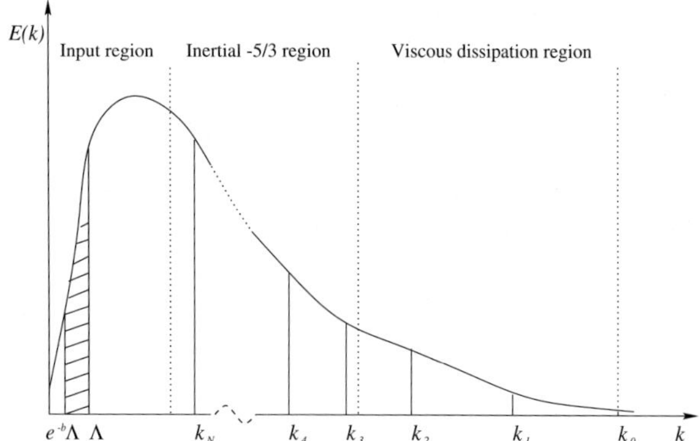

Fig. 2.19.   Wavenumber bands for FNS compared to those for Two-Field theory.

where h is the (inverse) spatial rescaling factor and η measures the bandwidth of the shell of modes being eliminated. Introduce the two-field decomposition:

$$u_k^+ = v_k^+ + \Delta_k^+, \tag{2.89}$$

where v_k^+ is another member of the turbulent ensemble, with the same statistical properties as u_k^+, and Δ_k^+ represents the phase difference between these fields.

Introduce as the basis for the *Conditional Average* (CA) a *biased subensemble*, the members of which have low-k modes which differ only slightly from u^-, this difference being represented by ϕ^-. Further define this subensemble to be such that

$$\langle \phi^- \rangle_c = 0. \tag{2.90}$$

From its definition, the CA has the properties:

$$\langle u_k^- \rangle_c = u_k^-, \quad \langle u_k^+ \rangle_c = \mathcal{O}\left(\langle \Delta_k^+ \rangle_c\right), \quad \langle u_k^- u_j^- \rangle_c = u_k^- u_j^- + \mathcal{O}\left(\langle \phi_k^- \phi_j^- \rangle_c\right). \tag{2.91}$$

$$\begin{aligned}
\langle u_k^- u_j^+ \rangle_c &= u_k^- \langle u_j^+ \rangle_c + \langle \phi_k^- u_j^+ \rangle_c \\
&= \mathcal{O}\left(\langle \Delta_j^+ \rangle_c, \langle \phi_k^- \Delta_j^+ \rangle_c\right),
\end{aligned} \tag{2.92}$$

where we have further assumed that the terms $\langle \Delta^+ \rangle_c, \langle \phi^- \phi^- \rangle_c$ and $\langle \phi^- \Delta^+ \rangle_c$ are small, and hence may be neglected as errors, and note that the $\langle \phi^- v^+ \rangle_c$ will be zero since ϕ^- and v^+ are independent.

Apply the CA to (2.63):

$$L_0 u_k^- = M_k^- \left\{ u_j^- u_{k-j}^- + \langle u_j^+ u_{k-j}^+ \rangle_c \right\}, \tag{2.93}$$

where we have dropped the terms $\langle \phi_k^- \phi_j^- \rangle_c$ and $2u_j^- \langle u_{k-j}^+ \rangle_c$ as these are assumed small. Next apply the CA to (2.64):

$$L_0 \langle u_k^+ \rangle_c = M_k^+ \left\{ u_j^- u_{k-j}^- + \langle u_j^+ u_{k-j}^+ \rangle_c \right\}, \qquad (2.94)$$

where we drop the same terms as in (2.93). Rearrange (2.94) to give an expression for $M^+ u^- u^-$ and then substitute into (2.64) to obtain

$$L_0 u_k^+ = M_k^+ \left\{ 2u_j^- u_{k-j}^+ + u_j^+ u_{k-j}^+ \right\} + H_k. \qquad (2.95)$$

The correction, H_k is given by

$$H_k = L_0 \langle u_k^+ \rangle_c - M_k^+ \left\{ \langle u_j^+ u_{k-j}^+ \rangle_c \right\} \qquad (2.96)$$

Use (2.95) to obtain a dynamical equation for $\langle u_j^+ u_{k-j}^+ \rangle_c$

$$\left(\frac{\partial}{\partial t} + \nu_0 j^2 + \nu_0 |\mathbf{k} - \mathbf{j}|^2 \right) \langle u_j^+ u_{k-j}^+ \rangle_c =$$

$$2M_j^+ \left\{ 2u_p^- \langle u_{j-p}^+ u_{k-j}^+ \rangle_c + \langle u_p^+ u_{j-p}^+ u_{k-j}^+ \rangle_c \right\} + \langle H_j u_{k-j}^+ \rangle_c. \qquad (2.97)$$

Since each term in (2.96) involves a conditional average, this implies

$$\langle H u^+ \rangle_c \sim \langle \cdot \rangle_c \langle u^+ \rangle_c \qquad (2.98)$$

and hence its neglect is consistent with earlier approximations.

To proceed further, introduce two *boundary layer* type approximations[a], assuming also that $k \sim k_d$ and $k_1 \sim k_0$:

(1) The velocity components in the high-k band are much smaller than those in the retained modes.
(2) The velocity components in the retained modes evolve very slowly on the time scales of the u^+ modes.

The first of these assumptions enables us to neglect the $\langle u^+ u^+ u^+ \rangle_c$ term in comparison to the $\langle u^+ u^+ \rangle_c u^-$ term. (In fact, the $\langle u^+ u^+ u^+ \rangle_c$ term can be treated by iteration and shown to be $\mathcal{O}(\eta)$, where η is the bandwidth of the u^+ region, in comparison to the $\langle u^+ u^+ \rangle_c u^-$ term; hence the assumption $k_1 \sim k_0$ implies this to be small in comparison). The second assumption allows us to take the operator involving the time derivative over to the right hand side to obtain:

$$\langle u_j^+ u_{k-j}^+ \rangle_c = 4M_j^+ \frac{\langle u_{j-p}^+ u_{k-j}^+ \rangle_c u_p^-}{\nu_0 j^2 + \nu_0 |\mathbf{k} - \mathbf{j}|^2}.$$

---

[a]These approximations were later replaced by consistent treatments to the appropriate order in perturbation theory: see section 4.6

Substituting into (2.63), this gives the equation for the retained modes:

$$L_0 u_k^- = M_k^- u_j^- u_{k-j}^- + \frac{4 M_k^- M_j^+ \langle u_{j-p}^+ u_{k-j}^+ \rangle_c u_p^-}{\nu_0 j^2 + \nu_0 |\mathbf{k} - \mathbf{j}|^2}. \tag{2.99}$$

The second term on the right hand side is linear in u^- and so may be interpreted in terms of an increment to the viscosity:

$$\delta \nu_0 = -\frac{4 M_k^- M_{k-j}^+ \langle u_{j-p}^+ u_{k-j}^+ \rangle_c}{k^2 (\nu_0 j^2 + \nu_0 |\mathbf{k} - \mathbf{j}|^2)}. \tag{2.100}$$

In order to evaluate the conditional average approximately, we decompose $\langle u^+ u^+ \rangle_c^+$, and retain only the leading order term $\langle v^+ v^+ \rangle_c$ $(= Q_{(v)}^+)$, then $Q_{(v)}^+$ may be represented by a Taylor series expansion of Q_k about $k = k_0$. That is,

$$Q_{(v)l}^+ = Q_l|_{l=k_0} + (l - k_0) \frac{\partial Q_l}{\partial l}\bigg|_{l=k_0} + \quad \text{higher order terms}, \tag{2.101}$$

where $l = k - j$ Q_l is determined by the assumption that the energy spectrum in the eliminated band is given by a power law of the form $E(k) = \alpha \epsilon^r k^s$. The calculation is then repeated iteratively until the viscosity converges to a fixed point. The general form for the viscosity increment is found to be

$$\delta \nu_0(k) = \frac{1}{k^2} \int \frac{d^3 j L(\mathbf{k}, \mathbf{j})}{\nu_0(j) j^2 + \nu_0(|\mathbf{k} - \mathbf{j}|)|\mathbf{k} - \mathbf{j}|^2}$$
$$\times \left[Q(l)|_{l=k_0} + (l - k_0) \frac{\partial Q(l)}{\partial l}\bigg|_{l=k_0} \right], \tag{2.102}$$

where $\mathbf{l} = \mathbf{k} - \mathbf{j}$.

Our general procedure is to eliminate modes progressively in shells defined by $k_{n+1} \leq k \leq k_n$, where

$$k_n = h k_{n-1} = h^n k_0 \qquad 0 \leq h \leq 1.$$

In this way we generate a recursion relationship for the effective viscosity

$$\nu_{n+1}(k) = \nu_n(k) + \delta \nu_n(k).$$

The form for the viscosity increment for the nth shell is found to be

$$\delta \nu_n(k) = \frac{1}{k^2} \int \frac{d^3 j L(\mathbf{k}, \mathbf{j})}{\nu_n(j) j^2 + \nu_n(|\mathbf{k} - \mathbf{j}|)|\mathbf{k} - \mathbf{j}|^2}$$
$$\times \left[Q(l)|_{l=k_n} + (l - k_n) \frac{\partial Q(l)}{\partial l}\bigg|_{l=k_n} \right]. \tag{2.103}$$

Step 2 of the RG algorithm is implemented as follows. Assume the spectrum in the band is given by $E(k) = \alpha \varepsilon^{2/3} k^{-5/3}$ and make the scaling transformation $k = k_{n+1} k'$. The effective viscosity becomes:

$$\nu_n(k) = \alpha^{1/2} \varepsilon^{1/3} k_n^{-4/3} \tilde{\nu}_n(hk').$$

The viscosity increment becomes:

$$\delta\tilde{\nu}_n(k') = \frac{1}{4\pi k'^2} \int d^3 j' \frac{L(\mathbf{k}',\mathbf{j}')Q'}{\nu_n(hj')j'^2 + \nu_n(hl')l'^2},$$

where

$$Q' = h^{11/3} - \frac{11}{3} h^{14/3}(l' - h^{-1}) + \quad \text{higher order terms.}$$

The recursion relation becomes

$$\tilde{\nu}_{n+1}(k') = h^{4/3}\tilde{\nu}_n(hk') + h^{-4/3}\delta\tilde{\nu}_n(k').$$

When these recursion relations are computed the result is as shown in Fig.2.20. . For a range of values of the (scaled) kinematic viscosity, the renormalized viscosity reaches a unique fixed point in four or five iterations. It should be noted that for a typical bandwidth of $h = 0.5$, four iterations reduces the volume of k-space (and correspondingly the number of degrees of freedom) by a factor of 16!

The local Reynolds number has been discussed in terms of its role as the 'coupling constant' and in Fig.2.21. we show that it reaches a maximum value of 0.4 at the fixed point, with this result holding for three different values of the bandwidth $\eta = (1 - k)$. The observable in this problem is the Kolmogorov prefactor α and in Fig.2.22. this is shown reaching a fixed point value of $\alpha = 1.60$ for three different initial values of the viscosity.

This insensitivity to arbitrarily chosen factors is an essential ingredient in RG and Fig.2.23. shows that the predicted value of $\alpha = 1.60 \pm 0.01$ holds for bandwidiths $0.25 \leq \eta \leq 0.45$. The breakdown of the theory for both lesser and greater values is readily understood in terms of the basic physics. For small η, the band of modes being eliminated is too narrow for asymptotic freedom to develop. Conversely, at large bandwidths we may expect the Taylor-series approximation to break down.

2.4.6. *Update of the two-field theory of turbulence*

A more formal treatment has been given by McComb and Johnston[46] with some support from numerical investigation.[47] Further developments can be found in the thesis of Khurom Kiyani.[36] We introduce a non-dimensional form of the NSE, based on:

$$\mathbf{u}(\mathbf{k}, t) \quad \text{defined on} \quad 0 \leq k \leq k_0, \tag{2.104}$$

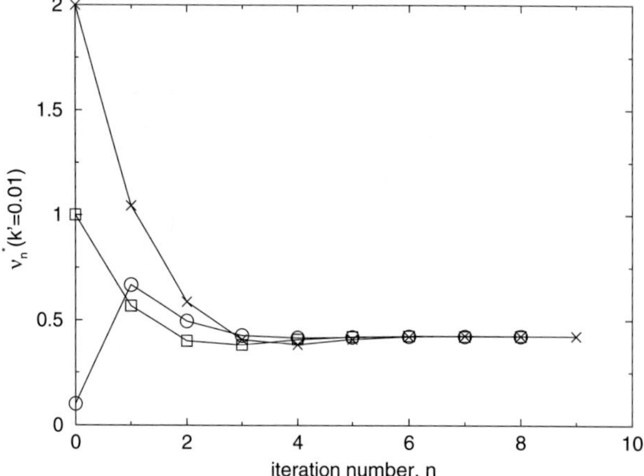

Fig. 2.20.   Effective viscosity reaching a fixed point.

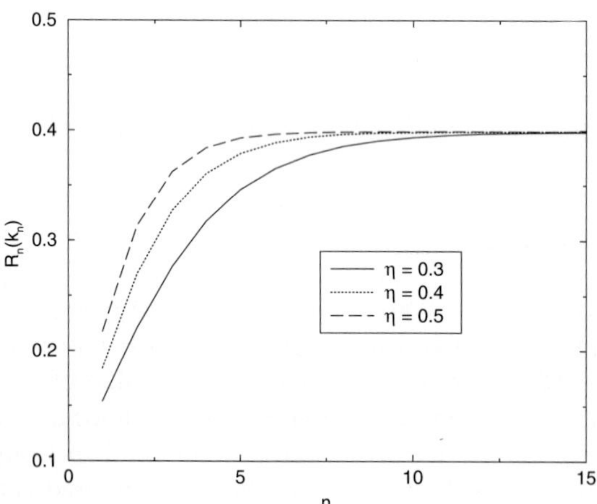

Fig. 2.21.   Local Reynolds number reaching a fixed point.

with scaled variables

$$\hat{k} = k/k_0; \qquad \hat{t} = t/\tau(k_0); \qquad \tau(k_0) = 1/\nu_0 k^2. \qquad (2.105)$$

$$\hat{\mathbf{u}}(\hat{\mathbf{k}}, \hat{t}) = \mathbf{u}(\mathbf{k}, t)/V(k_0); \qquad V^2(k) = \langle |\mathbf{u}(\mathbf{k}, t)|^2 \rangle. \qquad (2.106)$$

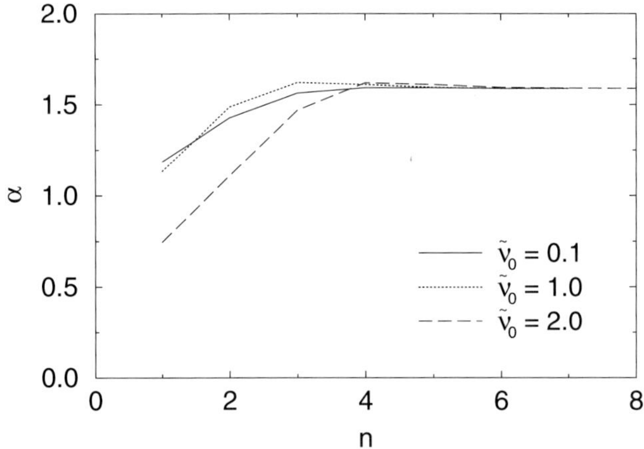

Fig. 2.22.   The Kolmogorov prefactor reaching a fixed point.

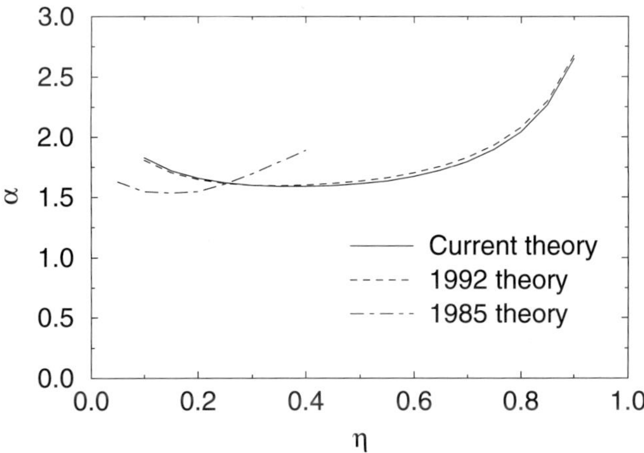

Fig. 2.23.   Variation of the Kolmogorov prefactor with choice of bandwidth.

Then the dimensionless NSE is:

$$\left(\frac{\partial}{\partial \hat{t}} + \hat{\nu}_0(\hat{k})\hat{k}^2\right)\hat{u}_\alpha(\hat{\mathbf{k}},\hat{t}) = R_0(k_0)M_{\alpha\beta\gamma}(\hat{\mathbf{k}})\int d^3\hat{j}\,\hat{u}_\beta(\hat{\mathbf{j}},\hat{t})\hat{u}_\gamma(\hat{\mathbf{k}}-\hat{\mathbf{j}},\hat{t})$$

$$+ \hat{f}_\alpha(\hat{\mathbf{k}},\hat{t}). \tag{2.107}$$

Here the dimensionless viscosity and forcing are given by:

$$\hat{\nu}_0(\hat{k}) = \tau(k_0)k_0^2\nu_0; \qquad \hat{f}_\alpha(\hat{\mathbf{k}},\hat{t}) = f_\alpha(\mathbf{k},t)\tau(k_0)/V(k_0). \tag{2.108}$$

We can also define the local Reynolds number

$$R_0(k_0) = \tau(k_0)V(k_0)k_0^4 = k_0^2 V(k_0)/\nu_0 = [E(k_0)]^{1/2}/\nu_0 k_0^{1/2}. \qquad (2.109)$$

Then the increment to viscosity (K.K. thesis, equation (4.43)):

$$\delta\nu_n(k) = \frac{1}{k^2}\int d^3j \frac{L(\mathbf{k},\mathbf{j})Q^+(|\mathbf{k}-\mathbf{j}|)}{\nu_n(j)j^2 + \nu_n(|\mathbf{k}-\mathbf{j}|)|\mathbf{k}-\mathbf{j}|^2 - \nu_n(k)k^2}. \qquad (2.110)$$

This result has been generalised from $n = 0$ to n and also $k' \to k$ etc.

2.4.7. *Non-Gaussian perturbation theory*

In two-field theory, the conditional average is evaluated in a sub-ensemble specified by

$$u_k^- = v_k^- + \phi_k^- \qquad (2.111)$$

where ϕ_k^- controls the error in the conditional average. Although both u_k^- and v_k^- must be solutions of the NSE (at least on $0 \le k \le k_1$) we may choose ϕ_k^- such that

$$\langle \phi_k^- \rangle_c = 0 \qquad (2.112)$$

and the CA is taken to have the properties

$$\langle u_k^- \rangle_c = v_k^- \qquad (2.113)$$

$$\langle u_k^- u_j^- \rangle_c = v_k^- v_j^- + \mathcal{O}\left(\langle \phi_k^- \phi_j^- \rangle_c\right). \qquad (2.114)$$

It also follows that we must have:

$$\langle v_k^- \rangle_c = v_k^-. \qquad (2.115)$$

We introduce a new soluble (or zero-order) model, as follows. A function v_k^+ is introduced as the solution of the dynamical equation:

$$L_{0k}v_k^+ = M_k^+ \left\{ v_j^- v_{k-j}^- + v_j^+ v_{k-j}^+ \right\}. \qquad (2.116)$$

This equation ensures that there is no direct coupling between v_k^- and v_k^+.

Equation (2.116) is not claimed to be a realizable physical system. In this respect it is no different from the Gaussian model in critical phenomena which does not undergo a phase transition in three dimensions and does not even possess a low-temperature fixed point.

The properties of the v^+ field under conditional average are:

$$\langle v^+ \rangle_c = \langle v^+ \rangle = 0 \qquad (2.117)$$

with

$$\langle v_k^+ v_{k'}^+ \rangle_c = \langle v_k^+ v_{k'}^+ \rangle = P_k \delta_{kk'} Q_v^+(k) \qquad (2.118)$$

where P_k is the transverse projector.

The problem then is to express $Q_v^+(k)$ in terms of $Q^+(k)$ formed from the u field. We do this by Taylor-series expansion as in two-field theory. Introduce a perturbation series for the u^+ field

$$u_k^+ = u_k^{+(0)} + \lambda u_k^{+(1)} + \lambda^2 u_k^{+(2)} + \ldots \tag{2.119}$$

where λ is a book-keeping parameter ($\lambda = 1$). Note: λ, not λ_0, to emphasise that this is non-Gaussian perturbation theory.

Then equating coefficients of each power of λ:

$$u_k^{+(0)} = v_k^+; \tag{2.120}$$

$$u_k^{+(1)} = R_{0k} M_k^+ \left\{ u_j^- u_{k-j}^- - v_j^- v_{k-j}^- + 2 u_j^- u_{k-j}^{+(0)} \right\}; \tag{2.121}$$

and so on.

Note: from (2.114), that when (2.121) is conditionally averaged, the cancellation of $u^- u^-$ and $v^- v^-$ terms removes the triple nonlinearity to $\mathcal{O}(\phi^- \phi^-)$.

This non-Gaussian perturbation theory leads to the same result as the two-field theory, thus:

$$\delta \nu_0(k) = \frac{1}{k^2} \int \frac{d^3 j \, L(\mathbf{k}, \mathbf{j})}{\nu_0(j) j^2 + \nu_0(|\mathbf{k} - \mathbf{j}|)|\mathbf{k} - \mathbf{j}|^2}$$
$$\times \left[Q(l)|_{l=k_0} + (l - k_0) \frac{\partial Q(l)}{\partial l} \bigg|_{l=k_0} \right]. \tag{2.122}$$

Compare this with the FNS result from Gaussian perturbation theory:

$$\delta \nu_0 = \frac{\lambda^2 D_0 A(d) S_d [\exp(\epsilon l) - 1]}{\nu_0^2 \Lambda^\epsilon (2\pi)^d \epsilon}, \tag{2.123}$$

where the arbitrary choice of stirring forces (i.e. D_0) is still present in the final result. In contrast, the two-field (or non-Gaussian result) depends on the actual spectral density Q of the turbulence velocity field.

Note that the two-field result is extended to further shells, with finite blocks of modes being eliminated recursively while the FNS result is immediately turned into a set of differential RG equations at $k = 0$.

A more general discussion of the new non-Gaussian perturbation theory can be found in.[48] For completeness we also mention that there has been some more general recognition of the need for a conditional average.[49–54]

2.5. Conclusion

In these lectures we have considered the turbulence problem as a branch of statistical physics and, in particular as part of what is known nowadays as *statistical field theory*. I believe that the renormalization methods described here have demonstrated their potential; either to solve the turbulence problem; or at least to be a significant ingredient in any such solution.

However, one must also be careful not to push the analogy between (say) the theory of critical phenomena (microscopic, random, thermal equilibrium, enormous number of degrees of freedom) with fluid turbulence (macroscopic, deterministic, far from equilibrium, large number of degrees of freedom) too far.

It is essential to recognise that a turbulent velocity field is a solution of a partial differential equation, with given initial and boundary conditions. Of course in practice, this solution will be irregular or chaotic, and it may well be permissible to treat it as if it were random. But this is always an approximation which requires examination. Also, with only about 10^4 degrees of freedom, as compared to 10^{24} for Avogadro-size systems, we cannot expect the 'large N limit' to operate so effectively; nor with microscales of $10^{-3}m$ as compared to $10^{-10}m$ for microscopic systems, can we expect 'finite-size effects' to be negligible.

In fact this is a characteristic of the experimental picture of turbulence: results are never quite as self-similar, self-preserving or scale-invariant as one might hope. Obviously this poses a problem for fundamental theories which, if only implicitly, are predicated on the existence of such universal behaviour.

Although such characteristics are part and parcel of renormalization methods, we should also bear in mind the existence of other fundamental approaches to turbulence which can be loosely subsumed under the twin headings of *vortex methods* and *dynamical systems theory*[a]. We have heard quite a few interesting talks on the latter subject in both the Workshop and the Summer School, and it is clear that this is a very active field of research. Yet mainstream turbulence researchers are inclined to see it as remote from their own concerns which involve large numbers of degrees of freedom and statistically averaged behaviour.

Nevertheless, it may be a mistake to see the situation in turbulence as being analogous to the statistical mechanics of ideal gases (say). There, dynamical systems theory aims to shed light on fundamental aspects such as ergodicity, but is unlikely to affect the successful theories which have existed for about a century. The problem of working out averages in both quantum and classical gases has been solved. Essentially, dynamical systems theory

---

[a]These two approaches are not necessarily mutually exclusive.

can only hope to give us a better understanding of how and why that solution is valid.

But this is not the case in turbulence, where we have many loose ends and, as I pointed out earlier, a much smaller number of degrees of freedom. The turbulence problem has long been characterized by the difficulty of handling the interplay between randomness and coherence, There seems to be a strong case for uniting the main fundamental approaches in order to separate out the random aspects from the deterministic. Possibly this could be done under the umbrella of some of the more eclectic approaches, such as multi-scale or multi-level methods? Certainly any initiative which produced a greater degree of cross-communication and a higher level of scholarship (as distinguished from 'research') would be a step in the right direction.

Bibliography

1. Laurie M Brown, editor. *Renormalization: From Lorentz to Landau (and Beyond)*. Springer-Verlag, 1993.
2. W.D. McComb. *Renormalisation Methods*. Oxford University Press, 2004.
3. A. N. Kolmogorov. The local structure of turbulence in incompressible viscous fluid for very large Reynolds numbers. *C. R. Acad. Sci. URSS*, 30:301, 1941.
4. A. N. Kolmogorov. Dissipation of energy in locally isotropic turbulence. *C. R. Acad. Sci. URSS*, 32:16, 1941.
5. W. D. McComb. *The Physics of Fluid Turbulence*. Oxford University Press, 1990.
6. G.K. Batchelor. *The theory of homogeneous turbulence*. Cambridge University Press, Cambridge, 2nd edn edition, 1971.
7. D.C. Leslie. *Developments in the theory of modern turbulence*. Clarendon Press, Oxford, 1973.
8. W. D. McComb. Theory of turbulence. *Rep. Prog. Phys.*, 58:1117–1206, 1995.
9. R. H. Kraichnan. The structure of isotropic turbulence at very high Reynolds numbers. *J. Fluid Mech.*, 5:497–543, 1959.
10. S.F. Edwards. The statistical dynamics of homogeneous turbulence. *J. Fluid Mech.*, 18:239, 1964.
11. J.R. Herring. Self-consistent field approach to turbulence theory. *Phys. Fluids*, 8:2219, 1965.
12. R. H. Kraichnan. Lagrangian-history closure approximation for turbulence. *Phys. Fluids*, 8(4):575–598, 1965.
13. R. H. Kraichnan and J.R. Herring. A strain-based Lagrangian-history turbulence theory. *J. Fluid Mech.*, 88:355, 1978.
14. Y. Kaneda. Renormalized expansions in the theory of turbulence with the use of the Lagrangian position function. *J. Fluid Mech.*, 107:131–145, 1981.
15. S. Kida and S.Goto. A Lagrangian direct-interaction approximation for ho-

78   *David McComb*

mogeneous isotropic turbulence. *J. Fluid Mech.*, 345:307–345, 1997.

16. W. D. McComb. A local energy transfer theory of isotropic turbulence. *J.Phys.A*, 7(5):632, 1974.

17. H.W Wyld, Jr. Formulation of the theory of turbulence in an incompressible fluid. *Ann.Phys*, 14:143, 1961.

18. J.R. Herring. Self-consistent field approach to nonstationary turbulence. *Phys. Fluids*, 9:2106, 1966.

19. L. L. Lee. A formulation of the theory of isotropic hydromagnetic turbulence in an incompressible fluid. *Ann.Phys*, 32:292, 1965.

20. R. H. Kraichnan. Decay of isotropic turbulence in the Direct-Interaction Approximation. *Phys. Fluids*, 7(7):1030–1048, 1964.

21. J.R. Herring and R.H. Kraichnan. *Comparison of some approximations for isotropic turbulence Lecture Notes in Physics*, volume 12, chapter Statistical Models and Turbulence, page 148. Springer, Berlin, 1972.

22. J.R. Herring and R.H. Kraichnan. A numerical comparison of velocity-based and strain-based Lagrangian-history turbulence approximations. *J. Fluid Mech.*, 91:581, 1979.

23. W. D. McComb and V. Shanmugasundaram. Numerical calculations of decaying isotropic turbulence using the LET theory. *J. Fluid Mech.*, 143:95–123, 1984.

24. W. D. McComb, V. Shanmugasundaram, and P. Hutchinson. Velocity derivative skewness and two-time velocity correlations of isotropic turbulence as predicted by the LET theory. *J. Fluid Mech.*, 208:91, 1989.

25. W. D. McComb, M. J. Filipiak, and V. Shanmugasundaram. Rederivation and further assessment of the LET theory of isotropic turbulence, as applied to passive scalar convection. *J. Fluid Mech.*, 245:279–300, 1992.

26. W. D. McComb and A.P. Quinn. Two-point, two-time closures applied to forced isotropic turbulence. *Physica A*, 317:487–508, 2003.

27. J.S. Frederiksen, A. G. Davies, and R.C. Bell. Closure theories with non-gaussian restarts for truncated two dimensional turbulence. *Phys. Fluids*, 6(9):3153, 1994.

28. J.S. Frederiksen and A. G. Davies. Eddy viscosity and stochastic backscatter parameterizations on the sphere for atmospheric circulation models. *J. Atmos. Sci.*, 54:2475–2492, 1997.

29. Jorgen S. Frederiksen and Antony G. Davies. Dynamics and spectra of cumulant update closures for two-dimensional turbulence. *Geophys. Astrophys. Fluid Dynamics*, 92:197, 2000.

30. Jorgen S. Frederiksen and Terence J. O'Kane. Inhomogeneous closure and statistical mechanics for Rossby wave turbulence over topography. *J. Fluid. Mech.*, 539:137–165, 2005.

31. A. P. Quinn. *Local Energy Transfer theory in forced and decaying isotropic turbulence*. PhD thesis, University of Edinburgh, 2000.

32. P.C. Martin, E.D. Siggia, and H.A. Rose. Statistical Dynamics of Classical Systems. *Phys. Rev. A*, 8(1):423–437, 1973.

33. W. D. McComb and K. Kiyani. Eulerian spectral closures for isotropic turbulence using a time-ordered fluctuation-dissipation relation. *Phys. Rev. E*,

72:016309-1–12, 2005.

34. W. D. McComb. Galilean invariance and vertex renormalization. *Phys. Rev. E*, 71:037301, 2005.

35. A. Berera and D. Hochberg. Galilean invariance and homogeneous anisotropic randomly stirred flows. *Phys. Rev. E*, 72:057301, 2005.

36. K. Kiyani. *An assessment of renormalization methods in the statistical theory of isotropic turbulence*. PhD thesis, University of Edinburgh, 2005.

37. K. Kiyani and W. D. McComb. Time-ordered fluctuation-dissipation relation for incompressible isotropic turbulence. *Physical Review E*, 70:066303-1–4, 2004.

38. J. C. Bowman, J. A. Krommes, and M. Ottaviani. The realizable Markovian closure. I. general theory, with application to three-wave dynamics. *Phys. Fluids B*, 5:3558–3589, 1993.

39. L.Ts. Adzhemyan, N.V. Antonov, and A.N. Vasiliev. *The Field Theoretic Renormalization Group in Fully Developed Turbulence*. Gordon and Breach Science Publishers, 1999.

40. D. Forster, D.R. Nelson, and M.J. Stephen. Large-distance and long-time properties of a randomly stirred fluid. *Phys. Rev. A*, 16(2):732–749, 1977.

41. W. D. McComb, W. Roberts, and A. G. Watt. Conditional-averaging procedure for problems with mode-mode coupling. *Phys. Rev. A*, 45(6):3507–3515, 1992.

42. D. Forster, D.R. Nelson, and M.J. Stephen. Long-time tails and the large-eddy behaviour of a randomly stirred fluid. *Phys. Rev. Lett.*, 36(15):867–869, 1976.

43. W. D. McComb and A. G. Watt. Conditional averaging procedure for the elimination of the small-scale modes from incompressible fluid turbulence at high Reynolds numbers. *Phys. Rev. Lett.*, 65(26):3281–3284, 1990.

44. W. D. McComb and A. G. Watt. Two-field theory of incompressible-fluid turbulence. *Phys. Rev. A*, 46(8):4797–4812, 1992.

45. W. D. McComb. Reformulation of the statistical equations for turbulent shear flow. *Phys. Rev. A*, 26(2):1078–1094, 1982.

46. W. D. McComb and C. Johnston. Conditional mode elimination and scale-invariant dissipation in isotropic turbulence. *Physica A*, 292:346, 2001.

47. W. D. McComb, A. Hunter, and C. Johnston. Conditional mode-elimination and the subgrid-modelling problem for isotropic turbulence. *Physics of Fluids*, 13:2030, 2001.

48. W. D. McComb. Asymptotic freedom, non-Gaussian perturbation theory, and the application of renormalization group theory to isotropic turbulence. *Phys. Rev. E*, 73:026303-1–7, 2006.

49. Y. Nagano and Y. Itazu. Renormalization group theory for turbulence: Eddy-viscosity type model based on an iterative averaging method. *Phys. Fluid*, 9(1):143–153, 1997.

50. M. K. Verma. Field theoretic calculation of renormalized viscosity, renomalized resistivity, and energy fluxes of magnetohydrodynamic turbulence. *Phys. Rev. E.*, 64(026305):1–4, 2001.

51. S. Sukoriansky, B. Galperin, and I. Staroselsky. Cross-term and ϵ-expansion

in RGE theory of turbulence. *Fluid Dynamics Research*, 33:319–331, 2003.

52. C.C. Chang, B-S. Lin, and C-T. Wang. Solvable model in renormalization group analysis for effective eddy viscosity. *Phys. Rev. E*, 67:1–4, 2003.

53. Y. Cao and W.K. Chow. Recursive renormalization-group calculation for the eddy viscosity and thermal eddy diffusity of incompressible turbulence. *Physica A*, 339:320–338, 2004.

54. V.L. Saveliev and M.A. Gorokhovski. Group-theoretical model of developed turbulence and renormalization of the Navier-Stokes equation. *Phys. Rev. E*, 72:1–6, 2005.

Chapter 3

Turbulence and Coherent Structures in the Ocean

Henk A. Dijkstra

Institute for Marine and Atmospheric research Utrecht (IMAU),
Department of Physics and Astronomy,
Utrecht University, Utrecht, the Netherlands
E-mail: dijkstra@phys.uu.nl

An introduction is given to the multi-scale processes that provide major contributions to the ocean's energy balance. Motivating problem is the issue to determine how the large-scale ocean will respond to variations in momentum and buoyancy exchange with the atmosphere. It appears that to obtain a clear answer to this problem, knowledge of turbulent flows at many different spatial and temporal scales is essential.

Contents

81

3.1. Introduction

On the large scale, the ocean circulation is driven by momentum fluxes (by the wind) and affected by fluxes of heat and freshwater at the ocean-atmosphere interface. The latter fluxes change the surface density of the ocean water and through mixing and advection, density differences are propagated horizontally and vertically.

A picture of the mainly wind-driven surface ocean circulation can be found in many texbooks (e.g., see Fig. 8.1 in[1]). In the North Pacific and North Atlantic, cellular type motions are found, with a large clockwise rotating subtropical gyre and a smaller anti-clockwise subpolar gyre. Currents at the western side of the basin are strongest, with the Gulf Stream in the Atlantic Ocean and the Kuroshio in the Pacific Ocean being the major currents. The Gulf Stream can be viewed as an eastward jet being part of both the North Atlantic subtropical and subpolar gyres. Typical horizontal velocities in the Gulf Stream are 1 ms^{-1}, whereas depth averaged velocities in the gyres are of order 0.01 ms^{-1}. The major current in the Southern Ocean is the Antarctic Circumpolar Current (ACC), which encircles the Antarctic continent from west to east. The ACC has an average volume transport of about 150 Sv ($1 \text{ Sv} = 10^6 \text{ m}^3\text{s}^{-1}$). Also in the Southern Hemisphere, the currents at the western side of each basin are the strongest, i.e., the East Australian Current in the South Pacific, the Brazil Current in the South Atlantic and the Agulhas Current in the South Indian Ocean.

The three-dimensional pattern of the time-mean flow, in particular the deep component, is much less known from observations. In addition to the forcing by the wind stress, this pattern is also affected by density differences, the latter mainly introduced through temperature and salinity gradients. When the surface layer is cooled and its density becomes larger than that of water just below, vigorous mixing will occur. The density change at depth affects horizontal density gradients and hence is able to affect the deep currents.

The last decade has seen a huge increase in the observational information available on the oceans' basin and global scales in particular through the World Ocean Circulation Experiment.[2] A profile of the temperature field[a] along a north-south section through the Atlantic (at 24°W) is plotted in Fig. 3.1. The ocean is stably stratified on the large scale as density

---

[a]formally this is the potential temperature field in which effects of compressibility of water at depth have been taken into account

decreases upwards everywhere. If we take z as a vertical coordinate, then a measure of the stratification is the buoyancy frequency, N^2 defined by

$$N^2 = -\frac{g}{\rho}\frac{\partial \rho}{\partial z} \tag{3.1}$$

Typical values of N are 5×10^{-4} s^{-1} in the deep ocean to 10^{-2} s^{-1} in the upper ocean.

Analysis of the WOCE section data combined with inversion studies have lead to estimates over the volume transports through the world ocean basins. In Fig. 2 in,[3] the boundaries between water masses are taken as certain so-called density surfaces. In the North Atlantic, about 16 Sv flows northwards in the upper layer. In the northern part of the basin, water mass transformations lead to sinking of about 15 Sv which is transported southwards at depth. Over the globe, about 30 Sv is upwelling from the abyss to upper layers in the ocean and this upwelling is balanced by sinking of surface waters and intermediate waters.

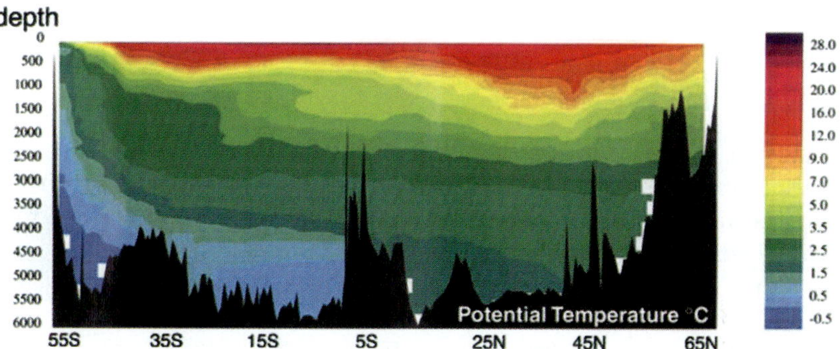

Fig. 3.1.   Temperature profile (°C) along a WOCE section in the Atlantic.[4]

The oceans take care of about 30%-40% of the total meridional heat transport of the combined ocean-atmosphere system. The meridional heat transport in the Atlantic is positive over the whole basin with a maximum of about 1.3 PW (1 PW $= 10^{15}$ W) at 25°N. In the Pacific, the heat transport is about a factor two smaller than in the Atlantic and it is southward in the South Pacific. The meridional heat transport in the Indian Ocean is mainly southward with a maximum of 1.8 PW near 20°S (see Fig. 1 in[3]).

Because of the role of the ocean heat transport in climate, there is much interest in how the large-scale ocean flow is affected by changes in the surface forcing. With an increasing concentration of greenhouse gases in the atmosphere, there is a potential to change the surface heat, freshwater and wind-stress forcing. How will the ocean circulation react to these changes in forcing? There may be quite some urgency to this problem as[5] have recently shown that the meridional heat transport at 25°N in the Atlantic has decreased by 15% (from 1.3 to 1.1 PW) over the period 1957-2004.

Recently, this problem has also gotten a new dimension due to the complexity of the energetics of ocean flows and it is now one of the major challenges in physical oceanography; it is the problem which we will address in this chapter. The approach to a possible answer will lead us along the multi-scale and multi-process (turbulent) flows in the ocean.

3.2. Specification of the problem

In the first subsection below, we start from the well-known equations of fluid dynamics. We next consider in more detail the energy balances for any ocean flow and discuss energy conversion mechanisms. In the last subsection, we indicate the challenges to understand how the large-scale ocean circulation is driven from an energetic point of view.

3.2.1. *Governing equations*

Consider a flow of water within a bounded region V on the earth. The equations of motion for a local volume of ocean water, as described from a reference frame moving along with the earth (with rotation vector $\boldsymbol{\Omega}$ and angular frequency $\Omega =| \boldsymbol{\Omega} |$) are[6]

$$\rho \left[\frac{D\mathbf{v}}{dt} + 2\boldsymbol{\Omega} \wedge \mathbf{v} \right] = -\nabla p - \rho\nabla\Phi + \rho\nu\nabla^2\mathbf{v} \qquad (3.2a)$$

$$\frac{D\rho}{dt} + \rho\nabla.\mathbf{v} = 0 \qquad (3.2b)$$

Here, $D/dt = \partial/\partial t + \mathbf{v}.\nabla$ is the material derivative and $\wedge$ is the vector product. The vector $\mathbf{v}$ is the velocity field of the flow, p is the pressure, and ρ is the density of the liquid. The quantity Φ is a potential, which represents the gravitational acceleration and the tide-generating accelerations. If we use z as the vertical coordinate then $\Phi = gz + \Phi_M$, where g is the gravitational acceleration and Φ_M is the tidal potential (see section 3.2 below). The quantity ν is the kinematic viscosity of water and represents the effects of changes in momentum due to random molecular motions.

As explained in appendix A, the salinity S is defined as the sum of the mass fraction of the ions in sea water. Let the local temperature of a volume of sea water be indicated by T, then conservation of heat and salt is formulated as

$$\rho C_p \frac{DT}{dt} = \nabla.\mathbf{q}_T + Q_T \tag{3.3a}$$

$$\rho \frac{DS}{dt} = \nabla.\mathbf{q}_S + Q_S \tag{3.3b}$$

In these equations, the terms $\mathbf{q}_T$ (Wm^{-2}) and $\mathbf{q}_S$ (kg m^{-2}s^{-1}) are the diffusive fluxes of heat and salt. The quantities Q_T (Wm^{-3}) and Q_S (kg m^{-3}s^{-1}) represent the internal sources and sinks of heat and salt. The quantity C_p is the heat capacity of the water.

As there are six equations and seven unknowns in the equations above, we need one more equation to close the system of equations. This is the equation of state, which for liquids such as salt water is given by

$$\rho = \rho(T, S, p) \tag{3.4}$$

For ocean water, the full equation of state can be found in.[7] A linear equation of state where

$$\rho = \rho_0(1 - \alpha_T(T - T_0) + \alpha_S(S - S_0)) \tag{3.5}$$

where T_0, S_0 and ρ_0 are reference values and α_T and α_S the volume expansion coefficients, is often an adequate approximation.

The flow in the ocean basin is driven by an atmospheric wind stress field, τ (in Nm^{-2}) and is affected by a downward heat flux Q_{oa} (in Wm^{-2}) and a downward freshwater flux $P - E$ (in ms^{-1}), where E and P represent evaporation and precipitation. These quantities enter the boundary conditions of the system of governing equations at the ocean-atmosphere interface. Boundary conditions at the continental boundaries and the bottom boundaries are relatively simple and consist of no-slip and no-flux conditions.

3.2.2. *The ocean energy balances*

Let the internal energy per unit of mass of the flow be indicated by E_I, the kinetic energy per unit of mass by E_K and the potential energy per unit mass by E_P. It is also convenient to introduce the stress tensor $\mathcal{T}$ of a Newtonian liquid as

$$\mathcal{T} = -p\mathcal{I} + 2\rho\nu(\mathcal{G} + \mathcal{G}^T) \rightarrow \nabla.\mathcal{T} = -\nabla p + \rho\nu\nabla^2\mathbf{v} \tag{3.6}$$

where $\mathcal{I}$ is the identity tensor and $\mathcal{G}_{ij} = \partial v_i/\partial x_j$ is the velocity gradient tensor.

The local form of the sum of the kinetic and internal energy balance (from the first law of thermodynamics) is

$$\rho\frac{D}{dt}(E_I + E_K) = -\nabla.\mathbf{q} - \rho\nabla\Phi.\mathbf{v} + \nabla(\mathcal{T}.\mathbf{v}) \tag{3.7}$$

where $\mathbf{q}$ is the total heat flux. Here the first term on the right hand side is the surface flux entering the fluid volume over the boundaries, the second term is the work by volume forces and the third term is the work by surface forces. The mechanical energy balance can be derived by multiplying the momentum balance (3.2a) by $\mathbf{v}$ to give

$$\rho\frac{DE_K}{dt} = -\rho\nabla\Phi.\mathbf{v} + \mathbf{v}.(\nabla.\mathcal{T}) \tag{3.8}$$

Subtracting the mechanical energy balance from the total energy balance gives, and the use of the identity

$$\mathbf{v}.(\nabla.\mathcal{T}) - \nabla(\mathcal{T}.\mathbf{v}) = \rho\varepsilon - p\nabla.\mathbf{v} \tag{3.9}$$

where $\varepsilon = \nu\nabla\mathbf{v}.\nabla\mathbf{v}$ is the viscous dissipation function, gives the internal energy balance

$$\rho\frac{DE_I}{dt} = -\nabla.\mathbf{q} + \rho\varepsilon - p\nabla.\mathbf{v} \tag{3.10}$$

With use of the identities $\mathbf{v}.(\mathbf{v}.\nabla\mathbf{v}) = \mathbf{v}.(\nabla E_K - \mathbf{v}\wedge(\nabla\wedge\mathbf{v}))$, $\mathbf{v}.\nabla^2\mathbf{v} = \nabla^2 E_K - \nabla\mathbf{v}.\nabla\mathbf{v}$ and the continuity equation (3.2b) multiplied by E_K, we can write the mechanical energy balance in the form

$$\frac{\partial(\rho E_K)}{\partial t} + \nabla.(\rho E_K\mathbf{v} - \rho\nu\nabla E_K + p\mathbf{v}) = -\rho\mathbf{v}.\nabla\Phi + p\nabla.\mathbf{v} - \rho\varepsilon \tag{3.11}$$

In the same way, we can write the internal energy balance as

$$\frac{\partial(\rho E_I)}{\partial t} + \nabla.(\rho E_I\mathbf{v} + \mathbf{q}) = -p\nabla.\mathbf{v} + \rho\varepsilon \tag{3.12}$$

The potential energy balance follows directly from the definition $E_P = \rho g z = \rho(\Phi - \Phi_M)$ and the continuity equation (multiplied by gz) and becomes an equation for Φ, i.e.

$$\frac{\partial(\rho\Phi)}{\partial t} + \nabla.(\rho\Phi\mathbf{v}) = \rho\mathbf{v}.\nabla\Phi + \rho\frac{\partial\Phi_M}{\partial t} \tag{3.13}$$

The global energy balances are obtained by integrating (3.11), (3.12) and (3.13) over the domain V. Care has to be taken that the upper (ocean-atmosphere) surface is a material deformable surface which satisfies the kinematic condition of no mass transport. When the outward normal on the boundary of V is indicated by $\mathbf{n}$, the global energy balances become

$$\frac{\partial}{\partial t}\int \rho E_K\, d^3x = -\int \rho E_K(\mathbf{v} - \mathbf{v}_s).\mathbf{n}\, d^2x$$

$$+ \int E_K(-p\mathbf{v} + \rho\nu\nabla E_K).\mathbf{n}\, d^2x - \int \rho\mathbf{v}.\nabla\Phi\, d^3x$$

$$+ \int p\nabla.\mathbf{v}\, d^3x - \int \rho\varepsilon\, d^3x \tag{3.14}$$

$$\frac{\partial}{\partial t} \int \rho \Phi \, d^3x = - \int \rho \Phi (\mathbf{v} - \mathbf{v}_s).\mathbf{n} \, d^2x$$

$$+ \int \rho \frac{\partial \Phi_M}{\partial t} \, d^3x + \int \rho \mathbf{v}.\nabla \Phi \, d^3x \qquad (3.15)$$

and

$$\frac{\partial}{\partial t} \int \rho E_I \, d^3x = - \int \rho E_I (\mathbf{v} - \mathbf{v}_s).\mathbf{n} \, d^2x$$

$$- \int \mathbf{q}.\mathbf{n} \, d^2x - \int p \nabla.\mathbf{v} \, d^3x + \int \rho \varepsilon \, d^3x \qquad (3.16)$$

In the equations above, each volume integral is indicated by d^3x, where, for example $d^3x = r^2 \cos\theta \, d\phi \, d\theta \, dz$ in spherical coordinates. Each surface integral is indicated by d^2x, i.e., $d^2x = r^2 \cos\theta \, d\phi \, d\theta$ in spherical coordinates.

The term involving the volume integral of $p\nabla.\mathbf{v}$ represents a conversion between kinetic and internal energy. Through the continuity equation, we find

$$p\nabla.\mathbf{v} = -\frac{p}{\rho} \frac{D\rho}{dt} \qquad (3.17)$$

and hence this term represents effects of density changes along streamlines (e.g., through compressibility and heat/salt changes).

The term involving the integral of $\rho \mathbf{v}.\nabla \Phi$ represents the conversion between kinetic and potential energy which is also referred to as the buoyancy production. For $\Phi = gz$, this term is written as $\rho g w$, where w is the vertical velocity of the liquid. Consider a situation in which $w > 0$ and where $N^2 > 0$ (stable stratification). When a fluid element (with higher density) moves upwards, it looses kinetic energy but it raises the potential energy of the system. In this case, the buoyancy production $\rho \mathbf{v}.\nabla \Phi$ is positive and it indeed shows up as a source term in (3.15) and a sink term in (3.14).

3.2.3. *A fundamental problem*

The basic equations as presented in the previous subsections have most of our confidence. The description depends on fluid parameters such as the kinematic viscosity which we know well for sea water. We know that these equations provide an accurate description of fluid phenomena as observed in laboratory experiments. Therefore, they are also expected to describe the small-scale (cm to m) processes in the ocean. Hence, when the surface forcing of the ocean is known it should in principle be possible to solve the governing equations and determine the sensitivity of the circulation with respect to strength of the momentum and buoyancy forcing.

However, typical values of the viscous dissipation ε in the ocean (as will be discussed in section 3.4.4. below) are $\varepsilon = 10^{-9}$ m^{2}s^{-3} and hence the Kolmogorov length scale $(\nu^3/\varepsilon)^{1/4} \approx 10^{-2}$ m. The flows in the ocean encompass a large range of scales, from centimeters and seconds on the very small scales towards thousands of kilometers and thousand of years on the very large scales. Flows on the basin scale such as the North Atlantic have a spatial dimension of 10^6 m and the ratio of the largest and smallest spatial scale of motion is therefore about a factor 10^8; the same holds for the ratio of temporal scales.

With current computational resources, we cannot use the fundamental equations to simulate large-scale ocean flows and hence have to resort to approximations of the effect of the small-scale processes. The usual step is to decompose the flow into a time-mean part and a fluctuating part, say for the velocity field

$$\mathbf{v} = \bar{\mathbf{v}} + \tilde{\mathbf{v}} \; ; \; \bar{\mathbf{v}} = \lim_{\Delta t \to \infty} \frac{1}{\Delta t} \int_{t_0}^{t_0 + \Delta t} \mathbf{v} \, dt \qquad (3.18)$$

Here, Δt is a time long compared to typical fluctuations of $\mathbf{v}$. When this decomposition is substituted into the momentum equations and the result is time-averaged, it is found that the mixing of momentum due to the fluctuations is much larger than that caused by molecular transport, i.e. for the first component of (3.2a), we find

$$\rho \, | \, \overline{\tilde{\mathbf{v}}.\nabla \tilde{u}} \, | \gg \nu \, | \, \nabla^2 u \, | \qquad (3.19)$$

To represent the effect of the fluctuations on the large-scale flow, closure assumptions have to be made which relate the turbulent exchange terms (here the Reynolds' stresses) to gradients in the time-mean flow. In the models of large-scale ocean flows, the effect of the small-scale processes are therefore incorporated through (very uncertain) closure assumptions. It appears ironic that we cannot represent the effects of the small-scale processes by making direct use of the well-known equations that govern them.

To understand how the large-scale flow is driven therefore requires a detailed study of processes leading to energy input in the ocean and of processes establishing the transport of energy towards its dissipation at the smallest scales.

3.3. Energy input

When the three global energy equations (3.14), (3.16) and (3.16) are added, we find

$$\frac{\partial}{\partial t}\int \rho(E_K + E_I + \Phi)\,d^3x + \int \rho(E_K + E_I + \Phi)(\mathbf{v} - \mathbf{v}_s).\mathbf{n}\,d^2x =$$

$$\int (-p\mathbf{v} + \rho\nu\nabla E_K).\mathbf{n}\,d^2x + \int \rho\frac{\partial \Phi_M}{\partial t}\,d^3x - \int \mathbf{q}.\mathbf{n}\,d^2x$$

The three terms on the right hand side represent the work due to normal and tangential stresses at the surface, the tidal energy input and the input due to buoyancy fluxes.

3.3.1. *Surface momentum forcing*

The winds act on the sea-surface to convert atmospheric kinetic energy to kinetic energy of the ocean. A rough estimate of the energy input can be deduced from a so-called bulk formula relating the wind stress magnitude τ_w (in Nm^{-2}) to the atmospheric wind speed u_a, say at 10 m height in the atmosphere, i.e.,

$$\tau_w = C_D \rho_a u_a^2 \tag{3.20}$$

where C_D is an semi-empirical drag coefficient and ρ_a is the density of the air. Using the estimate

$$\rho\nu\nabla E_K.\mathbf{n} \approx \tau_w u_o \tag{3.21}$$

where u_o is a typical surface velocity of the upper ocean, we obtain an estimate of the energy production as

$$P_w = s_e\,C_D \rho_a u_a^2 u_o \tag{3.22}$$

where s_e is the surface area of the Earth (about 3.5×10^{14} m^2). With $u_o = 0.3$ ms^{-1}, $u_a = 10$ ms^{-1}, $\rho_a = 1.25$ kgm^{-3} and $C_D = 1.25 \times 10^{-3}$, we then find an estimate $P_w = 16$ TW (1 TW $= 10^{12}$ W). With the uncertainties in each parameters taken into account a range of 10-60 TW results.

If we consider this input in more detail, we decompose the oceanic flow velocity field $\mathbf{v}$ into a low-frequency component $\bar{\mathbf{v}}$ and high-frequency fluctuations $\hat{\mathbf{v}}$, the latter, for example, associated with (breaking) surface waves. Similarly, we can decompose the wind stress field and the pressure field such that

$$\mathbf{v} = \bar{\mathbf{v}} + \hat{\mathbf{v}} \; ; \; \boldsymbol{\tau} = \bar{\boldsymbol{\tau}} + \hat{\boldsymbol{\tau}} \; ; \; p = \bar{p} + \hat{p} \tag{3.23}$$

The time averaged mechanical energy input P_w at the surface then becomes locally

$$P_w = \bar{\boldsymbol{\tau}}.\bar{\mathbf{v}} + \overline{\hat{\boldsymbol{\tau}}.\hat{\mathbf{v}}} + \overline{\hat{p}\hat{w}} \tag{3.24}$$

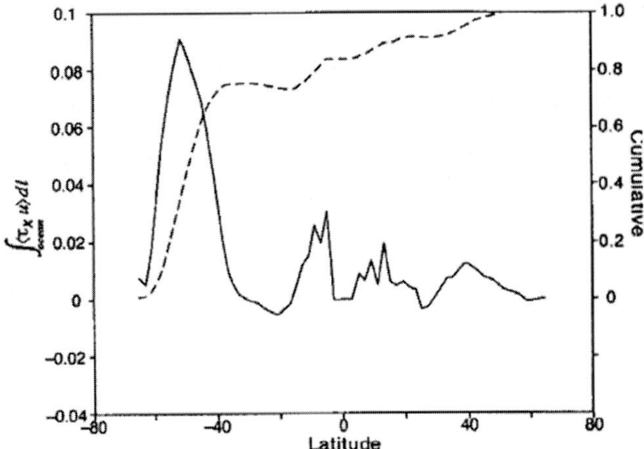

Fig. 3.2.   The drawn curve is the zonally integrated contribution of the term $\bar{\tau}^x \bar{u}_G$ over a 2° latitude band.   The dashed curve is the cumulative contribution starting in the Southern Ocean.   Values (in TW) are estimated from four years of satellite sea surface height data.[8]

The first term on the right hand side is the work of the mean wind stress on the quasi-steady ocean currents.   Its contribution can be split into a geostrophic and ageostrophic component, because

$$\bar{\mathbf{v}} = \bar{\mathbf{v}}_G + \bar{\mathbf{v}}_A \tag{3.25}$$

where the subscripts G and A indicate the geostrophic and ageostrophic part of the velocity, respectively.   An important contribution to the ageostrophic terms is that of the surface frictional boundary layer, the Ekman layer.   The different contributions have been estimated from observations in[8] and.[9,10]   In[8] the dominant term $\bar{\tau}^x.\bar{u}_G$ is plotted, having maxima of 2.5×10^{-2} Wm^{-2} in the Southern Ocean. The integrated result is shown in Fig. 3.2. where the dashed curve provides the total work; this leads to an estimate of about 1 TW input to the general circulation. This is a relatively small number considering that the total wind input is probably around 10-60 TW. The mechanical energy input into the Ekman layer has been estimated to be about 3 TW.[9,10]

3.3.2.   *Tidal forcing*

The Earth and Moon form a single mechanical system in which both rotate about a common center of mass, with a rotation period of 27.3 days. The Earth revolves eccentrically about the common centre of mass and all points within and on Earth follow circular paths, all of which have the same radius and the same angular velocity $2\pi/27.3$ days^{-1}. All points on Earth therefore experience an equal acceleration and hence an equal centrifugal force directed parallel to a line joining the centres of the Earth and the Moon. To keep the Moon in an orbit around the Earth, there is an exact balance between the gravitational attractive force (between them) and this centrifugal force at the center of mass.

To determine the tidal potential V_M, we consider first the idealized situation with the Moon above the equator and neglect the effect of the rotation of the Earth. At the Earth's surface, the gravitational attractive force by the Moon is not constant. It is greater than average on the side of the Earth facing the Moon (pulling the Earth away from its center) and less than average on the opposite site (Fig. 3.3.a). The resultant force (gravitational minus centrifugal) is called the tide-producing force. Consider the point E in Fig. 3.3., then the centrifugal force F_E^c is equal to the gravity force at the center of mass (which is approximately at the center of the Earth) and its magnitude is given by

$$F_E^c = \gamma \frac{M_E M_M}{R^2} \tag{3.26}$$

Here γ is the gravitational constant, M_E and M_M are the mass of the Earth and Moon, respectively and R is the distance between the centers of the Earth and the Moon.

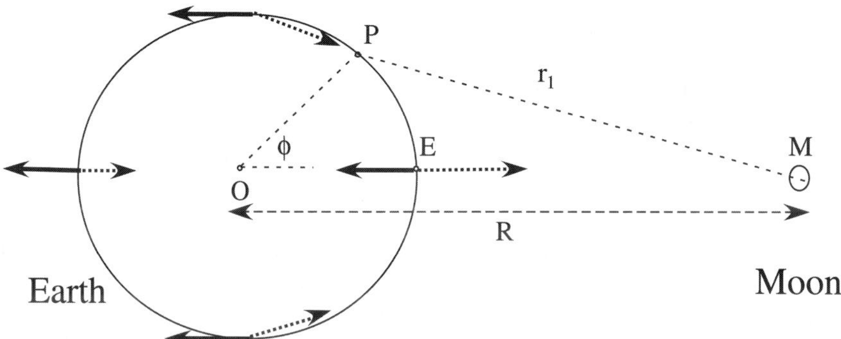

Fig. 3.3.   Sketch to show how the tide-generating force arises due to differences in centrifugal and gravitational forces and to compute the tidal potential at an arbitrary point P on the surface of the Earth. The solid arrows represent the centrifugal force and the dotted arrows the gravitational force.

The magnitude of the gravitational force F_E^g at E (Fig. 3.3.) is given by

$$F_E^g = \gamma \frac{M_E M_M}{(R - r)^2} \tag{3.27}$$

where r is the radius of the Earth. The tide-generating force F_E^t is then

$$F_E^t = F_E^g - F_E^c = \gamma \frac{M_M M_E r (2R - r)}{R^2 (R - r)^2} \approx \gamma \frac{2r M_M M_E}{R^3} \tag{3.28}$$

since $r \ll R$. The tide-generating force at E is directed towards the Moon and its magnitude scales with R^{-3}.

To compute the tidal force at an arbitrary point P on the surface of the Earth, we realize that the gravitational force is derived from the gravitational potential

$$V_g = -\gamma \frac{M_M}{r_1} \tag{3.29}$$

where r_1 is the distance between P and the Moon (Fig. 3.3.). Note that for point E, we have $r_1 = R - r$ and $F_E^g = -M_E \partial V_g / \partial r$. From the triangle OPM, we find

$$r_1^2 = r^2 + R^2 - 2rR \cos \phi \tag{3.30}$$

An expansion in powers of r/R then gives

$$V_g = -\gamma \frac{M_M}{R} \left(1 + \frac{r}{R} \cos \phi + \frac{1}{2} \left(\frac{r}{R} \right)^2 (3 \cos^2 \phi - 1) + \cdots \right) \tag{3.31}$$

The force associated with the second term of the potential exactly balances the centrifugal force and hence the tidal potential is given by the third term as

$$\Phi_M = -\gamma \frac{M_M r^2}{2R^3} (3 \cos^2 \phi - 1) \tag{3.32}$$

For $\phi = 0$, we find the tide-generating force in the radial direction as

$$F_E^t = -M_E \frac{\partial \Phi_M}{\partial r} = \gamma \frac{2r M_M M_E}{R^3} \tag{3.33}$$

which is exactly (3.28). The horizontal acceleration is given by

$$-\frac{1}{r} \frac{\partial \Phi_M}{\partial \phi} = \gamma M_M \frac{3r \sin 2\phi}{2R^3} \tag{3.34}$$

On the face of the Earth opposite to the Moon, the acceleration is positive whereas it is negative on the side facing the Moon. This causes two bulges on both sides on the ocean surface with a low sea surface in between. When the Earth rotates, an observer at the equator moves through the bulges twice (high tide) a day. This is the semi-diurnal or M_2 tide. As the Moon has a nonzero declination, there are also areas on Earth which experience a diurnal tide and a mixed tide.

While these are the basics of static tidal theory, reality is much more complicated. Also the Sun influences the tide-generating force. Moreover, the bulges induce tidal waves over Earth's surface which are influenced by continental geometry and cause a delay between the sea-surface height h and the tidal force. Hence, there is a constant non-equilibrium situation and a time-dependent tidal potential Φ_M, causing a net input of energy according to

$$\int \rho \frac{\partial \Phi_M}{\partial t}\, d^3x \approx \int \rho h \frac{\partial \Phi_M}{\partial t}\, d^2x \tag{3.35}$$

The energy input due to the M_2 tide has been estimated as 2.50 TW, while the total tidal energy input is about 3.75 TW.

3.3.3. *Buoyancy forcing*

An issue which is confusing at the moment is the role of the buoyancy forcing in the large-scale ocean circulation.[11] was the first to address this issue with laboratory experiments (see also[12]). Sandström studied the motion induced in a liquid by varying the height of the (equal) heating and cooling sources. In case the heating occurs at lower depth and hence lower pressure than the cooling, no flow was found. In the opposite case that the heating occurs at a higher pressure than the cooling, a flow develops in which warm fluid rises, moves towards the cooling area, sinks and returns at depth to the warm region. The vertical length scale of the flow in this case is about the vertical distance between the heat source and sink. These results lead Sandström to postulate that heating and cooling can maintain a circulation only when the heating is below the level of the cooling source. This has been interpreted to imply that heating and cooling at the same level cannot drive a circulation, which would only leave the wind and the tides as driving forces of the large-scale ocean circulation.[13]

Laboratory experiments and theoretical analyses[14] indicate, however, that it is incorrect to apply Sandström's postulate to the case of heating and cooling at the same geopotential surface. Here, a strong convective and persistent circulation appears in laboratory experiments both in rotating and non-rotating cases. Such flows are not in contradiction with recent theoretical work[15] which shows that a flow driven by surface buoyancy fluxes alone cannot generate an interior turbulence. As soon as the forcing is turned on, the differential heating in the surface will generate a horizontal pressure gradient in the fluid along the surface, which immediately forces a flow. As there will be a very small diffusion in the interior, there will be heat transport to the bottom. In the cooling area, a plume carries water to the bottom and in steady state both heat fluxes must match. In this way, the heating is distributed throughout the depth of the flow.

These results indicate that the buoyancy forcing can generate (available) potential energy which can do work against the friction in the system.

Reliable estimates of the energy input due to this buoyancy forcing are not available yet.

3.4.   Energetics of mixing

From the sketch of the three-dimensional circulation as given in the intro-duction, it appears that about 30 Sv of fluid is returning to the surface from deeper layers. For example, in the Atlantic warm and salty surface water is transported northward in the Atlantic Ocean where it is cooled. Vigorous mixing occurs in localized regions transporting fluid properties to about 2-3 km depths; by this process the potential energy is lowered. A deep current flows southward at depth which, for example, can be traced even in the Southern Ocean. Along this path, already part of this water is raised to the surface.

If there were no process to convert the heavy fluid to lighter fluid some-where at depth, the abyss would fill up with heavy density fluid. If the 30 Sv is transported upwards through upwelling with a constant upwelling velocity w_0, then continuity provides an estimate of w_0 from

$$w_0 s_e \approx 30 \times 10^6 \text{m}^3\text{s}^{-1} \rightarrow w_0 = 10^{-7} \text{ ms}^{-1} \tag{3.36}$$

where s_e is again the surface area of the ocean. Apparently there are mixing processes in the ocean that raise the center of mass and increase the potential energy of the flow. In this section, we discuss the energetics of this mixing and try to determine how much is needed to maintain the abyssal stratification as, for example, observed in Fig. 3.1..

3.4.1.   *A simple example*

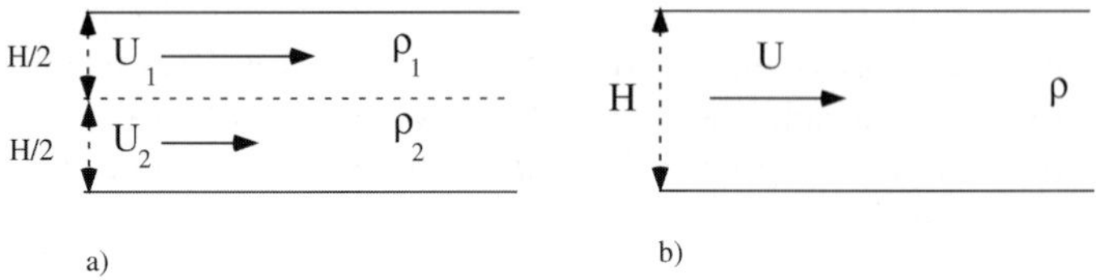

Fig. 3.4.   Sketch of a two layer situation with a stratified flow (a) which is completely mixed after some time (b).

To introduce the energetics of mixing, consider first the example of a two-layer fluid in which the layers (with densities $\rho_1 < \rho_2$ and thicknesses $H_1 = H_2 = H/2$) move at a different velocity with $U_2 > U_1$ (left panel of Fig. 3.4.). Because the liquid is stably stratified, the center of gravity falls below the middepth level (it is located in layer 2). Now suppose that mixing occurs (right panel of Fig. 3.4.) in which the heavier liquid is mixed upwards and the lighter liquid is mixed downwards. After mixing, the liquid moves with velocity U, its density is $\rho = (\rho_1 + \rho_2)/2$ and the center of mass is exactly at middepth. This mixing requires work to be done against the buoyancy forces and hence the potential energy of the system is increased. This follows from

$$
\Delta E_P = \int_0^H \rho g z \, dz - \left[\int_0^{H/2} \rho_2 g z \, dz + \int_{H/2}^H \rho_1 g z \, dz \right]
$$

$$
= \frac{1}{8}(\rho_2 - \rho_1) g H^2 > 0 \tag{3.37}
$$

Where does the energy to increase the potential energy comes from? It is clear here, that the energy must come from a change in kinetic energy of the flow. Indeed, as $U = (U_1 + U_2)/2$ by conservation of zonal momentum the kinetic energy change is

$$
\Delta E_K = \int_0^H \frac{1}{2}\rho U^2 \, dz - \left[\int_0^{H/2} \rho_2 U_2^2 \, dz + \int_{H/2}^H \rho_1 U_1^2 \, dz \right]
$$

$$
\approx -\frac{1}{8}\rho_0 (U_1 - U_2)^2 H < 0 \tag{3.38}
$$

where it is used that $\rho_1 \approx \rho_2 \approx \rho_0$, the latter being a reference density. Complete mixing is only possible when the kinetic energy loss is larger than the potential energy gain, i.e.,

$$
\frac{\Delta E_P}{-\Delta E_K} = \frac{(\rho_2 - \rho_1) g H}{\rho_0 (U_1 - U_2)^2} < 1 \tag{3.39}
$$

When this criterium is not fulfilled, there will be no complete but only partial mixing. This example shows clearly that an energy source is needed to mix density in a stably stratified liquid. In the next section, we consider a situation in which this is accomplished.

3.4.2. *Stability of stratified shear flows*

We now turn to a slightly more complex problem of a two-dimensional horizontally unbounded shear flow in a continuously stratified liquid bounded by two walls at $z = 0$ and $z = H$. The background density profile $\bar{\rho}(z)$ gives rise to a buoyancy frequency $N(z)$ and is assumed to be given. The

96   *Henk A. Dijkstra*

background velocity profile is given by $\bar{\mathbf{v}} = (\bar{u} = U(z), \bar{w} = 0)$. The two-dimensional equations of motion (neglecting rotation) are

$$\frac{\partial u}{\partial t} + u\frac{\partial u}{\partial x} + w\frac{\partial u}{\partial z} = -\frac{1}{\rho_0}\frac{\partial p}{\partial x} + \nu\nabla^2 u \qquad (3.40a)$$

$$\frac{\partial w}{\partial t} + u\frac{\partial w}{\partial x} + w\frac{\partial w}{\partial z} = -\frac{1}{\rho_0}\frac{\partial p}{\partial z} - \frac{\rho g}{\rho_0} + \nu\nabla^2 w \qquad (3.40b)$$

$$\frac{\partial u}{\partial x} + \frac{\partial w}{\partial z} = 0 \qquad (3.40c)$$

$$\frac{\partial \rho}{\partial t} + u\frac{\partial \rho}{\partial x} + w\frac{\partial \rho}{\partial z} = 0 \qquad (3.40d)$$

where we have used the traditional Boussinesq approximation[a]. The last equation (3.40d) follows from the temperature and salinity equations, in which molecular diffusion has been neglected, and the linear equation of state.

Consider first the inviscid case for which $\nu = 0$, then the boundary conditions at the vertical walls are only kinematic, i.e., $w = 0$ at $z = 0$ and $z = H$. Substituting the background velocity into the equations (3.40), we determine that the background pressure field $\bar{p}$ satisfies the hydrostatic balance

$$\frac{\partial \bar{p}}{\partial z} = -\bar{\rho} g \qquad (3.41)$$

We now assume that infinitesimally small perturbations on the background state occur, for example, $u = \bar{u} + \tilde{u}$, etc. Linearization of the equations with respect to the amplitude of the perturbations gives

$$\frac{\partial \tilde{u}}{\partial t} + \bar{u}\frac{\partial \tilde{u}}{\partial x} + \tilde{w}\frac{\partial \bar{u}}{\partial z} = -\frac{1}{\rho_0}\frac{\partial \tilde{p}}{\partial x} \qquad (3.42a)$$

$$\frac{\partial \tilde{w}}{\partial t} + \bar{u}\frac{\partial \tilde{w}}{\partial x} = -\frac{1}{\rho_0}\frac{\partial \tilde{p}}{\partial z} - \frac{\tilde{\rho} g}{\rho_0} \qquad (3.42b)$$

$$\frac{\partial \tilde{u}}{\partial x} + \frac{\partial \tilde{w}}{\partial z} = 0 \qquad (3.42c)$$

$$\frac{\partial \tilde{\rho}}{\partial t} + \bar{u}\frac{\partial \tilde{\rho}}{\partial x} + \tilde{w}\frac{\partial \bar{\rho}}{\partial z} = 0 \qquad (3.42d)$$

With introduction of a streamfunction ψ, with $\tilde{u} = \partial\psi/\partial z$ and $\tilde{w} = -\partial\psi/\partial x$ then elimination of the pressure perturbation from (3.42a-b) gives

$$\left(\bar{u}\frac{\partial}{\partial x} + \frac{\partial}{\partial t}\right)\nabla^2\psi - \frac{\partial\psi}{\partial x}\frac{\partial^2\bar{u}}{\partial z^2} = \frac{\partial\tilde{\rho}}{\partial x}\frac{g}{\rho_0} \qquad (3.43a)$$

$$\frac{\partial\tilde{\rho}}{\partial t} + \bar{u}\frac{\partial\tilde{\rho}}{\partial x} - \frac{\partial\psi}{\partial x}\frac{\partial\bar{\rho}}{\partial z} = 0 \qquad (3.43b)$$

---

[a]Here, this means that variations of the density are only considered in the buoyancy force, whereas the density is taken constant (ρ_0) elsewhere.

These equations admit separable solutions of the form

$$\psi = \Psi(z)e^{ik(x-ct)} \; ; \; \tilde{\rho} = R(z)e^{ik(x-ct)} \tag{3.44}$$

where k is the wavenumber and $c = c_r + ic_i$ is the complex growth factor. If $c_i > 0$ for a certain k then the background flow is unstable since this perturbation will grow exponentially in time. From (3.43b) we find that

$$(\bar{u} - c)R = \Psi \frac{\partial \bar{\rho}}{\partial z} \tag{3.45}$$

and using this in (3.43a), we obtain the Taylor-Goldstein equation

$$(\bar{u} - c)(\Psi'' - k^2\Psi) + \left(\frac{N^2}{\bar{u} - c} - \bar{u}''\right)\Psi = 0 \tag{3.46}$$

with $\Psi(0) = \Psi(H) = 0$. Just as for the stability of homogeneous shear flows (for which $N^2 = 0$), we can determine a necessary condition for instability. This is given by

$$Ri = \frac{N^2}{(\bar{u}')^2} < \frac{1}{4} \tag{3.47}$$

where Ri is the (local) Richardson number.

If this condition is satisfied, it is possible for perturbations to grow on the background flow which may lead to mixing. For the two-layer flow in section 3.4.1., we have

$$\frac{\partial \bar{u}}{\partial z} \approx \frac{(U_1 - U_2)}{H} \; ; \; N^2 \approx \frac{g(\rho_2 - \rho_1)}{\rho_0 H} \tag{3.48}$$

and the stability criterion (3.47) becomes

$$Ri = \frac{N^2}{(\bar{u}')^2} \approx \frac{gH(\rho_2 - \rho_1)}{\rho_0(U_1 - U_2)^2} < \frac{1}{4} \tag{3.49}$$

This provides the Richardson number with a physical interpretation: the numerator is the potential energy barrier to overcome whereas the denominator is the kinetic energy in the shear flow available for mixing.

3.4.3. *Turbulent stratified shear flows*

We consider now the stage in which the background flow is unstable and vigorous mixing occurs due to growth and interaction of perturbations. Again we write $u = \bar{u} + \hat{u}$, etc. but realize that the perturbations are now large compared to the background flow. Substituting these expressions into the viscous equations, multiplying the x-momentum equation by $\hat{u}$ and the z-momentum equation by $\hat{w}$ gives the equation for the kinetic energy $\hat{E}$ of the perturbations as

$$\frac{\partial \hat{E}}{\partial t} = \nu \nabla^2 \hat{E} - \bar{u}\frac{\partial \hat{E}}{\partial x} - \hat{\mathbf{u}}.\nabla \hat{E} - \frac{1}{\rho_0}\nabla.(\hat{\mathbf{u}}p)$$

$$- \hat{u}\hat{w}\frac{\partial \bar{u}}{\partial z} - \nu \mid \nabla \hat{\mathbf{u}} \mid^2 - \frac{g}{\rho_0}\hat{w}\hat{\rho} \qquad (3.50)$$

which we also could have derived directly from the general energy balance (3.11). When this equation is averaged over a long time interval (indicated by $<>$), experimental results show that there is a balance between the last three terms (between production and dissipation), i.e.,

$$< \hat{u}\hat{w}\frac{\partial \bar{u}}{\partial z} > + \varepsilon + \frac{g}{\rho_0} < \hat{w}\hat{\rho} > \approx 0 \qquad (3.51)$$

This balance motivates to define a mixing coefficient K_V (often called the diapycnal eddy diffusivity) as

$$K_V = -\frac{< \hat{w}\hat{\rho} >}{\frac{\partial \bar{\rho}}{\partial z}} = g\frac{< \hat{w}\hat{\rho} >}{\rho_0 N^2} \qquad (3.52)$$

It then follows from the balance (3.51) that

$$K_V = \frac{Ri_F}{1 - Ri_F}\frac{\varepsilon}{N^2} = \Gamma\frac{\varepsilon}{N^2} \qquad (3.53)$$

where Ri_F is the flux-Richardson number

$$Ri_F = -\frac{g}{\rho_0}\frac{< \hat{w}\hat{\rho} >}{< \hat{u}\hat{w}\frac{\partial \bar{u}}{\partial z} >} \qquad (3.54)$$

and Γ is called the mixing efficiency, often taken constant $\Gamma = 0.2$.

Also the flux Richardson number Ri_F has an interpretation in terms of an energy ratio. The denominator is the energy release from the background state to the kinetic energy of the perturbations through the Reynolds' stresses. The numerator is the potential energy change due to conversion of kinetic to potential energy. The importance of this result is that we have now an expression for the diapycnal eddy coefficient K_V in terms of the background stratification and the viscous dissipation.

3.4.4. *Mixing associated with the abyssal stratification*

This provides sufficient background to address the following problem: how much energy is needed to maintain the abyssal stratification as we observe it today? Consider the situation where water with a density ρ_N is sinking near a latitude θ_N towards the abyss and flows south to a certain latitude θ

at a depth H_0. The water column at θ has a constant buoyancy frequency N^2, which implies a linear density profile $\bar{\rho}(z) = \rho_N(1 - a(z + H_0))$ where $a = \rho_0 N^2/(g\rho_N)$. The work W_b required to lift this heavy water through the stratification to a certain height H by an upwelling velocity w_0 is given by

$$W_b = s_e \int_{-H_0}^{-H} (\rho_N - \bar{\rho})w_0 g \, dz = \rho_N \, a \, (H_0 - H)^2 w_0 g s_e \qquad (3.55)$$

With $a = 10^{-7}$ m^{-1}, $s_e = 3.5 \times 10^{14}$ m^2, $w_0 = 10^{-7}$ ms^{-1} and $H_0 - H = 1000$ m, we find that $W_b \approx 0.4$ TW.

From the definition of the flux Richardson number and the dominant turbulent kinetic energy balance equation, we find

$$g < \hat{w}\hat{\rho} > = \frac{Ri_F}{1 - Ri_F}\rho_0 \epsilon = \Gamma \rho_0 \varepsilon \qquad (3.56)$$

If an amount of 0.4 TW is needed to lift the heavy water upward, it follows for the volume averaged dissipation that

$$\int \rho_0 \varepsilon \, d^3 x = \frac{0.4}{\Gamma} \approx 2 \text{ TW} \rightarrow \varepsilon \approx 10^{-9} \text{ m}^2\text{s}^{-3} \qquad (3.57)$$

where $\Gamma = 0.2$ is used. With a typical value of $N = 10^{-3}$ s^{-1} in the abyssal ocean, this dissipation rate would give a background diapycnal mixing coefficient in the open ocean of about

$$K_V = \Gamma \frac{\varepsilon}{N^2} \approx 0.2 \times \frac{10^{-9}}{10^{-6}} = 2 \times 10^{-4} \text{ m}^2\text{s}^{-1} \qquad (3.58)$$

This value is a factor 10 too large than what is observed as basin averaged diapycnal diffusivity in the ocean.[16]

We must be careful, however, as we have totally neglected the effect of a possible buoyancy driven convective flow in the ocean. Maintenance of ocean stratification requires that the water column is continually overturned by taking dense water from the bottom and advecting water upward in the interior and this vertical exchange may be forced by buoyancy.[14] have shown that in a model where the interior flow is linked with surface fluxes and the regions of vertical sinking, a much smaller value of K_V is sufficient to maintain the stratification and the overturning circulation.

However, mixing is certainly important and the result in this section leads to the next question: how is transformation of energy from the input through the winds and the tides towards the viscous dissipation? This difficult issue will be addressed in the next section.

3.5.   Energy transformations

So far, we have learned that about 3.7 TW of energy enters the ocean through the tides, and that about 10-60 TW enters through the wind stress. Much of the wind stress input (about 90%) is, however, already directly dissipated in the surface layer. Phenomena such as breaking surface waves, Langmuir circulations and convection lead to a nearly uniform mixed layer in the upper ocean. Many of these processes occur at relatively small space and time scales in which horizontal gradients are insignificant. Similarly, much of the energy input from the tides (about 75% or 2.6 TW) is dissipated in shallow seas.

We have an estimate that about 1 TW of the wind stress input is available for the geostrophic (large-scale) ocean circulation and that 2 TW of abyssal mixing seems needed to maintain the observed stratification. There are two main pathways from the remaining surface energy input to the abyssal mixing energy. One pathway is a transfer of energy through successive instabilities of the general circulation, via meso-scale eddies and other coherent structures, towards smaller scales. The other pathway is through internal waves and the turbulence associated with the breaking of these waves. We discuss both pathways in the following subsections.

3.5.1.   *Internal wave pathway*

In section 3.4.2., we considered the problem of the linear stability of a shear flow in a stratified fluid. This problem serves nicely to introduce internal waves. We modify the problem slightly in that the domain is assumed unbounded in all directions and that the shear flow velocity is zero ($\bar{u} = 0$). The Taylor-Goldstein equation (3.46) then reduces to

$$c^2(\Psi'' - k^2\Psi) + N^2\Psi = 0 \tag{3.59}$$

for the function $\Psi(z)$. Substituting solutions $\Psi = e^{imz}\Psi_0$, leads with $\omega = kc$ and with (3.44) to the dispersion relation of travelling wave solutions (called internal waves, since they exist because the presence of the stratification)

$$\psi(x, z, t) = \Psi_0 e^{i(kx+mz-\omega t)} \Rightarrow \omega^2 = \frac{N^2 k^2}{k^2 + m^2} \tag{3.60}$$

To investigate the propagation of these waves , we determine their physical velocities (u, w) and density ρ. This gives

$$\psi = \Psi_0 \cos(kx + mz - \omega t) \tag{3.61a}$$

$$u = \frac{\partial \psi}{\partial z} = -\Psi_0 m \sin(kx + mz - \omega t) \tag{3.61b}$$

$$w = -\frac{\partial \psi}{\partial x} = \Psi_0 k \sin(kx + mz - \omega t) \tag{3.61c}$$

$$\rho = \frac{g}{\rho_0} N^2 \psi = \frac{g}{\rho_0} N^2 \Psi_0 \cos(kx + mz - \omega t) \tag{3.61d}$$

In Fig. 3.5., the vertical structure of an internal wave is plotted. For fixed z, we see the pattern of the streamfunction (and density). When the density increases through zero with increasing x the horizontal velocity is positive and the vertical velocity is negative. When the density decreases through zero with increasing x the horizontal velocity is negative and the vertical velocity is positive. This give the characteristic upward and downward movement parallel to lines of constant phase $kx + mz - \omega t$. In time, this leads to localized gradients in density. When the scale of the waves is so

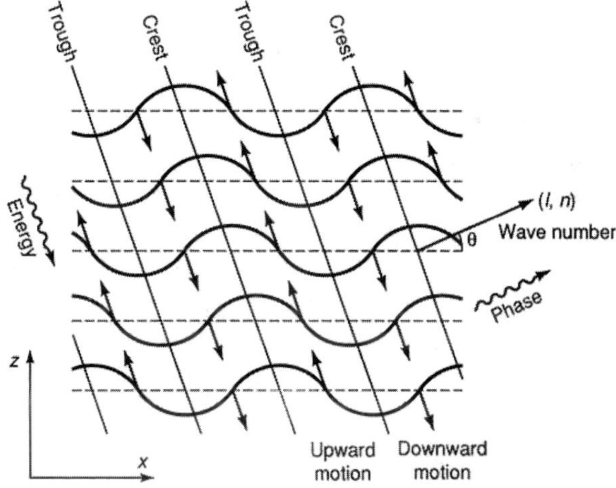

Fig. 3.5. Propagation of internal waves with z as vertical coordinate and x as horizontal coordinate. The drawn curves represent streamlines.

large that they are influenced by the Coriolis acceleration, the dispersion relation modifies to

$$\omega^2 = \frac{N^2 k^2 + f^2 m^2}{k^2 + m^2} \tag{3.62}$$

where $f = 2\Omega \sin \theta_0$ is the local Coriolis parameter and θ_0 the corresponding latitude.

Internal waves in the ocean can be generated by a large number of processes and nearly everywhere where a source of energy has some spatial and/or temporal variability. They are, for example, generated very efficiently by (tidal) flow over topography, by the instabilities of shear flows, and through mixing processes in the upper ocean. From the wind stress energy input, it is estimated that about 0.6 TW is directly transferred to force internal waves. From the (remaining) tidal input of 0.9 TW, most (0.8 TW) is used for the generation of internal tides, so-called internal waves with a tidal period. About 0.1 TW is transferred to the non-tidal internal wave field in the ocean. The energy spectrum of internal waves (the Garrett-Munk spectrum) has been carefully deduced from observations.[17] There is large energy at the inertial frequency. Integration of the wave spectrum provides an estimate of the energy contained in the internal wave field of about 1.4 EJ (1 EJ $= 10^{18}$ J).

The rate of decay of the internal wave field through breaking of these waves cannot be directly measured but it is estimated to be about 60 days. This leads to a dissipation rate ε_{IW} of

$$\int \rho \varepsilon_{IW} \, d^3x = \frac{1.4 \times 10^{18}}{60 \times 24 \times 3600} \approx 0.2 \text{ TW} \rightarrow \varepsilon_{IW} \approx 10^{-10} \text{ m}^2\text{s}^{-3} \quad (3.63)$$

This dissipation rate would give a background diapycnal mixing coefficient in the open ocean of about

$$K_V = \Gamma \, \frac{\varepsilon_{IW}}{N^2} \approx 0.2 \times \frac{10^{-10}}{10^{-6}} = 2 \times 10^{-5} \text{ m}^2\text{s}^{-1} \quad (3.64)$$

This value is in accordance with what is observed as a background diapycnal diffusivity in the ocean by direct microstructure measurements and tracer release experiments.[18,19] This background internal wave field in the open ocean appears not sufficient to provide the necessary 2 TW (it provides about 10% of it) to sustain the abyssal stratification and thermohaline circulation.

Mixing levels near sites of internal tide generation, however, are known to exceed this background value. Microstructure measurements have provided direct evidence that internal tides generate turbulence and mixing, in particular near abyssal topography.[20] Enhanced turbulence levels near internal tide generation sites are attributed to the dissipation of high wavenumber (large k and m) modes. About 2/3 of the total energy flux is, however, radiated away through low wavenumber modes. Using a barotropic model and a parameterization of the energy flux of the internal tide energy,[21] obtain an estimate of the resulting diapycnal diffusivity (Fig. 3.6.).

The interaction with internal tides with topography leads locally to intense mixing. At the 1000 m depth level (Fig. 3.6.a), the influence of the enhanced mixing due to internal tides is limited to a few locations. At 4000 m depth, however, about 10% of the map's area (Fig. 3.6.b) shows diffusivities larger than 10^{-4} m^{2}s^{-1} and maximum diffusivities (up to 3×10^{-3}) occur near the ocean bottom (Fig. 3.6.c). It is estimated that the

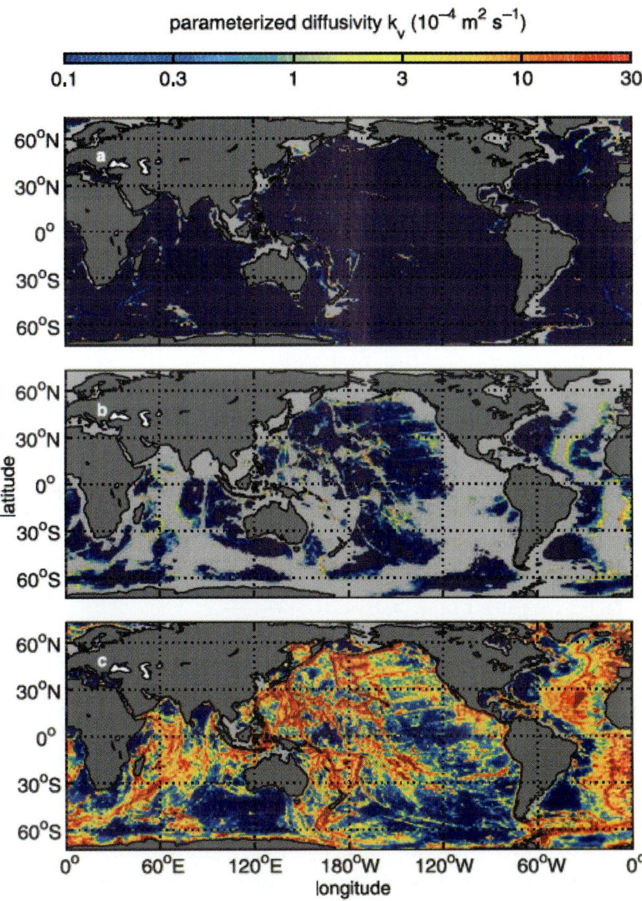

Fig. 3.6.   Estimates of the turbulent diapycnal diffusivity K_V along (a) a 1000m depth level (b) a 4000 m depth level and (c) the bottom boundary of the ocean.[21]

internal wave and internal tide pathway leads to an energy flux of about 1.3 TW which is available for abyssal mixing. This is, however, much more than that implied by the basin averaged values of K_V as deduced by.[16]

3.5.2. *The meso-scale eddy pathway*

At time scales longer than a day, the flows in the ocean are in near geostrophic equilibrium. On the large spatial scales of the general circulation, the effect of inertial accelerations are small with respect to the Coriolis acceleration. It is therefore possible to derive a simplified dynamical description on the long time scale, large spatial scale flows, the so-called quasi-geostrophic theory. We first turn to this theory in the next subsection and then study the origin of the meso-scale eddies in section 3.5.2.2. Specific coherent flows within the quasi-geostropic theory are addressed in section 3.5.2.3. and we conclude with a discussion of geostrophic turbulence in section 3.5.2.4..

3.5.2.1. *Quasi-geostrophic theory*

We extend the model in section 4.2 slightly by adding the lateral dimension (with coordinate y) and by including the Coriolis acceleration and its meridional variation on the sphere in the form of

$$f = f_0 + \beta_0 y \tag{3.65}$$

The governing equations then (with $\hat{\rho} = \rho - \bar{\rho}$ and $\hat{p} = p - \bar{p}$, where $\bar{p}$ and $\bar{\rho}$ are the hydrostatic fields), neglecting friction and diffusion altogether, become

$$\frac{Du}{dt} - (f_0 + \beta_0 y)v = -\frac{1}{\rho_0}\frac{\partial \hat{p}}{\partial x} \tag{3.66a}$$

$$\frac{Dv}{dt} + (f_0 + \beta_0 y)u = -\frac{1}{\rho_0}\frac{\partial \hat{p}}{\partial y} \tag{3.66b}$$

$$0 = -\frac{1}{\rho_0}\frac{\partial \hat{p}}{\partial z} - \frac{\hat{\rho}g}{\rho_0} \tag{3.66c}$$

$$\frac{\partial u}{\partial x} + \frac{\partial v}{\partial y} + \frac{\partial w}{\partial z} = 0 \tag{3.66d}$$

$$\frac{D\hat{\rho}}{dt} + w\frac{\partial \bar{\rho}}{\partial z} = 0 \tag{3.66e}$$

with D/dt being the material derivative $D/dt = \partial/\partial t + \mathbf{v}.\nabla$. The ratio of the inertial terms and the Coriolis acceleration is measured by

$$Ro = \frac{U}{2\Omega L} \tag{3.67}$$

where L is a typical horizontal length scale and U a horizontal velocity scale. For $U = 10^{-1}$ ms^{-1} and $L = 10^6$ m, we find typical values of $Ro = 10^{-3}$ and hence to first order inertia can be neglected. Under the additional condition that $\beta_0 y \ll f_0$, we obtain the dominant geostrophic balance

$$f_0 v_g = \frac{1}{\rho_0} \frac{\partial \hat{p}}{\partial x} \tag{3.68a}$$

$$f_0 u_g = -\frac{1}{\rho_0} \frac{\partial \hat{p}}{\partial y} \tag{3.68b}$$

where (u_g, v_g) are the geostrophic velocities. The geostrophic balance is degenerated as it leads immediately to the fact that the horizontal divergence $\partial u_g / \partial x + \partial v_g / \partial y = 0$. In a stratified flow, there can be no vertical velocity, $w_g = 0$, and hence no deformation of density surfaces, no pressure differences, and consequently no motion.

To investigate the impact of the ageostrophic effects (in this case, the inertia and the β_0 terms), we evaluate the ageostrophic terms using the geostrophic velocity to get

$$f_0 v = \frac{1}{\rho_0} \frac{\partial \hat{p}}{\partial x} + \frac{D_g u_g}{dt} - \beta_0 y v_g \tag{3.69a}$$

$$f_0 u = -\frac{1}{\rho_0} \frac{\partial \hat{p}}{\partial y} - \frac{D_g u_g}{dt} - \beta_0 y u_g \tag{3.69b}$$

where $D_g / dt = \partial / \partial t + \mathbf{v}_g . \nabla$. We now compute the horizontal divergence using the continuity equation as

$$\frac{\partial w}{\partial z} = -\left(\frac{\partial u}{\partial x} + \frac{\partial v}{\partial y}\right) = \frac{\beta_0}{f_0} v_g + \frac{1}{f_0} \frac{D_g \zeta}{dt} \tag{3.70}$$

where $\zeta = \partial u_g / \partial y - \partial v_g / \partial x$ is the vertical component of the geostrophic vorticity.

All terms in the density equation (3.66e) are small (on long time scale) and we approximate the advective transport of density by the geostrophic transport. This gives

$$w \frac{\partial \bar{\rho}}{\partial z} = -\frac{D_g \hat{\rho}}{dt} \rightarrow w = \frac{g}{\rho_0 N^2} \frac{D_g \hat{\rho}}{dt} \tag{3.71}$$

Using the hydrostatic equation (3.66c) to eliminate $\hat{p}$, the expression for w can be used in (3.70) to obtain (with the introduction of the geostrophic streamfunction such that $\hat{p} = \rho_0 f_0 \psi$) to give

$$\frac{\partial q}{\partial t} + J(\psi, q) = \frac{\partial q}{\partial t} + \frac{\partial \psi}{\partial x} \frac{\partial q}{\partial y} - \frac{\partial q}{\partial x} \frac{\partial \psi}{\partial y} = 0 \tag{3.72a}$$

$$q = \nabla^2 \psi + \beta_0 y + \frac{\partial}{\partial z}\left(\frac{f_0^2}{N^2} \frac{\partial \psi}{\partial z}\right) \tag{3.72b}$$

$$u = -\frac{\partial \psi}{\partial y} \; ; \; v = \frac{\partial \psi}{\partial x} \; ; \; \zeta = \nabla^2 \psi \tag{3.72c}$$

$$\hat{\rho} = \frac{\rho_0 f_0}{g} \frac{\partial \psi}{\partial z} \tag{3.72d}$$

where q is the potential vorticity (usually abbreviated with PV). This stratified quasi-geostropic (QG) model is one of the most important models to illustrate instabilities of large-scale flows in the atmosphere and ocean.

3.5.2.2. *Baroclinic instability*

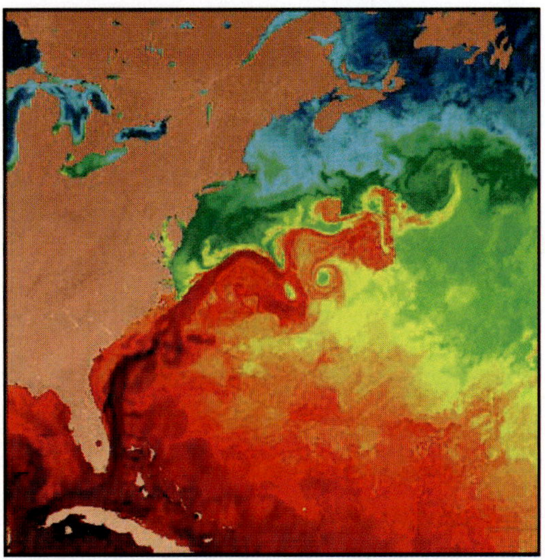

Fig. 3.7. Sea-surface temperature field in the Gulf Stream, showing the mesoscale eddies developing from mixed barotropic-baroclinic instabilities of the mean flow (Figure: courtesy of NASA).

Apart from instabilities associated with horizontal shear (called here barotropic instability), there is a particular new kind of instability of large-scale flows that is the main cause of the meso-scale eddies in the ocean. The nature of the baroclinic instability is most easily explained by looking at the stability of a zonal jet on an f-plane (where $\beta_0 = 0$) within a vertical domain $z \in [0, H]$. The background flow then becomes

$$\bar{\psi}(y, z) = -\frac{U}{H}yz \rightarrow \bar{u} = \frac{Uz}{H} \; ; \; \bar{v} = \bar{w} = 0 \; ; \; \bar{\rho} = -\frac{\rho_0 f_0}{g}\frac{U}{H}y \; ; \; \bar{q} = 0 \quad (3.73)$$

such that there is a meridional gradient in density associated with the zonal shear flow.

Linearization around the background flow and substitution of

$$\psi = \bar{\psi} + \Psi(z)e^{i(kx+ly-\sigma t)} \quad (3.74)$$

leads to an eigenvalue problem for Ψ and σ which together with the boundary conditions $w = 0$ at $z = 0$ and $z = H$ leads to an explicit expression for the complex growth factor $\sigma = \sigma_r + i\sigma_i$. The result is

$$\left(\frac{\sigma}{kU} - \frac{1}{2}\right)^2 = \frac{1}{4} + \frac{1}{\delta^2} - \frac{\coth\delta}{\delta} \quad (3.75)$$

where

$$\delta^2 = \frac{N^2}{f_0^2}(k^2 + l^2) \tag{3.76}$$

From this expression, we can deduce that for all wave numbers with

$$(k^2 + l^2)^{1/2} < 2.399\frac{f_0}{NH} \rightarrow \lambda > 2.169\frac{NH}{f_0} \tag{3.77}$$

the growth factor σ_i is positive. Maximum growth rates are found for wavelengths which are about $4NH/f$. The internal length scale NH/f is the internal Rossby radius of deformation, which is (with $H = 500$ m, $N = 10^{-3}$ s^{-1}) about 50 km at midlatitudes. We therefore expect strong growth of perturbation waves with a wavelength of about 200 km: these eventually lead to the mesoscale eddy field in the ocean as shown in Fig. 3.7. for the Gulf Stream region.

3.5.2.3.   *Coherent structures*

The interaction of the mixed barotropic-baroclinic perturbations with each other and with the mean flow leads to a complex turbulent flow, usually referred to as geostrophic turbulence. In such flows, also long-lived coherent flow patterns arise such as rings, dipoles, spirals and coherent vortices. Before we discuss in more detail the geostrophic turbulence flow regime, we give an example of a famous coherent flow structure, the modon.

Starting from (3.72), we first decompose the flow into a vertical component and a horizontal component by

$$\psi(x, y, z, t) = \Psi(x, y, t)\Xi(z) \tag{3.78}$$

As the equation is separable for constant N, the equation for the horizontal structure becomes

$$\frac{\partial Q}{\partial t} + J(\Psi, Q) = 0 \; ; \; Q = \nabla^2\Psi - F\Psi + \beta_0 y \tag{3.79}$$

where F is a constant which depends on the stratification and the vertical structure function Ξ. With an appropriate scaling, we can rewrite this equation as the so-called Charney-Hasegawa-Mima (CHM) equation

$$(\nabla^2 - 1)\frac{\partial \Psi}{\partial t} + \frac{\partial \Psi}{\partial x} + J(\Psi, \nabla^2\Psi) = 0 \tag{3.80}$$

The modon is a solution of (3.80) of the form

$$\Psi = \Phi(x - ct, y) \tag{3.81}$$

When substituted in (3.80) we get

$$J(\Phi + cy, \nabla^2\Phi - \Phi + y) = 0 \rightarrow \Phi + cy = G(\nabla^2\Phi - \Phi + y) \tag{3.82}$$

where G is an arbitrary function. To obtain a localized solution, the stream-function Ψ and its derivatives should vanish at infinity (for streamlines extending to infinity) and hence $cy = G(y)$, which implies that G is linear. This gives the equation for Φ as

$$\nabla^2 \Phi - (1 + \frac{1}{c})\Phi = 0 \tag{3.83}$$

The modon solution is obtained by invoking the ansatz that $c = -1/(1+\kappa^2)$ for all streamlines which do not extend to infinity (the interior region). The parameter κ is called the modon wave number. Also the modon radius a is introduced as the radius of the circle which is the boundary between streamlines which do and those which do not extend to infinity. In the exterior region, we still have $G(\xi) = c\xi$. Continuity and differentiability of the solution at $r = a$ leads to the conditions

$$r = a : \Phi + cy = 0 \; ; \; \lim_{r\uparrow a}\nabla\Phi = \lim_{r\downarrow a}\nabla\Psi \tag{3.84}$$

The solution of (3.83) can then be written in terms of Bessel functions and typical plots of modons can be found in.[22]

3.5.2.4. *Geostrophic turbulence*

Baroclinic instabilities draw their kinetic energy from the potential energy of the background flow (note the sloping isopycnals of the background flow (3.73)). Their presence and interaction lead to a flow state, which is usually referred to as geostrophic turbulence. This is one of the main reasons that oceanic kinetic energy is dominated (by approximately a factor 150) in transient phenomena rather than in the time-averaged flow. These instabilities cannot however, increase the potential energy by mixing since they lower the potential energy of the total flow. They provide, however, a pathway towards dissipation.

To understand this, consider flows in a domain V (which extends to infinity) described by (3.79), i.e.

$$\frac{\partial}{\partial t}(\nabla^2\Psi - F\Psi) + J(\psi, \nabla^2\Psi - F\Psi + \beta_0 y) = 0 \tag{3.85}$$

which hence conserve potential vorticity $Q = \nabla^2\Psi - F\Psi + \beta_0 y$ along streamlines and for which the velocity fields are in geostrophic balance. Multiplication of this equation by Ψ and integrating over the domain V, using kinematic boundary conditions gives

$$\frac{\partial}{\partial t}\int \frac{1}{2}(\nabla\Psi.\nabla\Psi + F\Psi^2)\, d^3x = 0 \tag{3.86}$$

which represents conservation of total energy $E = \frac{1}{2}(\nabla\Psi.\nabla\Psi + F\Psi^2)$. When (3.8.) is multiplied by $\nabla^2\Psi - F\Psi$, the result integrated over the volume V and use is made that $\Psi \to 0$ at the boundaries of the domain, we find

$$\frac{\partial}{\partial t} \int \frac{1}{2}(\nabla^2\Psi - F\Psi)^2 \, d^3x = 0 \tag{3.87}$$

which is the conservation of the potential enstrophy $Z = \frac{1}{2}(\nabla^2\Psi - F\Psi)^2$. Note that when $F = 0$, this quantity reduces to the usual enstrophy; in the discussion below, let $F = 0$ for simplicity.

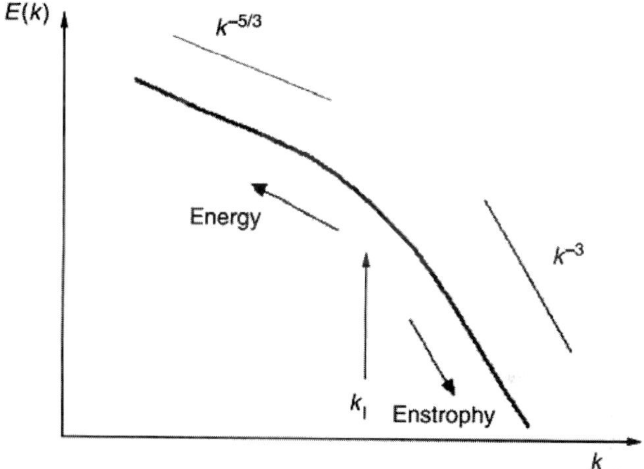

Fig. 3.8. Schematic spectrum for two-dimensional or quasi-geostrophic turbulence in its inertial ranges, given injection of energy and enstrophy at $k = k_1$.

For a particular flow, we can expand Ψ into a Fourier series and look at a wavenumber spectrum of the energy and enstrophy. Consider the flow components associated with three wavenumbers k_1, k_2 and k_3, which exchange energy and enstrophy in time. Define the energy change for the $i^t h$ wavenumber over a time interval $\Delta t = t_2 - t_1$ as $\Delta E_i = E(k_i, t_2) - E(k_i, t_1)$. Now note that in Fourier space, the Fourier components of the enstrophy are related by those of the energy through $Z_i = k_i^2 E_i$ (which can be seen by performing the Fourier transform). Conservation of energy and enstrophy implies

$$\Delta E_1 + \Delta E_2 + \Delta E_3 = 0 \tag{3.88a}$$
$$k_1^2 \Delta E_1 + k_2^2 \Delta E_2 + k_3^2 \Delta E_3 = 0 \tag{3.88b}$$

If we take, for example, $k_2 = 2k_1$ and $k_3 = 3k_1$, this will imply

$$\Delta E_1 = -\frac{5}{8}\Delta E_2 \ ; \ \ \Delta E_3 = -\frac{3}{8}\Delta E_2 \tag{3.89a}$$

$$\Delta Z_1 = -\frac{5}{32}\Delta Z_2 \ ; \ \ \Delta Z_3 = -\frac{27}{32}\Delta Z_2 \tag{3.89b}$$

If now mode k_2 looses energy to the other two modes ($\Delta E_2 < 0$) then more energy is transferred to smaller wavenumber k_1 than to the larger wavenumber k_3 ($\Delta E_1 > \Delta E_3$) and hence the energy moves to larger spatial scales. On the other hand, the enstrophy moves to smaller scales, since $\Delta Z_1 < \Delta Z_3$). This suggests that in freely decaying two-dimensional or quasi-geostrophic turbulence, the cascades in the inertial range will be upward for energy and downward for enstrophy. It can be shown that when k_1 is a typical wavenumber at which energy is supplied, the portion of the energy spectrum for $k > k_1$ follows a k^{-3} power law while that for $k < k_1$ follows a $k^{-5/3}$ power law (Fig. 3.8.).

How the energy is transferred from the large scale downwards is a major area of research at the moment. One of the possibilities is that the mesoscale structures become unstable to further instabilities for which the geostrophic equilibrium and potential vorticity constraints do not hold. These so-called unbalanced instabilities occur at smaller scales and for these flows the energy can cascade to high wavenumbers where it can eventually be dissipated. The part of the abyssal mixing that is provided by this meso-scale pathway is still very uncertain.

3.6.   Summary and conclusions

The ocean energy balance will be a hot topic in physical oceanography for the next few years to come. An overview of an interpretation from[13] of what is known at the moment is summarized in the box diagram in Fig. 3.9.. Of course, all the numbers in Fig. 3.9. have quite some uncertainty and a lot of future work will be directed towards better estimates. [13] are assuming, referring to Sandström's experiments, that the buoyancy forcing cannot drive the large-scale ocean circulation as it cannot provide the mixing field required. However, it is possible that through the buoyancy forcing the amount of energy needed to maintain the abyssal stratification is much smaller than 2 TW.[14]

Mixing is, however, essential to provide the increase in potential energy associated with the large-scale ocean circulation. To investigate the pathway of this energy, we have looked at the energy input, its pathways of transformation and its dissipation. The main energy input is provided by the wind and tides and a uncertain contribution from the buoyancy

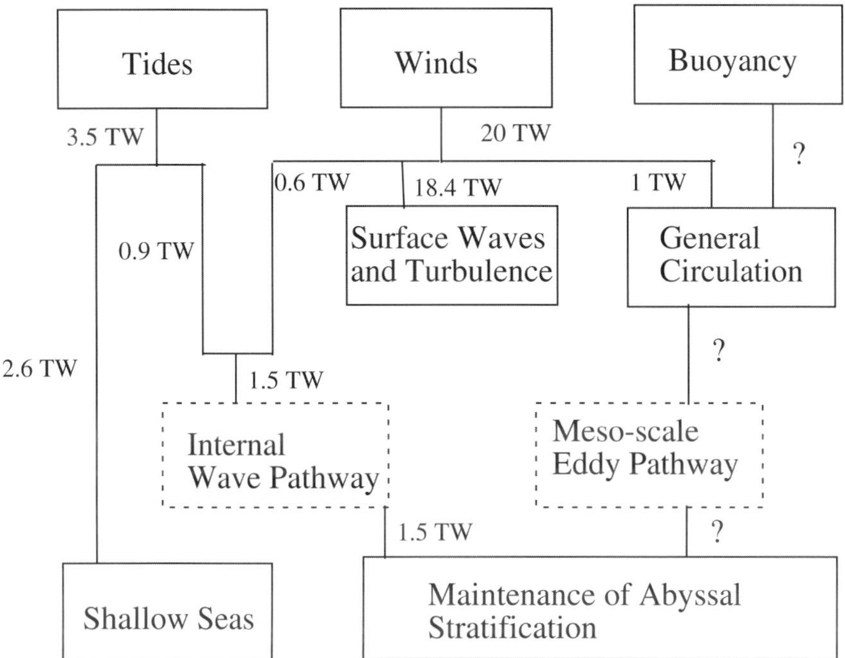

Fig. 3.9. Simplified picture of the status of knowledge on the energy budget of the global ocean circulation as in.[13] Fluxes from and to the reservoirs are in TW.

forcing. One pathway of energy transfer which could lead to the energy needed for the abyssal stratification is through internal tides and internal waves, which through interaction provide about 75% of the abyssal mixing required. The other 25% apparently is provided through the meso-scale eddy field, although the pathway is not completely clear.

The complexity of the processes which determine the energy balances in the ocean are striking and one can image how difficult it will be to establish reliable numbers. It is necessary to obtain good estimates, however, or otherwise our main question (how the ocean circulation will change with a changing atmospheric forcing) cannot be answered with much confidence.

3.7. Appendix A: Thermodynamics of sea water

Sea water consists of a dilute solution of ions, such as Cl^- and Mg^{2+}. In a certain volume element, let there be $n - 1$ of these ion types with masses $m_k, k = 1, \cdots, n - 1$ and indicate the mass of the water by m_n. The total mass m and the mass fractions $c_k, k = 1, \cdots, n-1$ are then given by

$$\sum_{k=1}^{n} m_k = m \; ; \; c_k = \frac{m_k}{m} \rightarrow \sum_{k=1}^{n} c_k = 1 \tag{3.90}$$

It is an experimental fact that the relative composition of the different ions is constant in sea water far from continental boundaries. This motivates to define the salinity S and the water fraction W as

$$S = \frac{1}{m} \sum_{k=1}^{n-1} m_k \; ; \; W = \frac{m_n}{m} \tag{3.91}$$

such that $c_k = \lambda_k S, k = 1, \cdots, n-1$ and

$$\sum_{k=1}^{n-1} \lambda_k = 1 \; ; \; S + W = 1 \tag{3.92}$$

In this way, sea water can be considered as a two-component liquid (water and salt) in which there is only one independent variable, the salinity S.

Let the chemical potential of each of the ion types be given by $\mu_k, k = 1, \cdots, n-1$ and that of the water by μ_n. With the entropy per unit mass indicated by η, the general Gibbs formula for the entropy change in the volume is

$$T d\eta = dE_I + pd(\frac{1}{\rho}) - \sum_{k=1}^{n} \mu_k dc_k \tag{3.93}$$

If we define

$$\mu_s = \sum_{k=1}^{n-1} \lambda_k \mu_k \; ; \; \mu_w = \mu_n \; ; \; \mu = \mu_s - \mu_w \tag{3.94}$$

then the last term in (3.93) can be written (using (3.90)) as

$$\sum_{k=1}^{n} \mu_k dc_k = \sum_{k=1}^{n-1} \mu_k dc_k + \mu_n dc_n = \sum_{k=1}^{n-1} (\mu_k - \mu_w) \, dc_k =$$

$$= \sum_{k=1}^{n-1} (\mu_k - \mu_w) \lambda_k \, dS = (\mu_s - \mu_w) \, dS = \mu \, dS \tag{3.95}$$

3.8.   Appendix B: The ocean entropy balance

The ocean is an open system which exchanges heat and matter with its surroundings. Let η be the entropy per unit mass in a certain volume of ocean water, then the entropy balance (using appendix A) is given by

$$T\frac{D\eta}{dt} = \frac{DE_I}{dt} + p\frac{D}{dt}(\frac{1}{\rho}) - \mu\frac{dS}{dt} \tag{3.96}$$

where μ is the difference in chemical potential between that of the salt ions (making up the salinity) and the water. For a dilute solution, the chemical potential μ is given by

$$\mu = \mu_0 + RT \ln S \tag{3.97}$$

where R J $(\text{kg K})^{-1}$ is a universal constant and μ_0 depends only on p and T.

The total entropy production of the total system can be written as that of the ocean system and that of its surroundings, i.e.

$$\dot{\sigma}_W = \dot{\sigma}_O + \dot{\sigma}_S \tag{3.98}$$

where

$$\dot{\sigma}_O = \frac{\partial}{\partial t} \int \rho\eta \, d^3x = \int \rho\frac{D\eta}{dt} \, d^3x \tag{3.99}$$

The entropy of the surrounding system will be changed by heat and material (salt) from the ocean through the ocean-atmosphere boundary and can be written as

$$\dot{\sigma}_S = \int \frac{\mathbf{q}_T.\mathbf{n}}{T} \, d^2x - \int \mu \, \mathbf{q}_S.\mathbf{n} \, d^2x \tag{3.100}$$

This finally leads to the entropy production for the total system as

$$\dot{\sigma}_W = \int \frac{1}{T} \left[\frac{D(\rho E_I)}{dt} + p\nabla.\mathbf{v} \right] d^3x + \int \frac{\mathbf{q}_T.\mathbf{n}}{T} \, d^2x$$

$$- \int \mu \, \frac{DS}{dt} \, d^3x - \int \mu \, \mathbf{q}_S.\mathbf{n} \, d^2x \tag{3.101}$$

Bibliography

1. J. P. Peixoto and A. H. Oort. *Physics of Climate.* AIP Press, New York, 1992.
2. WOCE. *Ocean Circulation and Climate: Observing and Modeling the Global Ocean [Siedler, G. and Church, J. and Gould, J. (eds)].* Academic Press, San Diego, USA, 2001.
3. A. Ganachaud and C. Wunsch. Improved estimates of global ocean circulation, heat transport and mixing from hydrographic data. *Nature*, 408, 453–457, 2000.
4. L. D. Talley. Some aspects of ocean heat transport by the shallow, intermediate and deep overturning circulations. In P. Clark, R. S. Webb, and L. D. Keigwin, editors, *Mechanisms of Global Climate Change.* AGU, Washington, DC, U.S.A., 1999.
5. H. L. Bryden, H. R. Longworth, and S. A. Cunningham. Slowing of the Atlantic meridonal overturning circulation at 25N. *Nature*, 438, 655–657, 2005.
6. J. Pedlosky. *Geophysical Fluid Dynamics. 2nd Edn.* Springer-Verlag, New York, 1987.

7. A. E. Gill. *Atmosphere-Ocean Dynamics*. Academic Press, New York, U.S.A., 1982.

8. C. Wunsch. The work done by the wind on the oceanic general circulation. *J. Phys. Oceanogr.*, 28, 2332–2340, 1998.

9. W. Wang and R. X. Huang. Wind Energy Input to the Ekman layer. *J. Phys. Oceanogr.*, 34, 1267–1275, 2004.

10. W. Wang and R. X. Huang. Wind Energy Input to the Surface Waves. *J. Phys. Oceanogr.*, 34, 1276–1280, 2004.

11. J. W. Sandström. Dynamische Versuche mit Meerwasser. *Annalen der Hydrographie und der Maritimen Meteorologie*, 1908.

12. A. Defant. *Physical Oceanography, vol I*. Pergamon, New York, 598 pp., 1961.

13. C. Wunsch and R. Ferrari. Vertical mixing, energy and the general circulation of the oceans. *Annual Review of Fluid Mechanics*, 36, 281–314, 2004.

14. G. Hughes and R. W. Griffiths. A simple convective model of the global overturning circulation, including effects of entrainment into sinking regions. *Ocean Modeling*, 12, 46–79, 2006.

15. F. Paparella and W. R. Young. Horizontal convection is non-turbulent. *J. Fluid Mech.*, 466, 205–214, 2002.

16. D. L. Rudnick and coauthors. From tides to mixing along the Hawaiian ridge. *Science*, 301, 355–357, 2003.

17. S. A. Thorpe. *The Turbulent Ocean*. Cambridge University Press, UK, 437 pp., 2005.

18. M. C. Gregg. Diapycnal mixing in the thermocline: a review. *J. Geophys. Res.*, 92, 5249–5286, 1987.

19. J. R. Ledwell, A. J. Watson, and C. S. Law. Mixing of a tracer in the pycnocline. *J. Geophys. Res.*, 103, 21,499–21,529, 2000.

20. K. L. Polzin, J. M. Toole, J. R. Ledwell, and R. W. Schmitt. Spatial variability of turbulent mixing in the abyssal ocean. *Science*, 276, 93–96, 1997.

21. L. C. StLaurent, H. L. Simmons, and S. R. Jayne. Estimating tidally driven mixing in the deep ocean. *Geophys. Res. Letters*, 29, doi:10.1029/2002GL015633, 2002.

22. G. E. Swaters. Spectral properties in modon stability theory. *Studies in Applied Mathematics*, 112, 235–258, 2004.

Chapter 4

Analytical Descriptions of Plasma Turbulence

John A. Krommes

Plasma Physics Laboratory, Princeton University
P.O. Box 451, MS 28
Princeton, NJ 08543–0451 USA
E-mail: krommes@princeton.edu

An overview of the current state of fundamental plasma turbulence theory as applied to magnetic confinement is presented. Various topics, both historical and contemporary, in the analytical description of plasma turbulence are discussed. Modern gyrokinetic formalism is described heuristically, and various important nonlinear equations derived therefrom are highlighted. Several examples of the transition to turbulence are mentioned. The venerable quasilinear theory is discussed as an important example of statistical closure. Resonance-broadening theory is motivated, and some plasma-physics uses of the direct-interaction approximation and its Markovian relatives are described. The Martin–Siggia–Rose formalism is introduced. The theory of zonal-flow generation is shown to provide interesting examples of field-theoretic functional techniques, Markovian closure theory, and adiabatic and wave-kinetic methods. As an example of intermittent phenomena, the formation and propagation of coherent plasma "blobs" are briefly discussed. Some outstanding issues and challenging research areas are enumerated.

Contents

115

4.1. LECTURE 1 – Introduction to Plasma Turbulence

A *plasma* is a collection of $N = \sum_s N_s$ charged particles of charge q_s (s is a species label: e for electrons and i for ions) and mass m_s in a volume V, with mean densities $\overline{n}_s \doteq N_s/V$. Frequently the plasma is overall charge-neutral, $\sum_s (\overline{n}q)_s = 0$. However, there are very interesting instances of *non-neutral plasmas* (O'Neil, 1995; Davidson, 2001). Although there are important examples of plasmas with just hundreds or thousands or particles, typically plasmas contain very many (e.g., Avagadro's number of) particles.

The many-body plasma has all of the complications of the neutral fluid, but has much richer behavior because of the presence of electromagnetic forces. Overall charge neutrality does not preclude charge fluctuations, either microscopic or macroscopic. Such fluctuations can conspire to produce a potpourri of nonlinear dynamics and turbulent motions. Those indeed occur in confined plasmas such as those used in magnetic fusion research because such systems are deliberately prevented from achieving thermal equilibrium. The inevitable mean gradients of density n, temperature T, and flow u serve as sources of "free energy" that can drive turbulence. That is unfortunate, since turbulence-induced losses of particles and heat shorten the energy confinement time and make the goal of achieving burning plasma substantially more difficult.

Other lecturers in this Summer School have spoken about nonlinear dynamics (P. Holmes), pattern formation (H. Swinney), fully developed turbulence (G. Falkovich, Chap. 1), renormalization techniques (W. D. McComb, Chap. 2), ocean circulation (H. Dijkstra, Chap. 3), atmospheric flows and instabilities (J. Frederiksen), numerical simulation (J. Jiménez, Chap. 6), optical nonlinearities (A. Desyatnikov, Chap. 8), and experimental techniques (M. Shats, Chap. 5; J. Soria, Chap. 7). Virtually all of those topics are represented in modern plasma research. In particular, there is a great commonality of nonlinear physics between the quasi-2D fluid flows of geophysics and of plasmas. Very much can be learned from the extensive research in these various other fields.

In these lectures I will give an introduction to some basic plasma turbulence phenomena and analysis, especially as motivated by some applications from fusion research. A typical modern experiment on the magnetic confinement of hot plasmas is pictured in Fig. 4.1.[a] It is impossible here to treat the myriad of practical details that any serious fusion researcher must confront: intrinsically complicated (toroidal) confinement geometry; inhomogeneous background profiles; an important population of magnetically trapped particles; magnetic shear; difficult edge boundary conditions, which include a transition from closed magnetic flux surfaces to open field lines; and more; entire books have been written on the subject (White, 2001; Wesson, 2004; Miyamoto, 2005). However, you will get a taste of what the issues are, and I will provide many entry points into the literature.[b] Unfortunately, space constraints preclude a complete discussion, and this is not a review; I apologize to the very many authors whose work could not be mentioned.

Not only can I not cover most of the technical details of toroidal confinement, I cannot do justice to many facets of plasma turbulence in other contexts. For example, I will say nothing about strong Langmuir turbulence [a consequence of nonlinearly coupled Langmuir (plasma) oscillations and ion acoustic waves], although that is very relevant to space and ionospheric plasma physics. For a basic introduction to that and other plasma-turbulence phenomena, a good place to start is the review article by Similon and Sudan (1990). It is interesting to contrast the state of fusion

---

[a]More pictures of contemporary fusion experiments as well as basic information about fusion research can be found at the web site of the Princeton Plasma Physics Laboratory, `http://www.pppl.gov`. The web site of the international ITER project, `http://www.iter.org`, may also be of interest.

[b]Course notes prepared by the leading experimentalist S. Zweben on various facets of plasma physics (including drift waves, plasma turbulence, and fusion) can be found at `http://www.pppl.gov/~szweben/Course/course.html`.

Fig. 4.1   The Tokamak Fusion Test Reactor at the Princeton Plasma Physics Laboratory (PPPL) as of 1989.   The plasma vacuum vessel is buried under a maze of support structures, diagnostics, neutral-beam injectors, *etc.* Courtesy of PPPL.

turbulence theory as described by that more than 15-year-old article with the emphases and recent developments described below.

In my oral lectures presented during the Summer School,[a] I attempted to provide rather elementary introductions to various interesting fusion-plasma issues. Most of that information will be included here, but these written lectures provide more details and focus somewhat more on systematic analytical techniques. Now sensible analytical approaches to turbulence that transcend basic dimensional analysis are very difficult to obtain, and if all researchers on plasma turbulence were to work exclusively on analytical methods, the field would probably not advance at all. Concerted experimental (and, more recently, numerical simulation) attacks on the challenging magnetic fusion problem have been absolutely essential in providing motivation,[b] identifying relevant problems, and determining key physical balances and effects. However, space constraints require me to limit the scope of the lectures in some way, and the complete absence of an analytical framework is also perilous; one can cite various examples where

---

[a]The orally presented version of these lectures contained considerably more figures than were feasible to include in the written manuscript. Slides of the oral talks can be obtained from `ftp://ftp.pppl.gov/pub/krommes/Talks/ANU06/talk*.pdf`.

[b](and, it must be said, research funding)

ill-founded intuition has led researchers astray.[a] One clearly needs all of theory, experiment, and numerical simulation. These lectures attempt to provide some feeling for the theoretical underpinnings and for the current analytical challenges. You will learn about some contemporary experimental issues in the accompanying lectures of M. Shats (Chap. 5). Numerical simulation of plasmas has some distinctive features relative to the fluid simulations described by J. Jiménez (Chap. 6), but is unfortunately not described here in any detail.

This introductory lecture will (very briefly!) cover

- the plasma as a many-body system;
- the ubiquity of linear plasma waves;
- the gyrokinetic description appropriate for low-frequency fluctuations in a strong magnetic field;
- the basic drift wave and the Hasegawa–Mima paradigm;
- other important nonlinear equations, including the Hasegawa–Wakatani equations and fluid equations for ion-temperature-gradient-driven (ITG) modes;
- some important physics phenomena such as entropy balances and Reynolds stress;
- and the transition to plasma turbulence.

In subsequent lectures, I will remark on the uses in plasma physics of renormalization methods (Lecture 2); some issues relating to the analytical description of zonal flows (Lecture 3); and various topics relating to intermittency and coherent structures, including a contemporary research problem involving the generation of plasma "blobs" (Lecture 4).

Let us begin with the fundamental difficulty that a modern fusion reactor contains an enormous number of particles, which can certainly not be treated individually.

4.1.1. *The Liouville and Klimontovich equations, the Vlasov–Poisson system, and plasma kinetic equations*

Plasmas are many-body systems; as such, they are governed by the *Liouville equation* for the N-particle probability density function (PDF)[b] $P_N(\Gamma, t)$, where[c] $\Gamma \doteq \{\boldsymbol{x}_i, \boldsymbol{v}_i \mid i = 1, \ldots, N\}$ denotes the set of all particle coordinates

---

[a]For some examples, see Krommes (2002).
[b]Statistics arise through an ensemble of initial conditions.
[c]I use the notation $\doteq$ for definitions.

120   *John A. Krommes*

and velocities:

$$\frac{\partial P_N}{\partial t} + \sum_{i=1}^{N} \left[v_i \cdot \frac{\partial P_N}{\partial x_i} \left(\frac{q}{m}\right)_i [E(x_i, t) + \frac{1}{c} v_i \times B(x_i, t)] \cdot \frac{\partial P_N}{\partial v_i} \right]$$
$$= 0. \qquad (4.1)$$

Thus, the study of plasma turbulence can be viewed as a special case of nonequilibrium statistical mechanics in the presence of electromagnetic forces. The Liouville equation states that the entire collection of N particles is conservative (particles are assumed to be neither created nor destroyed) and that phase-space volumes are conserved (Liouville's theorem). However, because N is a huge number, direct attacks on the Liouville equation are extremely complicated. Instead, reduced descriptions are usually used, as I will describe.

One reduction method is to derive equations for reduced n-particle distribution functions f_n (where $n = 1$, 2, or possibly 3) by integrating the Liouville equation over $N - n$ particle variables. That leads to the famous Bogoliubov–Born–Green–Kirkwood–Yvon (BBGKY) hierarchy of coupled PDF's. In that hierarchy one already sees the most fundamental technical issue in statistical physics, the *statistical closure problem*: distribution functions of order n are driven by ones of order $n + 1$, and no closed system for a small set of low-order distributions rigorously exists. This issue was remarked upon by both Prof. Falkovich (Chap. 1) and Prof. McComb (Chap. 2).

While one can develop approximate closures based on the traditional Liouville/BBGKY approach, it is more popular in plasma physics to use the equivalent, and for many purposes more physically appealing, *Klimontovich formalism* (Klimontovich, 1967). That begins with the *Klimontovich equation* for the singular plasma phase-space microdensity

$$\widetilde{N}_s(x, v, t) \doteq \frac{1}{\overline{n}_s} \sum_{i \in N_s} \delta(x - \widetilde{x}_i(t)) \delta(v - \widetilde{v}_i(t)), \qquad (4.2)$$

where s is a species label and the tildes denote random variables. The phase-space average (over an ensemble of initial conditions) of $\widetilde{N}$ is the one-particle distribution function[a] $f_1 \equiv f$: $\langle \widetilde{N}_s(x, v, t) \rangle = f_s(x, v, t)$. The *Klimontovich equation*, which can be obtained by straightforward time differentiation of Eq. (4.2), is

$$\frac{\partial \widetilde{N}_s}{\partial t} + v \cdot \nabla \widetilde{N}_s + \left(\frac{q}{m}\right)_s (\widetilde{E} + c^{-1} v \times \widetilde{B}) \cdot \frac{\partial \widetilde{N}_s}{\partial v} = 0, \qquad (4.3)$$

---

[a] f is normalized such that $V^{-1} \int dx \int dv\, f(x, v, t) = 1$, where V is the system volume.

where the Klimontovich microfields $\widetilde{\boldsymbol{E}}$ and $\widetilde{\boldsymbol{B}}$ are determined from Maxwell's equations, e.g., with $\widetilde{\rho}$ being the microscopic charge density,

$$\boldsymbol{\nabla} \cdot \widetilde{\boldsymbol{E}}(\boldsymbol{x}, t) = 4\pi \widetilde{\rho}(\boldsymbol{x}, t) = 4\pi \sum_s (\overline{n}q)_s \int d\boldsymbol{v}\, \widetilde{N}_s(\boldsymbol{x}, \boldsymbol{v}, t). \qquad (4.4)$$

Because Maxwell's equations are linear, the microfields are linear functionals of $\widetilde{N}$; thus the Klimontovich equation is *quadratically nonlinear*.[a] Of course, the Navier–Stokes equation (NSE) is also quadratically nonlinear because of the advective nonlinearity[b] $\boldsymbol{u} \cdot \boldsymbol{\nabla} \boldsymbol{u}$; that suggests both that there may be some commonality of nonlinear physical behavior between the neutral fluid and the plasma (at least at the macroscopic level) and that analytical methods[c] that have been used in neutral-fluid theory may also be applied to plasmas. I will discuss some aspects of renormalization for plasmas in Lectures 2 and 3.

Although the Klimontovich equation may have technical advantages over the Liouville equation, the physics contents of both equations are equivalent. In particular, both equations are time-reversible, so the derivation of dissipative, irreversible equations is entirely nontrivial.[d] Dissipation can manifest itself in both $\boldsymbol{v}$ space (collisional drag and velocity diffusion) and $\boldsymbol{x}$ space (Navier–Stokes-like fluid equations). These effects are usually treated sequentially. First one encapsulates the effects of particle discreteness in the dissipative *plasma collision operator* $C[f]$. Then moments of the collisional kinetic equation lead to dissipative fluid equations.[e]

$C[f]$ generalizes the Boltzmann collision operator (Cercignani, 1998) to deal with the long-ranged nature of the Coulomb force. The detailed form of C depends on whether the plasma is weakly- or strongly-coupled, terms that I will now define. A measure of the effects of particle discreteness is the *plasma parameter* $\epsilon_p \doteq 1/n\lambda_D^3 \sim n^{1/2}T^{-3/2}$, where λ_D is the Debye screening length[f] that defines the effective range of the Coulomb force around a *test particle* statistically shielded by all of the other plasma particles. A proper course on plasma statistical mechanics would discuss this

---

[a] In products such as $\widetilde{\boldsymbol{E}}\widetilde{N} \sim \sum_i \sum_j$, self-energy terms (i.e., $j = i$) are to be excluded.

[b] In the incompressible limit, the pressure is also a quadratically nonlinear functional of $\boldsymbol{u}$ (which enforces $\boldsymbol{\nabla} \cdot \boldsymbol{u} = 0$).

[c] [such as statistical renormalization; see the lectures of Prof. McComb (Chap. 2)]

[d] Some of the deep issues of nonlinear dynamics that arise here were discussed in the lectures of P. Holmes; for some references, see Holmes (1990) and Diacu and Holmes (1996).

[e] Without approximation, moment equations are not closed. In a collisional limit, the Chapman–Enskog procedure may be used to systematically effect closure and obtain formulas for the transport coefficients. More generally, one must incorporate the Landau damping due to the wave–particle resonance (Hammett and Perkins, 1990).

[f] The Debye length is defined by $\lambda_D^{-2} \doteq k_D^2 \doteq \sum_s (4\pi n q^2/T)_s$.

length and the dynamics of test particles in depth, but that is beyond the scope of these lectures. The *weakly coupled plasma* (Ichimaru, 1992a) is defined by $\epsilon_p \ll 1$; in this case, there are very many particles in a Debye sphere, the ratio of effective potential energy and thermal kinetic energy is small, and the plasma behaves almost like a continuous distribution of charge. Magnetically confined fusion plasmas are weakly coupled[a] (they are very hot[b]). *Strongly coupled plasmas* (Ichimaru, 1992b) satisfy the opposite limit. They are important in astrophysical contexts and in inertial fusion (high-density regimes).

Formally, the classical collision operator arises by averaging the Klimontovich equation. As a standard notation, I represent the fluctuation in any random quantity $\widetilde{A}$ as $\delta A \doteq \widetilde{A} - A$, where $A \doteq \langle \widetilde{A} \rangle$. The average of a product of two quantities $\widetilde{A}$ and $\widetilde{B}$ is then $\langle \widetilde{A}\widetilde{B} \rangle = AB + \langle \delta A\, \delta B \rangle$. Since one has $\langle \widetilde{N} \rangle = f$, one finds

$$\partial_t f + \boldsymbol{v} \cdot \boldsymbol{\nabla} f + (q/m)(\boldsymbol{E} + c^{-1}\boldsymbol{v} \times \boldsymbol{B}) \cdot \partial_{\boldsymbol{v}} f = -\partial_{\boldsymbol{v}} \cdot \langle \delta\boldsymbol{a}\, \delta N \rangle, \qquad (4.5)$$

where $\delta\boldsymbol{a}$ is the fluctuation of the random acceleration. The left-hand side of Eq. (4.5) involves the (time-reversible) Vlasov operator, which provides the mean-field description of the plasma. The right-hand side, which describes the effects of fluctuations, reduces in part to the time-irreversible plasma collision operator. However, it is important to recognize that fluctuations persist even in the limit of a continuous phase-space fluid ($\epsilon_p \to 0$) where the classical collision operators vanish. Thus, formally the right-hand side of Eq. (4.5) also includes the effects of turbulence. In this approach (Davidson, 1972), it makes no sense to average Eq. (4.5) again; since all quantities in it are already statistical means, the equation would merely be reproduced unchanged. Frequently, however, a different interpretation is used in which the first average is assumed to be performed over just the microscopic scales $k\lambda_D \gg 1$, so that the right-hand side reduces to *only* the plasma collision operator. Then macroscopic fluctuations remain in Eq. (4.5), which could be treated numerically. Alternatively, for analytical work a further average can be performed to obtain a "turbulent collision operator" from the nonlinear terms on the left-hand side of Eq. (4.5). This is analogous to the way

---

[a]For weak coupling, the Landau collision operator is generally used in practice. That operator is discussed in virtually all plasma-physics textbooks; see, for example, the early chapters of Helander and Sigmar (2002). A more fundamental description is given by the Balescu–Lenard operator (Ichimaru, 1992a, and references therein), which takes Debye shielding into account more systematically. Both the Landau and Balescu–Lenard operators are complicated because of their intrinsic integro-differential nature (they have Fokker–Planck form, but the Fokker–Planck coefficients depend on integrals over the entire particle distribution).

[b]Although fusion plasmas are hotter than the sun, their pressure $P \doteq nT$ is nevertheless small when compared to the effective pressure of the confining magnetic field: $\beta \doteq P/(B^2/8\pi) \ll 1$.

in which statistical descriptions of the NSE are usually obtained, except that here the averaging is done on a velocity-dependent kinetic equation. However, if the physics permits a fluid description, then an alternative procedure is to first derive dissipative fluid equations, then to average those. That is what is normally done for applications to turbulence in fusion plasmas.

4.1.2. *Basic concepts of linear theory*

Before one can discuss plasma turbulence (an intrinsically nonlinear phenomenon), one must have a solid understanding of the linear behavior of plasmas, which is much more complicated than that of the neutral fluid. For the incompressible NSE, in the absence of background flows the linear behavior is trivial, consisting of merely viscous damping described by the linear momentum diffusion equation $\partial_t \boldsymbol{u} = \mu \nabla^2 \boldsymbol{u}$, where μ is the (kinematic) viscosity. That is, to each Fourier wave number $\boldsymbol{k}$ is associated the stable eigenvalue[a] $\lambda_{\boldsymbol{k}} = -\mu k^2$.

The linear eigenvalue spectrum of a correspondingly homogeneous plasma is much richer. At the fluid level, such eigenvalues are determined by the 3×3 *dielectric tensor* $\mathsf{D}(\boldsymbol{k}, \omega)$, the determination of which is a standard exercise in introductory graduate-level courses on plasma waves (Stix, 1992). One begins with the linearized kinetic equation

$$\partial_t \delta f + \boldsymbol{v} \cdot \boldsymbol{\nabla} \delta f + (q/m)(\delta \boldsymbol{E} + c^{-1} \boldsymbol{v} \times \delta \boldsymbol{B}) \cdot \partial_v f = -\widehat{C}\delta f, \qquad (4.6)$$

where mean fields have been ignored for simplicity and $\widehat{C}$ is the linearized collision operator. One then seeks homogeneous, self-consistent equations for the fluctuating fields $\delta \boldsymbol{E}$ and $\delta \boldsymbol{B}$. The resulting dispersion relation[b]

$$\det\left[\mathsf{D}(\boldsymbol{k}, \omega) - \left(\frac{kc}{\omega}\right)^2 (\mathsf{I} - \widehat{\boldsymbol{k}}\,\widehat{\boldsymbol{k}})\right] = 0 \qquad (4.7)$$

has both high-frequency solutions — plasma-modified, transverse ($\boldsymbol{E} \perp \boldsymbol{k}$) light waves, with dispersion relation $\omega^2 = k^2 c^2 + \omega_p^2$, where ω_p is the plasma frequency[c] — and a variety of lower-frequency ones, often longitudinal ($\boldsymbol{E} \parallel \boldsymbol{k}$) and adequately treated in the electrostatic approximation $\boldsymbol{E} = -\boldsymbol{\nabla}\phi$. The familiar *Langmuir oscillation* $\omega^2 = \omega_{pe}^2 + 3k^2 v_{te}^2$

---

[a]I assume that perturbations vary as $e^{\lambda t} = e^{-i\omega t}$ with $\omega = \Omega + i\gamma$. λ is frequently used when referring to eigenspectra, especially in the context of nonlinear dynamics, while ω is generally used for wave dispersion relations.

[b]Frequently the collision operator is neglected or modeled by a constant ν (Krook form). The Krook collision operator does not conserve momentum or energy, which is sometimes a significant deficiency.

[c]For species s, $\omega_{ps} \doteq (4\pi n q^2/m)_s^{1/2}$; $\omega_p = (\sum_s \omega_{ps}^2)^{1/2}$.

$(k\lambda_D \ll 1)$ arises from intrinsically non-neutral charge fluctuations $\delta\rho$. Lower-frequency modes, however, can be *quasineutral*, i.e., $\delta\rho \approx 0$. We will be mostly concerned with such quasineutral normal modes.

In the presence of inhomogeneities (e.g., of density or temperature), the situation is yet more complicated. Not only can the homogeneous normal modes be strongly modified, new normal modes can appear. The study of normal modes in realistic confinement geometries is a full-time occupation for some researchers, sometimes requiring the use of supercomputers.

The basic qualitative fact to be gleaned from this cursory introduction to linear plasma theory is that it is easy for a plasma to support waves; they are the rule rather than the exception. The prevalence of linear plasma waves has strongly[a] influenced the development of plasma turbulence theory. Indeed, it is sometimes permissible to use a *weak turbulence* description, in which normal modes are taken to be independent to lowest order, then coupled perturbatively. However, in general turbulence leads to substantial *line broadening*, terms of all orders in perturbation theory are important, and one must use alternate (renormalization) techniques; see Lecture 2.

4.1.3. *The gyrokinetic description*

Although the derivation of the dielectric tensor of a magnetized plasma is standard, it is also tedious for an important, very basic physical reason. In a magnetic field, particles execute helical motion around the magnetic field lines. Polarization processes (the response to collective electromagnetic field fluctuations) must all be calculated with respect to that basic helical motion. The result is a slew of Bessel functions that describe the relative importance of every harmonic of the fundamental gyrofrequency $\omega_c \doteq qB/mc$. The resonance function $(\omega - k_\parallel v_\parallel - n\omega_c)^{-1}$ enters the formalism.

It is sometimes essential to describe physics at harmonics of the ion gyrofrequency ω_{ci}, for example when analyzing schemes for heating plasmas by injecting external rf power. However, normal modes of interest for microturbulence in magnetically confined plasmas typically satisfy $\omega \ll \omega_{ci}$.[b] For such modes, only the $n = 0$ harmonic needs to be considered. However, it is unduly complicated and unnecessary to calculate the entire dielectric tensor, only to then throw away most of its content. The *gyrokinetic formalism*[c] provides a methodology whereby the rapid gyration is analytically removed at the start, yielding a much cleaner and more physically intuitive

---

[a]Some would say "too strongly." Montgomery (1977) has made cogent criticisms of the overemphasis of the field on linear dynamics.

[b]This can be argued simply on the basis of the available free energy (Rosenbluth, 1965).

[c]What I call "gyrokinetics" here is referred to by some others as "*low-frequency* gyrokinetics." For some discussion of high-frequency gyrokinetics, see Qin et al. (2000).

description of the low-frequency plasma dynamics.

In general, gyrokinetics describes plasma physics in general configurations of electromagnetic fields that vary slowly in space and time. Although those are very important in practice, the main goal of these lectures is to discuss plasma turbulence, so I cannot here do complete justice to the formal derivation of the gyrokinetic formalism. Nowadays that is done with the aid of Hamiltonian methods (Dubin et al., 1983, and references therein; see Fig. 4.2) that exploit Lie perturbation theory (Cary, 1981) and techniques of differential geometry (Cary and Littlejohn, 1983), all of which are extensions of the pioneering work of Littlejohn.[a] Here, I will merely give a heuristic derivation that captures the essence of the physics. Consider a straight, spatially constant, externally imposed magnetic field $\boldsymbol{B} = B\widehat{\boldsymbol{b}}$. In the plane perpendicular to $\boldsymbol{B}$, a particle gyrates with gyrofrequency ω_c with gyration radius $\rho \doteq v_\perp/\omega_c$, where $v_\perp$ is the perpendicular velocity; the gyrocenter $\boldsymbol{R} \doteq \boldsymbol{x} - \boldsymbol{\rho}$ is fixed; see Fig. 4.3. Now it is "well known" and frequently stated that the quantity $\mu_0 \doteq \frac{1}{2}mv_\perp^2/B$ is an *adiabatic invariant*; that is, it remains constant if B is changed very slowly from one value to another. Actually, to be more precise, μ_0 is the lowest-order approximation (in small gradients, time variations, and electromagnetic fields) to a true adiabatic invariant μ called the *magnetic moment*. It is very much more convenient to use μ rather than $v_\perp$ or μ_0 as an independent velocity-space variable because μ is a constant of the motion.

Now introduce a temporally constant electric field $\boldsymbol{E}_\perp$ perpendicular to $\boldsymbol{B}$. This causes the gyrocenter to drift with an effective $\boldsymbol{E} \times \boldsymbol{B}$ velocity $\overline{\boldsymbol{V}}_E \doteq c\overline{\boldsymbol{E}} \times \widehat{\boldsymbol{b}}/B$ (Fig. 4.4). Here the overline means the effective electric field felt by the gyrocenter. In a Fourier representation, let $\boldsymbol{E}(\boldsymbol{x})$ vary as $\exp(i\boldsymbol{k}_\perp \cdot \boldsymbol{x}) = \exp(i\boldsymbol{k}_\perp \cdot \boldsymbol{R})\exp(i\boldsymbol{k}_\perp \cdot \boldsymbol{\rho})$. In a polar representation with $\angle\boldsymbol{k}_\perp = \alpha$ and $\angle\boldsymbol{\rho} = \theta$, it is easy to average the rapidly varying part $\exp[i k_\perp \rho \cos(\theta - \alpha)]$ over one gyration period to find that $\overline{\boldsymbol{V}}_{E,\boldsymbol{k}} = J_0(k_\perp\rho)\boldsymbol{V}_{E,\boldsymbol{k}}$; the zeroth-order Bessel function J_0 is ubiquitous in gyrokinetics.

Since the gyrocenter drifts with $\overline{\boldsymbol{V}}_E$, the magnetic moment generalizes to $\mu_0 \to \mu \doteq \frac{1}{2}m|\boldsymbol{v}_\perp - \overline{\boldsymbol{V}}_E|^2/B$; this expression is properly invariant to Galilean transformations of the velocity, describing the fact that (in the absence of any other drifts) the perpendicular motion is circular in the drifting frame. One can now write a plausible *gyrokinetic equation* for the distribution F of gyrocenters (in a constant $\boldsymbol{B}$):

$$\partial_t F(\boldsymbol{R}, \mu, v_\parallel, t) + v_\parallel\widehat{\boldsymbol{b}} \cdot \boldsymbol{\nabla}F + \overline{\boldsymbol{V}}_E \cdot \boldsymbol{\nabla}F + (q/m)\overline{E}_\parallel \partial_{v_\parallel} F = 0. \qquad (4.8)$$

This equation describes a five-dimensional (5D) phase space; the gyrophase θ does not appear since its effect on the gyrocenters has already

---

[a]Some of this work was reviewed by Krommes (2002), Appendix C, and further details can be found in the much more extensive review by Brizard and Hahm (2006).

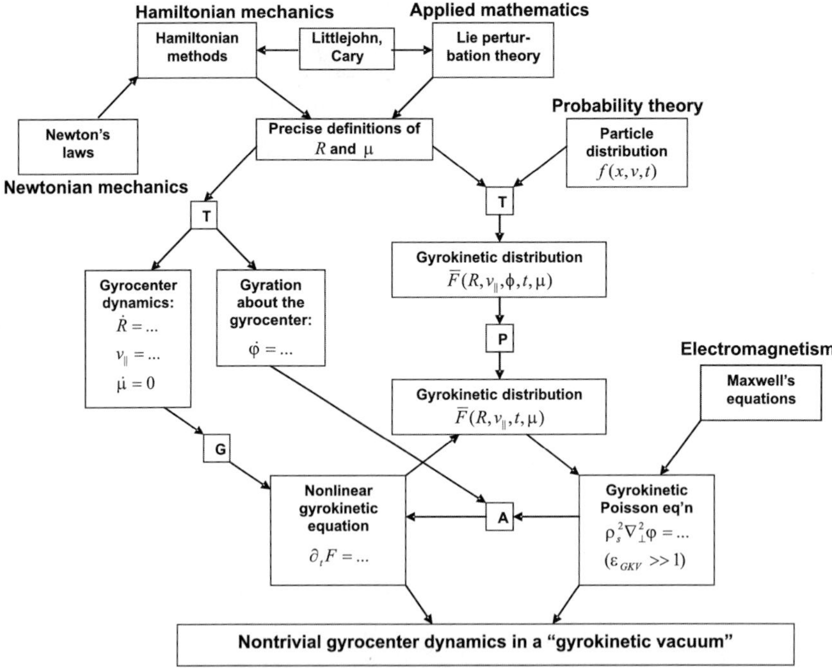

Fig. 4.2   The method of Dubin et al. (1983) for deriving the nonlinear gyrokinetic equation (GKE) was inspired by the seminal paper of Lee (1983), but used technology developed by Littlejohn and Cary. Precise definitions of the gyrocenter position R and the adiabatically conserved magnetic moment μ are determined asymptotically by Lie perturbation theory, which generates the gyrokinetic coordinate transformation T. In the new coordinates, the particle-coordinate distribution f becomes the gyroangle-dependent gyrokinetic distribution function $\widetilde{F}$. The gyrocenter distribution F is defined by the gyroangle average of $\widetilde{F}$ (i.e., $F=\mathsf{P}\widetilde{F}$, where P is the projection operator onto the $\partial/\partial\phi=0$ subspace). The fundamental gyrokinetic approximation G is to ignore the contributions to low-frequency evolution from fluctuations at or above the gyrofrequency. The gyrokinetic Poisson equation is the ordinary Poisson equation expressed in terms of F instead of f. The actual fields predicted by the gyrokinetic Poisson equation must be gyration-averaged (A) in order to determine the effective fields that act on the gyrocenters. Subsequent work (Hahm, 1988; Brizard and Hahm, 2006, and references therein) improved upon Dubin's method by employing differential one-forms.

been taken into account through the effective fields. Also, Eq. (4.8) contains no derivative with respect to μ, since μ is conserved for low-frequency motions with $\omega\ll\omega_c$. Thus, μ just appears as a parameter.

Next, allow $\boldsymbol{V}_E$ to vary slowly in time. It is well known[a] that this introduces a new drift, the particle *polarization drift* $\boldsymbol{V}^{\mathrm{pol}}\doteq\omega_c^{-1}\partial_t(c\boldsymbol{E}_\perp/B)$

---

[a]Elementary discussion of single-particle drifts can be found in Chandrasekhar (1960).

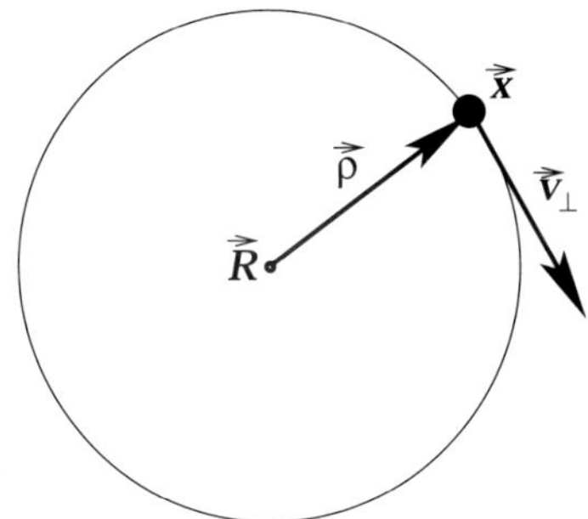

Fig. 4.3 Motion of a charged particle perpendicular to a constant magnetic field $\boldsymbol{B}$ out of the page. The particle position $\boldsymbol{x}$ and the gyrocenter position $\boldsymbol{R}$ are related by $\boldsymbol{x} = \boldsymbol{R} + \boldsymbol{\rho}$, where $\rho = v_\perp/\omega_c$.

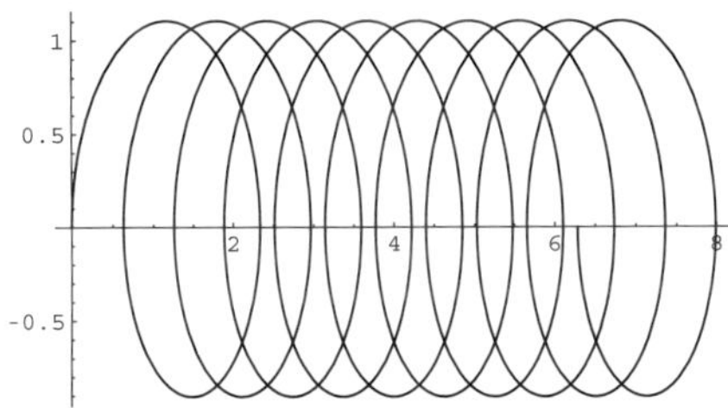

Fig. 4.4 Illustration of the $\boldsymbol{E} \times \boldsymbol{B}$ drift. $\boldsymbol{E} = E\widehat{\boldsymbol{y}} = \text{const}$; $\boldsymbol{B}$ is out of the page. $\boldsymbol{V}_E = c\boldsymbol{E} \times \widehat{\boldsymbol{b}}/B = (cE/B)\widehat{\boldsymbol{x}}$.

(Fig. 4.5). (Because ω_{ce} is large, electron polarization is negligible.) Note that polarization is not a finite-Larmor-radius (FLR) effect (involving the thermal gyroradius $\rho \doteq v_t/\omega_c$) because the polarization drift is independent of temperature and therefore survives in the limit $T \to 0$, for which the gyroradius vanishes. An important question is now, "Is it the particle or the gyrocenter that drifts with $\boldsymbol{V}^{\text{pol}}$?" If it is the gyrocenter, then one should incorporate $\boldsymbol{V}^{\text{pol}}$ into Eq. (4.8) as an additional term on the same footing as the $\boldsymbol{E} \times \boldsymbol{B}$ velocity. But that is not the case. Because the form of the Galilean-invariant magnetic moment μ is unchanged by the addition of time dependence to $\boldsymbol{V}_{E,\perp}$ and one strongly wants to use the adiabatically conserved μ as an independent variable, Eq. (4.8) *remains unchanged* under slow time dependence of $\boldsymbol{V}_E$. *The particle polarizes, not the gyrocenter.*[a] This results in a fluctuation in particle charge density relative to the gyrocenter charge density. Thus, *the effect of the polarization drift must appear in the gyrokinetic Poisson equation, not the kinetic equation.*

---

[a]Some further discussion of this point was given by Dubin and Krommes (1982).

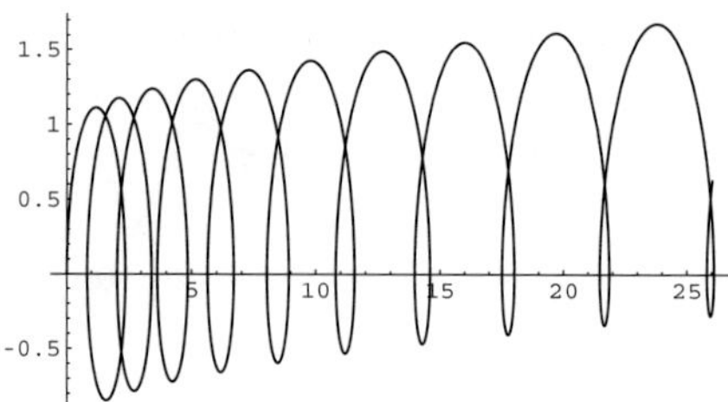

Fig. 4.5 Illustration of the *polarization drift* $\boldsymbol{V}_p = \omega_c^{-1}(c\dot{E}/B)\hat{\boldsymbol{y}}$ for $\dot{E} = \text{const}$. Here $\boldsymbol{V}_p$ is superimposed on the basic $\boldsymbol{E} \times \boldsymbol{B}$ drift of Fig. 4.4.

Specifically, the polarization charge can be calculated from the continuity equation $\partial_t \rho^{\text{pol}} + \boldsymbol{\nabla} \cdot (\rho \, \boldsymbol{V}^{\text{pol}}) = 0$. If this equation is linearized around a background density n, one obtains $\partial_t(\delta n^{\text{pol}}/n) = -\boldsymbol{\nabla} \cdot \delta \boldsymbol{V}^{\text{pol}}$. Because $\boldsymbol{V}^{\text{pol}}$ also involves a ∂_t, one can integrate immediately. In the electrostatic approximation, one has $\boldsymbol{E} = -\boldsymbol{\nabla}\phi$. It is convenient to introduce the dimensionless potential $\varphi \doteq e\phi/T_e$, the *sound speed* $c_s \doteq (ZT_e/m_i)^{1/2}$ (Z is the atomic number), and the ion *sound radius*[a] $\rho_s \doteq c_s/\omega_{ci}$. (This important length is very much smaller than characteristic macroscopic scale lengths such as the minor radius a of a tokamak: $\rho_s/a \ll 1$.) Then one readily finds

$$\delta n_i^{\text{pol}}/n_i = \rho_s^2 \nabla_\perp^2 \varphi. \tag{4.9}$$

An important physical connection can be obtained by noting that the right-hand side of Eq. (4.9) is just the $\hat{\boldsymbol{b}}$ component of the *vorticity* ϖ of the $\boldsymbol{E} \times \boldsymbol{B}$ flow, measured in units of ω_{ci}:

$$\varpi \doteq \hat{\boldsymbol{b}} \cdot (\boldsymbol{\nabla} \times \boldsymbol{V}_E) = \omega_{ci}(\rho_s^2 \nabla_\perp^2 \varphi). \tag{4.10}$$

A diagram illustrating this point is given by Krommes (2004b), Fig. 2. Thus, we will find that the important nonlinear equations that model plasma turbulence naturally involve ϖ as a dependent variable. Gyrokinetics brings the physics of (two-dimensional) vorticity to the fore.

Given a workable gyrokinetic formalism, we are now prepared to write the gyrokinetic Poisson equation. For particles at the laboratory coordinate position $\boldsymbol{x}$, Poisson's equation is

$$-\nabla^2 \phi = 4\pi\rho. \tag{4.11}$$

Upon introducing the electron Debye length $\lambda_{De} \doteq (4\pi n_e e^2/T_e)^{-1/2}$ and using the condition of overall charge neutrality $n_e = Zn_i$, Eq. (4.11) becomes

$$-\lambda_{De}^2 \nabla^2 \varphi = (Zen_i - en_e)/en_e = \delta n_i/n_i - \delta n_e/n_e. \tag{4.12}$$

---

[a]Note the distinction between the Roman s (sound) and the italic s (species). The sound radius ρ_s is the ion gyroradius evaluated at the electron temperature.

Let us now decompose the density fluctuations into gyrocenter and polarization contributions: $\delta n = \delta n^G + \delta n^{\mathrm{pol}}$. Electron polarization is negligible, so $\delta n_e \approx \delta n_e^G$. Slight rearrangement of Eq. (4.12) and use of Eq. (4.9) leads to

$$-\lambda_{De}^2 \widehat{\epsilon}_\perp \nabla^2 \varphi = \delta n_i^G / n_i - \delta n_e / n_e, \qquad (4.13)$$

where $\widehat{\epsilon}_\perp \nabla^2 \doteq \nabla^2 + (\rho_s^2/\lambda_{De}^2)\nabla_\perp^2$. If $k_\parallel$ is neglected in this formula,[a] then $\widehat{\epsilon}_\perp \nabla^2 \approx \epsilon_\perp \nabla_\perp^2$, where the *dielectric constant of the gyrokinetic vacuum* is

$$\epsilon_\perp \doteq 1 + \rho_s^2/\lambda_{De}^2. \qquad (4.14)$$

This quantity is so-named (Krommes, 1993) by analogy with the permittivity ϵ_0 of free space, which appears in the same location in Poisson's equation when MKS units are used. That is, we have derived a nonlinear dynamical system consisting of gyrocenters moving under the influence of $\boldsymbol{E} \times \boldsymbol{B}$ drifts in a "vacuum" state that embodies through $\epsilon_\perp$ the effect of particle polarization.

The *gyrokinetic regime* is defined (Krommes et al., 1986) by $\rho_s^2/\lambda_{De}^2 \gg 1$. In this limit, ion polarization is important, $\epsilon_\perp \gg 1$, and the 1 in Eq. (4.14) is generally neglected (the fluctuations are taken to be quasineutral). This is the regime of interest for tokamak microturbulence. The gyrokinetic Poisson equation

$$-\rho_s^2 \nabla_\perp^2 \delta\varphi = \delta n_i^G / n_i - \delta n_e / n_e \qquad (4.15)$$

then shows, in conjunction with Eq. (4.10), that *a deficiency of ion gyrocenters implies positive plasma vorticity*; thus an evolution equation for vorticity is ubiquitous in gyrokinetics.

The alternate limit $\rho_s^2/\lambda_{De}^2 \to 0$ ($B \to \infty$) is called the *guiding-center limit*. Guiding-center models of plasmas have proven to be very useful for elucidating some basic nonlinear statistical properties of strongly magnetized plasmas (Taylor and McNamara, 1971; Taylor and Thompson, 1973; Taylor, 1974; Joyce and Montgomery, 1973; Vahala, 1974); however, it would take us too far afield to discuss those here.

The GKE (4.8) is essentially[b] correct (in the absence of electromagnetic effects) for spatially uniform $\boldsymbol{B}$; otherwise, the effects of the magnetic ∇B and curvature drifts must be included. This must be done with care in order to preserve the Hamiltonian nature of the dynamics (which, for example, guarantees that phase-space volume is conserved under the evolution).

---

[a]This is a common approximation for tokamak microturbulence. If $k_\parallel$ is replaced by the inverse connection length of the magnetic lines [$k_\parallel \to (Rq)^{-1}$] and one assumes $k_\perp \rho_s = O(1)$, then $k_\parallel/k_\perp \approx \rho_s/Rq = (\rho_s/a)(a/Rq) \ll 1$, where a is the minor radius.
[b]In detail, the effective potentials are also modified by second-order ponderomotive-type effects (Dubin et al., 1983; Brizard and Hahm, 2006). Those are generally small.

Modern methods of differential geometry are ideal for this purpose. General variational techniques based on differential one-forms were described by Cary and Littlejohn (1983), and their application to the problem of inhomogeneous magnetic geometry was carried out by Hahm (1988) and reviewed by Brizard and Hahm (2006). The result is (approximately)

$$\frac{\partial F}{\partial t} + \boldsymbol{\nabla} \cdot [(v_\parallel \widehat{\boldsymbol{b}} + \overline{\boldsymbol{V}}_E + \boldsymbol{V}_d)F]$$
$$+ \frac{\partial}{\partial v_\parallel} \left\{ \left[\left(\frac{q}{m} \right) \overline{E}_\parallel - \mu \widehat{\boldsymbol{b}} \cdot \boldsymbol{\nabla} B + v_\parallel \boldsymbol{\kappa} \cdot \overline{\boldsymbol{V}}_E \right] F \right\} = 0, \qquad (4.16)$$

where the *magnetic drift velocity* is defined by

$$\boldsymbol{V}_d \doteq \left(\frac{\mu B}{\omega_c} \right) \widehat{\boldsymbol{b}} \times \boldsymbol{\nabla} \ln B + \left(\frac{v_\parallel^2}{\omega_c} \right) \widehat{\boldsymbol{b}} \times \boldsymbol{\kappa} \approx \left(\frac{\mu B + v_\parallel^2}{\omega_c} \right) \widehat{\boldsymbol{b}} \times \boldsymbol{\kappa}, \qquad (4.17)$$

with

$$\boldsymbol{\kappa} \doteq \widehat{\boldsymbol{b}} \cdot \boldsymbol{\nabla} \widehat{\boldsymbol{b}} \qquad (4.18)$$

being the *curvature vector*; the last approximation in Eq. (4.17) is valid for vacuum fields.[a] Also, for $T_i \neq 0$ the gyrocenter density generalizes in $\boldsymbol{k}$ space to[b]

$$\delta n_{\boldsymbol{k}}^G = \int d\boldsymbol{Z} J_0(k_\perp \rho) F_{\boldsymbol{k}}(\boldsymbol{Z}) \qquad (4.19)$$

and the polarization density generalizes (for a Maxwellian background) to

$$\delta n_i^{\mathrm{pol}} = -(1 - \Gamma_0) \left(\frac{Ze\phi}{T_i} \right) n_i, \qquad (4.20)$$

where $\Gamma_0(b) \doteq e^{-b} I_0(b)$, I_0 is the modified Bessel function of the first kind, and $b \doteq k_\perp^2 v_\perp^2 / \omega_c^2$. With the aid of $\Gamma_0(b) \approx 1 - b$ for small b, one can readily verify that Eq. (4.20) properly reduces to Eq. (4.9) as $T_i \to 0$ (T_i cancels out in this limit).

---

[a][more generally, for $\beta \doteq P/(B^2/8\pi) \ll 1$]. To prove this, one notes that for a general vector $\boldsymbol{A}$, one has the identity $\boldsymbol{A} \times \boldsymbol{\nabla} \times \boldsymbol{A} = \boldsymbol{\nabla}(\frac{1}{2}A^2) - \boldsymbol{A} \cdot \boldsymbol{\nabla} \boldsymbol{A}$. When applied to the unit vector $\widehat{\boldsymbol{b}}$ ($\widehat{\boldsymbol{b}} \cdot \widehat{\boldsymbol{b}} = 1$), one has $\widehat{\boldsymbol{b}} \cdot \boldsymbol{\nabla} \widehat{\boldsymbol{b}} \doteq \boldsymbol{\kappa} = -\widehat{\boldsymbol{b}} \times \boldsymbol{\nabla} \times \widehat{\boldsymbol{b}} = -\widehat{\boldsymbol{b}} \times \boldsymbol{\nabla} \times (\boldsymbol{B}/B) = -\widehat{\boldsymbol{b}} \times [(4\pi/c)(\boldsymbol{j}/B) + \widehat{\boldsymbol{b}} \times \boldsymbol{\nabla} \ln B]$; thus $\widehat{\boldsymbol{b}} \times \boldsymbol{\kappa} = (4\pi/c)\boldsymbol{j}_\perp/B + \widehat{\boldsymbol{b}} \times \boldsymbol{\nabla} \ln B$. In the absence of perpendicular currents, the $\boldsymbol{\nabla} B$ and curvature drifts thus combine to give formula (4.17).
[b]Here $\boldsymbol{Z} = \{\boldsymbol{R}, \mu, v_\parallel, \phi\}$ is the "proper" set of gyrokinetic coordinates perturbatively constructed such that μ is adiabatically conserved through all orders. The gyrocenter PDF F is independent of ϕ. The J_0 arises because Poisson's equation holds in particle coordinates but the GKE evolves the gyrocenter PDF. The succession of coordinate transformations that leads systematically to Eq. (4.19) was first given by Dubin et al. (1983); for a review, see Krommes (2002), Appendix C.

The electrostatic approximation $\boldsymbol{E} = -\boldsymbol{\nabla}\phi$ is appropriate for sufficiently small plasma pressures, $\beta \doteq nT/(B^2/8\pi) \ll m_e/m_i$.[a] This critical value is frequently exceeded in modern devices, and more generally one must include electromagnetic effects (Hahm et al., 1988) *via* a vector potential $\boldsymbol{A}$. For $\beta \ll 1$, it is adequate to consider just $A_\parallel$, which generates magnetic perturbations perpendicular to the strong background field. Those are responsible for *field-line bending* effects, which emerge from the variation in direction of the unit vector $\widehat{\boldsymbol{b}}$ that defines the direction of the *total* magnetic field. Although electromagnetic effects are quite important in practice,[b] I will not discuss them here.

4.1.4. *Drift waves and the Hasegawa–Mima equation*

We are now prepared to discuss the basic low-frequency normal mode of a magnetically confined plasma, the electrostatic *drift wave* or *universal mode*. It is "universal" because it is driven by the gradient of the background density profile, which is inevitably present in a confined plasma. I will use gyrokinetic techniques to derive the basic nonlinear evolution equation for the electrostatic potential of the drift wave, namely the Hasegawa–Mima equation.[c] Further perspectives are given by Horton and Hasegawa (1994). A detailed review of drift waves and their role in anomalous transport is by Horton (1999).

4.1.4.1. *Derivation of the Hasegawa–Mima equation*

For simplicity, let us consider cold ions $(T_i \to 0)$ and *adiabatic electrons*. By the *adiabatic limit*, one means that the parallel wave phase velocity is much smaller than the parallel particle velocity: $\omega/k_\parallel v_\parallel \ll 1$. In this case, the electrons have time to move rapidly along the field lines and achieve a statistical (frequently called "Boltzmann") equilibrium based on the instantaneous potential: $\delta n_e/n_e \approx e\delta\phi/T_e \equiv \delta\varphi$. [This approximation, called *adiabatic response*, cannot be true for fluctuations that do not vary along the field lines, i.e., $k_\parallel = 0$ (*convective cells*). I will return to this important observation in Sec. 4.1.6.1. below and in the later discussion (Lecture 3) of zonal flows.] Since the ions are cold, they can be fully

---

[a]The electrostatic criterion can be written as $c_A \gg v_{te}$, where $c_A \doteq (B^2/4\pi n_i m_i)^{1/2}$ is the Alfvén speed.

[b]In the National Spherical Torus Experiment (NSTX) at the Princeton Plasma Physics Laboratory, β may be as high as 40%.

[c]The original derivation of this equation was by Hasegawa and Mima (1978). Those authors, as have many others subsequently, began with the *particle* fluid equations of Braginskii (1965). The present derivation, based on the GKE, is substantially cleaner and more intuitive.

described by *fluid equations*. The basic gyrocenter continuity equation

$$\partial_t N_i^G + \boldsymbol{\nabla} \cdot (\boldsymbol{V}_E N_i^G) + \nabla_\parallel (u_{\parallel i} N_i^G) = 0 \qquad (4.21)$$

can be written on physical grounds or derived rigorously by integrating the (collisionless) GKE over velocity. To lowest order in $k_\parallel u_{\parallel i}/\omega$, one can neglect the ion parallel flow. Since $\boldsymbol{\nabla} \cdot \boldsymbol{V}_E = 0$ for constant B,[a] one has

$$\partial_t N_i^G + \boldsymbol{V}_E \cdot \boldsymbol{\nabla} N_i^G \approx 0, \qquad (4.22)$$

which exhibits the basic role of the $\boldsymbol{E} \times \boldsymbol{B}$ drift velocity in advecting the ion density (this is the "drift" in "drift wave"[b]). Decompose $N_i^G = \langle N_i^G \rangle + \delta N_i^G$, where $\langle N_i^G \rangle$ is a background or statistical mean profile and δN_i^G is the fluctuation. Such a decomposition seems reasonably well motivated in cases where the fluctuations are small: $\delta N_i^G / \langle N_i^G \rangle \ll 1$, as is typically the case in the hot core of a tokamak. Because fluctuations will couple to $\boldsymbol{\nabla} \langle N_i^G \rangle$, the resulting model can be called *gradient-driven* (to be distinguished from *flux-driven models* to be discussed in Lecture 4). Also assume that the mean electric field vanishes: $\langle \boldsymbol{V}_E \rangle = \boldsymbol{0}$. For simplicity, let us work in a slab geometry in which the y and z directions are periodic; slight inhomogeneity is allowed in the x direction (the radial direction in a torus). The statistical average of Eq. (4.21) becomes

$$\partial_t \langle N_i^G \rangle + \partial_x \Gamma = 0, \qquad (4.23)$$

where the turbulent flux of ion gyrocenters is

$$\Gamma \doteq \langle \delta V_{E,x} \delta N_i^G \rangle. \qquad (4.24)$$

Equation (4.23) introduces the basic *transport problem*; I will have more to say about that later.

The reader used to similar manipulations of the NSE may worry that classical dissipation does not appear in Eq. (4.23). The simple reason is that we have employed a *collisionless* GKE, but the justification for that is dubious. I will return to this point in Sec. 4.1.5.3.. For now, I continue with collisionless gyrokinetics. The fluctuations obey[c]

$$\partial_t \delta N_i^G + \delta \boldsymbol{V}_E \cdot \boldsymbol{\nabla} \langle N_i^G \rangle + \boldsymbol{\nabla} \cdot (\delta \boldsymbol{V}_E \delta N_i^G) - \partial_x \Gamma = 0; \qquad (4.25)$$

they are, of course, coupled to the evolution of the background profile. To make further progress, let us define the *density gradient scale length L_n* by

---

[a]One should not be lulled into a false sense of simplicity; this is not true in a torus. See, for example, the discussion of geodesic acoustic modes in Sec. 4.3.5..

[b]Some people believe that "drift" refers to "diamagnetic drift," but they are mistaken; see footnote a, p. 134.

[c]Very frequently, the $\partial_x \Gamma$ term is dropped from Eq. (4.25). That is simply not correct for inhomogeneous statistics.

$L_n^{-1} \doteq -\partial_x \ln\langle N\rangle$; in the simplest approximation (or to lowest order in a multiple-space-scale expansion), this is assumed to be constant. Then

$$\boldsymbol{V}_E \cdot \boldsymbol{\nabla}\langle N_i^G\rangle = \left(\frac{cT_e}{eB}\right) \hat{\boldsymbol{z}} \times \boldsymbol{\nabla}\varphi \cdot (-L_n^{-1}\langle N_i^G\rangle\hat{\boldsymbol{x}}) = (V_{*e}\partial_y\varphi)\langle N_i^G\rangle, \quad (4.26\text{a,b})$$

where

$$V_{*e} \doteq cT_e/(eBL_n) = \kappa c_{\mathrm{s}}, \tag{4.27}$$

is the so-called electron diamagnetic velocity[a] and[b]

$$\kappa \equiv \kappa_n \doteq \rho_{\mathrm{s}}/L_n \ll 1. \tag{4.28}$$

If one assumes that a statistically steady state has been achieved, $\partial_t\langle N_i^G\rangle = 0$, then $\partial_x\Gamma$ must vanish according to Eq. (4.23). Then dividing Eq. (4.25) by $\langle N_i^G\rangle$ gives

$$\partial_t(\delta N_i^G/\langle N_i^G\rangle) + V_{*e}\partial_y\varphi + \boldsymbol{V}_E \cdot \boldsymbol{\nabla}(\delta N_i^G/\langle N_i^G\rangle) \approx 0. \tag{4.29}$$

(A small term $\propto \delta N_i^G/L_n$ was neglected.) Finally, one may close the system by using the quasineutral gyrokinetic Poisson equation [the left-hand side of Eq. (4.12) is neglected] together with adiabatic electron response:

$$\delta N_i^G/N_i^G \approx -(\rho_{\mathrm{s}}^2\nabla_\perp^2\delta\varphi - \delta\varphi). \tag{4.30}$$

When one inserts this into Eq. (4.29), one notices that $\boldsymbol{V}_E \cdot \boldsymbol{\nabla}\varphi = 0$. (That is, the so-called $\boldsymbol{E} \times \boldsymbol{B}$ nonlinearity $\boldsymbol{V}_E\cdot\boldsymbol{\nabla}n_e$ vanishes for adiabatic electron response.) Finally, then, one obtains the *Hasegawa–Mima equation*,

$$\partial_t(\underbrace{1}_{\substack{\text{adiabatic}\\ \text{electron}\\ \text{response}}} - \underbrace{\rho_{\mathrm{s}}^2\nabla_\perp^2}_{\substack{\text{ion}\\ \text{polarization}}})\delta\varphi + \underbrace{V_{*e}\partial_y\delta\varphi}_{\substack{\text{advection of}\\ \text{background}\\ \text{profile}}} + \underbrace{\delta\boldsymbol{V}_E \cdot \boldsymbol{\nabla}(-\rho_{\mathrm{s}}^2\nabla_\perp^2\delta\varphi)}_{\substack{\text{advection of vorticity,}\\ \text{or the}\\ \textit{polarization drift}\\ \textit{nonlinearity}}} = 0.$$

$$(4.31)$$

As indicated, this equation describes *fluctuations* in the electrostatic potential. In practice, the δ notation is frequently dropped: $\delta\varphi \to \varphi$.

In linear order, one neglects the advective nonlinearity and obtains the dispersion relation

$$\omega = \frac{\omega_{*e}}{1 + k_\perp^2\rho_{\mathrm{s}}^2}, \tag{4.32}$$

where $\omega_{*e} \doteq k_y V_{*e}$ is called the electron diamagnetic frequency. This is the linear drift wave, the mechanism of which is described in Fig. 4.6. It is dispersive because of ion polarization. In the *gyrokinetic ordering*,

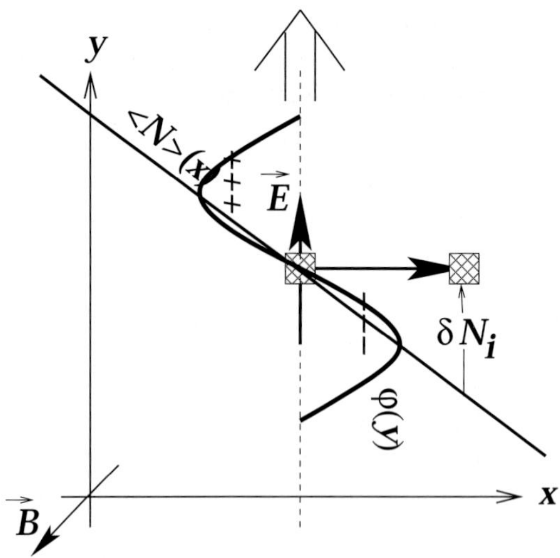

Fig. 4.6 Basic physics of the linear drift wave. Assume a background density gradient in the $-x$ direction with scale length $L_n \doteq -d\ln\langle N\rangle/dx$, a magnetic field $\boldsymbol{B}$ in the z direction, and a sinusoidal potential $\varphi(y,t) = \cos(k_y y - \Omega t)$. Ion polarization is neglected. Electrons instantaneously move along $\boldsymbol{B}$ to neutralize the potential according to the adiabatic response $\delta n_e/n_e = \delta\varphi$. The resulting electric field creates an $\boldsymbol{E} \times \boldsymbol{B}$ drift, which carries a parcel of ion density. In order that the potential be consistent with the resulting density fluctuation, the wave must propagate upward (in the electron diamagnetic direction). Quantitatively, the density fluctuation in time $\Delta t = \Omega^{-1}$ is $\delta N_i = \Delta t\,\delta V_{Ex}(N_i/L_n)$, or $\delta N_i/N_i = (cT_e/eB)k_y\delta\varphi/\Omega L_n$. By quasineutrality, $\delta N_i/N_i = \delta n_e/n_e = \delta\varphi$, leading to the dispersion relation $\Omega = \omega_{*e}$.

one assumes $k_\perp\rho_\mathrm{s} = O(1)$; thus the polarization shielding is not a small correction. Again, that correction is not an FLR effect.

The Hasegawa–Mima equation, with its polarization-drift nonlinearity, serves as a fundamental nonlinear paradigm for drift-wave dynamics. It and its generalizations can exhibit chaotic solutions. Because in large devices $\kappa \ll 1$, the fluctuations are of microscopic character[a] and one refers to drift-wave *microturbulence*.[b]

---

[a] In Braginskii fluid theory, the diamagnetic flow velocity for species s is defined by $\boldsymbol{u}_{*s} \doteq (nq)_s^{-1}(c/B)\widehat{\boldsymbol{b}} \times \boldsymbol{\nabla}P_s$. This is identical to formula (4.26b) for constant temperature; however, the diamagnetic flow is not the origin of the V_{*e} in the drift wave. $\boldsymbol{u}_{*s}$ is an FLR effect (it vanishes for $T \to 0$); the $\boldsymbol{E} \times \boldsymbol{B}$ velocity does not so vanish. Also, for constant $\boldsymbol{B}$ the divergence of the diamagnetic flux $n\boldsymbol{u}_*$ vanishes, so does not contribute to the continuity equation.

[b] The scalar κ is easily distinguishable in context from the magnetic curvature vector $\boldsymbol{\kappa}$.

[a] In small (especially linear) experimental devices, drift-wave eigenmodes are frequently excited with scale of the order of the radius of the plasma column.

[b] Tokamaks do not exhibit steady-state macroturbulence; macroscopic MHD instabilities

When both the background density gradient and the adiabatic electron response are neglected, the Hasegawa–Mima equation reduces to the 2D Euler equation in the vorticity representation:

$$\partial_t \varpi + \boldsymbol{V}_E \cdot \boldsymbol{\nabla} \varpi = 0. \tag{4.33}$$

More generally, Eq. (4.31) is actually identical in form to the Charney equation of geophysics (Charney and Stern, 1962); it is thus frequently called the Charney–Hasegawa–Mima equation.[a] It is time-reversible, so is properly viewed as a generalization of the Euler equation of neutral-fluid theory. As such, this equation left to its own devices will achieve statistical equilibrium (Kraichnan and Montgomery, 1980, and references therein), which is not a sensible description of the forced, dissipative turbulence encountered in practice. Nevertheless, we know on general principles that the quadratic invariants of the nonlinear term importantly constrain the spectral evolution (see the lectures of Prof. Falkovich in Chap. 1), so the Hasegawa–Mima equation as it stands is already nontrivial. Under periodic boundary conditions, the equation conserves the two quadratic invariants

$$\mathcal{E} = \frac{1}{2} \int d\boldsymbol{x} \,(\delta\varphi^2 + |\boldsymbol{\nabla}_\perp \delta\varphi|^2) = \frac{1}{2} \sum_{\boldsymbol{k}} (1 + k_\perp^2)|\delta\varphi_{\boldsymbol{k}}|^2, \tag{4.34a}$$

$$\mathcal{W} = \frac{1}{2} \int d\boldsymbol{x} \,[|\boldsymbol{\nabla}_\perp \delta\varphi|^2 + (\nabla_\perp^2 \delta\varphi)^2] = \frac{1}{2} \sum_{\boldsymbol{k}} k_\perp^2 (1 + k_\perp^2)|\delta\varphi_{\boldsymbol{k}}|^2. \tag{4.34b}$$

$\mathcal{E}$ is called the *energy*,[b] while $\mathcal{W}$ is called the *potential enstrophy*. Their conservation implies the possibility of a *dual cascade* (Kraichnan, 1967) in $\boldsymbol{k}$ space of energy to the left and enstrophy to the right.

I will describe dissipative embellishments of the Hasegawa–Mima equation in Sec. 4.1.6. below.

4.1.4.2. *Dimensional analysis of the Hasegawa–Mima equation; Bohm vs gyro-Bohm scaling*

Difficult nonlinear problems require complicated techniques and analysis for their full solution. But frequently the fundamental *scaling properties*

---

typically lead to global field-line stochastization and catastrophic loss of confinement (called *disruption*).

[a](especially when geophysicists are known to be in the audience). The equation may also be known to some workers as the "barotropic vorticity equation with long-wavelength stabilization" (Frederiksen, 2006a).

[b]Each of the two terms in $\mathcal{E}$ has a simple physical interpretation. The term in $|\boldsymbol{\nabla}_\perp \delta\phi|^2$ is the kinetic energy associated with $\boldsymbol{E} \times \boldsymbol{B}$ motion of the ions. The term in $\delta\varphi^2$ is associated with parallel compression of the electrons as they adiabatically neutralize the potential fluctuations. To see this, let an overbar denote the spatial average. Then the work done on the plasma due to parallel electron flow is $\partial_t \overline{\mathcal{E}} = \overline{\delta j_\parallel \delta E_\parallel} = -\overline{\delta j_\parallel \nabla_\parallel \delta\phi} = \overline{(\nabla_\parallel \delta j_\parallel)\delta\phi} = -\overline{(\partial_t \delta\rho_e)\delta\phi} = en_e \overline{\partial_t(\delta n_e/n_e)\delta\phi} = (n_e T_e)\partial_t(\tfrac{1}{2}\overline{\delta\varphi^2}).$

of the physics can be established straightforwardly, as I will demonstrate with the Hasegawa–Mima equation. The general subject of dimensional analysis and scaling has been beautifully treated by Barenblatt (1996) and was briefly reviewed by Krommes (2002), Appendix B.[a] However, for the Hasegawa–Mima problem, it is possible to proceed more intuitively. First, let us put the Hasegawa–Mima equation into dimensionless form. I now write $\delta\varphi \to \varphi$. Upon displaying all dimensional coefficients, one has

$$(1-\rho_{\mathrm{s}}^2\nabla_\perp^2)\frac{\partial\varphi}{\partial t}+\left(\frac{cT_e}{eBL_n}\right)\frac{\partial\varphi}{\partial y}+\left(\frac{cT_e}{eB}\right)\widehat{\boldsymbol{z}}\times\boldsymbol{\nabla}\varphi\cdot\boldsymbol{\nabla}(-\rho_{\mathrm{s}}^2\nabla_\perp^2\varphi)=0. \quad (4.35)$$

The form of the polarization-drift ($\nabla_\perp^2$) terms indicates that the natural spatial scale is ρ_{s}. Upon noting that $cT_e/eB = \rho_{\mathrm{s}}c_{\mathrm{s}} = \rho_{\mathrm{s}}^2\omega_{ci}$, it is clear that in terms of the dimensionless spatial scales $\boldsymbol{x}/\rho_{\mathrm{s}}$ and times $\omega_{ci}t$ the Hasegawa–Mima equation becomes

$$(1 - \nabla_\perp^2)\partial_t\varphi + \kappa\,\partial_y\varphi + \widehat{\boldsymbol{z}}\times\boldsymbol{\nabla}\varphi\cdot\boldsymbol{\nabla}(-\nabla_\perp^2\varphi)=0, \quad (4.36)$$

where κ is defined by Eq. (4.28). This dimensionless equation contains a single dimensionless parameter κ on which all physical consequences of the equation depend. Note, however, that the dimensionless time $\omega_{ci}t$ is "unnatural" because the gyration time scale was analytically removed with the derivation of the GKE. Indeed, we will see that κ sets the characteristic time scale for all balances between linear and nonlinear terms and thus for all of Hasegawa–Mima physics.

In fact, κ can be removed from the problem altogether by a further scaling transformation. Write $t = T\bar{t}$ and $\varphi = P\overline{\varphi}$. The terms in Eq. (4.36) then stand in the balance $T^{-1}P : \kappa P : P^2$, the unique solution of which is $P = \kappa$, $T = \kappa^{-1}$. That is, in terms of $\bar{t} \doteq \kappa\omega_{ci}t$ and $\overline{\varphi} \doteq \kappa^{-1}\varphi$, the Hasegawa–Mima equation becomes

$$(1 - \overline{\nabla}_\perp^2)\partial_{\bar{t}}\overline{\varphi} + \partial_{\bar{y}}\overline{\varphi} + \widehat{\boldsymbol{z}}\times\overline{\boldsymbol{\nabla}}\overline{\varphi}\cdot\overline{\boldsymbol{\nabla}}(-\overline{\nabla}_\perp^2\overline{\varphi})=0. \quad (4.37)$$

This equation contains no parameters, it is *form-invariant* against any further rescaling, and all terms are of order one. Furthermore, the natural time variations are slow relative to the gyration time: $\partial/\partial(\omega_{ci}t) = \kappa\,\partial/\partial\bar{t}$. Equation (4.37) would be the natural form to use for a numerical attack on Hasegawa–Mima dynamics.

Now suppose that the nonlinear dynamics of Eq. (4.37) somehow determine a turbulent diffusion coefficient D. [This coefficient will actually vanish for strict (i.e., conservative) Hasegawa–Mima physics, but the basic arguments are valid more generally.] According to elementary random-walk

---

[a] Plasma physicists are probably most familiar with the work of Connor and Taylor (1977), reviewed by Connor (1988). An important early paper on the dimensional analysis of fusion plasmas was by Kadomtsev (1975).

theory, $D = \Delta x^2/2\Delta t$, where Δx and Δt are the fundamental step size and time. Thus, the units of D are

$$D \sim \frac{x^2}{t} \sim \left(\frac{(x/\rho_s)^2}{\omega_{ci} t} \right) D_B, \tag{4.38}$$

where

$$D_B \doteq \rho_s c_s = \rho_s^2 \omega_{ci} = cT_e/eB. \tag{4.39}$$

D_B is the so-called *Bohm diffusion coefficient*, named after the well-known observation of Bohm (1949), who suggested that the cross-field transport coefficient in a certain arc plasma appeared to obey $D \approx \frac{1}{16} D_B$. It exhibits the *Bohm scaling* $D_B \sim B^{-1}$. Although this decreases with increasing magnetic field, it does so only weakly. If Bohm scaling prevailed, it would be difficult to design an economical fusion reactor.

But the physically dimensionless D also scales with κ. Upon rescaling κ, one sees that the space scale is unchanged but the time scales with κ^{-1}. Since $D \sim t^{-1}$, we conclude that

$$D \sim \kappa D_B = \left(\frac{\rho_s}{L_n} \right) \left(\frac{cT_e}{eB} \right) \doteq D_{\mathrm{gB}} \sim \frac{1}{B^2}. \tag{4.40}$$

This is called *gyro-Bohm scaling* and depends more strongly on magnetic field than does Bohm scaling. Accordingly, gyro-Bohm scaling is much more amenable to the construction of a viable fusion reactor.

The extra factor of κ in Eq. (4.40) can be interpreted as the natural rms fluctuation level,[a] since $\varphi \sim \kappa$. This amplitude is frequently called the *mixing-length level*, as it follows from the following heuristic argument. Consider a fluid element beginning at $x = 0$ and wandering perpendicularly with the random $\boldsymbol{E} \times \boldsymbol{B}$ velocity. For distances shorter than the *autocorrelation length* or *mixing length* L_{ac} of the turbulence, a Taylor expansion is useful: $n(x) \approx n(0) + L_{\mathrm{ac}}\partial_x n(0)$. Thus the density fluctuation that develops over L_{ac} is $\delta n/\langle n \rangle \approx L_{\mathrm{ac}}/L_n \sim \rho_s/L_n = \kappa$; one must have $L_{\mathrm{ac}} \sim \rho_s$ since ρ_s is the only natural spatial scale. The φ scaling $\varphi \sim \kappa$ then follows from the adiabatic relation $\delta n/\langle n \rangle \sim \varphi$.[b]

---

[a] An important early model of cross-field transport in a strong magnetic field is due to Taylor and McNamara (1971), who considered the statistical dynamics of charged rods in thermal equilibrium. They derived Bohm diffusion for the cross-field transport coefficient D, but their formula for D also contained the rms fluctuation level. That takes a specific value in thermal equilibrium (with no dependence on B), but the basic form is more general.

[b] The associated $\boldsymbol{E} \times \boldsymbol{B}$ velocity is then $\boldsymbol{V}_E \sim V_* \doteq (\rho_s/L_n)c_s$, so the mixing-length balance can be interpreted as balancing the linear term $V_*\partial_y$ and the nonlinear term $\boldsymbol{V}_E \cdot \boldsymbol{\nabla}$. However, statistical steady-state balance arguments really should be applied to the spectral balance equation for mean-squared fluctuations, not to the primitive amplitude equation. When that is done, the reactive diamagnetic frequency $k_y V_*$ disappears in favor of the linear growth rate; see further discussion in Sec. 4.2.3. on resonance-broadening theory. Note, however, that an additional ratio of γ/Ω does not necessarily affect the magnetic-field scaling very much because γ usually scales with Ω.

In summary, for Hasegawa–Mima dynamics dimensional/scaling analysis determines all basic transport quantities to within a constant. In a random walk, the characteristic steps are $\Delta x = \rho_{\mathrm{s}}$ and $\Delta t = (\kappa\omega_{ci})^{-1}$, leading to the transport estimate[a] $D \sim \Delta x^2/\Delta t$ or

$$D \sim \rho_{\mathrm{s}}^2/(\kappa\omega_{ci})^{-1} = \kappa\rho_{\mathrm{s}}c_{\mathrm{s}} = D_{\mathrm{gB}}. \qquad (4.41)$$

This result is uncertain only to within an undetermined constant C that is nominally of order unity: $D = CD_{\mathrm{gB}}$.

4.1.5. *The gyrokinetic transport problem*

Unfortunately, one order-unity coefficient is $C = 0$, and I will show that, in fact, the Hasegawa–Mima density transport coefficient vanishes identically. This could have been easily anticipated from the fact that the Hasegawa–Mima equation is time-reversible; irreversibility is required for transport. But let us attack the transport problem somewhat more generally by deriving some consequences of formula (4.24) for the turbulent flux of ion gyrocenters, which is a very basic result for low-frequency fluctuations in strongly magnetized plasmas.

4.1.5.1. *Ambipolar transport*

First I will prove that the mean fluxes of ion and electron gyrocenters are equal, i.e., that ambipolar transport ensues. To do so, multiply the gyrokinetic Poisson equation (4.13) by $V_{E,x}$ $(= -\partial_y\varphi)$:

$$(-\partial_y\varphi)\nabla_\perp^2\varphi = \overline{n}_i^{-1}V_{E,x}\delta N_i^G - \overline{n}_e^{-1}V_{E,x}\delta n_e. \qquad (4.42)$$

Upon using $\overline{n}_e = Z\overline{n}_i$, rearranging, and statistically averaging, one obtains

$$\overline{n}_e\langle(-\partial_y\varphi)\nabla_\perp^2\varphi\rangle = Z\Gamma_i - \Gamma_e. \qquad (4.43)$$

In a slab model, one has translational invariance, so one can integrate by parts under the average: $\langle(-\partial_y\varphi)\nabla^2\varphi\rangle = \langle(\partial_y\boldsymbol{\nabla}\varphi)\cdot\boldsymbol{\nabla}\varphi\rangle = \frac{1}{2}\partial_y\langle|\boldsymbol{\nabla}\varphi|^2\rangle = 0$. Thus one concludes that the gyrokinetic Poisson equation constrains the gyrocenter fluxes to be *ambipolar*[b]:

$$Z\Gamma_i = \Gamma_e. \qquad (4.44)$$

This is true even though the ion dynamics are highly nonlinear.

Now consider the Hasegawa–Mima electron flux, which is constrained by the adiabatic response $\delta n_e/\langle n_e\rangle = \delta\varphi$: $\Gamma_e = \langle(-\partial_y\delta\varphi)\delta n_e\rangle = \langle(-\partial_y\delta\varphi)\delta\varphi\rangle\langle n_e\rangle = -\frac{1}{2}\langle n_e\rangle\partial_y\langle\delta\varphi^2\rangle = 0$. For purely adiabatic response, the electron particle flux vanishes. But since the fluxes are ambipolar, we conclude that the ion flux vanishes as well.

---

[a] This result follows equally well from the scaling $D \sim \Delta v^2\Delta t$. Note that $\Delta t^{-1} = \kappa\omega_{ci} = c_{\mathrm{s}}/L_n$, $\omega_* = k_y V_* = (k_y\rho_{\mathrm{s}})(c_{\mathrm{s}}/L_n) \sim \Delta t^{-1}$, and $V_* = \rho_{\mathrm{s}}c_{\mathrm{s}}/L_n = \rho_{\mathrm{s}}/\Delta t$. Thus the estimates $\Delta v \sim V_*$ and $\Delta t \sim \omega_*^{-1}$ recover the gyro-Bohm scaling.

[b] Ambipolarity holds on the average, but not microscopically.

4.1.5.2. *Nonadiabatic response and transport scaling*

Clearly one must consider nonadiabatic electron response in order to obtain nonzero particle flux. The simplest way of doing this is to postulate the so-called $i\delta$ *model*,[a]

$$\delta n_{e,k} = (1 - i\delta_{\bm k})\delta\varphi_{\bm k}. \tag{4.45}$$

Here $\delta_{\bm k}$ is a wave-number-dependent constant, presumably determined from linear gyrokinetic theory. In dimensionless units, one now has

$$\Gamma_e = \langle(-\partial_y\varphi)\delta n_e\rangle \tag{4.46a}$$

$$= \sum_{\bm k}\langle(-ik_y\delta\varphi_{\bm k})(1 + i\delta_{\bm k}^*)\delta\varphi_{\bm k}^*\rangle \tag{4.46b}$$

$$= \sum_{\bm k}k_y\,\mathrm{Re}\,\delta_{\bm k}\langle|\delta\varphi_{\bm k}|^2\rangle. \tag{4.46c}$$

The flux is therefore positive if $k_y\,\mathrm{Re}\,\delta_{\bm k} > 0$. Because the linear dispersion relation is now $\omega = \omega_*/(1 + k_\perp^2 - i\delta_{\bm k})$, one sees that positive $k_y\delta_{\bm k}$ corresponds to linear instability (driven by the electrons[b]).

From the linear dispersion relation, one trivially determines that $\gamma_{\bm k}/\Omega_{\bm k} = \delta_{\bm k}/(1 + k_\perp^2)$, or $\delta_{\bm k} \sim \gamma_{\bm k}/\Omega_{\bm k}$. A consequence is that if the turbulence saturates at the mixing level, then the turbulent diffusion coefficient has the scaling[c]

$$D \sim \left(\frac{\gamma}{\Omega}\right)D_{\mathrm{gB}}. \tag{4.47}$$

This result was anticipated by footnote b, p. 137 and can be found in the monograph of Kadomtsev (1965), p. 107. It can also be obtained from the random-walk scaling $D \sim \Delta v^2\tau_{\mathrm{ac}}$ provided one uses the appropriate autocorrelation time τ_{ac}, defined by $\tau_{\mathrm{ac}} \doteq \int_0^\infty d\tau\, C_{vv}(\tau)$, where $C_{vv}(\tau)$ is the Lagrangian[d] correlation function of the $\bm E \times \bm B$ velocity normalized to $C_{vv}(0) = 1$. That function should capture the essential ingredients of a wave-like oscillation (frequency Ω) damped by turbulence at rate η: $\tau_{\mathrm{ac}} = \mathrm{Re}\int_0^\infty d\tau\,\exp(-i\Omega\tau - \eta\tau) = \eta/(\Omega^2 + \eta^2) \approx (\eta/\Omega)\Omega^{-1}$ for $\eta/\Omega \ll 1$.[e] In a statistical steady state, the rate of turbulent damping η should scale with the rate of turbulent forcing γ, so one finds $\tau_{\mathrm{ac}} \sim (\gamma/\Omega)\Omega^{-1}$. With $\Omega \sim \omega_*$ and $\Delta v \sim V_*$, one recovers the scaling (4.47).

---

[a](frequently attributed to Waltz)

[b]Kinetically, the instability drive results from inverse electron Landau damping.

[c]This follows systematically by writing Fick's Law $\Gamma_e = -D\,\partial_x n_e$ as $\Gamma_e = DL_n n_e$ and identifying D by restoring dimensions to Eq. (4.46c), or less formally by noting that if Fick's Law holds, then for fixed background gradient any scaling exhibited by Γ_e must be inherited by D.

[d]Lagrangian means that the function is evaluated along the actual trajectories of the fluid elements: $C_{vv}(\tau) \doteq \langle\delta V(\bm x(\tau),\tau)\delta V(\bm x(0),0)\rangle$. The major difficulty of analytical transport theory is relating the Lagrangian correlation function to the Eulerian one $C_{vv}(\bm x,t,\bm x',t')$.

[e]Physically δ is small; however, one can contemplate the limit $\delta \to \infty$. That does *not*

4.1.5.3. *Entropy balance, dissipation, and steady-state flux*

In the previous section, positive δ corresponds to positive γ, i.e., antidissipation (physically, due to inverse Landau damping on the electrons). In the presence of such forcing, the system will not saturate unless some dissipative terms are included. There are interesting and subtle relationships between steady-state flux and dissipation.

Recall that $\Gamma_i = Z^{-1}\Gamma_e$ due to the ambipolarity constraint. Note, however, that one cannot calculate Γ_i by applying an $i\delta$ model directly to the ions. The ions obey predominantly *fluid*,[a] not adiabatic, response, and their motion is substantially nonlinear because of the polarization-drift nonlinearity. However, one can form an equation for ion flux by manipulating the ion continuity equation. Because dissipation is important, let us generalize that equation to

$$\partial_t \delta N_i^G = -V_* \partial_y \delta\varphi - \delta\boldsymbol{V}_E \cdot \boldsymbol{\nabla} \delta N_i^G - \widehat{\nu}\, \delta N_i^G, \qquad (4.48)$$

where $\widehat{\nu}$ is a positive-definite dissipative operator.[b] Upon multiplying by δN_i^G and averaging, one obtains (with $V_* \to \kappa$ in dimensionless units)

$$\partial_t[\tfrac{1}{2}\langle(\delta N_i^G)^2\rangle] = \kappa\Gamma_i - \mathcal{D}, \qquad (4.49)$$

where $\mathcal{D}$ is the dissipation function $\mathcal{D} \doteq \langle \delta N_i^G \mid \widehat{\nu} \mid \delta N_i^G \rangle > 0$. Equation (4.49) provides one example of what is commonly called an *entropy balance equation*.[c] If a steady, dissipative state is achieved, then positive flux results inevitably, being in balance with the dissipation: $\kappa\Gamma_i = \mathcal{D}$.

The interpretation of entropy balances like Eq. (4.49) was discussed at length by Krommes and Hu (1994), who were motivated by earlier considerations of entropy evolution by Lee and Tang (1988). It is important to note that typically *one does not obtain the value of Γ by calculating the dissipation directly*. Rather, the flux saturates first, and the dissipation subsequently adjusts to absorb that amount of flux. This is completely analogous to the way the dissipation scale is determined in the K41 description of homogeneous, isotropic, 3D Navier–Stokes turbulence. The

---

correspond to the limit of arbitrarily large γ; for the DW dispersion relation, it is easy to check that $\gamma_{\max} = \omega_*/[2(1 + k_\perp^2)]$ at $\delta = 1 + k_\perp^2$ and $\gamma/\Omega = 1$. At larger δ, the form of the nonlinearity changes from vorticity advection (polarization-drift nonlinearity) to the so-called $\boldsymbol{E} \times \boldsymbol{B}$ nonlinearity and the random-walk arguments must be reconsidered. Incidently, the equation with $\delta \neq 0$ is called the Terry–Horton equation (Terry and Horton, 1982; Horton, 1986).

[a]For any species, fluid response is defined by the limit $\omega/k_\parallel v_t \gg 1$.

[b]In a fluid description, $\widehat{\nu}$ can model ion Landau damping. Kinetically, an example of $\widehat{\nu}$ is the linearized plasma collision operator $\widehat{C}$. The reconciliation of fluid and kinetic models was discussed by Krommes and Hu (1994).

[c]The quadratic quantity that is singled out in the balance equation need not be literally an entropy, although it is sometimes a second-order approximation to it.

ideas of Krommes and Hu were verified by Sugama et al. (2003) and, more recently, by Candy and Waltz (2005), and below.

Recently entropy balances have been intensely discussed in the context of gyrokinetic particle simulation. One approach to solving the GKE is to apply the method of characteristics to the deviation δf of the distribution function from some reference distribution f_0 such as a Maxwellian. This so-called δf *method*, originally due to Kotschenreuther (1991) and Parker and Lee (1993), has low noise since sampling error associated with the background distribution is eliminated. One unwelcome byproduct, however, is that for entirely collisionless simulations with nonzero turbulent flux the entropy-like quantity $\int d\boldsymbol{x}\, d\boldsymbol{v}\, \delta f^2/f_M$ grows indefinitely with time[a] [i.e., there is no dissipation term in the equation analogous to Eq. (4.49)]. In the context of particle simulation, this is referred to as the problem of *growing weights*.[b] Nevins et al. (2005) have argued and verified by computer simulation that at very long times the growing weights can contaminate the signal-to-noise ratio and substantially suppress the measured flux.

Any other approach to strictly collisionless gyrokinetics should, in principle, also suffer from this difficulty. However, in practice certain algorithms have built-in numerical dissipation that may either be larger than or altogether replace physical collisional dissipation. This issue was thoroughly discussed by Candy and Waltz (2005) for the GYRO gyrokinetic simulation code based on conventional (non-Monte Carlo) methods for discretizing the GKE. The balance between steady-state flux and dissipation was verified, as was the insensitivity of the results to the actual value of the (sufficiently small) dissipation); these results argue in favor of the robustness of the numerical algorithm employed in GYRO. However, it is fair to say that at the time of writing the relative merits of the various schemes for solving the nonlinear GKE were still intensely disputed.

4.1.5.4. *Reynolds stresses and transport*

In Sec. 4.1.5., statistically averaging the GKE led to a transport equation for particle density. In Navier–Stokes theory, the analogous procedure is to average the Navier–Stokes momentum equation. As described in more detail in such references as Tennekes and Lumley (1972) and briefly reviewed by Krommes (2002), that leads to

$$\partial_t \boldsymbol{U} + \boldsymbol{U} \cdot \boldsymbol{\nabla} \boldsymbol{U} = \rho_m^{-1} \boldsymbol{\nabla} \cdot (-P\mathsf{I} + 2\mu\mathsf{S} + \boldsymbol{\tau}), \qquad (4.50)$$

---

[a]This was called the *entropy paradox* by Krommes and Hu (1994) because the growth of mean-squared fluctuations is incompatible with the assertion of a strict steady state.
[b]The *particle weight* w is defined by $w \doteq \delta f/f_0$. A thorough discussion of the interpretation of the δf method was given by Hu and Krommes (1994), who also showed how to calculate the associated sampling noise.

where $\mathsf{S} \doteq \frac{1}{2}[(\boldsymbol{\nabla U}) + (\boldsymbol{\nabla U})^T]$ is the *rate-of-strain tensor* and $\boldsymbol{\tau} \doteq -\rho_m\langle \delta\boldsymbol{u}\,\delta\boldsymbol{u}\rangle$ is called the *Reynolds stress* (Reynolds, 1895). $\boldsymbol{\tau}$ describes the effects of the fluctuations on the mean flow. The analogy between the turbulent momentum flux $\langle \delta\boldsymbol{u}(\rho_m\delta\boldsymbol{u})\rangle$ and the turbulent gyrokinetic particle flux $\langle \delta\boldsymbol{V}_E\delta N\rangle$ is clear.

The traditional interpretation of Reynolds stress according to Eq. (4.50) involves the generation and evolution of macroscopic, statistically mean flows in an inhomogeneous medium. However, the nomenclature is frequently extended to include the generation of long-wavelength *fluctuations* from the nonlinear interaction of short-wavelength fluctuations. Thus, consider the Hasegawa–Mima polarization-drift nonlinearity $\delta\boldsymbol{V}_E \cdot \boldsymbol{\nabla}\nabla_\perp^2 \delta\varphi = \delta\boldsymbol{V}_E \cdot \boldsymbol{\nabla}\delta\varpi$. For homogeneous turbulence, the complete statistical average of this term vanishes. However, if the term is averaged over just y and z [equivalent to extracting the $(k_y, k_z) = (0,0)$ component], one obtains[a]

$$\langle \delta\boldsymbol{V}_E \cdot \boldsymbol{\nabla}\delta\varpi\rangle_{y,z} = \partial_x\langle \delta V_{E,x}\delta\varpi\rangle_{y,z}(x), \qquad (4.51)$$

so this "Reynolds stress" can drive the *zonal fluctuation* $\delta\varphi_{k_y=0,k_z=0}(x)$, which can be interpreted as a fluctuating shear flow. Such arguments figure importantly in discussions of the "L–H" transition between low- and high-confinement regimes in tokamaks. I will return to this subject in Lecture 3.

4.1.6. *Some other important equations*

In the next few sections, I will briefly describe some other equations of importance to modern plasma nonlinear theory.

4.1.6.1. *The generalized Hasegawa–Mima equation*

Let us begin by questioning the foundations of the Hasegawa–Mima equation. In Sec. 4.1.4. I pointed out that the assumption of adiabatic electron response fails for fluctuations with $k_\parallel = 0$. Instead, such fluctuations are in the *fluid limit* $\omega/k_\parallel v_\parallel \gg 1$. Since the physics hinges on the relative importance of the parallel electron current, $j_{\parallel e} = (-e)u_{\parallel e}n_e$, one may begin with the continuity equation for electrons,

$$\partial_t\delta n_e + \delta\boldsymbol{V}_E \cdot \boldsymbol{\nabla}\langle n_e\rangle + \boldsymbol{\nabla}\cdot(\delta\boldsymbol{V}_E\delta n_e) + \nabla_\parallel[\delta j_{\parallel e}/(-e)] = 0. \qquad (4.52)$$

---

[a]An interesting identity is (Krommes, 2004a) $\boldsymbol{V}_E \cdot \boldsymbol{\nabla}\nabla^2\varphi = (\partial_{xx} - \partial_{yy})(V_{Ex}V_{Ey}) + \partial_{xy}(V_{Ey}^2 - V_{Ex}^2)$. This demonstrates two points. First, for homogeneous, isotropic turbulence both $\langle V_{Ex}V_{Ey}\rangle = 0$ and $\langle V_{Ex}^2\rangle = \langle V_{Ey}^2\rangle$, showing that the complete statistical average of the Reynolds stress vanishes. Second, if this identity is averaged over y using periodic boundary conditions, the right-hand side reduces to $\partial_{xx}\langle V_{Ex}V_{Ey}\rangle_y$. The x integral of the y-averaged vorticity equation $\partial_t\varpi + \boldsymbol{V}_E \cdot \boldsymbol{\nabla}\varpi = 0$ then gives $\partial_t\langle V_y\rangle_y = -\partial_x\langle V_{Ex}V_{Ey}\rangle_y$, a conventional Reynolds equation that describes the generation of zonal flow from inhomogeneous velocity fluctuations.

The perpendicular part of the electron continuity equation can be processed in the same way as for the ions [Eq. (4.29)]. Then, upon taking the time derivative of $\delta N_i^G = \delta n_e - \nabla_\perp^2 \delta\varphi$, one obtains

$$\underbrace{\{-V_* \partial_y \delta\varphi - \delta \boldsymbol{V}_E \cdot \boldsymbol{\nabla} n_e - \nabla_\parallel [\delta j_{\parallel e}/(-e)]\}}_{\partial_t \delta n_e}$$
$$- \nabla_\perp^2 (\partial_t \delta\varphi) + \underbrace{[V_* \partial_y \delta\varphi + \delta \boldsymbol{V}_E \cdot \boldsymbol{\nabla}(\delta n_e - \nabla_\perp^2 \delta\varphi)]}_{-\partial_t \delta N_i^G} = 0. \quad (4.53)$$

Note that the diamagnetic terms and the $\boldsymbol{V}_E \cdot \boldsymbol{\nabla} n_e$ terms cancel, therefore giving the vorticity equation[a]

$$\partial_t \varpi + \boldsymbol{V}_E \cdot \boldsymbol{\nabla}\varpi = \nabla_\parallel j_{\parallel e}/e. \quad (4.54)$$

For modes with $k_\parallel = 0$, $\nabla_\parallel j_{\parallel e}$ vanishes, giving rise to the 2D Euler equation for vorticity; that is, convective cells are hydrodynamic. For other modes, adiabatic response is the asymptotically dominant response. Both of these limits are captured in the *generalized Hasegawa–Mima equation*, defined by

$$(\widehat{\alpha} - \nabla_\perp^2)\partial_t \varphi + \widehat{\alpha} V_* \partial_y \varphi + \boldsymbol{V}_E \cdot \boldsymbol{\nabla}(\widehat{\alpha} - \nabla_\perp^2)\varphi = 0, \quad (4.55)$$

where $\widehat{\alpha} = 0$ for convective cells ($k_\parallel = 0$) and $\widehat{\alpha} = 1$ otherwise. This equation captures the essence of the interaction between drift waves and convective cells, including zonal flows. I will return to it in Lecture 3.

4.1.6.2. *Collisional drift waves and the Hasegawa–Wakatani equation*

Adiabatic response will also fail if the plasma is sufficiently collisional, since collisions impede the rapid motion of the electrons along the field lines. One can use the general vorticity equation derived in the last section, but now one calculates the electron current from the parallel component of the momentum equation:

$$m_e n_e du_{\parallel e}/dt = -en_e E_\parallel - \nabla_\parallel (n_e T_e) - \nu_{ei} m_e n_e u_{\parallel e}, \quad (4.56)$$

where viscous stresses and the ion parallel flow have been ignored. Electron inertia is also negligible, and if temperature fluctuations are ignored, one ultimately obtains

$$\nabla_\parallel [\delta j_{\parallel e}/n_e(-e)] = D_\parallel \nabla_\parallel^2 (\delta\varphi - \delta n_e/n_e), \quad (4.57)$$

---

[a]This same equation is used in "reduced" plasma MHD (Strauss, 1976).

144   *John A. Krommes*

where $D_\| \doteq v_{te}^2/\nu_{ei}$ is the classical collisional diffusion coefficient for parallel motion. Upon substituting into Eq. (4.54), one obtains[a]

$$\frac{d\delta\varpi}{dt} = -D_\| \nabla_\|^2 \left(\delta\varphi - \frac{\delta n_e}{n_e} \right) + \mu \nabla_\perp^2 \delta\varpi, \tag{4.58a}$$

$$\frac{d}{dt} \left(\frac{\delta n_e}{n_e} \right) = -D_\| \nabla_\|^2 \left(\delta\varphi - \frac{\delta n_e}{n_e} \right) - V_* \frac{\partial\delta\varphi}{\partial y} + D\nabla_\perp^2 \left(\frac{\delta n_e}{n_e} \right) \tag{4.58b}$$

These are called the *Hasegawa–Wakatani equations*, originally discussed by Hasegawa and Wakatani (1983) and Wakatani and Hasegawa (1984). They provide a paradigm for collisional drift-wave fluctuations, such as might occur in a cold plasma edge. (Because they have been derived in slab geometry, they are missing magnetic-curvature-drive effects that are likely to be important in practice; see Sec. 4.1.6.4..)

One can recover the Hasegawa–Mima equation from the Hasegawa–Wakatani equations by passing to the limit $\widehat{\alpha} \doteq -D_\| \nabla_\|^2 \to \infty$. (This is not possible for modes with $k_\| = 0$.) In order that other terms remain finite, one must have $\delta n_e/n_e = \delta\varphi$ (i.e., adiabatic response) to lowest order. Then subtracting Eq. (4.58a) from Eq. (4.58b) cancels the parallel terms and leads to (a dissipative version of) the Hasegawa–Mima equation.

The Hasegawa–Wakatani system possesses four quadratic invariants, as reviewed by Krommes (2002), Sec. 2.4.5. The thermal-equilibrium statistical mechanics of the dissipationless limit was discussed by Koniges et al. (1991); it provides useful information about cascade directions. Numerical simulations of the full system have been done by Biskamp et al. (1994), Biskamp and Zeiler (1995), and Camargo et al. (1995). Comparisons of statistical theory with direct numerical simulations were done by Hu et al. (1997), as briefly described in Sec. 4.2.5.2..

4.1.6.3. *Ion-temperature-gradient-driven modes*

Both the universal drift wave (as described by the Hasegawa–Mima equation) and the Hasegawa–Wakatani equations are very useful for elucidating various basic principles. However, as models of reality, they are deficient in one crucial aspect: they have been derived under the assumption of cold ions, $T_i \to 0$; in a fusion reactor, however, the ions are required to be hot. Although it is not difficult to use linear kinetic theory to embellish the drift-wave dispersion relation with FLR effects, one must not overlook the possibility that an entirely new mode of oscillation arises when $T_i \neq 0$. Indeed, that turns out to be the case, in the form of the *ion-temperature-gradient-driven* (ITG) *mode.* It is widely believed that the ITG mode is

---

[a]Here perpendicular dissipation (the μ and D terms) has been added by hand. Those terms can be derived rigorously from collisional gyrokinetics.

primarily responsible for the turbulent ion heat losses in modern toroidal devices.

In the absence of magnetic curvature (see the next section), it is straightforward to linearize the GKE in the presence of a background ion temperature gradient and find the dispersion relation for normal modes. With the usual fluid ordering $\omega/k_\parallel v_{ti} \gg 1$, one is led to the cubic dispersion relation

$$\omega^3 - \Omega_{*n}\omega^2 - \Omega_{\mathrm{s}}^2\omega - \Omega_T^3 = 0, \tag{4.59}$$

where the coefficients are defined by

$$\Omega_{*n} \doteq \frac{\omega_*^n}{1 + k_\perp^2 \rho_{\mathrm{s}}^2}, \quad \Omega_{\mathrm{s}} \doteq \frac{k_\parallel c_{\mathrm{s}}}{(1 + k_\perp^2 \rho_{\mathrm{s}}^2)^{1/2}}, \quad \Omega_T \doteq \left(\Omega_{\mathrm{s}}^2 \widehat{\tau}\omega_*\right)^{1/3} = \left(\Omega_{\mathrm{s}}^2 |\omega_{*i}|\right)^{1/3}. \tag{4.60a,b,c}$$

Here $\widehat{\tau} \doteq T_i/T_e$ and $\omega_* \doteq \omega_*^n + \omega_*^T$. Ω_{*n} is the eigenfrequency of the universal drift wave discussed in Sec. 4.1.4., while Ω_{s} is the frequency of the ion sound wave in a strong magnetic field. The effects of ∇T_i enter through the hybrid frequency Ω_T. If that is set to zero, the dispersion relation factors into $0 = \omega(\omega^2 - \Omega_{*n}\omega - \Omega_{\mathrm{s}}^2)$. The two nontrivial roots describe the co- and counter-propagating ion sound waves modified by the density gradient; for $k_\parallel \to 0$ the copropagating sound wave becomes the drift wave. Of more interest, a third, zero-frequency mode has also appeared. That evolves into the (slab) ITG mode as Ω_T is increased.

As $\Omega_T \to \infty$, the dispersion relation reduces to

$$0 = \omega^3 - \Omega_T^3, \tag{4.61}$$

the solution of which introduces the three cube roots of unity: $\omega = \{1, e^{2\pi i/3}, e^{4\pi i/3}\}\Omega_T$. The unstable root at angle $\theta = \frac{2}{3}\pi$ is the slab ITG mode. The root at $\theta = 0$ has evolved from the drift-wave branch, while the last root at $\theta = \frac{4}{3}\pi$ is connected to the counter-propagating sound-wave branch. This is illustrated in Fig. 4.7.[a]

To understand the physics of the slab ITG mode, it is helpful to derive a fluid description (by taking appropriate moments of the GKE) and closing the hierarchy appropriately. For example, a fluid system corresponding to the dispersion relation (4.61) is

$$\frac{\partial \delta n}{\partial t} = -\frac{\partial \delta u}{\partial z}, \quad \frac{\partial \delta u}{\partial t} = -c_{\mathrm{s}}^2\frac{\partial \delta T}{\partial z}, \quad \frac{\partial \delta T}{\partial t} = -\widehat{\tau}V_*^T\frac{\partial \delta\varphi}{\partial y}. \tag{4.62a,b,c}$$

(Nonlinear effects can be incorporated by changing $\partial_t \to d/dt$.) Such systems can be used to show that the basic instability mechanism is associated with *negative compressibility*. The argument and an instructive diagram are

---

[a]Because the coefficients of the dispersion relation (4.59) are all real, the complex conjugate of a root is also a root.

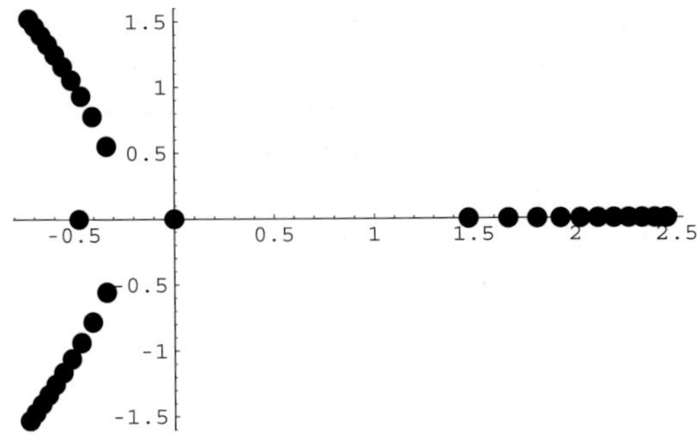

Fig. 4.7   The three branches of the slab ITG–DW dispersion relation. The branch in the right half-plane is the drift wave; the branch in the upper left quadrant is the unstable ITG mode.

given in detail by Cowley et al. (1991). The basic physics of this and other important normal modes are also discussed by Weiland (2000).

Note that if an ITG mode can exist, so can an electron-temperature-gradient-driven (ETG) mode. ETG modes are believed to be important for electron heat transport. Further discussion is given in Lecture 3.

4.1.6.4.   *The effects of magnetic curvature*

In confinement devices, the magnetic field is inevitably curved and inhomogeneous. For example, a purely toroidal magnetic field $\boldsymbol{B} = B\widehat{\boldsymbol{\zeta}}$ has circular field lines in a plane $Z = 0$ orthogonal to the major symmetry axis and thus has a curvature vector $\boldsymbol{\kappa} \doteq \widehat{\boldsymbol{b}} \cdot \boldsymbol{\nabla}\widehat{\boldsymbol{b}}$ of magnitude R^{-1} pointing toward the center (R is the horizontal distance to the center line[a]). A vacuum toroidal field is produced by external poloidal currents flowing in solenoidal windings wrapped on the surface of the confinement chamber; the surface integral of Ampere's law $\boldsymbol{\nabla} \times \boldsymbol{B} = (4\pi/c)\boldsymbol{j}$ then shows that $B \propto R^{-1}$.

In a weakly inhomogeneous magnetic field, single particles exhibit the ∇B *drift* $\boldsymbol{V}_{\nabla B} \doteq (\tfrac{1}{2}v_{\perp}^2/\omega_c)\widehat{\boldsymbol{b}} \times \boldsymbol{\nabla} \ln B$. If the field lines are curved, particles also exhibit the *curvature drift* $\boldsymbol{V}_{\boldsymbol{\kappa}} \doteq (v_{\parallel}^2/\omega_c)\widehat{\boldsymbol{b}} \times \boldsymbol{\kappa}$. Because ω_c is signed, $\boldsymbol{V}_d \doteq \boldsymbol{V}_{\nabla B} + \boldsymbol{V}_{\boldsymbol{\kappa}}$ depends on the sign of charge; if a toroidal $\boldsymbol{B}$ is out of the page, ions drift down and electrons drift up at (mass-independent) rate $2(\rho_s/R)v_{ts}$. Such charge separation produces a vertical electric field, which then causes an outward $\boldsymbol{E} \times \boldsymbol{B}$ drift. This basic effect precludes confinement by a purely toroidal field.[b] It also provides an important driving mechanism

---

[a]In a familiar model of circular magnetic flux surfaces nested concentrically around a magnetic axis of distance R_0, one has $R = R_0 + r\cos\theta$, where r and θ are polar coordinates in the poloidal cross section with origin at the magnetic axis.

[b]In the tokamak, this problem is solved by driving toroidal currents to produce a poloidal

for both macroscopic and microscopic instabilities.

The particle drift shows up as an additional term in the GKE (4.16): $\partial_t F + \boldsymbol{\nabla}\cdot(\boldsymbol{V}_d F) + \cdots = 0$. When this equation is integrated over velocity, one finds a mass-independent contribution to the gyrocenter continuity equation of the form $\partial_t N_s + 2(c/q_s B)\widehat{\boldsymbol{b}}\times\boldsymbol{\kappa}\cdot\boldsymbol{\nabla}P_s$, where $P_s \doteq (nT)_s$ is the pressure of species s. When the ion and electron gyrocenter charge densities are added to form the plasma vorticity according to the gyrokinetic Poisson equation (4.13), the drift terms combine to give a contribution driven by the total pressure gradient:

$$\partial_t \varpi \propto 2\widehat{\boldsymbol{b}}\times\boldsymbol{\kappa}\cdot\boldsymbol{\nabla}P + \cdots . \tag{4.63}$$

This effect is frequently called *curvature drive*, although it includes (through the factor of 2) the $\boldsymbol{\nabla}B$ drift as well.

Curvature drive is responsible for many significant effects in modern tokamaks, including the so-called ideal and resistive *ballooning modes*.[a] One important example relevant to microturbulence is the *curvature-driven ITG mode*, which is robustly unstable even for $k_\parallel \to 0$. In the limit of large T_i, the basic equations describe the evolution of ion gyrocenter density (due to the particle drifts associated with ion temperature perturbations) coupled with the linear advection equation for ion temperature:

$$\partial_t \delta N_i = -V_d \partial_y \delta T_i, \quad \partial_t \delta T_i = -V_*^T \partial_y \delta\varphi, \tag{4.64a,b}$$

where $V_d \doteq 2\rho_s/R$. This describes a nonpropagating instability with growth rate $\gamma = (\omega_d \omega_*^T)^{1/2}/(1 + k_\perp^2 \rho_s^2)$, where $\omega_d \doteq k_y V_d$. The physical mechanism is described in Fig. 4.8, and further discussion can be found in Ottaviani et al. (1997). It is this curvature-driven mode that is important in practice.

The particle drifts are also responsible for the so-called *geodesic acoustic mode* (GAM) (see Sec. 4.3.5.) and figure prominently in the formation of quasicoherent objects known as *blobs* that form near the plasma edge, as I will discuss in Lecture 4.

4.1.7. *The transition to plasma turbulence*

In the next several lectures, we will consider forced, dissipative, steady states of plasma turbulence, for which statistical descriptions are appropriate. But what about the *transition* to that turbulence?

Over the history of the fusion program, much discussion has focused on states of fully developed plasma microturbulence, with relatively little attention paid to transition. There are several elementary reasons for

---

field; the total field then connects regions of good curvature (pointing away from the plasma) and bad curvature (pointing toward the plasma).

[a]Introductory discussions and physical mechanisms can be found in Miyamoto (1978).

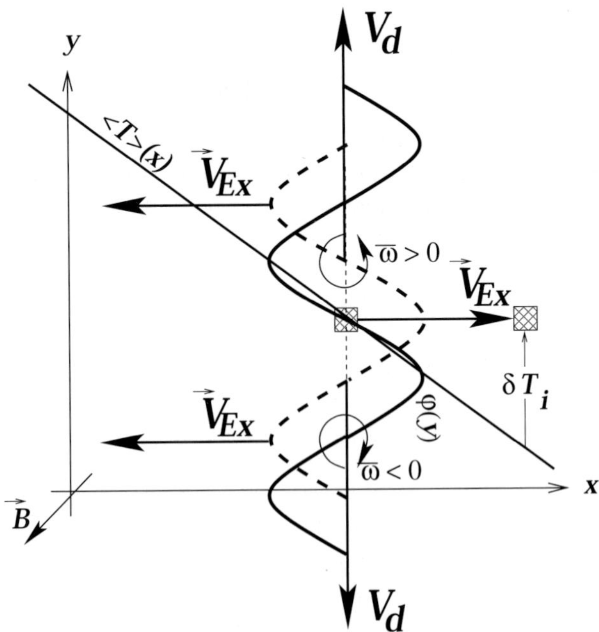

Fig. 4.8   Physical mechanism of the curvature-driven ITG instability.   Assume a background ion temperature gradient in the $-x$ direction with scale length $L_T \doteq -d\ln\langle T_i\rangle/dx$, a magnetic field $\boldsymbol{B}$ in the z direction, and a sinusoidal potential $\varphi(y) \propto \cos(k_y y)$. $\boldsymbol{E} \times \boldsymbol{B}$ advection creates an ion temperature (pressure) fluctuation δT_i. In regions of positive (negative) δT_i, ions drift downward (upward). The charge imbalance creates an electric field $(\delta\varphi - \nabla_\perp^2 \delta\varphi = \delta N_i)$ that reinforces the original advection.

this. First, starting up a large tokamak discharge is a rather violent, only semi-controlled process. When a quasi-steady state is finally established, the profiles of density, temperature, flow, magnetic field, *etc.* are controlled by considerations of macroscopic MHD stability, magnetic geometry, and transport effects in the presence of external sources. Because all variables are coupled, it may not be possible to exercise fine control over, say, all details of the global temperature profile without changing some other variable in an undesirable way.

Second, microturbulence by definition involves fluctuations of very short scale with respect to the macroscopic scale lengths or system dimensions. Therefore, the Fourier mode spacing $\delta k \doteq 2\pi/L$, where L is a periodicity length such as the minor circumference, is very small (i.e., the mode number is large). For example, let fluctuations vary poloidally as $\exp(im\theta) = \exp(ik_y y)$, with $k_y \doteq m/r$. The poloidal circumference is

$L = 2\pi r$, and one has

$$\frac{\delta k}{k} = \frac{2\pi}{kL} = \frac{2\pi}{(m/r)(2\pi r)} = \frac{1}{m}. \tag{4.65}$$

Now for microfluctuations with $k_y\rho_s = O(1)$, one has $(m/r)\rho_s = O(1)$ or $m = O(r/\rho_s) \gg 1$. Thus, if mode m is unstable the neighboring mode at $m + 1$ is also likely to be unstable and one expects an essentially continuous *band* of unstable modes.[a] This complicates discussion of discrete bifurcations as an order parameter such as the temperature gradient is tuned.

This difficulty is a well-known property of spatially extended systems (Newell and Whitehead, 1969; Manneville, 1990); it can be sensibly handled with the aid of multiple-scale analysis. For now, it is merely important to understand why careful studies of the transition to turbulence are typically not done on large tokamaks. However, such studies have been performed on small laboratory devices such as, early on, the so-called Q machines (Q stands for quiescent). Parameters in such small, weak-field, low-magnetic-shear devices are such that macroscopic normal modes can be excited and studied in a controlled manner. It is natural to analyze systems with small numbers of degrees of freedom using dynamical systems techniques (Guckenheimer and Holmes, 1983).

Because the small machines are generally of relatively low temperature, they are usually collision-dominated. That is unlike the highly collisionless hot core of a fusion reactor, but may be relevant to the cold edge of such devices. Most of the bifurcation analyses have been done on such collisional systems. Because variants of the drift wave are generally involved, the typical scenario is for a Hopf bifurcation to occur at the point of linear instability of the most unstable mode.[b] What happens next is somewhat model-dependent, but one would not be surprised to observe some variant of the scenario of Ruelle and Takens (1971). This was nicely verified in the experiments of Klinger et al. (1997), who studied the transition to drift-wave chaos and turbulence in a laboratory triple-plasma device.

Although such fine details may not be highly relevant to large, fusion-scale devices, the transition to turbulence in those devices is still of interest, particular if it suggests methods to suppress the turbulence altogether. Large-scale computer simulations are very useful in this context, and it

---

[a]The details of this discussion are correct for fluctuations in systems with cylindrical symmetry, such as many of the small laboratory devices. In toroidal geometry, azimuthal symmetry is broken and m is no longer a good "quantum number." Poloidal harmonics are coupled even linearly by the poloidal variation of the toroidal magnetic-field strength, $B = B_0/(1 + \epsilon\cos\theta)$, where $\epsilon \doteq r/R^{(0)} < 1$ is the inverse aspect ratio and θ is the poloidal angle. However, the basic qualitative points about mode spacing remain valid.

[b]An early and important calculation was by Hinton and Horton (1971), although the word "bifurcation" does not appear in that paper.

is now becoming possible to perform reasonably realistic simulations of collisionless confined plasmas. That subject is too extensive to be discussed here, but I will mention one important example of the kinds of results that can be obtained. For a certain set of standard parameters (called the "Cyclone base case"), Dimits et al. (2000) summarized various numerical calculations of the ion heat flux as a function of temperature gradient κ_T for collisionless ITG turbulence. Let κ_c denote the threshold for linear ITG instability. Dimits *et al.* found that ITG fluctuations were suppressed for an order-unity range $\kappa_c < \kappa_T < \kappa_*$, unlike a standard bifurcation scenario. The difference $\kappa_* - \kappa_c$ is called the *Dimits shift*. It is due to the excitation of poloidally symmetric *zonal flows* that exert a strongly stabilizing shearing effect on the ITG modes. The calculation of the collisionless Dimits shift presents a difficult problem in nonlinear plasma dynamics. I will return to it in Lecture 3, which focuses on zonal flows.

Although the transition to turbulence is quite interesting and I have here merely scratched the surface, of more practical importance are the substantial levels of quasi-steady-state (micro)turbulence experienced by plasma discharges typical of contemporary confinement experiments. Beginning with the pioneering work of Taylor (1915) [see also Taylor (1921)], statistical methods have been used to describe such turbulence. In the next lecture, I will turn to some aspects of statistical closure theory for plasmas.

4.2. LECTURE 2 – Statistical Closures and Plasma Turbulence

Nowhere is the dichotomy between the theoretical and experimental approaches to understanding the nonlinear, turbulent behavior of plasmas more evident than in the subject of statistical closure theory. Theorists are perhaps inexorably drawn to analytical closure formalism (Orszag, 1977) because it presents very difficult challenges in applied mathematics, even for the most physically idealized models. Experimentalists are understandably skeptical of the relevance of very idealized models; they are interested in scaling trends (which as we saw in Lecture 1 can follow from basic dimensional and scaling analysis), the direct identification of physical mechanisms, qualitative spectral features, *etc.* Hopefully the perspectives of both camps influence the others in a useful way.[a]

In any event, in this and the next lecture I will provide an overview of some applications of statistical closures to plasma turbulence. Because the

---

[a]One example where the two camps meet is in the theory and use of the *bicoherence* (the square of a normalized triplet correlation function) for diagnosing important mode couplings (Itoh et al., 2005b; White et al., 2006, and references therein).

foundations of statistical plasma turbulence theory, a discipline approximately 40 years old, were recently reviewed by Krommes (2002), I will discuss here only a few highlights. Although logically one can discuss the present state of plasma closure theory without much reference to its historical development, pedagogically it is very instructive to review the thinking and methodology of the early workers. Their research led to nomenclature and approximate techniques that are still in use today, and both the deficiencies and successes of the various formalisms shed important light on the difficulties that are to be surmounted. Specifically, I will remark on

- quasilinear theory (the first paradigm for the nonlinear consequences of wave–particle interaction);
- weak-turbulence theory for wave–wave interactions (a systematic description of mode–mode coupling when the spectrum is composed of weakly damped or growing waves);
- resonance-broadening theory (the attempt of Dupree and others to formulate a description of strong plasma turbulence);
- the Martin–Siggia–Rose formalism for renormalized classical statistical dynamics;
- the direct-interaction approximation (the thoroughly studied formulation by Kraichnan of a self-consistent and realizable statistical closure);
- and certain Markovian closures, including the first successful quantitative comparison between direct numerical simulations of nontrivial plasma dynamics and statistical closure theory.

These topics do not begin to exhaust the myriad of practical applications that pay homage to statistical methods. But one must start somewhere.

4.2.0.1. *Background on renormalization and statistical closures*

All of the above topics involve *renormalization* in one way or another. Nice introductions to renormalization can be found in the lectures of Prof. McComb (Chap. 2) and, in more detail, in the book by McComb (2004); see also the review by Krommes (2002). However, for continuity I offer a few remarks here as well.

In quantum field theory (Zinn-Justin, 1996; Amit, 2005), *renormalization* implies the replacement of an unobservable "bare" quantity (mass or charge) that may be formally infinite by a physically observable one that is finite. Quantum renormalization procedures are necessary in order to tame singularities that arise because of a lack of knowledge of the detailed physical processes that operate at very large momenta (very short scales). That is not the situation for the classical problems, such as the NSE, in which we will be interested. Such equations are assumed to describe the physics exactly at all scales, and the presence of dissipation provides a natural large-k cutoff. Renormalization is then more easily understood as a procedure for taking into account certain effects from all orders in perturbation theory in cases in which low-order perturbation theory is physically misleading or ill-defined. For example, consider the diffusion Green's function, which in $\boldsymbol{k}$ space is $G_{\boldsymbol{k}}(\tau) = H(\tau)\exp(-k^2 D\tau)$. This is a bounded function. However, its small-argument Taylor expansion, $G_{\boldsymbol{k}}(\tau) = H(\tau)[1 - k^2 D\tau + \frac{1}{2}(k^2 D\tau)^2 + \cdots]$, is secular at any finite order of truncation. In cases like this, regular perturbation theory applied to the observable itself is inadequate. If the rule for generating the nth-order term in such an expansion is sufficiently well understood, renormalization techniques can be used to sum the series. Alternatively, a properly renormalized theory need never generate perturbative expansions in the first place. Examples include Dupree's resonance-broadening theory (Sec. 4.2.3.) and Kraichnan's direct-interaction approximation (Sec. 4.2.4.2.), both of which are subsumed by the MSR formalism (Sec. 4.2.4.1.).

It is possible to establish connections between (i) the classical procedures involving, for example, diagrammatic resummation, and (ii) the quantum-mechanical introduction of a large-momentum cutoff that regularizes perturbation theory and on which the physical observables are ultimately shown to not depend. One elementary but instructive example, which in part illustrates how *anomalous exponents* (Barenblatt, 1996) arise, is discussed by Krommes (2002), Sec. 6.1.2. Another is the calculation of static Debye shielding, already cited by Prof. McComb in Chap. 2 as a physically important example of renormalization. I shall discuss that problem in the next few paragraphs.

The static Debye shielding problem is to solve Poisson's equation

$$-(\nabla^2 - k_D^2)\phi(\boldsymbol{x}) = 4\pi\delta(\boldsymbol{x}) \tag{4.66}$$

(the Debye wave number k_D is defined in footnote f, p. 121). With $r \doteq |\boldsymbol{x}|$, the well-known solution is $\phi(\boldsymbol{x}) = r^{-1}\exp(-k_D r)$, which results from evaluating the inverse Fourier transform[a]

$$\phi(\boldsymbol{x}) = \int \frac{d\boldsymbol{k}}{(2\pi)^3}\, e^{i\boldsymbol{k}\cdot\boldsymbol{x}}\left(\frac{4\pi}{k^2 + k_D^2}\right) = \frac{1}{r}\left(\frac{2}{\pi}\,\mathrm{Im}\int_0^\infty d\bar{k}\,\frac{\bar{k}e^{i\bar{k}}}{\bar{k}^2 + \delta^2}\right), \tag{4.67}$$

where $\delta \doteq k_D r$ is the unique dimensionless parameter of the problem. Suppose instead that one attempts to solve the problem by regular perturbation theory, treating k_D as a small parameter. That is not well motivated because δ depends on scale, but if one tries anyway, one must iterate

$$G^{(0)-1}\phi = \delta(\boldsymbol{x}) - \Sigma\phi, \tag{4.68}$$

---

[a]The last parenthesized integral in Eq. (4.67) can be performed by contour integration by writing it as $\pi^{-1}\,\mathrm{Im}\int_{-\infty}^\infty d\bar{k}\,\bar{k}e^{i\bar{k}}/(\bar{k}^2 + \delta^2)$ and evaluating the single residue at $\bar{k} = i\delta$.

where $\Sigma \doteq k_D^2/4\pi$ and the bare Green's function (the potential of a unit point charge) obeys[a] $-(4\pi)^{-1}\nabla^2 G^{(0)}(\boldsymbol{x};\boldsymbol{x}') = \delta(\boldsymbol{x} - \boldsymbol{x}')$ or $G^{(0)}(\boldsymbol{x};\boldsymbol{x}') = G^{(0)}(\boldsymbol{x} - \boldsymbol{x}') = |\boldsymbol{x} - \boldsymbol{x}'|^{-1}$. At lowest order, one finds $\phi^{(0)}(\boldsymbol{x}) = G^{(0)}(\boldsymbol{x}) = r^{-1}$. At first order, $\phi^{(1)}(\boldsymbol{x}) = -\Sigma \int d\overline{\boldsymbol{x}}\, G^{(0)}(\boldsymbol{x} - \overline{\boldsymbol{x}})G^{(0)}(\overline{\boldsymbol{x}})$. It can be easily verified that this convolution is *divergent* at large distances, so one cannot proceed without introducing a large-distance cutoff $r_{\max}$ or small-k cutoff $k_{\min}$. With a finite $r_{\max}$ or nonzero $k_{\min}$, all terms in the perturbation expansion are finite, and one finds formally $\phi(\boldsymbol{x}) = G(\boldsymbol{x})$, where $G = G^{(0)} \star \sum_{n=0}^{\infty}(-\Sigma G^{(0)}\star)^n$ and $\star$ denotes convolution. One can sum the geometric series to give $G = G^{(0)} \star (1 + \Sigma G^{(0)})^{-1} = (G^{(0)-1} + \Sigma)^{-1}$. *After* the summation, the dependence on the cutoff is benign and can be ignored. For example, with a $\boldsymbol{k}$-space cutoff, one has $G^{(0)}(\boldsymbol{x}) = A(k_{\min})/r$, where $A(k_{\min}) \doteq (2/\pi) \int_{k_{\min}}^{\infty} d\overline{k}\, \sin(\overline{k})/\overline{k} = (2/\pi)\,\mathrm{Si}(k_{\min})$ is an *analytic* function of $k_{\min}$ that smoothly approaches 1 as $k_{\min} \to 0$.

In this calculation, renormalization involved the introduction of a cutoff, summation of more and more complicated physical effects through all orders, and the ultimate replacement of the bare Green's function $G^{(0)}$ by the renormalized function G that obeys

$$G^{-1} = G^{(0)-1} + \Sigma. \tag{4.69}$$

Here Σ generalizes the renormalized mass of quantum field theory; Eq. (4.69) is an example of a *Dyson equation* (Dyson, 1949; Krommes, 2002).[b]

Now the result (4.69) is utterly trivial in this example because it merely restates the original problem (4.66).[c] However, in turbulence problems the value of Σ is not given but must be calculated from statistical averages[d] of complicated nonlinear terms.[e] If there is a natural physical cutoff, perturbative evaluations of the effective Σ will be finite, but the physics may

---

[a]If $\widehat{C} = \widehat{A}\widehat{B}$, where $\widehat{A}$, $\widehat{B}$, and $\widehat{C}$ are linear operators represented in $\boldsymbol{x}$ space by the kernels $A(\boldsymbol{x};\boldsymbol{x}')$, etc., then $C(\boldsymbol{x};\boldsymbol{x}') = \int d\overline{\boldsymbol{x}}\, A(\boldsymbol{x};\overline{\boldsymbol{x}})B(\overline{\boldsymbol{x}};\boldsymbol{x}')$. The inverse of the operator $\widehat{A}$ obeys $\int d\overline{\boldsymbol{x}}\, A^{-1}(\boldsymbol{x};\overline{\boldsymbol{x}})A(\overline{\boldsymbol{x}};\boldsymbol{x}') = \delta(\boldsymbol{x} - \boldsymbol{x}')$. Thus $G^{(0)-1}(\boldsymbol{x};\boldsymbol{x}') = -(4\pi)^{-1}\nabla^2\delta(\boldsymbol{x} - \boldsymbol{x}')$; in $\boldsymbol{k}$ space, $G_{\boldsymbol{k}}^{(0)-1} = k^2/4\pi$.

[b]The form of the Dyson equation shows that it is best to approximate, by perturbation theory or otherwise, the *inverse* of the Green's function. See the discussion by Martin et al. (1973) and related remarks by Krommes (2002).

[c]However, it should be noted that diagrammatic summation techniques were also used in the early discussions by Balescu (1963) of the linearized Vlasov equation, which arises in the calculation of the dynamic screening of a moving test charge. Nowadays one would calculate the Vlasov Green's function directly (Ichimaru, 1973) by Fourier transformation techniques that generalize the static calculation, but the diagrammatic solution is instructive if only to show that it should be avoided if at all possible.

[d]Indeed, the static shielding problem itself is ultimately derived from statistical considerations of the many-body plasma physics.

[e]A very instructive model that has been used to demonstrate statistical renormalization techniques, both successful and unsuccessful, is the *stochastic oscillator* of Kubo (1959,

not be represented properly unless terms are summed through all orders. Of course, explicit summations may not be possible in practice, so alternate, nonperturbative approaches are desirable. One such is the procedure of Martin et al. (1973), to be discussed briefly in Sec. 4.2.4.1..

Renormalizations effect *statistical closures*, which are closed equations for a finite (hopefully small) subset of statistical observables. In the remainder of this lecture I will discuss various examples of plasma closures. Since plasmas are collections of charged particles, it was inevitable that an early focus on particle (as opposed to fluid) dynamics developed. The behavior of the plasma as a dielectric medium, with its many modes of collective oscillation, was appreciated early on. It was then natural that the consequences of wave–particle interaction would be pursued. Now the many aspects of the interaction of charged particles with waves merit an entirely separate treatise; I can only scratch the surface here. Specifically, I will in the next section discuss basic aspects of the quasilinear description of stochasticity. That serves as the simplest illustration of a statistical closure for plasmas (which includes a quite subtle renormalization), highlights connections to modern understanding of nonlinear dynamics, and motivates Dupree's development of resonance-broadening theory (RBT), the first statistical theory that addressed issues of strong plasma turbulence. I describe the RBT in Sec. 4.2.3..

4.2.1. *Quasilinear theory*

The venerable *quasilinear theory* (QLT) of the wave–particle interaction was first described by Vedenov et al. (1962) and Drummond and Pines (1962). In essence, the quasilinear mechanism for the saturation of a kinetic instability driven by inverse Landau damping exploits velocity-space diffusion to flatten the background distribution function in the vicinity of the linear resonance, thereby turning off the source of free energy. This mechanism can be generalized to include spatial diffusion as well, so in principle might figure in the saturation of spatial-profile-driven instabilities such as the universal drift instability or ITG modes. However, since macroscopic spatial gradients are inevitable in magnetically confined plasmas (they are maintained by sources or boundary conditions), quasilinear flattening cannot be effective in removing such gradients; one must look for other saturation mechanisms, notably nonlinear mode coupling. Nevertheless, study of the original quasilinear theory is very instructive, and

---

1962b) and Kraichnan (1961), $\partial_t \psi + i\widetilde{\omega}\psi = 0$, where $\widetilde{\omega}(t)$ is a Gaussian time series. For details, see the seminal paper of Kraichnan (1961) and the review by Krommes (2002). An interesting recent use of that model to clarify subtle issues in MHD turbulence is by Lithwick and Goldreich (2003).

the basic physics (stochasticity of particle motion) figures importantly in several practical applications such as current drive (Fisch, 1987).

4.2.1.1. *Heuristic estimate of the quasilinear velocity-space diffusion coefficient*

The basic ingredients of the QLT are as follows. For simplicity, I will work in one dimension in this and the next subsection. We are given a linear dispersion relation $\omega_k = \Omega_k + i\gamma_k$; in order to speak about well-defined eigenmodes, it is assumed that $|\gamma_k/\Omega_k| \ll 1$. The modes are assumed to be unstable, $\gamma_k > 0$, over a certain wave-number range Δk. Random noise will then excite the unstable eigenmodes, which will grow to nontrivial levels within Δk. In elementary QLT, it is *assumed* that the resulting electric field induces particle stochasticity, i.e., velocity-space diffusion. The resulting diffusion coefficient can be estimated, using straightforward random-walk arguments,[a] from

$$D_v(v) = [(q/m)^2 \langle \delta E^2 \rangle]\tau_{\mathrm{ac}}(v), \tag{4.70}$$

where the bracketed term is the mean-square acceleration experienced during the *autocorrelation time* $\tau_{\mathrm{ac}}(v)$, which is interpreted as the time a particle moving with velocity v "stays in sync" with the wave packet. Now that wave packet moves with the group velocity $v_{\mathrm{gr}} \doteq \partial\Omega_k/\partial k$. A spectral spread Δk corresponds to a wave-packet half-width $\Delta x = \pi/\Delta k$. After a Galilean frame change to the velocity of the wave packet, one then estimates $\tau_{\mathrm{ac}} \approx \Delta x/|v - v_{\mathrm{gr}}|$, or

$$\tau_{\mathrm{ac}}(v) \approx \pi/\Delta k|v - v_{\mathrm{gr}}|. \tag{4.71}$$

4.2.1.2. *Quasilinear diffusion from random walk*

It is instructive to derive the results (4.70) and (4.71) more rigorously. I continue with the diffusion ansatz. A basic result from random-walk theory is that the diffusion coefficient of some random variable z is the time integral

---

[a]A basic result is that if in 1D a step Δx is taken with equal probabilities to the left and right in a characteristic time Δt (the *Drunkard's Walk*), then on the average position will diffuse according to $\langle x^2 \rangle = 2Dt$, where $D = \frac{1}{2}\Delta x^2/\Delta t = \frac{1}{2}\Delta v^2\Delta t$. More generally, $D = \int_0^\infty d\tau \langle \delta v(\tau)\delta v(0) \rangle$, where δ denotes the fluctuation from the mean and $v(\tau)$ denotes the Lagrangian time dependence $v(x(\tau), \tau)$ (Taylor, 1921). Variables other than position can diffuse as well. In general, if a derivative of a variable z is driven by Gaussian white noise f, $\dot{z} = f(t)$ with $\langle \delta f(t)\delta f(t') \rangle = 2D_z\delta(t - t')$, then z will diffuse with diffusion coefficient D_z (see further, more explicit discussion in Sec. 4.2.3.1.). This is the case for an unmagnetized charged particle, which obeys $\dot{v} = qE(t)/m$. To the extent that $E(x(t), t)$ can be idealized as a Gaussian process, diffusion will occur in v space.

of the Lagrangian correlation function of $\delta\dot{z}$:

$$D_z = \int_0^\infty d\tau \, \langle \delta\dot{z}(\tau)\delta\dot{z}(0)\rangle, \qquad (4.72)$$

where $A(t)$ means that the quantity $A(x,t)$ is to be evaluated at the random position: $A(t) = A(x(t),t)$; this is called *Lagrangian time dependence*. To discuss velocity diffusion, one takes $z = v$, the time derivative of which is the acceleration due to the electric field: $\dot{v} = (q/m)E(x(t),t)$. Thus

$$D_v = \left(\frac{q}{m}\right)^2 \int_0^\infty d\tau \, \langle \delta E(x(\tau),\tau)\delta E(x(0),0)\rangle. \qquad (4.73)$$

The basic dimensional scaling (4.70) is already apparent in formula (4.73). To obtain more detail, let us introduce the Fourier decomposition $\delta E(x,t) = \sum_k \delta E_k(t)\exp(ikx - i\Omega_k t)$; slow growth or damping is included in the time dependence of the Fourier amplitudes. Thus

$$D_v = \left(\frac{q}{m}\right)^2 \int_0^\infty d\tau \, \left\langle \sum_{k,k'} \delta E_k \delta E_{k'} e^{ikx(\tau)-i\Omega_k\tau} e^{ik'x(0)} \right\rangle. \qquad (4.74)$$

Over what does one average? In the most general instance, each of δE_k, $x(\tau)$, and $x(0)$ could be random. Here I will treat the case in which the Fourier amplitudes are given; that is, we study the passive *stochastic acceleration* problem. With regard to position, one may change variables and work with[a] $x(0)$ and $\Delta x(\tau) \doteq x(\tau) - x(0)$. Here $\Delta x(\tau)$ measures the distance traveled in time τ: $\Delta x = v\tau + \int_0^\tau d\tau' \, \delta a(\tau')(\tau - \tau')$. I will deal with the influence of the random acceleration when we study the RBT in Sec. 4.2.3.. If one ignores it, i.e., if one assumes that the trajectories are linear, then $\Delta x(\tau) \approx v\tau$ is no longer random (since we are studying diffusion conditional on the current value of velocity). However, statistical homogeneity dictates that $x(0)$ must be distributed uniformly. Thus

$$D_v = \left(\frac{q}{m}\right)^2 \int_0^\infty d\tau \sum_{k,k'} \delta E_k \delta E_{k'} e^{i(kv-\Omega_k)\tau} \langle e^{i(k+k')x(0)}\rangle \qquad (4.75a)$$

$$= \left(\frac{q}{m}\right)^2 \int_0^\infty d\tau \sum_k |\delta E_k|^2 e^{i(kv-\Omega_k)\tau}, \qquad (4.75b)$$

where the result $\langle e^{i(k+k')x(0)}\rangle = \delta_{k,-k'}$ was used. One may now attempt to perform the time integral; however, as it stands, that integral is not convergent as $\tau \to \infty$. That is a consequence of the neglect of nonlinear corrections

---

[a]Formally, it is frequently more convenient to work with the formula $D_z = \int_0^\infty d\tau \, \langle \delta\dot{z}(0)\delta\dot{z}(-\tau)\rangle$ [equivalent to formula (4.72) for stationary turbulence] and to consider propagation backwards in time; see the discussion by Krommes (2002), Appendix E. However, that is unnecessary for the present introductory discussion.

to the free-streaming trajectory, which will be studied in Sec. 4.2.3. Here, one may simply define the integral by adding the convergence factor $e^{-\epsilon\tau}$. Then

$$\int_0^\infty d\tau\, e^{i(kv-\Omega_k)\tau} e^{-\epsilon\tau} = \frac{1}{-i(kv - \Omega_k + i\epsilon)} \equiv 2\pi\delta_+(kv - \Omega_k) \quad (4.76a)$$

$$= \pi\delta(kv - \Omega_k) + iP\left(\frac{1}{kv - \Omega_k}\right). \quad (4.76b)$$

Symmetry dictates that only the real part contributes to (4.75b). Thus

$$D_v = \pi\left(\frac{q}{m}\right)^2 \sum_k |\delta E_k|^2 \delta(kv - \Omega_k). \quad (4.77)$$

Although we are making progress, the result (4.77) is somewhat ill-behaved. Although one expects that $D_v(v)$ should be a smooth function of v, Eq. (4.77) presents the velocity-space diffusion as a superposition of spikes centered on the phase velocities of the waves. This unphysical result is also a consequence of the neglect of nonlinear effects, which we will rectify shortly. Meanwhile, one way out is to assume that the mode spacing $\delta k \doteq 2\pi/L$ is small enough that the discrete wave-number summation may be replaced by an integral: $\sum_k \to \int_{-\infty}^\infty dk/\delta k$. If one also defines the spectral density by $\mathcal{E}(k) \doteq L|\delta E_k|^2$, then

$$D_v = \pi\left(\frac{q}{m}\right)^2 \int_{-\infty}^\infty \frac{dk}{2\pi}\, \mathcal{E}(k)\delta(kv - \Omega_k). \quad (4.78)$$

This is the form in which the quasilinear diffusion coefficient is frequently presented. One can, however, go further by performing the integral. Now

$$\delta(kv - \Omega_k) = \frac{\delta(k - k_0)}{|v - \partial\Omega_k/\partial k|_{k_0}}, \quad (4.79)$$

where $k_0(v)$ is the resonant wave number for given v: $k_0 v = \Omega_{k_0}$. A reality condition dictates that this equation has two solutions $\pm|k_0|$. Then

$$D_v = \frac{1}{2}\left(\frac{q}{m}\right)^2 \sum_{k_0} \mathcal{E}(k_0)\frac{1}{|v - v_{\mathrm{gr}}|}. \quad (4.80)$$

To demonstrate the essential equivalence of this result to Eq. (4.70), I rearrange formula (4.80) as

$$D_v = \left(\frac{q}{m}\right)^2 \left(\frac{\sum_{k_0} \Delta k \mathcal{E}(k_0)}{2\pi}\right)\left(\frac{\pi}{\Delta k|v - v_{\mathrm{gr}}|}\right) \quad (4.81a)$$

$$\approx (q/m)^2 \delta E^2 \tau_{\mathrm{ac}}(v). \quad (4.81b)$$

We have recovered the heuristic estimate (4.70)!

Basic QLT was already discussed in 1962,[a] more than ten years before modern ideas about nonlinear dynamics were widely known in the plasma-physics community. Russian research was key. The important publications by Zaslavskii and Sagdeev (1967) and Chirikov (1969); the influential book by Sagdeev and Galeev (1969), which contained extensive discussion of QLT; and the later paper of Zaslavskii and Chirikov (1972) were forerunners of the plasma applications of stochasticity in the middle 1970's by Smith and Kaufman (1975) and Karney (1978). Nowadays, we understand (Lichtenberg and Lieberman, 1992) that velocity-space diffusion is a consequence of the overlap of phase-space islands in a Hamiltonian system (the *Chirikov criterion*), which leads to the destruction of Kolmogorov–Arnold–Moser (KAM) surfaces and the onset of trajectory stochasticity (extreme sensitivity to initial conditions). Such stochasticity is intrinsically a nonlinear effect that cannot be treated perturbatively. In Sec. 4.2.3., we will see how to approximately take that nonlinearity into account with a simple renormalization, and we will pursue the connection between the quasilinear diffusion formula and stochastic Hamiltonian dynamics.

4.2.1.3. *Quasilinear theory as a statistical closure*

Let us now consider QLT from the point of view of standard statistical closure theory. I begin with Eq. (4.5) for the one-particle distribution function f, specialized to a spatially homogeneous, unmagnetized medium:

$$\partial_t f = -(q/m)\partial_{\boldsymbol{v}} \cdot \langle \delta \boldsymbol{E} \delta N \rangle. \tag{4.82}$$

Without approximation, the fluctuations obey

$$\partial_t \delta N + \boldsymbol{v} \cdot \boldsymbol{\nabla} \delta N + (q/m)\boldsymbol{E} \cdot \partial_{\boldsymbol{v}} \delta N + (q/m)\delta \boldsymbol{E} \cdot \partial_{\boldsymbol{v}} f$$
$$+ (q/m)\partial_{\boldsymbol{v}} \cdot (\delta \boldsymbol{E} \delta N - \langle \delta \boldsymbol{E} \delta N \rangle) = 0. \tag{4.83}$$

The mean field $\boldsymbol{E}$ can be taken to vanish. The simplest closure approximation is to drop the nonlinear terms altogether; that is the "linear" in "quasilinear". The quadratic term on the right-hand side of Eq. (4.82) is the "quasi". Thus

$$\partial_t \delta N + \boldsymbol{v} \cdot \boldsymbol{\nabla} \delta N \approx -(q/m)\delta \boldsymbol{E} \cdot \partial_{\boldsymbol{v}} f. \tag{4.84}$$

Equations (4.82) and (4.84) form a closed system, since $\delta \boldsymbol{E}$ is linearly related to δN *via* Poisson's equation. One may solve for δN by introducing the zeroth-order Green's function $g^{(0)}$ for particle motion, which obeys

$$(\partial_t + \boldsymbol{v} \cdot \boldsymbol{\nabla})g^{(0)}(z,t;z',t') = \delta(t - t')\delta(z - z'), \tag{4.85}$$

[a]It is amazing that only a few years later Kadomtsev (1965) discussed not only QLT and weak-turbulence theory but also the sophisticated direct-interaction approximation (which he called the *weak-coupling approximation*).

where $z \equiv \{x, v\}$. The solution of Eq. (4.85) is

$$g^{(0)}(z, t; z', t') = H(t - t')\delta(x - x' - v'(t - t'))\delta(v - v'). \qquad (4.86)$$

Then

$$\delta N(z, t) = \int d\bar{z}\, g^{(0)}(z, t; \bar{z}, 0)\delta N(\bar{z}, 0)$$

$$- \int_0^t d\bar{t} \int d\bar{z}\, g^{(0)}(z, t; \bar{z}, \bar{t}) \left(\frac{q}{m}\right) \delta \boldsymbol{E}(\bar{x}, \bar{t}) \cdot \frac{\partial f(\bar{v}, \bar{t})}{\partial \bar{v}}. \qquad (4.87)$$

For collisionless theory ($\epsilon_p \to 0$), one assumes that the initial conditions are smooth[a]; those contribute an initial transient that will phase-mix away and can be ignored. Thus the right-hand side of Eq. (4.82) becomes

$$\left(\frac{q}{m}\right)^2 \frac{\partial}{\partial \boldsymbol{v}} \cdot \int_0^t d\bar{t} \int d\bar{z}\, \langle \delta \boldsymbol{E}(x, t) g^{(0)}(z, t; \bar{z}, \bar{t})\delta \boldsymbol{E}(\bar{x}, \bar{t})\rangle \cdot \frac{\partial f(\bar{v}, \bar{t})}{\partial \bar{v}}. \qquad (4.88)$$

Because $g^{(0)}$ is not random, it can be removed from the angular brackets if desired. The non-Markovian form (4.88) is sometimes called the *Bourret approximation* (Bourret, 1962). With the aid of the specific result for $g^{(0)}$, formula (4.88) reduces to

$$\left(\frac{q}{m}\right)^2 \frac{\partial}{\partial \boldsymbol{v}} \cdot \int_0^t d\tau\, \langle \delta \boldsymbol{E}(x, t)\delta \boldsymbol{E}(x - v\tau, t - \tau)\rangle \cdot \frac{\partial f(v, t - \tau)}{\partial v}. \qquad (4.89)$$

We recognize in the integrand the lowest-order (streaming) approximation to the Lagrangian correlation function $\mathsf{C}(\tau)$ of the electric field.

So far the formalism is still time-reversible. This is in accord with the general discussion of Orszag (1977), who stressed that the infinite hierarchy of cumulants, of which Eq. (4.82) is the first member, is formally time-reversible. One must therefore pay close attention, for at the next step irreversibility will be introduced. When dealing with issues involving weak spatial inhomogeneity or temporal nonstationarity, it is conventional to write two-point Eulerian correlation functions in the form $C(\boldsymbol{x}, t, \boldsymbol{x}', t') = C(\boldsymbol{\rho}, \tau \mid \boldsymbol{X}, T)$, where $\boldsymbol{\rho} \doteq \boldsymbol{x} - \boldsymbol{x}'$, $\boldsymbol{X} \doteq \frac{1}{2}(\boldsymbol{x} + \boldsymbol{x}')$, $\tau \doteq t - t'$, and $T \doteq \frac{1}{2}(t + t')$. Thus the kernel of formula (4.89) involves $\mathsf{C}(\boldsymbol{v}\tau, \tau \mid \boldsymbol{x} - \frac{1}{2}\boldsymbol{v}\tau, t - \frac{1}{2}\tau) \equiv \mathsf{C}(\tau \mid t - \frac{1}{2}\tau)$. We are assuming homogeneous statistics, so C does not depend on its $\boldsymbol{X}$ argument; it does, however, retain T dependence. Let us now assume that as a function of τ $\mathsf{C}(\tau \mid t)$ decays on a characteristic autocorrelation time τ_{ac}. [This will turn out to be the same $\tau_{\mathrm{ac}}(\boldsymbol{v})$ we have discussed previously.] Then for $t \gg \tau_{\mathrm{ac}}$, the *Markovian approximation* is justifiable:

$$\int_0^t d\tau\, \mathsf{C}(\tau \mid t - \tfrac{1}{2}\tau)f(t - \tau) \approx \left(\int_0^\infty d\tau\, [\mathsf{C}(\tau \mid t) - \tfrac{1}{2}\tau\, \partial_t \mathsf{C}]\right) f(t). \qquad (4.90)$$

---

[a]If singular initial conditions were permitted and the fields were evaluated for the microfluctuations at scales shorter than λ_D, this result when inserted into Eq. (4.82) would ultimately yield the plasma collision operator.

I have retained the nonstationary correction term $\tau \partial_t \mathsf{C} = O(\gamma \tau_{\mathrm{ac}})$, but ignored a correction involving $\partial_t f$. That latter term can be shown to be small in the quasilinear limit.[a] The first term can be readily shown to produce the diffusion operator

$$\frac{\partial}{\partial \boldsymbol{v}} \cdot \mathsf{D}(\boldsymbol{v}) \cdot \frac{\partial f}{\partial \boldsymbol{v}}, \quad \text{where } \mathsf{D}(\boldsymbol{v}) \doteq \left(\frac{q}{m}\right)^2 \int \frac{d\boldsymbol{k}}{(2\pi)^3} \langle \delta \boldsymbol{E} \delta \boldsymbol{E} \rangle(\boldsymbol{k}) \delta(\boldsymbol{k} \cdot \boldsymbol{v} - \Omega_{\boldsymbol{k}}).$$

$$(4.91\mathrm{a,b})$$

This form, the multidimensional version of the D_v calculated in the previous section, is irreversible! That is a consequence of the Markovian approximation, which in turn is predicated on the assumption that the Lagrangian correlation function decays as $\tau \to \infty$. However, in a finite-sized box (in which the Fourier amplitudes are discrete) it turns out that the linear approximation to $\mathsf{C}(\tau \mid t)$ does *not* decay[b] as $\tau \to \infty$; as we discussed in Sec. 4.2.1.2., one must add at least a convergence factor ϵ, which is a subtle appeal to the effects of nonlinear physics that we have formally dropped. Thus irreversibility is inexorably tied up with the long-time behavior of the Lagrangian correlation function. In the discussion below of resonance-broadening theory, I will estimate the omitted nonlinear effects and show that the onset of dynamical stochasticity justifies the present calculation.

There is one more interesting twist to the formalism, which relates to the nonstationary correction term in Eq. (4.90). Clearly irreversible velocity-space diffusion applies to the resonant particles: $\boldsymbol{k} \cdot \boldsymbol{v} = \Omega_{\boldsymbol{k}}$. What about the nonresonant particles? Suppose we consider nonstationary evolution of f. As the fields grow (with the linear growth rate, in this model), the boundaries of the resonant region will shift, and certainly the particles outside of that region will shift as well. The nonresonant particles will "slosh" in response to the oscillating fields; that is, the nonresonant particles carry mechanical momentum and energy that will grow along with the fields. If the process is now reversed, so that the fields damp, the nonresonant momentum and energy will decay along with the fields; the nonresonant dynamics are *reversible*.

---

[a] This exercise is best done after the material in Sec. 4.2.3. on resonance-broadening theory is studied. The dimensionless correction is of the order of $\tau_{\mathrm{ac}} D_v / \Delta v^2$, where Δv is the characteristic width of the resonant region. The characteristic nonlinear decorrelation time is $\omega_d \sim (k^2 D_v)^{1/3}$, with $k \Delta v \sim \omega_d$. Thus the correction is $O(\tau_{\mathrm{ac}} \omega_d)$. The condition $\tau_{\mathrm{ac}} \omega_d < 1$ defines the upper boundary of the quasilinear regime.

[b] It suffers quasirecurrences on a recurrence time $\tau_r \doteq N \tau_{\mathrm{ac}}$, where N is the number of modes in the spectrum. In an infinite box (Fourier integral representation), $\tau_r \to \infty$ and the Markovian approximation is apparently justified even in the absence of nonlinear effects. That can be seen from formula (4.75b) by changing the wave-number summation to an integration and performing that integration first (which is the natural order of the two improper integrals). That gives a $\delta(\tau)$, which can then be integrated over τ; one is led to formula (4.80) without the need for an intermediate convergence factor. Some of the ideas mentioned here were addressed in the PhD thesis of Tetreault (1976).

In fact, the nonresonant effects are contained in the nonstationary term in Eq. (4.90), which can be readily seen to be time-reversible. It is therefore suggested to split off the fluctuation-dependent nonresonant particles from f by writing $f = F + f_\text{nres}$, where $\partial_t f_\text{nres} \sim \partial_t \delta E^2$, which can be integrated to obtain f_nres in terms of the fluctuation level. F is taken to obey the irreversible diffusion equation

$$\frac{dF}{dt} = \frac{\partial}{\partial \boldsymbol{v}} \cdot \mathsf{D}(\boldsymbol{v}) \cdot \frac{\partial F}{\partial \boldsymbol{v}}, \tag{4.92}$$

while

$$f_\text{nres} \doteq -\frac{1}{2}\left(\frac{q}{m}\right)^2 \frac{\partial}{\partial \boldsymbol{v}} \cdot \left[\sum_{\boldsymbol{k}} \widehat{\boldsymbol{k}} |\delta \boldsymbol{E}_{\boldsymbol{k}}|^2 \widehat{\boldsymbol{k}} \frac{\partial}{\partial \Omega_{\boldsymbol{k}}} \mathrm{P}\left(\frac{1}{\Omega_{\boldsymbol{k}} - \boldsymbol{k} \cdot \boldsymbol{v}}\right) \right] \cdot \frac{\partial F}{\partial \boldsymbol{v}} \tag{4.93}$$

describes the nonresonant sloshing. It can be shown (Kaufman, 1972) that the resulting system of resonant and nonresonant particles properly conserves momentum and energy. Specifically, the nonresonant momentum and energy densities can be calculated from Eq. (4.93) to be

$$\sum_s \int d\boldsymbol{v} \begin{pmatrix} (m\overline{n})_s \boldsymbol{v} \\ \frac{1}{2}(m\overline{n})_s v^2 \end{pmatrix} f_{\text{nres},s}(\boldsymbol{v}) = \sum_{\boldsymbol{k}} \begin{pmatrix} \boldsymbol{k} \\ \Omega_{\boldsymbol{k}} \end{pmatrix} \mathcal{N}_{\boldsymbol{k}} - \sum_{\boldsymbol{k}} \begin{pmatrix} 0 \\ 1 \end{pmatrix} \frac{\mathcal{E}_{\boldsymbol{k}}}{8\pi}, \tag{4.94}$$

where

$$\mathcal{N}_{\boldsymbol{k}} \doteq \frac{\partial \operatorname{Re} \mathcal{D}(\boldsymbol{k}, \Omega_{\boldsymbol{k}})}{\partial \Omega_{\boldsymbol{k}}} \frac{\mathcal{E}_{\boldsymbol{k}}}{8\pi}, \tag{4.95}$$

can be interpreted as the *wave action density*,[a] $\mathcal{E}_{\boldsymbol{k}} \doteq |\delta \boldsymbol{E}_{\boldsymbol{k}}|^2$, and $\mathcal{D}(\boldsymbol{k}, \omega)$ is the electrostatic dielectric function. Then it can be shown that

$$\frac{d}{dt} \sum_s \int d\boldsymbol{v} \begin{pmatrix} (m\overline{n})_s \boldsymbol{v} \\ \frac{1}{2}(m\overline{n})_s v^2 \end{pmatrix} F_s(\boldsymbol{v}) = -2 \sum_{\boldsymbol{k}} \gamma_{\boldsymbol{k}} \begin{pmatrix} \boldsymbol{k} \\ \Omega_{\boldsymbol{k}} \end{pmatrix} \mathcal{N}_{\boldsymbol{k}}. \tag{4.96}$$

So far, our calculations [which have loosely followed the seminal paper of Kaufman (1972)] have been systematic, although not necessarily inspired. However, the split into resonant plus nonresonant parts of the distribution function is so intuitively pleasing that one is led to search for a more profound interpretation. That was done in a beautiful paper by Dewar (1973). Dewar considered a canonical transformation from the particle coordinates $(\boldsymbol{x}, \boldsymbol{p})$ to *oscillation-center variables*, defined such that the coordinate transformation is nonsecular. He showed that F is properly interpreted as the *oscillation-center PDF*, and that the nonresonant sloshing is hidden in the transformation of variables. The idea of the oscillation center is a core concept that has generalizations and analogies to many situations in which nonresonant nonlinear physics is at work. For example,

---

[a]Fundamental and elegant discussion of wave action density is given by Whitham (1974).

the well-known ponderomotive force due to inhomogeneous high-frequency electromagnetic fields acts on the oscillation center of a particle and can be calculated by various nonresonant averaging procedures such as Lie perturbation theory (Cary and Kaufman, 1977; Cary and Kaufman, 1981). A gyrocenter is also an example of an oscillation center.

This discussion shows that even the simplest statistical closure for collisionless Vlasov physics contains nontrivial physics content. One must be careful to avoid applying incompatible approximations to the resonant and nonresonant particles, lest important conservation laws be violated. More complicated closures that model truly nonlinear effects will face all of these difficulties and more.

The basic mechanism of quasilinear saturation is that an unstable feature in the distribution function is smoothed out by the diffusion, thus removing the source of free energy that produces a positive linear growth rate. This scenario has been verified for the one-dimensional[a] bump-on-tail problem in the beautiful experiments of Roberson et al. (1971) and Roberson and Gentle (1971). It is well worth the time of a budding plasma physicist to study those papers in order to gain appreciation of the difficulties involved in setting up a clean experiment in bounded geometry and comparing its results against calculations originally performed in an infinite, homogeneous medium.

Although quasilinear flattening can be applied to x-space problems as well [the drift-wave problem is discussed by Sagdeev and Galeev (1969)], it is rarely possible to saturate instabilities by flattening spatial profiles because in most cases those profiles are maintained by external sources. For example, in confined toroidal plasmas there will inevitably be nontrivial, essentially stationary profiles of mean ("background") density and temperature. To understand the physics of saturation in such circumstances, one must look to *nonlinear mode coupling*.

4.2.2. *Weak-turbulence theory*

The next step beyond QLT is weak-turbulence theory (WTT), which is a perturbative development of the fluctuation amplitudes φ in a small coupling parameter, followed by statistical averaging based on the *random-phase approximation*. As applied to wave–wave coupling, the statistical ansatz requires that a threshold for wave stochasticity be exceeded. That can be studied using Hamiltonian techniques in which at least three oscilla-

---

[a]In more than one dimension, a simple argument of Sagdeev and Galeev (1969) shows that a truly stationary plateau in the distribution function cannot form because an infinite portion of phase space would have to be flattened. A quasiplateau may form at finite times, but ultimately the fluctuations decay to zero.

tors (waves) are coupled together. For exactly three oscillators, transformation to action–angle variables followed by elimination of nonresonant terms leads to an integrable Hamiltonian for resonant three-wave interactions.[a] However, when multiple wave-number triads are considered, resonance overlap and stochasticity can ensue (Zakharov, 1984). For sufficiently weak interactions, stochasticity occurs predominantly in the wave phases θ, where fluctuations are written in the form $\delta\varphi_{\bm{k}} = a_{\bm{k}} e^{i\theta_{\bm{k}}}$. n-wave interactions with $n > 3$ can also be studied, as can wave–wave–particle interactions.

I will not show the relatively straightforward details of the derivation of the wave kinetic equation (WKE) for resonant three-wave interactions. In essence, one exploits the random-phase approximation to perturbatively determine an expression for the coarse-grained time derivative $dC_{\bm{k}}/dt \approx [C_{\bm{k}}(t + \Delta t) - C_{\bm{k}}(t)]/\Delta t$, where $C_{\bm{k}}(t) \equiv C_{\bm{k}}(t, t)$ [$C_{\bm{k}}(t, t')$ being the two-time correlation function] and Δt is the stochastization time. If the basic wave amplitude is of order ϵ, one must expand $\varphi = \sum_{n=1}^{\infty} \epsilon^n \varphi^{(n)}$ and work through fourth order according to[b] $C_{\bm{k}} \approx \langle \delta\varphi_{\bm{k}}^{(1)} \delta\varphi_{\bm{k}}^{(1)*} \rangle + \langle \delta\varphi_{\bm{k}}^{(1)} \delta\varphi_{\bm{k}}^{(3)*} \rangle + \langle \delta\varphi_{\bm{k}}^{(3)} \delta\varphi_{\bm{k}}^{(1)*} \rangle + \langle \delta\varphi_{\bm{k}}^{(2)} \delta\varphi_{\bm{k}}^{(2)*} \rangle$. (To obtain the third-order terms, one must iterate twice on the nonlinearity.) For a nonlinear fluid equation described in Fourier space by the mode-coupling equation

$$\partial_t \varphi_{\bm{k}} - L_{\bm{k}} \varphi_{\bm{k}} = \frac{1}{2} \sum_{\Delta} M_{\bm{k},\bm{p},\bm{q}} \varphi_{\bm{p}}^* \varphi_{\bm{q}}^*, \tag{4.97}$$

where $\sum_{\Delta}$ means the sum over all $\bm{p}$'s and $\bm{q}$'s such that $\bm{k} + \bm{p} + \bm{q} = \bm{0}$, the final result for homogeneous statistics is the *spectral balance equation*

$$\partial_t C_{\bm{k}} - 2\gamma_{\bm{k}} C_{\bm{k}} + 2\,\mathrm{Re}(\eta_{\bm{k}}^{\mathrm{nl}}) C_{\bm{k}} = 2 F_{\bm{k}}^{\mathrm{nl}}, \tag{4.98}$$

where $\gamma_{\bm{k}} \doteq \mathrm{Re}\, L_{\bm{k}}$ is the linear growth rate and the nonlinear effects are represented by

$$\eta_{\bm{k}}^{\mathrm{nl}} \doteq -\sum_{\Delta} M_{\bm{k},\bm{p},\bm{q}} M_{\bm{p},\bm{q},\bm{k}}^* \theta_{\bm{k},\bm{p},\bm{q}}^* C_{\bm{q}}, \tag{4.99a}$$

$$F_{\bm{k}}^{\mathrm{nl}} \doteq \frac{1}{2} \sum_{\Delta} |M_{\bm{k},\bm{p},\bm{q}}|^2 \,\mathrm{Re}(\theta_{\bm{k},\bm{p},\bm{q}}) C_{\bm{p}} C_{\bm{q}}, \tag{4.99b}$$

and the *triad interaction time*

$$\theta_{\bm{k},\bm{p},\bm{q}} \doteq \pi\delta(\Delta\Omega), \tag{4.100}$$

$\Delta\Omega \doteq \Omega_{\bm{k}} + \Omega_{\bm{p}} + \Omega_{\bm{q}}$ being the frequency mismatch due to wave dispersion. $F_{\bm{k}}^{\mathrm{nl}}$ (a positive-definite quantity) is called the nonlinear forcing; $\eta_{\bm{k}}^{\mathrm{nl}}$ (typically positive) is called the nonlinear damping.

---

[a]The resonant three-wave problem is described, although not in Hamiltonian terms, by Sagdeev and Galeev (1969) and Davidson (1972).

[b]Odd-order terms vanish under the random-phase approximation.

Since these expressions originate in systematic perturbation theory, they inherit the energy-conservation properties of the primitive amplitude equations. The quantity $\widetilde{\mathcal{I}} = \sum_k \sigma_k |\delta\varphi_k|^2$ is conserved for each realization of a statistical ensemble if $\sigma_k M_{k,p,q} + \text{c.p.} = 0$, where c.p. denotes the cyclic permutation $k \to p \to q$; for the spectral balance equation, that constraint can easily be shown to guarantee the nonlinear conservation of $\mathcal{I} = \sum_k \sigma_k C_k$. Moreover, the perturbation-theoretic derivation implies that the general forms hold as well for situations of strong turbulence. The assumption of weak turbulence shows up most clearly in the form of $\theta_{k,p,q}$, which is here approximated by $\pi\delta(\Delta\Omega)$. In strong-turbulence theory, one expects that delta function to be broadened. I will discuss the proper generalization in Sec. 4.2.5. on Markovian closures.

Although much can be and has been said about weak-turbulence theory,[a] I will not do so here. Some basic ideas were mentioned in the lectures by Prof. Falkovich (Chap. 1), and a significant review article is by Zakharov (1984). For now, the important ideas are the underlying assumption of stochasticity, which can be justified from Hamiltonian wave theory, and the systematic perturbative development, which gives the basic form of the nonlinear terms.

The most direct path to strong-turbulence generalizations of the weak-turbulence spectral balance equation is *via* formal closure techniques, especially the direct-interaction approximation. However, in view of its historical importance, I will first take a detour and discuss Dupree's resonance-broadening theory.

4.2.3. *Resonance-broadening theory*

The quasilinear formalism rests heavily on the linear wave–particle resonance. It is obvious that nonlinearity will modify that resonance in some way (indeed, we have noted that nonlinearity is required to regularize the theory), and Dupree mounted a serious attack. His seminal resonance-broadening theory (RBT) was developed in the latter half of the 1960's; the formalism was also pursued by Weinstock in that same period. Although the calculations of Dupree and Weinstock were clothed with rather heavy-duty mathematical apparatus, the fundamental intuition is firmly grounded in the classical theories of Langevin equations and Brownian motion (Uhlenbeck and Ornstein, 1930; Wang and Uhlenbeck, 1945). Therefore, let us begin by briefly reviewing those topics, which are also useful background for various other analytical formalisms as well.

---

[a]One elementary prediction (Kadomtsev, 1965, p. 107) is that saturation levels scale like γ/Ω relative to the mixing-length level. Some discussion of this point in the context of an exactly solvable model is given by Krommes (2002), Appendix J.

4.2.3.1. *Classical Langevin theory*

Consider an unmagnetized, nonrelativistic plasma for simplicity. Newton's second law of motion for a charged particle is then

$$\dot{\boldsymbol{v}} = \left(\frac{q}{m}\right) \boldsymbol{E}(\boldsymbol{x}(t), t). \tag{4.101}$$

Note the *Lagrangian time dependence*; $\boldsymbol{E}$ is measured at the position of the moving particle. If one considers only the internal fields generated by all of the other $N - 1$ particles, then $\boldsymbol{E} = \widetilde{\boldsymbol{E}}$, the Klimontovich microfield. Equation (4.101) simply describes the characteristic acceleration that appears in the Klimontovich equation (4.3).

Rigorously, one should solve the N-body problem to find $\widetilde{\boldsymbol{E}}(t)$. The method of Langevin (1908) circumvents that extreme difficulty by postulating a decomposition into (i) a smooth, linear, frictional-drag term; and (ii) a rapidly varying and random acceleration:

$$\dot{\boldsymbol{v}} + \underbrace{\nu\boldsymbol{v}}_{\substack{\text{coherent} \\ \text{drag}}} = \underbrace{\delta\boldsymbol{a}(t)}_{\substack{\text{random} \\ \text{acceleration}}}. \tag{4.102}$$

Furthermore, $\delta\boldsymbol{a}$ is taken to be *Gaussian white noise*[a] with zero mean. It is thus entirely specified by its two-time correlation function:

$$\langle\delta\boldsymbol{a}(t)\delta\boldsymbol{a}(t')\rangle \equiv \mathsf{F}(t, t') = 2D_v\delta(t - t')\mathsf{I} \tag{4.103}$$

(I is the unit tensor). F is called the (covariance of the) noise. In near-equilibrium plasmas, the physics and the value of the drag coefficient ν can be understood by detailed considerations of the polarization process that ensues when a moving test particle is inserted into the plasma. However, I do not require those results here; I just need the general form and properties of the linear Langevin equation.

Langevin's model is a linear *stochastic differential equation*, well discussed by van Kampen (1976). By definition, "solution" of such an equation means the determination of the multivariate PDF of the independent variables.[b] For nonlinear equations, that is very difficult; however, the linearity of Eq. (4.102) allows great progress. Let us introduce the Green's function $R(t; t')$ for the coherent operator on the left-hand side. It obeys

$$\partial_t R(t; t') + \nu R(t; t') = \delta(t - t'), \tag{4.104}$$

---

[a]Of course, this is not a rigorous consequence of the nonlinear dynamical equations. In the context of the Langevin model, it can be justified by coarse-graining the time axis into segments Δt much longer than the autocorrelation time τ_{ac}. The total acceleration during Δt is then the sum of a large number of random velocity increments. If those are taken to be statistically independent (an assumption stronger than that of being uncorrelated), then the central limit theorem predicts essentially Gaussian statistics.

[b]In most cases, this is equivalent to the determination of all multi-point cumulants.

166   *John A. Krommes*

which is easily solved to give

$$R(\tau) = H(\tau)e^{-\nu\tau}, \tag{4.105}$$

where $\tau \doteq t-t'$ and $H(\tau)$ is the Heaviside unit step function that guarantees causality. In terms of $R(\tau)$, the formal solution of Eq. (4.102) for $t > 0$ is

$$v(t) = R(t)v_0 + \int_0^t d\bar{t}\, R(t;\bar{t})\delta a(\bar{t}). \tag{4.106}$$

Here v_0 is an initial condition imposed at some arbitrary time $t = 0$. It can either be assumed to be constant or to be a Gaussian random variable statistically independent of δa. We have now expressed $v(t)$ as a linear superposition of the Gaussian random variable δa. An important theorem is, *The sum of two Gaussian variables is Gaussian.* Upon interpreting the time integration as Riemann summation,[a] one concludes that $v(t)$ is also a Gaussian random variable and is therefore specified entirely by its mean and covariance. For constant v_0, one has $\langle v(t)\rangle = R(t)v_0$, so the fluctuating velocity is $\delta v(t) = \int_0^t d\bar{t}\, R(t;\bar{t})\delta a(\bar{t})$. One can now calculate the velocity covariance $C(t,t') \doteq \langle \delta v(t)\delta v(t')\rangle$:

$$C(t,t') = \int_0^t d\bar{t}\int_0^{t'} d\bar{t}'\, R(t;\bar{t})F(\bar{t},\bar{t}')R(t';\bar{t}'). \tag{4.107a}$$

Amazingly, the form of this equation remains correct even for extremely sophisticated renormalizations, as we will see later. In general, the detailed forms of F and R are difficult to determine self-consistently; however, for the heuristic linear Langevin model we have the white-noise approximation (4.103) for F and the explicit form (4.105) for R, so the integrations required in Eq. (4.107a) can be performed:

$$C(t;t') = \left(\frac{D_v}{\nu}\right)\left(e^{-\nu|\tau|} - e^{-\nu(t+t')}\right). \tag{4.108}$$

As an important special case, consider equal times: $t' = t$. This gives the velocity dispersion from the statistically sharp initial condition:

$$C(t,t) = \langle \delta v^2(t)\rangle = \left(\frac{D_v}{\nu}\right)(1 - e^{-2\nu t}) \tag{4.109a}$$

$$\rightarrow \left(\frac{D_v}{\nu}\right)\begin{cases} 2\nu t & (\nu t \ll 1), \\ 1 & (\nu t,\ \nu t' \rightarrow \infty). \end{cases} \tag{4.109b}$$

The short-time limit displays velocity-space diffusion: $\langle \delta v^2(t)\rangle = 2D_v t$. At long times, the fluctuation level saturates. If we assume kinetic-energy

---

[a]Mathematicians will no doubt cringe at this cavalier discussion, but for our purposes there is no problem.

equilibration with a Gibbsian thermal bath at temperature T, then it must be true that $\frac{1}{2}M\langle\delta v^2\rangle \to \frac{1}{2}T$, or $D_v/\nu = T/M$; this is the famous *Einstein relation* (Einstein, 1905; Einstein, 1908).

Of course, random velocity induces a random position. Given our results so far, it is not hard to show that for $\nu t \ll 1$ one has

$$\langle\delta x^2(t)\rangle = \frac{2}{3}D_v t^3. \tag{4.110}$$

This result figures importantly in the resonance-broadening theory to be discussed shortly.

4.2.3.2. *Resonance broadening for $B = 0$*

In an unmagnetized plasma, the basic linear wave–particle interaction occurs when the component of particle velocity in the direction of the wave propagation is equal to the wave phase velocity: $\hat{\boldsymbol{k}}\cdot\boldsymbol{v} = \omega/k$, or $\omega - \boldsymbol{k}\cdot\boldsymbol{v} = 0$. Mathematically, this *Landau resonance* appears as either the argument of a Dirac delta function, $\delta(\omega - \boldsymbol{k}\cdot\boldsymbol{v})$ (if two-sided time correlations are under discussion), or as a resonant denominator, $(\omega - \boldsymbol{k}\cdot\boldsymbol{v} + i\epsilon)^{-1}$ (if one-sided, causal response functions are involved). The latter is nothing but the Fourier transform of the Green's function for a streaming particle:

$$R_{\boldsymbol{k}}^{(0)}(\boldsymbol{v},\tau;\boldsymbol{v}') = H(\tau)e^{-i\boldsymbol{k}\cdot\boldsymbol{v}\tau}\delta(\boldsymbol{v} - \boldsymbol{v}'), \tag{4.111a}$$

$$\int_{-\infty}^{\infty} d\tau\, e^{i\omega\tau} R_{\boldsymbol{k}}^{(0)}(\boldsymbol{v},\tau;\boldsymbol{v}') = \lim_{\epsilon\to 0}\frac{\delta(\boldsymbol{v} - \boldsymbol{v}')}{-i(\omega - \boldsymbol{k}\cdot\boldsymbol{v} + i\epsilon)}. \tag{4.111b}$$

In Eq. (4.111a), the phase factor $\boldsymbol{k}\cdot\boldsymbol{v}\tau$ is the lowest-order approximation to $\boldsymbol{k}\cdot\widetilde{\boldsymbol{x}}_b(-\tau)$, where[a] $\widetilde{\boldsymbol{x}}_b(-\tau) \approx \boldsymbol{v}\tau$ is the positive distance measured back along the trajectory from the current position $\boldsymbol{x}$ at time t. However, under the influence of a fluctuating Langevin electric field, that distance becomes uncertain. Thus it is useful to define the *mean response function* by a statistical average over the fluctuating field. In order to avoid technical complications about correlations between the Langevin position and velocity (Benford and Thomson, 1972), let us consider the reduced response function obtained by integrating over $\boldsymbol{v}'$:

$$\int d\boldsymbol{v}'\, R_{\boldsymbol{k}}(\boldsymbol{v},\tau;\boldsymbol{v}') = \langle e^{-i\boldsymbol{k}\cdot\widetilde{\boldsymbol{x}}_b(-\tau)}\rangle. \tag{4.112}$$

Within the (very approximate) Langevin framework, $\widetilde{\boldsymbol{x}}_b$ is assumed to be a Gaussian random variable. The required average can therefore be evaluated by cumulant expansion [see Kubo (1962a) and the beginning of Sec. 4.2.4.1.]. With the result (4.110), one obtains

$$R_{\boldsymbol{k}}(\boldsymbol{v},\tau) = H(\tau)\exp(-i\boldsymbol{k}\cdot\boldsymbol{v}\tau - \tfrac{1}{3}k^2 D_v\tau^3). \tag{4.113}$$

---

[a]$\widetilde{\boldsymbol{x}}_b(-\tau)$ has the same numerical value as the function $\Delta\boldsymbol{x}(\tau)$ introduced previously, but it is expressed in different variables. For a discussion, see Krommes (2002), Appendix E.

This introduces the new nonlinear rate $\nu_D \doteq (\frac{1}{3}k^2 D_v)^{1/3}$. Although the Fourier transform of Eq. (4.113) can be obtained either analytically[a] or (much more simply) numerically, the essential qualitative idea emerges by considering an exponential form with the characteristic decay rate ν_D:

$$\int_0^\infty d\tau \, e^{i\omega\tau} e^{-i\boldsymbol{k}\cdot\boldsymbol{v}\tau - \nu_D\tau} = \frac{1}{-i(\omega - \boldsymbol{k}\cdot\boldsymbol{v} + i\nu_D)}. \tag{4.114}$$

The real part of this is a Lorentzian that clearly displays a *resonance broadening* of width ν_D:

$$\operatorname{Re} R_{\boldsymbol{k},\omega} = \frac{\nu_D}{(\omega - \boldsymbol{k}\cdot\boldsymbol{v})^2 + \nu_D^2}. \tag{4.115}$$

Clearly ν_D renormalizes the zeroth-order response.

Dupree asserted that the formula (4.114) should be used to replace the zeroth-order result (4.111b) that appears in the usual linear approximation to the plasma dielectric function. Whereas in linear theory the wave growth rate $\gamma_{\boldsymbol{k}}$ is determined by particles moving exactly at the phase velocity $\boldsymbol{v}_{\mathrm{ph}}$ of the wave (and thus ultimately by the slope of the background distribution function at $\boldsymbol{v}_{\mathrm{ph}}$), the resonance broadening implies that in nonlinear theory a group of particles centered on the phase velocity can interact with the wave. To the extent that those particles can sample a range of background slopes, this raises the possibility of a stabilizing nonlinear correction to the effective growth rate, $\gamma_{\boldsymbol{k}}^{\mathrm{eff}} \approx \gamma_{\boldsymbol{k}} - \Delta\gamma_{\boldsymbol{k}}$, even when the background distribution is held fixed. This mechanism was suggested already in Dupree's first paper on the RBT (Dupree, 1966).

4.2.3.3. *Trapping, stochasticity, and resonance broadening*

It is obvious that ν_D replaces the convergence factor ϵ employed in Sec. 4.2.1.2.. I will now show that when the fields are large enough that the Chirikov criterion for stochasticity is satisfied, the resonance broadening smooths the resonances sufficiently to justify a Fourier integral representation and, equivalently, to guarantee that the Lagrangian correlation function does indeed decay rapidly, on the τ_{ac} scale, with no recurrences. Now a single Fourier amplitude plus its complex conjugate corresponds to the particle Hamiltonian $H(\boldsymbol{p}, \boldsymbol{x}) = p^2/2m - q\phi_{\boldsymbol{k}} \cos(\boldsymbol{k}\cdot\boldsymbol{x} - \Omega_{\boldsymbol{k}}t)$. A frame change to the wave phase velocity reduces this to the pendulum Hamiltonian $\frac{1}{2}p^2 - \epsilon\cos\theta$, where $\epsilon \doteq mq\phi_{\boldsymbol{k}}$. The island width can be readily calculated to be $\Delta p = 4\sqrt{\epsilon}$ or, in terms of velocity, $\Delta v = 4v_{\mathrm{tr}}$, where the *trapping velocity* is $v_{\mathrm{tr}} \doteq (q\phi_{\boldsymbol{k}}/m)^{1/2}$. In the absence of any other Fourier components, resonant particles will therefore be trapped inside this island. If all of the amplitudes are roughly equal, then a Chirikov island overlap

---

[a] It is an Airy function of complex argument.

parameter can be estimated according to $S = \Delta v / \delta v$, where δv is the separation between adjacent resonances due to the fact that wave numbers are quantized with spacing δk: in 1D, $\delta v = \delta k |\partial(\omega_k/k)/\partial k| = |v_{\mathrm{ph}} - v_{\mathrm{gr}}|(\delta k/k)$.

Clearly one either has trapping (for very small amplitudes) or diffusion (when the Chirikov criterion is satisfied). One of the confusing points of Dupree's early papers is that he sometimes referred to the resonance broadening as describing "trapping." That is incorrect because the resonance broadening is couched in terms of D_v and velocity-space diffusion only exists in the stochastic regime in which islands are destroyed. In Dupree's strong defense, however, his basic work in 1966 and 1967 preceded the widespread and detailed understanding of nonlinear stochastic Hamiltonian dynamics (Smith and Kaufman, 1975; Karney, 1978; Treve, 1978).

An integral representation is justified when the change in effective linear frequency $kv - \omega_k$ between adjacent wave numbers is smaller than the nonlinear resonance broadening: $|v_{\mathrm{ph}} - v_{\mathrm{gr}}|\delta k < (k^2 D_v)^{1/3}$. (I will drop all numerical coefficients in the following estimates.) The left-hand side can be written as $k\,\delta v$. We have $D_v = (q/m)^2 \langle \delta E^2 \rangle \tau_{\mathrm{ac}}$. One can estimate $\langle \delta E^2 \rangle \approx N \delta E_k^2$, where N is the number of modes in the spectrum ($N \delta k = \Delta k$) and δE_k is a typical Fourier amplitude. Note from formula (4.71) that $N \tau_{\mathrm{ac}} = (k\,\delta v)^{-1}$, since $\Delta k / \delta k = N$ and one can take $v = v_{\mathrm{ph}}$. Also, $(q/m)^2 \delta E_k^2 = k^2 v_{\mathrm{tr}}^4 \to k^2 \Delta v^4$. Thus, a continuum representation is justified if $k\,\delta v < [k^2 (k\,\delta v)^{-1} k^2 \Delta v^4]^{1/3}$, which is readily seen to reduce to $\delta v < \Delta v$ or $S > 1$. One concludes that stochasticity (a nonlinear phenomenon) justifies the Fourier integral representation. This is a beautiful consistency; had it not been obtained, important physics would have been overlooked.

4.2.3.4. *Resonance broadening for $B \neq 0$*

Both Dupree (1967) and Weinstock (1968) noted that a resonance-broadening formalism can be applied as well to strongly magnetized situations in which one considers the cross-field motion of fluid elements rather than single particles; it is that work that affords the most immediate connection to the more modern studies of plasma turbulence, which have focused mostly on fluid models. Assume that gyrocenters move cross-field with the $\boldsymbol{E} \times \boldsymbol{B}$ drift. If the electric field is again idealized to be Gaussian white noise, we now have the more traditional picture of a Gaussian random *velocity* (rather than acceleration), so perpendicular position is predicted to diffuse according to $\langle \delta x^2(\tau) \rangle = 2 D_\perp \tau$. As I discuss further below, the $D_\perp$ that appears here is the diffusion coefficient of a test fluid element.

In Dupree's original work, he appealed to intuition to justify the assumption of stochasticity. However, the transition to perpendicular stochasticity can be put on firm footing, as was shown later by Ching (1973), Hirshman (1980), and, in much more detail, Isichenko et al. (1992). In the first two references, the single wave $\phi = \phi_0 \cos(k_x x)\cos(k_y y + k_z z - \Omega t)$ was considered.[a] After transforming to the wave frame, one is led to the conserved and integrable Hamiltonian

$$K(x, y_w) = -\alpha x + \epsilon \cos(k_x x)\cos(k_y y_w), \tag{4.116}$$

with x and y_w playing the roles of conjugate momentum and coordinate and where $\alpha \doteq (\Omega - k_z v)/k_y$ and $\epsilon \doteq c\phi_0/B$. K plays for $\boldsymbol{E} \times \boldsymbol{B}$ motion the role of the reference pendulum Hamiltonian of unmagnetized theory. Representative phase contours are shown in Fig. 4.9, which demonstrates the existence of both *spatial trapping*[b] as well as unbounded motion. Fig. 4.10 shows that the addition of a second wave with incommensurate wavevector and frequency can destroy the separatrices and lead to global stochasticity. This is the fundamental nonlinear dynamical mechanism that underlies Dupree's basic assumption of stochastic diffusion.

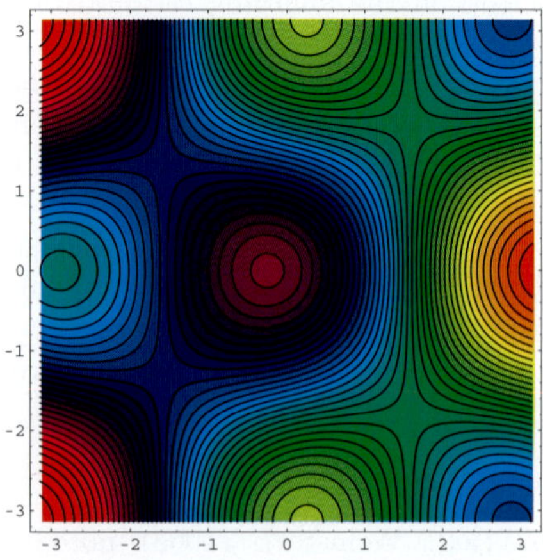

Fig. 4.9 Phase contours of the integrable $\boldsymbol{E} \times \boldsymbol{B}$ Hamiltonian (4.116) for a single wave demonstrate the possibility of spatial trapping; here $\epsilon/\alpha = 0.5$.

In the stochastic regime, one must now consider

$$\langle e^{-i\boldsymbol{k}\cdot\widetilde{\boldsymbol{x}}_b(-\tau)}\rangle \approx e^{-k_\perp^2 D_\perp \tau}, \tag{4.117}$$

where the assumption of Gaussian statistics was used to truncate the cumulant expansion. The characteristic broadening rate is now $\nu_D = k_\perp^2 D_\perp$;

---

[a]The assumption of a standing wave in the x direction can be justified on the basis of nontrivial boundary conditions in the direction of macroscopic inhomogeneity.

[b]The frequency of small oscillations around the elliptic fixed points is $\omega_{\mathrm{tr}} = |\epsilon k_x k_y| = |ck_x k_y \phi_0/B|$.

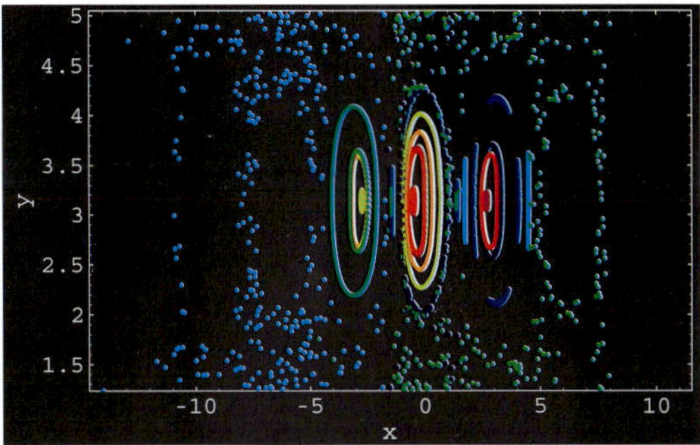

Fig. 4.10   Poincaré plot showing that regions of stochasticity develop for $\boldsymbol{E} \times \boldsymbol{B}$ motion in the presence of a second wave.

this can be seen to be a consequence of *random Doppler shifts* of the perpendicular velocity. By assuming that the turbulence should grow until the plasma achieves marginal stability, Dupree asserted that nonlinear saturation should occur when the resonance-broadening rate becomes as large as the linear growth rate: $\nu_D = \gamma_{\boldsymbol{k}}$. This can be rearranged into a *prediction* for the steady-state test-particle diffusion coefficient:

$$D_\perp = \gamma_{\boldsymbol{k}} / k_\perp^2. \tag{4.118}$$

In its dependence on γ, this result agrees with the earlier prediction of Kadomtsev (1965), discussed here as Eq. (4.47). Somewhat anomalously, it seems that widespread comprehension of the basis and limitations of this important formula has diminished rather than increased with time.

The prediction of Kadomtsev and Dupree that $D_\perp$ involves the dissipative growth rate rather than the real frequency[a] was clearly an advance in understanding, since dissipation and transport are inevitably linked according to random-walk theory. Nevertheless, although I wish to stress the positive contributions of RBT, it should also be noted that there are several (closely related) difficulties with formula (4.118), which will be discussed in turn after they are enumerated:

(1) $D_\perp$ is supposed to be independent of wave number; however, the right-hand side of formula (4.118) in general depends on $\boldsymbol{k}$.

(2) $D_\perp$ appears to represent a dissipative effect (the diffusive rate $k_\perp^2 D_\perp$ appears), yet it models the effect of a conservative advective nonlinearity. Thus, it is unclear that energy is properly conserved.

(3) The RBT does not clearly distinguish between the passive diffusion of a test fluid element and the self-consistent diffusion of the entire plasma.

---

[a]The gyro-Bohm diffusion coefficient (4.40) involves no dissipative quantity.

Difficulty 1: Wave-number dependence, or not. Because $D_\perp$ entered from the beginning as a $\boldsymbol{k}$-independent parameter, one must not take the right-hand side of Eq. (4.118) literally; the $\boldsymbol{k}$ and $\gamma_{\boldsymbol{k}}$ must be understood to represent some *typical* wave number and growth rate, respectively.[a] Thus the formula is essentially dimensional in nature,[b] in spite of the possibly formidable mathematics from which it was derived.

Difficulty 2: *Energy conservation*. The RBT does not conserve energy (Orszag and Kraichnan, 1967; Dupree and Tetreault, 1978; Krommes, 2002). If the nonlinear terms are modeled by a diffusion operator $k^2 D_\perp$, a basic spectral evolution equation is $\partial_t C_{\boldsymbol{k}} = -k^2 D_\perp C_{\boldsymbol{k}}$. This predicts that all Fourier amplitudes except for $\boldsymbol{k} = \boldsymbol{0}$ damp (and $C_{\boldsymbol{0}}$ can be taken to vanish). Because the right-hand side is negative-definite, the energy evolves according to $\partial_t \mathcal{E} < 0$, which is incompatible with the assumption of a conservative nonlinearity.

To better understand the difficulty with energy conservation, it is useful to interpret the general form (4.98) of the wave kinetic equation in terms of a Langevin representation. Consider the nonlinear Langevin equation

$$\partial_t \psi_{\boldsymbol{k}} + \eta_{\boldsymbol{k}} \psi_{\boldsymbol{k}} = \widetilde{f}_{\boldsymbol{k}}, \tag{4.119}$$

where $\widetilde{f}_{\boldsymbol{k}} \propto \widetilde{w}(t)$, $\widetilde{w}(t)$ being white noise with unit intensity. Thus, we assume that $\langle \widetilde{f}_{\boldsymbol{k}}(t) \widetilde{f}_{\boldsymbol{k}}^*(t') \rangle = 2F_{\boldsymbol{k}}^{\mathrm{nl}} \delta(t - t')$. By introducing the Green's function of the left-hand side of Eq. (4.119), which obeys $(\partial_t + \eta_{\boldsymbol{k}}) R_{\boldsymbol{k}}(t; t') = \delta(t - t')$, and solving for the forced response according to $\psi_{\boldsymbol{k}}(t) = \int_{-\infty}^{t} dt'\, R(t; t') \widetilde{f}_{\boldsymbol{k}}(t')$, it is not hard to show that the general form (4.98) of the Markovian wave kinetic equation is reproduced provided that $F_{\boldsymbol{k}}^{\mathrm{nl}}$ has the form (4.99b) (that is, it describes the beating of two amplitudes $\boldsymbol{p}$ and $\boldsymbol{q}$ to drive a fluctuation at $\boldsymbol{k}$). The Green's function R is said to describe the *coherent response*, while $\widetilde{f}_{\boldsymbol{k}}$ is called *incoherent noise*. Statistical steady states evidently arise by a balance between the incoherent noise (intrinsically nonlinear) and the coherent response (both linear and nonlinear).

The balance $\nu_D \sim \gamma$ is a dimensionally correct approximation to $\mathrm{Re}\,\eta = \gamma$, which would hold on a wave number by wave number basis in steady state if the nonlinear forcing on the right-hand side of the spectral balance equation were neglected. Thus, the fundamental approximation of the RBT is that *the incoherent noise is neglected*. Energy conservation is lost immediately, whether or not the diffusive approximation is made on $\eta_{\boldsymbol{k}}^{\mathrm{nl}}$. That is actually more or less appropriate for small-k fluctuations,

---

[a] Note that some growth rates are negative, whereas $D_\perp$ is positive.

[b] For the drift-wave problem with $k_\perp \rho_s = O(1)$, both of the estimates $D \sim \gamma/k_\perp^2$ [Eq. (4.118)] and $D \sim (\gamma/\Omega) D_{\mathrm{gB}}$ [Eq. (4.47)] reduce to $D \sim \gamma \rho_s^2$.

although the physics of the large k's is quite different. But the neglect of the incoherent noise is fundamental and unjustifiable.[a]

Difficulty 3: Fluid vs test-particle diffusion. The distinction between test-particle and fluid diffusion coefficients is well known to students of classical (collisional) cross-field transport in a magnetic field.[b] An analogous distinction exists for turbulence. In Dupree's 1966 paper that introduced his "test wave" formalism for assessing the effects of resonance broadening, he stated, "The method we employ for solving the Vlasov–Maxwell equation consists of two distinct pieces. First, we assume knowledge of the electric field $\boldsymbol{E}$ .... As a second step, we must ... require that the f so determined does ... produce the assumed $\boldsymbol{E}$ [via Poisson's equation]." Later he asserted, "The fact that the initial phases of the background waves in the subsidiary [test wave] problem are uncorrelated ... does not prevent the [Fourier coefficients] so calculated from being used to describe an actual system in which all the initial phases have some precise relation to each other and to f." But by initially assuming that the electric field is known, Dupree lost the self-consistency between the waves and the particles. Thus, his $D_\perp$ is a test-particle quantity. This is at least a quantitative deficiency; however, there can be more serious consequences. Unfortunately, the single $D_\perp$ of the RBT is frequently invoked in multiple contexts (cross-field transport,

---

[a]Dupree later attempted to include the incoherent noise with his theory of *phase-space granulations* or *clumps* (Dupree, 1972). Some aspects of that theory are controversial; see Krommes (1997), and references therein.

[b]Classical transport theory is a consequence of random microfluctuations on sub-Debye scales. According to standard random-walk considerations, a characteristic step size for a gyrospiraling particle should be ρ_s, where s denotes species (either e for electrons or i for ions); a characteristic step time should be the inverse of the species-dependent collision rate ν_s. One has $\nu_e = \nu_{ee} + \nu_{ei} \approx \nu_{ee}$ (both rates are comparable), and $\nu_i = \nu_{ii} + \nu_{ie} \approx \nu_{ii}$. One then estimates that the cross-field test-particle diffusion coefficients scale as $D_e^{(\text{test})} \sim \rho_e^2 \nu_e$ and $D_i^{(\text{test})} \sim \rho_i^2 \nu_i$, with ratio

$$D_i^{(\text{test})}/D_e^{(\text{test})} \sim \left(\frac{\rho_i}{\rho_e}\right)^2 \left(\frac{\nu_i}{\nu_e}\right) \sim \left(\frac{m_i}{m_e}\right)^{1/2} \gg 1. \qquad (4.120)$$

(The mass scaling $\nu_s \sim m_s^{-1/2}$ follows from the standard estimate $\nu \sim n\sigma v_t$, where σ is the mass-independent Coulomb scattering cross section.) It would appear that cross-field transport is dominated by the ions. In fact, however, the cross-field fluxes of the entire plasma fluid are equal. Because momentum is conserved during a collision, the *net* fluid transport due to like-particle collisions vanishes and one must exclude like-species collision rates in the random-walk estimates. Thus, one estimates $\nu_e \to \nu_{ei}$ and $\nu_i \to \nu_{ie}$ and finds $D_e^{(\text{fluid})} \sim \rho_e^2 \nu_{ei}$ and $D_i^{(\text{fluid})} \sim \rho_i^2 \nu_{ie}$, with ratio

$$D_i^{(\text{fluid})}/D_e^{(\text{fluid})} \sim \left(\frac{\rho_i}{\rho_e}\right)^2 \left(\frac{\nu_{ie}}{\nu_{ei}}\right) = 1. \qquad (4.121)$$

This last result is known as *ambipolar diffusion*.

spectral balance, renormalized response). In a more complete theory, all of those physical processes are distinct, and in some cases (notably spectral transfer) the distinction between test-particle and fluid diffusion can be qualitatively significant (Krommes, 2002). Given modern understanding, it is perhaps better to be guided by more systematic formalisms such as the DIA (Sec. 4.2.4.2.) or Markovian closures (Sec. 4.2.5.) rather than to attempt to untangle the threads of Dupree's various courageous attempts.

Although I am stressing here the conceptual problems with the RBT, it is instructive to also note one practical difficulty with Dupree's early work on drift waves: effects related to the polarization drift were lost [see, for example, the equations studied by Dupree and Tetreault (1978)]. The situation would no doubt be different had nonlinear gyrokinetics existed at the time.

Nevertheless, the RBT was seminal and inspirational. It stressed the importance of nonperturbative treatments of stochasticity (especially of the $\boldsymbol{E} \times \boldsymbol{B}$ variety), showed how to make practical estimates of nontrivial saturation levels, and demonstrated that, at least approximately, an analytical description of complicated plasma nonlinearities may not be impossible. I wlll now turn to some of the later, more systematic developments.

4.2.4. *"Systematic" renormalization and the direct-interaction approximation*

Resonance-broadening theory obviously involves renormalization. But just how is that accomplished, and when should one trust it? Dupree (1966) initially suggested a formalism involving the successive introduction of more and more *test waves*. The method was somewhat reminiscent of Kraichnan's original procedure for deriving his direct-interaction approximation (DIA) (see the lectures of Prof. McComb in Chap. 2), yet Dupree did not recover the DIA but rather (quite approximately) renormalized equations for passive advection. For self-consistent situations, Kraichnan's approach, involving an assessment of the role of an elementary triad *removed* from the sea of all interacting modes, was superior. It took some time for this to be thoroughly appreciated. Orszag and Kraichnan (1967) critiqued the formalism of Dupree (1966) soon after Dupree's work was published. They discussed several variants of a *Vlasov DIA* and enumerated a variety of cogent insights into the difficulties of a proper theory. However, the intrinsic physical complications of a turbulent system of coupled fields and charged particles precluded immediate practical work on the Vlasov DIA, and plasma turbulence theory developed independently of the important work of Orszag and Kraichnan.

Interest in systematically renormalized plasma turbulence theory was rekindled by the seminal paper of Martin, Siggia, and Rose (Martin et al., 1973) on the statistical dynamics of classical nonlinear systems. That generating-functional method made explicit contact with the powerful non-perturbative techniques of quantum field theory, for example as successfully applied to quantum electrodynamics. It gave a new perspective on the DIA and clarified issues of self-consistency. The emphasis of MSR was on the development of closed, renormalized equations for low-order, multipoint cumulants; that aspect of the formalism was thoroughly reviewed by Krommes (2002) and will be described only very briefly here. However, other methods of calculation are also possible, including direct calculation of the generating functional under certain approximations. That line of attack, which has been pursued quite recently by Spineanu and Vlad (2005) in the context of coherent structures intermixed with random turbulence, will be briefly described in Lecture 4.

4.2.4.1. *MSR formalism and the Dyson equations for turbulence*

The MSR formalism can be described as either a cumulant generating-functional procedure or, equivalently, a *path-integral representation* of the statistical dynamics. What do these mean?

In probability theory, it is well known that the Fourier transform $Z(k)$ of a PDF $P(x)$ (Z is called the *characteristic function*) is a moment generating function, and that $W \doteq \ln Z$ is a *cumulant generating function*: $Z(k) \doteq (2\pi)^{-1} \int_{-\infty}^{\infty} dk \exp(-ikx)P(x)$; $\langle x^n \rangle = \partial^n Z(k)/\partial(-ik)^n_{|k=0}$; $\langle\langle x^n \rangle\rangle = \partial^n W(k)/\partial(-ik)^n_{|k=0}$. The key property of the cumulant (denoted by double angle brackets) of a set of n random variables is that if any variable in the set is statistically independent of all of the others, then the cumulant vanishes. Cumulants thus describe intrinsic statistical correlations (Kubo, 1962a) and can be small.[a] A Gaussian distribution is described by just its first cumulant (the mean) and its second cumulant (the variance).

To describe the application of these results to physics, I first remind the reader of some well-known results from equilibrium statistical mechanics that relate to the determination of extensive ensemble averages by certain variations of intensive parameters. In terms of the $6N$-dimensional phase-space point Γ, the Gibbs PDF (canonical ensemble) for thermal-equilibrium states is $P(\Gamma) = e^{-\beta H(\Gamma)}/Z$, where $\beta \doteq T^{-1}$ and the *partition function* is

---

[a]Smallness is not sufficient. In general, truncated cumulant expansions are *nonrealizable*; that is, the predicted PDF has negative regions. Furthermore, if certain cumulants have the wrong sign, the PDF may not even be normalizable. Kraichnan (1980) has stressed the importance of realizability in formulating sensible statistical closures.

$Z(\beta) \doteq \int d\Gamma \, e^{-\beta H}$. If $f(H)$ is some polynomial function of the random energy, then the ensemble-averaged f is

$$\langle f(H) \rangle = Z^{-1} \int d\Gamma \, f\left(\frac{\partial}{\partial(-\beta)}\right) e^{-\beta H}. \tag{4.122}$$

Cumulants can be generated by

$$\langle\!\langle H^n \rangle\!\rangle = \frac{\partial^n \ln Z}{\partial(-\beta)^n}. \tag{4.123}$$

For example,

$$\langle H \rangle = \frac{\partial \ln Z}{\partial(-\beta)}, \quad \langle\!\langle H^2 \rangle\!\rangle \equiv \langle \delta H^2 \rangle = \frac{\partial^2 \ln Z}{\partial \beta^2}. \tag{4.124a,b}$$

Summarizing, the statistics of energy fluctuations can be obtained by varying the canonical partition function with respect to $\beta \doteq T^{-1}$.

In the canonical ensemble, the number N of particles is fixed. If instead one is dealing with an open system and is interested in the statistics of particle number, one must consider the *grand canonical ensemble*

$$P(\Gamma; \beta, \eta) = \frac{1}{Z} e^{-\beta H_N + \eta N}, \quad Z(\beta, \eta) \doteq \sum_{N=0}^{\infty} \int d\Gamma \, e^{-\beta H_N + \eta N}. \tag{4.125a,b}$$

(These formulas are conventionally written in terms of the chemical potential $\mu \doteq \eta/\beta$, but the η form is more useful for the interpretation and developments to be described next.) Then

$$\langle\!\langle N^n \rangle\!\rangle = \frac{\partial \ln Z(\beta, \eta)}{\partial \eta^n}. \tag{4.126}$$

For example, $\langle N \rangle = \partial \ln Z / \partial \eta$ and

$$\frac{\partial^2 \ln Z}{\partial \eta^2} = \sum_{N=0}^{\infty} \int d\Gamma \, N^2 \left(Z^{-1} e^{-\beta H_N + \eta N}\right)$$

$$- \left[\sum_{N=0}^{\infty} \int d\Gamma \, N \left(Z^{-1} e^{-\beta H_N + \eta N}\right)\right]^2 \tag{4.127a}$$

$$= \langle N^2 \rangle - \langle N \rangle^2 = \langle\!\langle N^2 \rangle\!\rangle \equiv \langle \delta N^2 \rangle. \tag{4.127b}$$

Thus, the mean density and indeed all of the higher-order cumulants are related to the change in the equilibrium statistical distribution due to the addition of an extra particle. In thermal equilibrium, the logarithm of the grand canonical partition function is a *generating function* for the cumulants of particle number.[a]

---

[a] In the above expressions, the operation $\sum_{N=0}^{\infty} e^{\eta N}$ takes the place of Fourier transformation, with $\eta \equiv -ik$.

Now consider the possibility of obtaining a cumulant generating functional[a] for turbulence. The fundamental difficulty is that, unlike the case of thermal equilibrium, *for turbulence one does not know the form of the steady-state probability density functional of ψ*. Thus it is not clear how to write a useful explicit form of a partition or generating functional.[b]

This was the problem attacked and solved in the seminal paper by Martin, Siggia, and Rose (MSR) (Martin et al., 1973). They generalized to classical dynamics[c] powerful functional techniques, originally due to Schwinger (1951a, 1951b, 1951c) ,[d] that had been previously used successfully in quantum field theory (de Dominicis and Martin 1964a, 1964b) . Note that it may be asking too much to obtain the explicit form of a nonequilibrium cumulant generating functional. However, within a variational formulation in which all cumulants vary in concert as an external source η is changed, it seems plausible that one may be able to deduce useful relationships between the cumulants that go beyond the trivial fact, known from simple probability theory, that a cumulant of order $n+1$ is the derivative with respect to η of a cumulant of order n. Indeed, what ultimately emerges is a (matrix) *Dyson equation* for turbulence in which closure difficulties are concealed in a *turbulent collision operator* Σ. In a certain natural way, the lowest-order approximation to Σ leads to Kraichnan's direct-interaction approximation.

The MSR formalism has been thoroughly reviewed by Krommes (2002); most of that discussion will not be repeated here. However, I will discuss the path-integral representation of the MSR generating functional, both because of its intrinsic interest and because there have been recent attempts to evaluate that functional directly (see further discussion in Lecture 4). Now in any theory that describes statistics involving multiple times, it is crucial to embed the constraint that the dynamics evolve according to a specified equation. I therefore focus on a time-dependent function $\widetilde{q}(t)$ (again, tildes denote random variables) and consider the random dynamics $\partial_t \widetilde{q} = \widetilde{A}(\widetilde{q}, t)$. Statistics enter through either $\widetilde{A}$ (the tilde emphasizes that $\widetilde{A}$ may contain a random parameter or possess random functional dependence) or a random initial condition $\widetilde{q}_0$. If one considers the one-sided function

---

[a]It must be a functional rather than a function because one must consider statistics involving an infinite number of space-time points.

[b]Even in thermal equilibrium, it is not clear how to generate time-lagged correlations; the time-independent grand canonical ensemble does not contain enough information.

[c]It may seem peculiar to "generalize" quantum-mechanical techniques to the classical domain, since the former would seem to be the more fundamental. The difficulty has to do with the fact that the states in a classical system are distributed continuously, so certain commutators that are nontrivial quantum-mechanically vanish in the classical limit. See the MSR paper for further, very clear discussion.

[d]Some entertaining references related to Schwinger are by Martin (1979) and Mehra and Milton (2000).

$\widetilde{q}_+(t) \doteq H(t - t_0)\widetilde{q}(t)$, the initial condition can be incorporated into $\widetilde{A}$ at the price of a delta function:

$$\partial_t \widetilde{q}_+ = \widetilde{A}(\widetilde{q}, t) + \delta(t - t_0)\widetilde{q}_0 \tag{4.128}$$

[thus $\widetilde{q}_+(0_+) = \widetilde{q}_0$]. An example of a random $\widetilde{A}$ is $\widetilde{A}(q, t) = i\widetilde{\Omega}q + \widetilde{f}(t)$, where $\widetilde{\Omega}$ and $\widetilde{f}$ are random; this form would be appropriate for the description of forced, passive advection.

The initial goal is to obtain statistical averages of functions F of the $\widetilde{q}$'s evaluated at n discrete times:

$$\langle F \rangle(t_1, t_2, \ldots, t_n) \doteq \langle F(\widetilde{q}(t_1), \widetilde{q}(t_2), \ldots, \widetilde{q}(t_n)) \rangle \equiv \langle \widetilde{F} \rangle. \tag{4.129}$$

For example, one can obtain the n-point multitime moment as $M_n = \langle \widetilde{F} \rangle$, where $\widetilde{F} \doteq \widetilde{q}(t_1) \ldots \widetilde{q}(t_n)$. Now

$$F(\widetilde{q}(t_1) \ldots \widetilde{q}(t_n)) = \int dq_0 \, dq_1 \, \ldots \, dq_n \, F(q_1, \ldots, q_n)$$
$$\times \, \delta(q_0 - \widetilde{q}_0)\delta(q_1 - \widetilde{q}(t_1)) \ldots \delta(q_n - \widetilde{q}(t_n)). \tag{4.130}$$

If one were to perform the average at this point, one would just represent $\langle F \rangle$ in terms of the standard representation of an n-fold, discretely multivariate PDF, i.e., the n-dimensional generalization of $P(q) = \langle \delta(q - \widetilde{q}) \rangle$. However, that quantity is unknown and is difficult to calculate because the value of $\widetilde{q}(t)$ depends on *all* previous times t' according to the formal integration of the equation of motion. It is technically helpful to incorporate all of those times into the representation. Therefore, let us imagine that the continuous time axis is discretized according to $t \to t_k \doteq k\Delta t$ for $0 \le k \le N$ with $N \to \infty$, and define $D[q] \doteq dq_0 \, dq_1 \, \ldots \, dq_N$. Thus

$$\widetilde{F} = \int D[q] F(q_1, \ldots, q_n) \prod_{i=0}^{N} \delta(q_i - \widetilde{q}(t_i)). \tag{4.131}$$

Next, we incorporate the constraint that $\widetilde{q}(t)$ is governed by the equation of motion. If we use an explicit time differencing, we have, for example,

$$\frac{\widetilde{q}(t_i) - \widetilde{q}(t_{i-1})}{\Delta t} = \widetilde{A}(\widetilde{q}(t_{i-1}), t_{i-1}). \tag{4.132}$$

Consider factors i and $i - 1$ of the product in Eq. (4.131):

$$\delta(q_i - \widetilde{q}(t_i))\delta(q_{i-1} - \widetilde{q}(t_{i-1}))$$
$$= \delta(q_i - \widetilde{q}(t_{i-1}) - \Delta t \widetilde{A}(q_{i-1}, t_{i-1}))\delta(q_{i-1} - \widetilde{q}(t_{i-1})) \tag{4.133a}$$
$$= \Delta t^{-1}\delta(\Delta t^{-1}(q_i - q_{i-1}) - \widetilde{A}(q_{i-1}, t_{i-1}))\delta(q_{i-1} - \widetilde{q}(t_{i-1})) \tag{4.133b}$$

The random fields at t_{i-1} were replaced by the observer coordinate q_{i-1} due to the presence of the second delta function. At this point, we therefore have formally

$$\widetilde{F} = \Delta t^{-N} \int D[q] F(q_1, \ldots, q_n) \delta[\dot{q}(t) - \widetilde{A}(q(t), t)] \delta(q_0 - \widetilde{q}_0). \qquad (4.134)$$

We now represent the delta functions as Fourier integrals, e.g., $\delta(q-\widetilde{q}) = (2\pi)^{-1} \int dp \, e^{ip(q-\widetilde{q})}$. Define $\widehat{q} \doteq p/(i\Delta t)$, the *random Lagrangian* $\widetilde{\mathcal{L}}(t) \doteq \widehat{q}(t)[\dot{q}(t) - \widetilde{A}(q(t), t)]$, and the *random action* $\widetilde{S} \doteq \lim_{T\to\infty} \int_{0_-}^{T} dt' \, \widetilde{\mathcal{L}}(t')$. We then ultimately obtain

$$\widetilde{F} = \mathcal{N} \int D[q] D[\widehat{q}] F[q] e^{-\widetilde{S}}, \qquad (4.135)$$

where one generalized to an arbitrary functional dependence of F on $q(t)$ and $\mathcal{N}$ is a normalization coefficient.

Although the $\widehat{q}$'s were originally introduced for technical convenience, they have a profound interpretation: they are related to *infinitesimal response functions*. If the dynamical equation is augmented to $\dot{\widetilde{q}} = \widetilde{A}(\widetilde{q}, t) + \widetilde{f}^{(\text{ext})}(t)$, where $\widetilde{f}^{(\text{ext})}$ is an external source functionally independent of $\widetilde{q}$, then the random two-point infinitesimal response function is defined by

$$\widetilde{R}(t; t_1) \doteq \frac{\delta \widetilde{q}(t)}{\delta f^{(\text{ext})}(t_1)}. \qquad (4.136)$$

By causality, this function vanishes for $t < t_1$. As you know from the lectures of Prof. McComb (Chap. 2), the mean response function $R \doteq \langle \widetilde{R} \rangle$ figures importantly in renormalized theories of turbulence, such as the DIA or the LET. Now the presence of $f^{(\text{ext})}$ adds to the random Lagrangian the term $-\widehat{q}(t) f^{(\text{ext})}(t)$, so it is easy to verify that $\widehat{q}$ is the momentum-space representation of a functional derivative with respect to $f^{(\text{ext})}$:

$$\frac{\delta^n \widetilde{F}}{\delta f^{(\text{ext})}(t_1) \ldots \delta f^{(\text{ext})}(t_n)} = \mathcal{N} \int D[q] D[\widehat{q}] F[q] \widehat{q}(t_1) \ldots \widehat{q}(t_n) e^{-\widetilde{S}}. \qquad (4.137)$$

For $\widetilde{R}(t; t_1)$ [Eq. (4.136)], replace $F[q]$ by $q(t)$ and set $n = 1$.

In the path-integral representation, statistics reside only in the random action $\widetilde{S}$. One can therefore perform the average that is ultimately required for either correlation functions or mean response functions by "merely" defining an *effective action* S according to $\langle e^{-\widetilde{S}} \rangle = e^{-S}$. Then one can define the *generating functional*

$$Z[\eta, \eta] \doteq \mathcal{N} \int D[q] D[\widehat{q}] \, e^{-S} e^{\int dt' \, [\eta(t')q(t') + \eta(t')\widehat{q}(t')]}. \qquad (4.138)$$

As discussed above, η can be interpreted as a statistically sharp source term added to the equation of motion; similarly, η plays the role of a source term added to an appropriate *adjoint equation*. In the end, $\boldsymbol{\eta} \doteq (\eta, \eta)^T$ can be set to zero. Any Taylor-expandable function of $\boldsymbol{\psi} \doteq (q, \widehat{q})^T$ can therefore be obtained from functional differentiation. For example, the two-point moment follows from $\langle q(t_1)q(t_2)\rangle = \delta^2 Z/\delta\eta(t_1)\delta\eta(t_2)|_{\boldsymbol{\eta}=0}$, and the two-point mean response function follows from $R(t_1;t_2) = \langle q(t_1)\widehat{q}(t_2)\rangle = \delta^2 Z/\delta\eta(t_1)\delta\eta(t_2)|_{\boldsymbol{\eta}=0}$. It is left as an exercise to verify that $R(t_1;t_2)$ is appropriately causal.

Finally, one can generate cumulants from $W[\boldsymbol{\eta}] \doteq \ln Z[\boldsymbol{\eta}]$. It is another exercise to verify that[a] $\langle\langle\widehat{q}\rangle\rangle_{|\boldsymbol{\eta}=0} = 0$. Thus $R = \langle q\rangle\langle\widehat{q}\rangle + \langle\langle q\,\widehat{q}\rangle\rangle = \langle\langle q\,\widehat{q}\rangle\rangle$, $C = \langle q\,q\rangle - \langle q\rangle\langle q\rangle = \langle\langle q\,q\rangle\rangle$, and the formalism naturally generates the correlation–response matrix

$$\mathsf{C}(1,2) \doteq \langle \delta\boldsymbol{\psi}(1)\delta\boldsymbol{\psi}^T(2)\rangle = \begin{pmatrix} C(1,2) & R(1;2) \\ R(2;1) & 0 \end{pmatrix} = \frac{\partial^2 W[\boldsymbol{\eta}]}{\partial\boldsymbol{\eta}(1)\delta\boldsymbol{\eta}(2)}\Big|_{\boldsymbol{\eta}=0}.$$

$$(4.139\text{a,b,c})$$

The generating-functional representation of the nonlinear dynamics is useful for direct attacks on the correlation and response functions. However, it does not in itself solve the closure problem, since further functional differentiations of $W[\boldsymbol{\eta}]$ with respect to $\boldsymbol{\eta}$ merely generate higher-order cumulants. To close the cumulant hierarchy, a further idea is required. The basic step is the introduction of a Legendre transformation from $\boldsymbol{\eta}$ (an external source) to the mean field $\langle\langle\boldsymbol{\psi}\rangle\rangle[\boldsymbol{\eta}] = \delta W[\boldsymbol{\eta}]/\delta\boldsymbol{\eta}$ (an internal quantity). That mean field is necessarily nonzero for nonzero $\boldsymbol{\eta}$. The MSR approach, which generalizes seminal techniques of quantum field theory (de Dominicis and Martin, 1964a), is to formulate, in an $\boldsymbol{\eta}$-independent way, functional integral equations relating the statistical observables. Once those equations are in hand, one can trivially pass to the limit $\boldsymbol{\eta} \to \mathbf{0}$, thereby obtaining information about the statistical dynamics of the original system. The basic idea is illustrated in Fig. 4.11.

Of course, the functional integral equations are extremely complex, and I shall not review here the details of their development; see the review by Krommes (2002) or the textbook by Amit (2005). In brief, statistical closure is effected by the introduction of so-called *vertex functions*, which are essentially suitably normalized three-point correlation or response functions. At lowest order, only one vertex function is nonzero, and that is approximated by the bare coupling coefficient. What results is the *direct-interaction approximation* (see the next section).

To what extent MSR vertex renormalizations can be useful[b] is a matter of some debate. Certainly the MSR formalism is "not a panacea," as Martin

---

[a]This result is a generalization of $(2\pi)^{-1}\int dx\,dk\,k e^{ikx} = \int dk\,k\delta(k) = 0$. Note that although the exponent in this example is symmetrical in x and k, the path-integral formalism is not symmetrical in q and $\widehat{q}$.

[b]In the presence of symmetries one can deduce *Ward identities* (Amit, 2005).

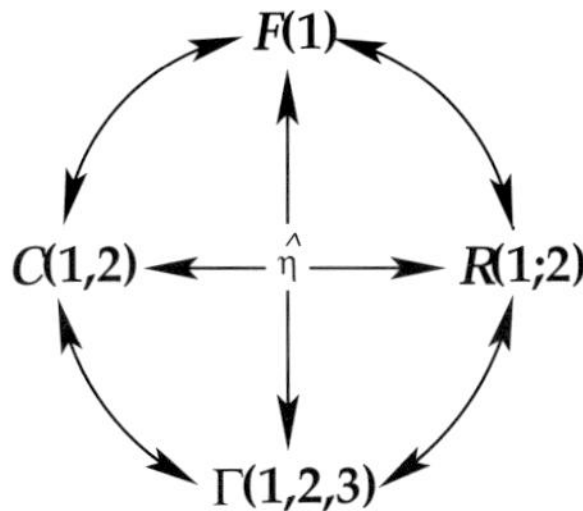

Fig. 4.11   The basic idea of the MSR formalism. When the nonlinear system is probed with $\boldsymbol{\eta}$, all statistical observables change functionally in concert. Here the mean field is represented by $\boldsymbol{F}$ (not to be confused with the arbitrary functional employed in the development of the path-integral representation), and a set of three vertex functions is represented by Γ. Variations with respect to $\boldsymbol{\eta}$ can be replaced by variations with respect to $\boldsymbol{F}$; $\boldsymbol{\eta}$ can then be set to zero, leaving functional equations relating the observables.

(1976) has emphasized to this author privately. One important difficulty is that the formalism is Eulerian-based; therefore, no finite-order truncation of the vertex renormalizations are random-Galilean-invariant. This was stressed by Kraichnan (1964), who studied the first vertex renormalization in some detail. I will not pursue vertex renormalizations here.

4.2.4.2. *General form of the direct-interaction approximation*

Even though the direct-interaction approximation of Kraichnan (1959) omits vertex renormalization, it is still a sophisticated theory. You have learned about the DIA in the context of the NSE from Prof. McComb (Chap. 2), and further details are given by McComb (1990) and Krommes (2002). It can be viewed as a non-Markovian, two-time generalization of the wave kinetic equation of weak-turbulence theory. In addition to the exact equation for the mean field, it consists (for homogeneous statistics) of the following coupled system of equations for the (two-sided) correlation function $C_{\boldsymbol{k}}(t,t')$ and (one-sided) mean response function $R_{\boldsymbol{k}}(t;t')$:

$$(\partial_t - L_{\boldsymbol{k}})C_{\boldsymbol{k}}(t,t') + \int_0^t d\bar{t}\,\Sigma_{\boldsymbol{k}}^{\mathrm{nl}}(t;\bar{t})C_{\boldsymbol{k}}(\bar{t},t')$$

$$= \int_0^{t'} d\bar{t}\,F_{\boldsymbol{k}}^{\mathrm{nl}}(t,\bar{t})R_{\boldsymbol{k}}^*(t';\bar{t}), \tag{4.140a}$$

$$(\partial_t - L_{\boldsymbol{k}})R_{\boldsymbol{k}}(t;t') + \int_{t'}^t d\bar{t}\,\Sigma_{\boldsymbol{k}}^{\mathrm{nl}}(t;\bar{t})R_{\boldsymbol{k}}(\bar{t};t') = \delta(t-t'), \tag{4.140b}$$

where

$$\Sigma_{\boldsymbol{k}}^{\mathrm{nl}}(t;t') \doteq -\sum_{\Delta} M_{\boldsymbol{k},\boldsymbol{p},\boldsymbol{q}} M_{\boldsymbol{p},\boldsymbol{q},\boldsymbol{k}}^{*} R_{\boldsymbol{p}}^{*}(t;t') C_{\boldsymbol{q}}^{*}(t,t'), \qquad (4.141\mathrm{a})$$

$$F_{\boldsymbol{k}}^{\mathrm{nl}}(t,t') \doteq \frac{1}{2} \sum_{\Delta} |M_{\boldsymbol{k},\boldsymbol{p},\boldsymbol{q}}|^{2} C_{\boldsymbol{p}}^{*}(t,t') C_{\boldsymbol{q}}^{*}(t,t'). \qquad (4.141\mathrm{b})$$

The similarities of Eq. (4.140a) and Eqs. (4.141) to their Markovian wave kinetic equation counterparts (4.98) and (4.99) are evident. Roughly speaking, the intermediate $\bar{t}$ integrals in Eqs. (4.140) determine an effective mode–mode correlation time that in Eqs. (4.99) has been called $\theta_{\boldsymbol{k},\boldsymbol{p},\boldsymbol{q}}$. I will be more specific about the connection in Sec. 4.2.5. below.

4.2.4.3. *Applications of the direct-interaction approximation to plasmas*

The DIA for plasma physics was pursued by DuBois and coworkers, Krommes, and others in the period roughly spanning 1975–1985. Much of the history was reviewed by Krommes (2002), and it is unnecessary to repeat that lengthy discussion here. Suffice it to say that, in the end, the DIA for plasmas has so far proven to be useful more as an organizing template for conceptual discussion and further approximations rather than as a tool for explicit calculations. One important technical difficulty relates to anisotropy. For neutral fluids, it is sometimes not unreasonable to contemplate situations of homogeneous, isotropic turbulence. Then the DIA can be dramatically simplified by analytically performing the integrals over wavevector angles. For magnetically confined plasmas with background profile gradients, however, the fluctuations are intrinsically anisotropic because of (i) the dependence of the diamagnetic frequency on k_y, and (ii) the fundamental physics distinction between the parallel and perpendicular directions. Inhomogeneity is also an issue if one is attempting to simultaneously treat fluctuations in the entire confinement vessel. By working in sum and difference coordinates and applying WKB techniques, it is possible to develop (Yoshizawa et al., 2001) a two-scale specialization of the general inhomogeneous DIA, and that has been used recently (Gürcan et al., 2006) for very approximate investigation of the tendency of turbulence to spread from linearly unstable to linearly stable zones. However, rigorous (even numerical) solutions of the full DIA for problems of interest to plasmas are few (LoDestro et al., 1991).

One promising approach to numerical solution of non-Markovian closures closely related to the DIA involves the use of non-Gaussian restarts or *cumulant updates* (Rose, 1985) that simplify the calculation of the time-history integrals. Detailed investigations of this technique for 2D turbulence have been done by Frederiksen and coworkers [see, e.g., Frederiksen et al.

(1994) and Frederiksen and Davies (2000)]; the most recent account can be found in the Workshop paper of Frederiksen (2006b).

Finally, one should note McComb's local-energy-transfer (LET) theory, which addresses the issue of random Galilean invariance that plagues the DIA. Although that issue is frequently considered to be unimportant for plasma microturbulence, which typically does not have a well-developed inertial range, the theory is worth appreciating on its own. For details and references, see the lectures of Prof. McComb (Chap. 2).

4.2.5. *Markovian closures*

Although the full DIA is quite complicated in practice, its general structure is important. Its nonlinear terms properly conserve quadratic invariants, and its time-history integrals afford a plausible self-consistent determination of nonlinear time scales (modulo important issues about random Galilean invariance and the interactions of disparate scales). In this section, I will show how the DIA can be used to motivate a Markovian spectral evolution equation that generalizes to strong turbulence the weak-turbulence WKE; then I will describe some plasma applications of Markovian closures.

4.2.5.1. *The DIA-based EDQNM*

The basic procedure is to (i) assert the fluctuation–dissipation relation $C_{k,+}(t,t') = R_k(t;t')C_k(t,t)$, where the subscript $+$ denotes one-sided in time; and (ii) assume that the time-lagged response function possesses the Markovian dynamics

$$\partial_t R_k(t;t') + \eta_k R_k(t;t') = \delta(t - t'), \tag{4.142}$$

where $\eta_k \doteq -L_k + \eta_k^{\mathrm{nl}}$. If the fluctuation–dissipation ansatz is inserted into the equal-time version of Eq. (4.140a), one encounters the construction

$$\theta_{k,p,q} \doteq \int_0^t dt' \, G_k(t,t')G_p(t,t')G_q(t,t'), \tag{4.143}$$

where $R_k(t;t') \doteq H(t - t')G(t,t')$. $\theta_{k,p,q}$ is called the *triad interaction time*.[a] One is led to a renormalized spectral balance equation of the same form as Eqs. (4.98) and (4.99), but with $\theta_{k,p,q}$ defined not by Eq. (4.100) but rather by the evolution equation that follows by differentiating Eq. (4.143),

$$\partial_t \theta_{k,p,q} + \Delta\eta\,\theta_{k,p,q} = 1 \quad [\theta_{k,p,q}(0) = 0]. \tag{4.144}$$

The steady-state solution of Eq. (4.144) is

$$\theta_{k,p,q} = \Delta\eta^{-1}, \tag{4.145}$$

---

[a]More precisely, it is $\mathrm{Re}\,\theta$ that ought to be called the triad interaction time; however, frequently people do not bother to make the distinction. See further discussion below.

showing how the relaxation rates of all three Fourier amplitudes participate symmetrically in defining the overall coherence time of the renormalized three-wave interaction.

The equations (4.98), (4.99), (4.142), and (4.144) define what is sometimes called (Bowman et al., 1993) the *DIA-based eddy-damped quasinormal Markovian approximation* (EDQNM). It is a plausible generalization of the wave kinetic equation of weak-turbulence theory to predict self-consistently determined nonlinear damping times η_k^{nl} while preserving energy-conserving spectral dynamics. But is it well-behaved? For example, does $C_k(t)$ remain non-negative for all k's and all times?

That the true $C_k(t)$ is non-negative is one example of a *realizability constraint*. An infinite set of such constraints follow from the basic fact that any probability density function(al) is non-negative; they can be presented as inequalities relating moments of various orders (Kraichnan, 1980). I will not pause to describe those here. However, a basic argument about realizability is very important: *If one can demonstrate a well-posed stochastic differential equation for a primitive random amplitude such that the resulting statistics are described exactly by a statistical closure, then that closure must be realizable.* Let us see how that works for the EDQNM. Consider the Langevin equation

$$\partial_t \psi_k - L_k \psi_k + \eta_k^{\mathrm{nl}} \psi_k = \widetilde{f}_k(t), \qquad (4.146)$$

where $\widetilde{f}_k$ is a centered white-noise process such that

$$\langle \widetilde{f}_k(t) \widetilde{f}_k^*(t') \rangle = 2F_k^{\mathrm{nl}} \delta(t - t'). \qquad (4.147)$$

It is not difficult to show that the resulting spectral balance equation is just Eq. (4.98). Realizability [specifically, $C_k(t) \geq 0$] is thus assured if one can construct an $\widetilde{f}_k$ that reproduces Eq. (4.147). One possibility is

$$\widetilde{f}_k(t) = \frac{1}{\sqrt{2}} \widetilde{w}(t) \sum_\Delta M_{k,p,q} [\mathrm{Re}\, \theta_{k,p,q}]^{1/2} \widetilde{\xi}_p^*(t) \widetilde{\xi}_q^*(t), \qquad (4.148)$$

where $\widetilde{w}(t)$ is a white-noise process with unit intensity [i.e., $\langle \widetilde{w}(t)\widetilde{w}(t') \rangle = \delta(t - t')$] and the ξ_k's are auxiliary Gaussian variables with variance $C_k(t)$. Here $\mathrm{Re}\, \theta_{k,p,q}$ is interpreted as the nonlinear correlation time; its presence is required to ensure dimensional consistency [due to the presence of $\widetilde{w}(t)$].

This construction can be shown to be well-founded for situations in which linear waves are absent[a] ($L_k = \gamma_k$). Then $\theta_{k,p,q}$ is real, and the explicit solution of Eq. (4.144),

$$\theta_{k,p,q}(t) = \int_0^t dt' \exp\left(-\int_{t'}^t d\bar{t}\, \Delta\eta(\bar{t}) \right), \qquad (4.149)$$

---

[a]One example is the NSE for homogeneous turbulence.

shows that always $\theta_{k,p,q} \geq 0$. The square root in Eq. (4.148) is then well defined, and the entire Langevin construction makes sense; *the wave-free EDQNM is realizable.*

The situation is not so clear when linear waves are allowed (and, as we have seen, such waves are ubituitous in plasma physics). Now $\theta_{k,p,q}$ and $\Delta\eta$ become complex and, although the general formula (4.149) still holds, there is no guarantee that $\mathrm{Re}\,\theta_{k,p,q} \geq 0$. The prospect of a negative correlation time is disquieting,[a] and the interpretation of the square root in Eq. (4.148) is unclear. If F_k itself becomes negative, realizability is certainly violated because a negative coefficient in Eq. (4.147) is incompatible with a positive PDF. Indeed, direct numerical solutions of the EDQNM reveal that the closure is not realizable in the presence of wave physics. This result, due to Bowman et al. (1993), was a major and unwelcome surprise at the time.

Eventually, Bowman was able to cure this realizability difficulty by developing what he called the *realizable Markovian closure* (RMC). In essence, he modified the transient evolution of $\theta_{k,p,q}$ such that its real part was not allowed to develop negative regions, while ensuring that the time-asymptotic steady state reproduced that of the original EDQNM. While this is not hard to do for scalar amplitudes, it presents major technical difficulties for multi-field problems in which θ becomes a tensor in the field indices. Modifications must be done with care in order that nonlinear invariants are preserved. In the end, however, Bowman was successful.

4.2.5.2. *Plasma-physics applications of Markovian closures*

Specifically, the RMC was applied Hu et al. (1995, 1997) to the 2D version of the Hasegawa–Wakatani equations, in which the operator $\widehat{\alpha} \doteq -D_\parallel \nabla_\parallel^2$ is replaced by the constant adiabaticity parameter α. The RMC was solved numerically and the predicted particle flux was compared with that of direct numerical simulations. The agreement was excellent. A quasilinear calculation of the flux was quantitatively poor,[b] showing that nonlinear effects were important and were being correctly modeled by the RMC.

It should be noted that closures such as the DIA-based EDQNM or the RMC derived therefrom are not invariant to random Galilean transformations. Bowman and Krommes (1997) have explored realizable test-field models that generalize the RMC.

The results of Bowman and Hu represent the most quantitatively suc-

---

[a]Note that, for any complex variable z, $\mathrm{Re}(z)$ is not an analytic function of z.

[b]In this context, quasilinear means that the cross-correlation $\langle \delta\varphi\, \delta n \rangle$ that enters in the definition of the flux is reduced to a k-dependent weighting of $C_k \doteq \langle |\delta\varphi_k|^2 \rangle$ by using the relationship between the variables obtained from linear theory; the resulting formula is evaluated with the measured C_k. Such a calculation necessarily preserves dimensional and scaling constraints.

cessful achievements of statistical closure theory for plasmas. They demonstrate that, for a sufficiently simple nonlinear paradigm, analytical theory can make detailed predictions that go beyond the consequences of simple scaling theory. The RMC is certainly not the last word; a Markovian version of McComb's LET theory (Chap. 2) would be interesting to study, and the apparently successful, though numerically challenging, non-Markovian approach of Frederiksen (2006b) ought to be explored in plasma contexts. However, even Markovian closures for homogeneous statistics are quite complicated for multifield models; applications to inhomogeneous statistics such as are encountered in confinement devices may be too difficult to be worth the effort, and alternate approaches may be required. From the point of view of basic theory, the work of Kraichnan (1985) on statistical decimation deserves to be further explored. But it is not surprising that most effort on the spectral dynamics and associated transport for realistic situations focuses nowadays on increasingly sophisticated numerical simulations. One partial exception in which analytical methods have recently given useful insights is the theory of zonal flows, to be discussed in the next lecture.

4.3.   LECTURE 3 – Zonal Flows in Plasmas

We have thus seen that a statistical treatment (Markovian closure) of mode coupling can provide a sensible and quantitatively accurate description of two-point spectral functions and turbulent flux. However, the constant-k_z approximation used in the Hasegawa–Wakatani calculations of Hu et al. (1995, 1997) precludes study of one potentially very important ingredient of plasma turbulence: the *convective cells* (CC's) possessing, by definition, $k_z = 0$. As further special cases of CC's, one defines a *zonal mode* as a CC with $k_y = 0$ (and $k_x \neq 0$), and a *streamer* as a CC with $k_x = 0$ (and $k_y \neq 0$). Later, we will also encounter (Sec. 4.3.5.) the *geodesic acoustic mode*, which is a normal mode of oscillation of a CC (with both k_x and k_y nonzero) of importance in toroidal geometry.

A zonal potential, possessing only a $k_x \neq 0$, is associated with a *zonal flow* (ZF) in the y or poloidal direction: $V_y(x) = \partial_x \varphi(x)$. Such flows, having no component in the radial direction, do not directly contribute to transport.[a] However, because zonal flows can interact *via* nonlinear mode coupling with other Fourier amplitudes such as drift waves (DW's), they can affect the saturation level of the entire system and thereby regulate the overall transport level. The usual argument is that the presence of ZF's *reduces* the drift-wave fluctuation level, therefore inhibits the transport. That is because such flows are typically sheared, $\partial_x V_y(x) = \partial_x^2 \varphi(x)$, and

---

[a]This can also be understood spectrally because formulas such as (4.46b) involve k_y.

it is expected that shear flows should destroy DW eddies. This has been verified in a variety of computer simulations such as those of Lin et al. (1998).

Streamers, on the other hand, may be expected to contribute importantly to transport because of their large radial extent. They have been observed in numerical simulations (Drake et al., 1988; Jenko et al., 2000) and are expected to be important in ETG turbulence. Although space considerations preclude an extensive discussion of streamer physics (and the topic is in any event far from closed), see Sec. 4.3.6. for brief remarks on ETG transport and Sec. 4.4.4.3. for the role of streamers in the generation of blobs in edge turbulence.

In this lecture, I will describe some of the analytical methods that have proven to be useful in discussing the DW–ZF problem (still very much an area of active research). I will leave the present state of experimental understanding to the accompanying lectures of M. Shats (Chap. 5). A much lengthier review of ZF physics has been given by Diamond et al. (2005); see also the written version (Itoh et al., 2006) of the 2005 review talk by K. Itoh. However, the present emphases are somewhat different and my discussion is not a proper subset of the topics covered by those reviews. I will address

- the generation of zonal flows *via* modulational instability;
- the role of zonal flows in the transition to collisionless ITG turbulence;
- simple ideas about spectral bifurcations and the transition between low- and high-confinement modes in tokamaks;
- disparate-scale interactions of drift waves and zonal flows in steady-state turbulence;
- and a few remarks on geodesic acoustics modes and on transport due to ETG modes.

4.3.1. *Modulational instability and zonal-flow generation*

In discussing the generation of ZF's, it is important to distinguish between (i) an *initial-value problem,* in which DW's but no ZF's are present initially; and (ii) a *statistically steady turbulent state,* in which DW's and ZF's coexist. It is traditional to state that zonal flows are generated by the modulational instability of drift waves, and I will sketch such a derivation in this section.[a] However, that initial-value problem would not seem to be as relevant as an analysis of a statistically steady state, such as described by Markovian closure theory. That will be discussed in Sec. 4.3.4.

---

[a]Further insights and references on the generation of ZF's by modulational instability can be found in the Workshop paper of Dewar and Abdullatif (2006).

                    John A. Krommes

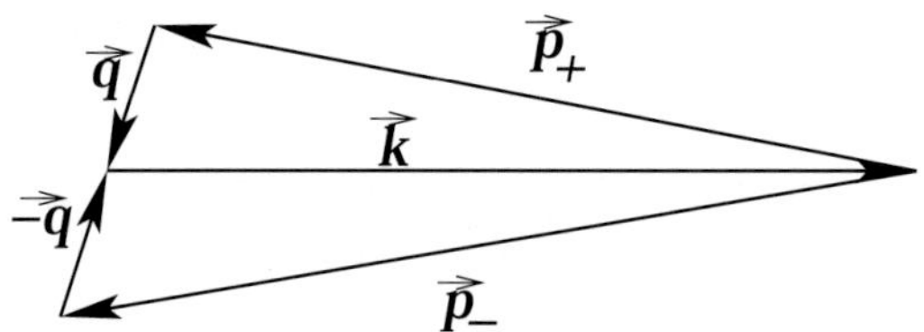

Fig. 4.12 Wavevector triads for the basic modulational instability calculation. The basic process is built from two three-wave interactions, not (as is sometimes asserted) a four-wave interaction.

I shall illustrate the modulational instability of drift waves by working with a general scalar-field model for quadratic nonlinearity, such as would be appropriate for the electrostatic Hasegawa–Mima equation.[a] Thus one begins with the general mode-coupling representation (4.97). For simplicity, let us ignore all linear growth or dissipation. One postulates a fixed-amplitude "pump" wave at wavevector $\boldsymbol{k}$ and inquires into the fate of a small-amplitude perturbation at wavevector $\boldsymbol{q}$. Now quadratic nonlinearity requires that wavevectors interact in triads, so *sidebands* are generated at wavevector $\boldsymbol{p}_\pm = -(\boldsymbol{k} \pm \boldsymbol{q})$; see Fig. 4.12. Let $\varphi_{\boldsymbol{p}_\pm} \equiv \varphi_\pm$. The small perturbations (denoted by Δ) then evolve according to

$$\partial_t \Delta\varphi_+ + i\Omega_+ \Delta\varphi_+ = M_{\boldsymbol{p}_+,\boldsymbol{q},\boldsymbol{k}} \Delta\varphi_{\boldsymbol{q}}^* \varphi_{\boldsymbol{k}}^*, \tag{4.150a}$$

$$\partial_t \Delta\varphi_- + i\Omega_- \Delta\varphi_- = M_{\boldsymbol{p}_-,-\boldsymbol{q},\boldsymbol{k}} \Delta\varphi_{-\boldsymbol{q}}^* \varphi_{\boldsymbol{k}}^*, \tag{4.150b}$$

$$\partial_t \Delta\varphi_{\boldsymbol{q}} + i\Omega_{\boldsymbol{q}} \Delta\varphi_{\boldsymbol{q}} = M_{\boldsymbol{q},\boldsymbol{k},\boldsymbol{p}_+} \varphi_{\boldsymbol{k}}^* \Delta\varphi_+^* + M_{\boldsymbol{q},-\boldsymbol{k},-\boldsymbol{p}_-} \varphi_{-\boldsymbol{k}}^* \Delta\varphi_{-\boldsymbol{p}_-}^*. \tag{4.150c}$$

By using reality conditions such as $\varphi_{-\boldsymbol{k}} = \varphi_{\boldsymbol{k}}^*$, one finds that a closed linear system results for the amplitudes $\Delta\varphi_+$, $\Delta\varphi_-^*$, and $\Delta\varphi_{\boldsymbol{q}}^*$. For simplicity, I will ignore $\Omega_{\boldsymbol{q}}$. That would be rigorously correct if $\varphi_{\boldsymbol{q}}$ were a zonal flow, and is reasonable in general if $q \ll k$. One then must consider the equations

$$\partial_t \Delta\varphi_+ + i\Omega_+ \Delta\varphi_+ = [M_{\boldsymbol{p}_+,\boldsymbol{q},\boldsymbol{k}} \varphi_{\boldsymbol{k}}^*(t)]\Delta\varphi_{\boldsymbol{q}}^*, \tag{4.151a}$$

$$\partial_t \Delta\varphi_-^* - i\Omega_- \Delta\varphi_-^* = [M_{\boldsymbol{p}_-,-\boldsymbol{q},\boldsymbol{k}}^* \varphi_{\boldsymbol{k}}(t)]\Delta\varphi_{\boldsymbol{q}}^*, \tag{4.151b}$$

$$\partial_t \Delta\varphi_{\boldsymbol{q}}^* = [M_{\boldsymbol{q},\boldsymbol{k},\boldsymbol{p}_+}^* \varphi_{\boldsymbol{k}}(t)]\Delta\varphi_+$$
$$+ [M_{\boldsymbol{q},-\boldsymbol{k},-\boldsymbol{p}_-}^* \varphi_{\boldsymbol{k}}^*(t)]\Delta\varphi_-^*. \tag{4.151c}$$

This system is nontrivial because of the time dependence of the drift-wave $\varphi_{\boldsymbol{k}}(t)$ on the right-hand side. However, it can be converted into a conventional eigenvalue problem by the transformation $\Delta\varphi_\pm = e^{i\Omega_{\boldsymbol{k}}t}\Delta\widehat{\varphi}_\pm$. This

---

[a]Much more elaborate calculations have been done. See, for example, the analysis by Guzdar et al. (2001) of the modulational instability of finite-β drift waves (asserted to be relevant to the L–H transition in tokamaks; see further discussion in Sec. 4.3.3.) and the calculations by Singh et al. (2005) of the modulational instability of drift-resistive ballooning modes with electromagnetic effects, relevant to edge physics.

removes the rapid time dependence by introducing the frequency shifts $\Delta\Omega_\pm \doteq \Omega_k + \Omega_\pm = \Omega_k - \Omega_{k\pm q}$. Then

$$\partial_t \Delta\widehat{\varphi}_+ + i\Delta\Omega_+ \Delta\widehat{\varphi}_+ = (M_{\boldsymbol{p}_+,\boldsymbol{q},\boldsymbol{k}}|\varphi_{\boldsymbol{k}}|)\Delta\varphi_{\boldsymbol{q}}^*, \tag{4.152a}$$

$$\partial_t \Delta\widehat{\varphi}_-^* - i\Omega_- \Delta\widehat{\varphi}_-^* = (M_{\boldsymbol{p}_-,-\boldsymbol{q},\boldsymbol{k}}^*|\varphi_{\boldsymbol{k}}|)\Delta\varphi_{\boldsymbol{q}}^*, \tag{4.152b}$$

$$\partial_t \Delta\varphi_{\boldsymbol{q}}^* = (M_{\boldsymbol{q},\boldsymbol{k},\boldsymbol{p}_+}^*|\varphi_{\boldsymbol{k}}|)\Delta\widehat{\varphi}_+$$
$$+ (M_{\boldsymbol{q},-\boldsymbol{k},-\boldsymbol{p}_-}^*|\varphi_{\boldsymbol{k}}|)\Delta\widehat{\varphi}_-^*. \tag{4.152c}$$

Eigenvalues λ $(\partial_t \to \lambda)$ are thus determined from

$$\det \begin{pmatrix} -i\Delta\Omega_+ - \lambda & 0 & M_{\boldsymbol{p}_+,\boldsymbol{q},\boldsymbol{k}}|\varphi_{\boldsymbol{k}}| \\ 0 & i\Delta\Omega_- - \lambda & M_{\boldsymbol{p}_-,-\boldsymbol{q},\boldsymbol{k}}^*|\varphi_{\boldsymbol{k}}| \\ M_{\boldsymbol{q},\boldsymbol{k},\boldsymbol{p}_+}^*|\varphi_{\boldsymbol{k}}| & M_{\boldsymbol{q},-\boldsymbol{k},-\boldsymbol{p}_-}^*|\varphi_{\boldsymbol{k}}| & -\lambda \end{pmatrix} = 0. \tag{4.153}$$

This is a cubic equation in λ, for which exact analytical solutions are available. Those can be used to show that if $|\varphi_{\boldsymbol{k}}|^2$ is large enough (relative to the stabilizing dispersion term $\Delta\Omega_+\Delta\Omega_-$), then there is for $\omega = i\lambda$ one real root and two complex conjugate ones (one of which is the modulational instability). To get a flavor for the result, one can proceed approximately by introducing the multiscale ordering parameter $\epsilon \doteq q/k \ll 1$ and assuming that $M|\varphi_{\boldsymbol{k}}| = O(1)$.[a] Asymptotic balance using the method of Kruskal (1965) shows that the dominant balances are between either the terms of $O(\lambda^0)$ and $O(\lambda^1)$ (giving an oscillatory root) or $O(\lambda^1)$ and $O(\lambda^3)$, giving the approximate dispersion relation[b]

$$\lambda^2 = -\Delta\Omega_+\Delta\Omega_- + (MM)|\varphi_{\boldsymbol{k}}|^2, \tag{4.154a}$$

$$(MM) \doteq M_{\boldsymbol{p}_+,\boldsymbol{q},\boldsymbol{k}}M_{\boldsymbol{q},\boldsymbol{k},\boldsymbol{p}_+}^* + M_{\boldsymbol{p}_-,-\boldsymbol{q},\boldsymbol{k}}^*M_{\boldsymbol{q},-\boldsymbol{k},\boldsymbol{p}_-}^*. \tag{4.154b}$$

Wave dispersion is thus stabilizing, but for sufficiently large DW amplitude [and assuming that $(MM) > 0$], one finds an instability with growth rate[c]

$$\lambda = (MM)^{1/2}|\varphi_{\boldsymbol{k}}|. \tag{4.155}$$

This is the modulational instability for the initial growth of convective cells (especially zonal flows) driven by drift waves. Dimensionally, λ is the sweeping frequency associated with advection of a long wavelength passive scalar by a drift-wave $\boldsymbol{E} \times \boldsymbol{B}$ velocity: $\lambda = O(\boldsymbol{q} \cdot \boldsymbol{V}_{E,\boldsymbol{k}})$.

---

[a]This is not quite correct. If the M's describe $\boldsymbol{E} \times \boldsymbol{B}$ advection, then they involve the factor $\widehat{\boldsymbol{z}} \cdot \boldsymbol{q} \times \boldsymbol{k} = O(\epsilon)$. Other factors may offset that, however, and at least for generalized Hasegawa–Mima dynamics it is consistent to assume that $(MM) = O(\epsilon)$. This does not change the dominant balances.

[b]This result is not entirely consistent (it omits an imaginary term arising from iteration); however, it captures the spirit of the solution and the proper threshold.

[c]As an example, one may consider generalized Hasegawa–Mima dynamics, for which $M_{\boldsymbol{k},\boldsymbol{p},\boldsymbol{q}} = \widehat{\boldsymbol{z}} \cdot (\boldsymbol{p} \times \boldsymbol{q})[(\alpha_q + q_\perp^2) - (\alpha_p + p_\perp^2)]/(\alpha_k + k_\perp^2)$, with α vanishing for the zonal mode. Through lowest order in q/k and for $\boldsymbol{q} = q\widehat{\boldsymbol{x}}$, one finds $(MM) = 2|\boldsymbol{k} \times \boldsymbol{q}|^2[1 - 4k_x^2/(1 + k_\perp^2)]$, which is positive for $k_x^2 < \frac{1}{3}(1 + k_y^2)$.

4.3.2. *Zonal flows, the Dimits shift, and the transition to ITG turbulence*

In Lecture 1, I briefly mentioned the numerical simulations of Dimits et al. (2000), which indicated that for collisionless ITG modes the transition to turbulence occurs, as a function of the temperature gradient κ_T, not at the threshold for linear instability κ_c but rather at a larger value κ_*. This phenomenon is called the *Dimits shift*. Dimits *et al.* believed and Rogers et al. (2000) argued convincingly that the phenomenon is due to the generation of ZF's that wipe out the DW's[a] for $\kappa_T < \kappa_*$. Only with the recent work of Kolesnikov and Krommes (2005a, 2005b), however, has an attempt been made to address the transition from the point of view of systematic bifurcation theory. The calculation has some unusual aspects that I will briefly describe.

From the previous section, we know that ZF generation from a DW pump requires a DW sideband (SB) as well. Kolesnikov and Krommes therefore considered a fluid model that contained just one DW and one ZF. To naturally capture a linear instability of the curvature-driven ITG type, at least two coupled fields are required; Kolesnikov and Krommes considered vorticity ϖ and temperature T. With the assumed boundary conditions,[b] the zonal modes are real, so the number of independent variables is

$$
\begin{aligned}
\text{DW:}\ & 2 \times 2 = 4 && \text{(complex, two fields)} \\
\text{DW sideband:}\ & 2 \times 2 = 4 && \text{(complex, two fields)} \\
\text{ZF:}\ & \underline{1 \times 2 = 2} && \text{(real, two fields)} \\
& 10 && \text{(total real degrees of freedom).}
\end{aligned}
$$

The resulting model will not be written down here; consult Kolesnikov and Krommes (2005a) for the details.[c] Fig. 4.13 shows the result of a numerical solution of the 10D model for $\kappa_c < \kappa_T < \kappa_*$, projected into the 3D subspace whose axes are the real part of ITG vorticity (x), the imaginary part of ITG vorticity (y), and the zonal vorticity (z). Colors correspond to different segments of the temporal evolution. The figure shows that from an arbitrary initial condition, a burst of DW activity (and a resulting driven ZF fluctuation) ensues. At later times, however, the dynamics are attracted to

---

[a]Following conventional, if sloppy, usage, I refer to generic "DW's" rather than the more specific "ITG modes."

[b]Fluctuations are assumed to vanish at $x = 0$ and $x = \pi/k_x$. This boundary condition is a very crude attempt to model the localizing effect of magnetic shear on radial eigenfunctions (Horton et al., 1996).

[c]Additional remarks and background on the calculations of Kolesnikov and Krommes can be found in the Workshop paper of Krommes (2006) included in the accompanying volume.

a stable fixed point in the phase space for which only the zonal amplitudes are nonzero. Simulation also reveals that the fixed point becomes unstable at a certain value κ_*; that value is identified with the upper boundary of the Dimits-shift regime.[a]

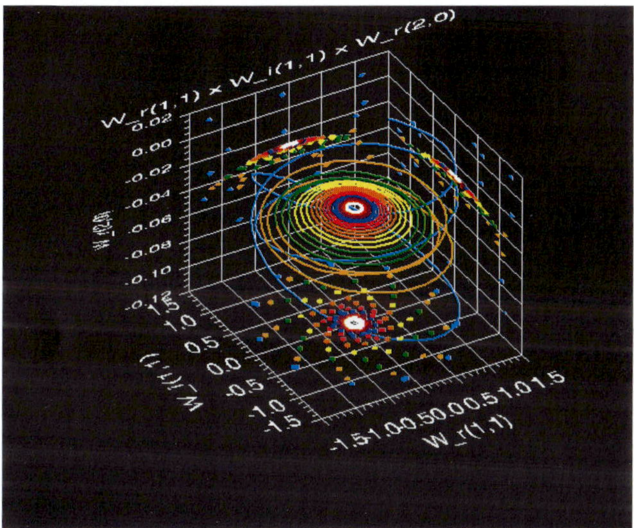

Fig. 4.13   Projection of the 10D ITG dynamics onto the 3D subspace composed of $\mathrm{Re}\,\varpi_1$, $\mathrm{Im}\,\varpi_1$, and ϖ_2. The colors indicate successive intervals of the time evolution. The dynamics evolve through an initial transient toward a fixed point with vanishing ITG activity but nonzero zonal flow. Several color movies of such dynamics can be found at `ftp://ftp.pppl.gov/pub/krommes/Misc/Dimits_shift/movie*.mpg`.

Analytically, it is of interest to calculate the position of the fixed point and the value of κ_*. That can be addressed either by constructing a local center manifold (Guckenheimer and Holmes, 1983) or by (somewhat tedious) direct analysis of the 10D system. One important point is that undamped ZF's must be counted in the center eigenspace along with the bifurcating DW. Thus, in the present model the center eigenspace and associated center manifold are 4D. This must be distinguished from the case of order-unity ZF damping, for which the center eigenspace is just 2D and a conventional Hopf bifurcation occurs at linear threshold.

Although much can be said about the bifurcation analysis (and the deficiencies of the severely truncated[b] slab model), I shall not do so here.

---

[a]The 10D model does not saturate for $\kappa_T > \kappa_*$. That is a deficiency; however, higher-order truncations do saturate.

[b]Recall from the discussion in Sec. 4.1.7. that we are dealing with a spatially extended system. The fundamental three modes that are considered in the present bifurcation

The central point is that the model illustrates the importance of ZF's is suppressing DW's. In the Dimits-shift regime, the suppression is complete. Above the turbulence threshold, that is not the case; DW's and ZF's coexist. I will discuss aspects of that self-consistent interaction in Sec. 4.3.4.. First, however, I will introduce a toy model of interacting DW and ZF spectra.

4.3.3. *Spectral bifurcations and the L–H transition*

In 1982 a spontaneous transition between states of poor confinement (the *low mode* or *L mode*) and relatively better confinement (the *high mode* or *H mode*) was discovered on the ASDEX tokamak (Wagner et al., 1982). Such transitions have now been observed on many devices; their physics remains a topic of contemporary interest. A very brief introduction to current thinking, together with key references, is given by White et al. (2006). One possible mechanism is the spontaneous generation of zonal flows (sheared poloidal flows), which then suppress the turbulence. This amounts to a bifurcation of the spectral intensity.

The three-wave DW–SB–ZF system requires the presence of the ZF in order to define the third leg of the wave-number triad. It therefore does not support the conventional solution in which only DW's are excited. Because many-mode models are very difficult to study analytically, Diamond et al. (1994) considered the toy model of coupled DW intensity $\mathcal{D}$ and ZF intensity $\mathcal{Z}$ (both assumed to be non-negative)

$$\dot{\mathcal{D}} = (\gamma - \alpha\mathcal{D})\mathcal{D} - \mathcal{Z}\mathcal{D}, \tag{4.156a}$$

$$\dot{\mathcal{Z}} = -\nu\mathcal{Z} + \mathcal{D}\mathcal{Z}. \tag{4.156b}$$

Here γ represents the linear DW growth rate, α is a measure of the non-linear DW self-interaction, and ν represents linear collisional damping on the ZF's. Total fluctuation energy $\mathcal{E}$ is defined by $\mathcal{E} = \mathcal{D} + \mathcal{Z}$. The coefficient of the energy-conserving cross-coupling terms $\mathcal{Z}\mathcal{D}$ has been set to one for simplicity.[a] This model supports two fixed points (statistically steady states), which I shall call the $\mathcal{D}$ *state* and the $\mathcal{Z}$ *state*:

$$\mathcal{D}\ \text{state:}\ \mathcal{D} = \gamma/\alpha, \quad \mathcal{Z} = 0, \tag{4.157a}$$

$$\mathcal{Z}\ \text{state:}\ \mathcal{D} = \nu, \quad \mathcal{Z} = \gamma - \alpha\nu. \tag{4.157b}$$

Linear stability analysis of these fixed points readily reveals that the pure $\mathcal{D}$ state with no zonal activity is stable (and the $\mathcal{Z}$ state unstable) for

---

analysis are microscopic in scale. The possibility then exists of the modulational instability of an essential continuum of long-wavelength excitations that couple to the fundamental modes. A first attempt at describing that phenomenon with the aid of multiple-scale analysis is described by Kolesnikov and Krommes (2005a).

[a]This amounts to a normalization of the time scale.

sufficiently large ν; the $\mathcal{D}$ state is destabilized when $\gamma \geq \alpha\nu$, in which regime the $\mathcal{Z}$ state is stable (with positive ZF fluctuation intensity $\mathcal{Z}$). Clearly the $\mathcal{D}$ state (with high DW activity and transport) is a proxy for the low confinement mode; similarly, the $\mathcal{Z}$ state represents the high confinement mode. The model therefore demonstrates the possibility of an L–H *spectral bifurcation* controlled by the values of physical parameters.[a]

At the time of writing, experimental verification of such a scenario appears to be mixed. Bicoherence studies on the DIII-D tokamak (Moyer et al., 2001) appear to be consistent with a Reynolds-stress-driven sheared flow scenario. However, more recent bicoherence studies on the NSTX spherical tokamak were not in obvious accord with such a scenario (White et al., 2006), although the diagnostics employed were difficult and preliminary. Further work obviously remains, but it seems plausible that as experimental techniques are refined the detailed physics of the L–H transition will be uncovered within the next few years. In the accompanying lectures by M. Shats (Chap. 5), you will learn more about the current experimental situation.

4.3.4. *Interactions of disparate scales in steady-state turbulence*

Such spectral-bifurcation models make no attempt to open up the details of the nonlinear physics on a mode-by-mode basis. That is by no means easy to do and remains a subject of current interest. However, analytical treatments of the interactions between CC's and DW's are considerably simplified if the modes are assumed to be well separated in spatial scale (Diamond et al., 1998). If the DW's have wave number $\boldsymbol{k}$ and the CC's have wave number $\boldsymbol{q}$, then the assumption of relatively long-wavelength CC's corresponds to the inequality $\epsilon \doteq q/k \ll 1$. Note that we do not need to specialize to ZF's at this time.

Early work on the ZF problem (Diamond et al., 1998) relied on various intuition about the proper form of the equations for disparate-scale interactions, such as use of a wave kinetic equation to describe the propagation of a wave packet of DW's in the long-wavelength field of a ZF. While many such ideas are heuristically correct, certain subtleties can be appreciated and mistakes avoided only with detailed understanding of the structure and content of (at least) Markovian closure theory. Krommes and Kim (2000) attacked that problem directly by expanding in small ϵ the EDQNM approximation for generalized Hasegawa–Mima dynamics. That permits one to make contact with traditional discussions of eddy viscosity,

---

[a]The present model is extremely heuristic, and certain implications drawn from it can be misleading. For further discussion and extensions of the model, see Ball et al. (2002) and Ball (2006).

as we will see.

First, however, let me sketch a heuristic derivation of the nonlinear ZF growth rate for the specific case of generalized Hasegawa–Mima dynamics. The idea is to relate the growth of ZF energy to the corresponding loss of DW energy when the short-wavelength drift wave packet is modulated by the long-wavelength ZF. To describe that modulation, one needs an appropriate wave kinetic equation for the DW's.

4.3.4.1.   *Wave kinetic equations for weakly inhomogeneous dynamics*

The reduction of spectral balance equations for weakly inhomogeneous dynamics to wave kinetic equations (WKE's) was addressed early on in an important paper by Carnevale and Martin (1982), who considered the equal-time version of the Dyson equation (4.140a) for the two-point spectrum. A subtle mathematical error in their calculations, very relevant to the proper form of the WKE's discussed below, was discussed in detail by Krommes and Kim (2000). If we concentrate just on the DW–ZF interactions, the correct result has the form

$$\partial_T \mathcal{N}_{\boldsymbol{k}} - \{\Omega_{\boldsymbol{k}}, \mathcal{N}_{\boldsymbol{k}}\} = 0. \tag{4.158}$$

Here $\mathcal{N}_{\boldsymbol{k}}$ is proportional,[a] in a way that I will describe in the next section, to the basic two-point spectrum $C_{\boldsymbol{k}}(T)$ introduced in Lecture 2. Also, $\Omega_{\boldsymbol{k}} \doteq \boldsymbol{k} \cdot \underline{\boldsymbol{V}}$ is the sweeping frequency of a DW of wave number $\boldsymbol{k}$ in the presence of a CC velocity field $\underline{\boldsymbol{V}}$. Finally, the notation $\{A, B\}$ means the Poisson bracket of A and B:

$$\{A, B\} \doteq \left(\frac{\partial A}{\partial \boldsymbol{X}} \cdot \frac{\partial B}{\partial \boldsymbol{k}} - \frac{\partial A}{\partial \boldsymbol{k}} \cdot \frac{\partial B}{\partial \boldsymbol{X}}\right) + [(\boldsymbol{X}, \boldsymbol{k}) \Leftrightarrow (T, -\omega)] \tag{4.159a}$$

$$= \frac{\partial}{\partial \boldsymbol{k}} \cdot \left(\frac{\partial A}{\partial \boldsymbol{X}} B\right) - \frac{\partial}{\partial \boldsymbol{X}} \cdot \left(\frac{\partial A}{\partial \boldsymbol{k}} B\right) + [(\boldsymbol{X}, \boldsymbol{k}) \Leftrightarrow (T, -\omega)] \tag{4.159b}$$

where X and T describe the slow temporal and long spatial CC variations while $\boldsymbol{k}$ and ω describe Fourier components of the rapid DW variations. Thus Eq. (4.158) (which arises by integrating over all ω) is explicitly

$$\partial_T \mathcal{N}_{\boldsymbol{k}}(\boldsymbol{X}, T) + \partial_{\boldsymbol{k}}\Omega_{\boldsymbol{k}} \cdot \nabla \mathcal{N}_{\boldsymbol{k}} - \nabla \Omega_{\boldsymbol{k}} \cdot \partial_{\boldsymbol{k}} \mathcal{N}_{\boldsymbol{k}} = 0. \tag{4.160}$$

The conservative form (4.159b) guarantees that the spatially averaged plasmon number[b] $\overline{\mathcal{N}} \doteq \int d\boldsymbol{X} \sum_{\boldsymbol{k}} \mathcal{N}_{\boldsymbol{k}}(\boldsymbol{X}, T)$ is conserved, even under the interaction between the DW's and the ZF's.

---

[a] $\mathcal{N}_{\boldsymbol{k}}$ has sometimes been called the *plasmon density* or *plasmon action*. However, it *differs* in both form and interpretation from the wave action density defined in Eq. (4.95). Wave action can be discussed within a linear context, while the interesting properties of $\mathcal{N}_{\boldsymbol{k}}$ are associated with nonlinear interactions.

[b] Here and elsewhere I am mixing discrete wavevector sums and the continuum approximation ($\partial/\partial \boldsymbol{k}$ appears in the WKE). It is easier to keep track of dimensions this way, since $\sum_{\boldsymbol{k}}$ is dimensionless; however, in the actual evaluation of the formulas wave-number integration is implied.

In the calculations to follow on the ZF growth rate, it is important to understand that Ω_k is a nonlinear (fluctuation-dependent) advection frequency, not the linear mode frequency. To understand how such a quantity arises, consider the generalized Hasegawa–Mima dynamics (4.55). Among other things, that equation describes the advection of drift waves (φ'_k) by ZF's ($\underline{\varphi}_q$) according to $0 = \partial_t \varphi' + (1 - \nabla_\perp^2)^{-1} \underline{V}_E \cdot \boldsymbol{\nabla}(1 - \nabla_\perp^2)\varphi'$; to lowest order in q/k, the Fourier transform of the advection term is just $i\boldsymbol{k} \cdot \underline{V}_{E,q} \doteq i\Omega_k$.

4.3.4.2. *Casimir invariants and the plasmon density*

To identify a quantity $\overline{\mathcal{N}}$ that is conserved under the DW–ZF interaction, one may turn to the theory of noncanonical (functional) Hamiltonian dynamics, as reviewed by Morrison (1998). Quantities that are conserved regardless of the Hamiltonian are called *Casimir invariants*; they are properties merely of the Poisson brackets. Krommes and Kolesnikov (2004) have developed a general formalism for systems of coupled PDE's containing the $\boldsymbol{E} \times \boldsymbol{B}$ advective nonlinearity; those can be written in the general form (I assume the summation convention for repeated indices)

$$\partial_t \psi^i = \widehat{L}^i{}_j \psi^j + \{\psi^i, \mathcal{H}\}, \quad \text{where} \quad \{A, B\} \doteq \overline{S^{ij}[\psi]\left[\frac{\delta A}{\delta \psi^i}, \frac{\delta B}{\delta \psi^j}\right]}, \quad \text{(4.161a,b)}$$

$\widehat{L}^i{}_j$ is an arbitrary linear operator, $S^{ij}[\psi] \doteq S^{ij}{}_k \psi^k$ is assumed to be symmetric in i and j, and $[A, B] \doteq \widehat{z} \cdot \boldsymbol{\nabla} A \times \boldsymbol{\nabla} B$. It can then be shown (Krommes and Kolesnikov, 2004) that a Casimir invariant conserved by the nonlinear term is $\mathcal{C} = \frac{1}{2}\overline{\psi^i \psi^i}$. For example, when one specializes to a single scalar equation, the Poisson bracket becomes proportional to $\overline{\psi[\delta A/\delta\psi, \delta B/\delta\psi]}$. As pointed out by Morrison and Greene (1980), the equation $\partial_t \varpi = \{\varpi, \mathcal{H}\}$ reproduces the 2D Euler equation if $\mathcal{H} = \frac{1}{2}\overline{\varpi(-\nabla^{-2})\varpi} = \frac{1}{2}\overline{|\boldsymbol{\nabla}\varphi|^2}$; then $\mathcal{C}$ is the enstrophy $\frac{1}{2}\overline{\varpi^2}$. When ϖ is replaced by[a] $N \doteq (\widehat{\alpha} - \nabla^2)\varphi$, one obtains (Krommes and Kolesnikov, 2004) the generalized Hasegawa–Mima dynamics (4.55), with Casimir

$$\mathcal{C} = \frac{1}{2}\overline{N^2}. \quad \text{(4.162)}$$

The Casimirs discussed so far are conserved by the primitive dynamics. However, we will be most interested in statistical formalisms, in which Casimirs will be mean-square fluctuation quantities linearly proportional (through wave-number weightings or gradients) to the fluctuation intensity $\mathcal{E}$; I will denote the statistical Casimirs by $\mathcal{N}$ rather than $\mathcal{C}$. Now consider a disparate-scale statistical decomposition into short-wavelength

---

[a] N is just the ion gyrocenter density.

DW's and long-wavelength CC's. It was shown quite generally by Krommes and Kolesnikov (2004) that in mean square *the short-wavelength contribution to the Casimir remains conserved under long-wavelength modulation* to lowest order in the scale-separation parameter ϵ. Fundamentally, this is a consequence of the fact that Casimir conservation is independent of the Hamiltonian. That the specific $\mathcal{N}$ given by Eq. (4.162)[a] is conserved under the DW–ZF interaction was first proven with different means by Smolyakov and Diamond (1999); those authors did not appreciate the connection to Hamiltonian dynamics and Casimir invariants.

4.3.4.3. *Modulation and random ray propagation*

Although the Casimir is conserved under the evolution of the wave packet, the central wave number $\boldsymbol{k}$ of that packet is not conserved under long-wavelength modulation. Instead, the characteristics of Eq. (4.160) show that $\boldsymbol{k}$ evolves (in a Lagrangian sense) according to the ray equation

$$\dot{\boldsymbol{k}} = -\boldsymbol{\nabla}\Omega_{\boldsymbol{k}}. \tag{4.165}$$

(This equation describes refraction of the ray by the ZF, which modulates the density seen by the DW.) Because of the prevalence of $\overline{k}^2 \doteq 1 + k^2$ in the theory, let us dot Eq. (4.165) with $\boldsymbol{k}$ and multiply and divide the right-hand side by $\overline{k}^2$ to obtain

$$(\overline{k}^2)^{\cdot} = -2\gamma_{\boldsymbol{k}}^{(1)}\overline{k}^2, \quad \text{where} \quad \gamma_{\boldsymbol{k}}^{(1)} \doteq \overline{k}^{-2}\boldsymbol{k}\cdot\boldsymbol{\nabla}\Omega_{\boldsymbol{k}}. \tag{4.166a,b}$$

The superscript 1 indicates an effect of first order in the inhomogeneity. $\gamma_{\boldsymbol{k}}^{(1)}$ should not be confused with the ZF growth rate we will calculate shortly; the latter describes mean-square effects. Recall that $\Omega_{\boldsymbol{k}} = \boldsymbol{k}\cdot\underline{\boldsymbol{V}}$. For turbulent steady states with $\langle\underline{\boldsymbol{V}}\rangle = \boldsymbol{0}$, $\langle\gamma_{\boldsymbol{k}}^{(1)}\rangle = 0$ through first order.

$\gamma_{\boldsymbol{k}}^{(1)}$ will be a crucial ingredient in our subsequent calculation of the ZF growth rate. It has a simple heuristic interpretation. Note that although

<hr>

[a] It is interesting to express the statistical version of Eq. (4.162) in terms of the decomposition into short and long scales. Let the fundamental spectral quantity be $C^{\varphi\varphi}(\boldsymbol{x},\boldsymbol{x}') \doteq \langle\delta\varphi(\boldsymbol{x})\delta\varphi(\boldsymbol{x}')\rangle \equiv C(\boldsymbol{\rho}\mid\boldsymbol{X})$. Then

$$\langle\delta N(\boldsymbol{x})\delta N(\boldsymbol{x})\rangle = (\widehat{\alpha} - \nabla^2)(\widehat{\alpha} - \nabla'^2)C(\boldsymbol{x}-\boldsymbol{x}'\mid\tfrac{1}{2}(\boldsymbol{x}+\boldsymbol{x}'))\big|_{\boldsymbol{x}'=\boldsymbol{x}} \tag{4.163a}$$

$$= [(\widehat{\alpha} - \nabla_{\boldsymbol{\rho}}^2) + O(\nabla_{\boldsymbol{X}}^2)]C(\boldsymbol{\rho}\mid\boldsymbol{X})\big|_{\substack{\boldsymbol{\rho}=\boldsymbol{0}\\\boldsymbol{X}=\boldsymbol{x}}} \tag{4.163b}$$

$$= \sum_{\boldsymbol{k}}[(1+k^2)^2 + O(\nabla^2)]C_{\boldsymbol{k}}(\boldsymbol{x}). \tag{4.163c}$$

Thus the DW Casimir is

$$\tfrac{1}{2}\overline{\langle\delta N^2\rangle} = \tfrac{1}{2}\int d\boldsymbol{X}\sum_{\boldsymbol{k}}(1+k^2)^2 C_{\boldsymbol{k}}^{\varphi\varphi}(\boldsymbol{X}). \tag{4.164}$$

$\mathcal{N}$ is conserved, DW energy $\mathcal{E} \doteq \frac{1}{2}\sum_{\boldsymbol{k}} \overline{k}^2 C_{\boldsymbol{k}}$ is not. We have $\mathcal{N}_{\boldsymbol{k}} = \overline{k}^2 \mathcal{E}_{\boldsymbol{k}}$. In a Lagrangian frame moving with the wave packet (and ignoring sums over wave numbers for the moment), one has $\dot{\mathcal{N}}_{\boldsymbol{k}} = 0 = (\overline{k}^2)\dot{}\mathcal{E}_{\boldsymbol{k}} + \overline{k}^2 \dot{\mathcal{E}}_{\boldsymbol{k}}$, or, upon using Eq. (4.166b), $\dot{\mathcal{E}}_{\boldsymbol{k}} = 2\gamma_{\boldsymbol{k}}^{(1)}\mathcal{E}_{\boldsymbol{k}}$. A more precise version of this result that properly takes into account the spectral content of the wave packet is

$$\partial_T \overline{\mathcal{E}} = 2 \sum_{\boldsymbol{k}} \overline{\gamma_{\boldsymbol{k}}^{(1)}\mathcal{E}_{\boldsymbol{k}}}. \tag{4.167}$$

Indeed, this is exactly the result that follows from rewriting Eq. (4.158) in terms of energy:

$$\partial_T \mathcal{E}_{\boldsymbol{k}} = \overline{k}^{-2}\{\Omega_{\boldsymbol{k}}, \overline{k}^2 \mathcal{E}_{\boldsymbol{k}}\} = \{\Omega_{\boldsymbol{k}}, \mathcal{E}_{\boldsymbol{k}}\} + 2\gamma_{\boldsymbol{k}}^{(1)}\mathcal{E}_{\boldsymbol{k}}, \tag{4.168a,b}$$

where the last term arises from

$$2\gamma_{\boldsymbol{k}}^{(1)} \doteq \{\Omega_{\boldsymbol{k}}, \ln \overline{k}^2\} = 2\boldsymbol{k} \cdot \nabla\Omega_{\boldsymbol{k}}/\overline{k}^2. \tag{4.169}$$

The interpretation of this energy variation is simply that the spectral content of the drift wave packet is changed by the refractive effect of the modulating zonal flow. Again, at first order the effect vanishes upon averaging over a homogeneous ensemble of ZF's.

4.3.4.4. *Heuristic calculation of the zonal flow growth*

We are finally prepared to calculate the ZF growth rate $\gamma_{\boldsymbol{q}}$. By definition, that describes the mean-square energy gain of the CC's due to their modulation of the DW's; equivalently, it is the loss of energy of the DW's due to that modulation. Because we have a sensible description of the DW's in the form of the modulated WKE, we will use the latter formulation. Let $\underline{\mathcal{E}}$ ($\mathcal{E}'$) denote the long- (short-)wavelength energy,[a] and (temporarily) define an effective CC growth rate $\underline{\gamma}$ independent of wave number. Now the CC and/or DW energies can change for various reasons; here we are interested in the specific effects of DW–ZF interaction. Consider an energy increment $\Delta\mathcal{E}$ due to that interaction. Then $\partial_T \Delta\underline{\mathcal{E}} = 2\underline{\gamma}\Delta\underline{\mathcal{E}} = -\partial_T \Delta\mathcal{E}'$, which leads to the formula $\underline{\gamma} = -(\partial_T \Delta\mathcal{E}')/2\Delta\underline{\mathcal{E}}$. This formula is rigorous,[b] but it does not address the more ambitious goal of calculating $\gamma_{\boldsymbol{q}}$ separately for each $\boldsymbol{q}$. However, the following generalization is plausible. Let us write a diagonal representation for the CC energy as $\underline{\mathcal{E}} = \sum_{\boldsymbol{q}} |\delta\varepsilon_{\boldsymbol{q}}|^2$. Then it can be shown that

$$\gamma_{\boldsymbol{q}} = -\frac{1}{2}\left(\frac{\delta^2\partial_T\mathcal{E}'}{\delta\langle\varepsilon_{\boldsymbol{q}}\rangle\delta\langle\varepsilon_{\boldsymbol{q}}\rangle^*}\right)\Big|_{\langle\varepsilon_{\boldsymbol{q}}\rangle=0}. \tag{4.170}$$

---

[a]In Fourier space, I will refer to $\underline{\mathcal{E}}$ and $\mathcal{E}'$ as $\mathcal{E}_{\boldsymbol{q}}$ and $\mathcal{E}_{\boldsymbol{k}}$, respectively, avoiding the redundant notation $\underline{\mathcal{E}}_{\boldsymbol{q}}$ and $\mathcal{E}'_{\boldsymbol{k}}$.

[b](although it is useful only if one knows how to calculate the indicated increments)

198   *John A. Krommes*

Here $\langle \varepsilon_q \rangle$ is a mean "potential" induced by a small perturbation away from the statistically steady state. Further remarks on this quantity will be given below. Basically, we see that the CC growth rate can be obtained from the second-order variation of the DW energy due to the CC modulation. It is plausible that the effect is of second order because first-order effects should average away.

A formula similar although not identical to Eq. (4.170) was first postulated and evaluated by Diamond et al. (1998), differences lying in (i) the interpretation of just what quantity (i.e., $\langle \varepsilon_q \rangle$) participates in the functional variations, and (ii) important technical issues relating to random Galilean invariance. The detailed derivation within a systematically renormalized statistical theory requires the MSR generating-functional apparatus (Sec. 4.2.4.1.), as described by Krommes and Kim (2000). The utility of the MSR formalism here is that it provides a technically efficient way of probing the system and thus to examine the dynamical details of ZF's or DW's separately even in a state of homogeneous, steady-state turbulence.

I will now proceed step by step through the implementation of the functional variations required in formula (4.170). Formula (4.167) is more explicitly

$$\partial_T \mathcal{E}' = 2 \sum_k \overline{\gamma_k^{(1)}(\boldsymbol{X}; [\langle \underline{\varphi} \rangle]) \mathcal{E}_k(\boldsymbol{X}; [\langle \underline{\varphi} \rangle])}, \qquad (4.171)$$

where I have indicated the dependence on the modulation scale $\boldsymbol{X}$ and the functional dependence on the mean potential. $\gamma_k^{(1)}$ is linear in $\langle \underline{\varphi} \rangle$, so one needs to calculate the energy variation only through first order as well. The barring operation denotes an integration over all of the long scales. That can be expressed in terms of the large-scale Fourier amplitudes $\boldsymbol{q}$ by means of Parseval's theorem:

$$\partial_T \mathcal{E}' = 2 \sum_k \sum_q \gamma_{k;q}[\langle \varphi_q \rangle]^{(1)} \mathcal{E}_{k;q}[\langle \varphi_q \rangle]^*. \qquad (4.172)$$

Explicitly, $\gamma_k^{(1)}(\boldsymbol{X}; [\langle \underline{\varphi} \rangle]) = \overline{k}^{-2} \boldsymbol{k} \cdot \boldsymbol{\nabla}(\boldsymbol{k} \cdot \hat{\boldsymbol{z}} \times \boldsymbol{\nabla}\langle \underline{\varphi} \rangle)$, so, upon replacing $\boldsymbol{\nabla} \to i\boldsymbol{q}$, $\gamma_{k;q}^{(1)}[\langle \varphi_q \rangle] = -\overline{k}^{-2}(\boldsymbol{k} \cdot \boldsymbol{q})(\hat{\boldsymbol{z}} \cdot \boldsymbol{q} \times \boldsymbol{k})\langle \varphi_q \rangle$. The first functional variation in Eq. (4.170) will just remove the last $\langle \varphi_q \rangle$.

Now we need to evaluate the first-order part of $\mathcal{E}_{k;q}[\langle \varphi_q \rangle]$. Because we know that the naturally conserved quantity is the Casimir $\mathcal{N}$, it is convenient (although not necessary) to work with $\mathcal{N}_k$ instead of $\mathcal{E}_k$:

$$\partial_T \mathcal{E}' = 2 \sum_k \sum_q \gamma_{k;q}[\langle \varphi_q \rangle] \overline{k}^{-2} \mathcal{N}_{k;q}[\langle \varphi_q \rangle]^*. \qquad (4.173)$$

The modulational contribution to $\mathcal{N}_k$ follows from the WKE (4.160):

$$\partial_T \mathcal{N}_k + (\partial_k \Omega_k[\langle \underline{\varphi} \rangle]) \cdot \boldsymbol{\nabla}\mathcal{N}_k - (\boldsymbol{\nabla}\Omega_k[\langle \underline{\varphi} \rangle]) \cdot \partial_k \mathcal{N}_k = 0. \qquad (4.174)$$

The first-order change in $\mathcal{N}_k$ due to the presence of $\langle\varphi\rangle$ obeys

$$\partial_T \Delta\mathcal{N}_k[\langle\underline{\varphi}\rangle] + (\partial_k \Delta\Omega_k[\langle\underline{\varphi}\rangle]) \cdot \underbrace{\nabla\mathcal{N}_k}_{(a)} + \underbrace{(\partial_k\Omega_k)}_{(b)} \cdot \nabla\Delta\mathcal{N}_k[\langle\underline{\varphi}\rangle]$$

$$- (\nabla\Delta\Omega_k[\langle\underline{\varphi}\rangle]) \cdot \partial_k\mathcal{N}_k - \underbrace{(\nabla\Omega_k)}_{(c)} \cdot \partial_k\Delta\mathcal{N}_k[\langle\underline{\varphi}\rangle] = 0. \quad (4.175)$$

Here the crossed-out terms vanish because the modulation is made to a statistically homogeneous (unmodulated) background. Term (a) vanishes because there are no mean gradients in that background, while terms (b) and (c) vanish because $\langle\varphi\rangle = 0$ in the unmodulated state. Thus, at this point we have, upon Fourier transforming the large-scale variations,

$$\partial_T \Delta\mathcal{N}_{k;q}[\langle\varphi_q\rangle] = (iq\Delta\Omega_{k;q}[\langle\varphi_q\rangle]) \cdot \partial_k\mathcal{N}_{k;q}. \quad (4.176)$$

Now we must integrate Eq. (4.176) to find the variation $\Delta\mathcal{N}_{k;q}$. Here one must appeal to one more heuristic argument. The effect of the modulation must build up over some characteristic time ΔT. Because γ_q characterizes a statistically steady state, ΔT must be a property of that state as well (specifically, of the steady-state interactions between the DW's and the ZF's). Therefore, I assert that $\Delta T = \mathrm{Re}\,\theta_{q,k,-k}$, where $\theta_{q,k,p}$ is an appropriate triad interaction time deduced from Markovian closure theory. As discussed by Krommes and Kim (2000), *this θ must respect random Galilean invariance because it describes the interaction between disparate scales!*[a] Since in general $p = -k - q$, one has $p \approx -k$ to lowest order in ϵ.

Thus, with $\theta^r_{q,k,-k} \equiv \mathrm{Re}\,\theta_{q,k,-k}$, we have

$$\Delta\mathcal{N}_{k;q}[\langle\varphi_q\rangle] = i\theta^r_{q,k,-k}\Delta\Omega_{k;q}[\langle\varphi_q\rangle]q \cdot \partial_k\mathcal{N}_{k;q}. \quad (4.177)$$

The complex conjugate of this expression is needed, and the required functional variation with respect to $\langle\varphi_q\rangle^*$ removes the $\langle\varphi_q\rangle$ dependence from the right-hand side of Eq. (4.177). The only remaining task is to relate $\langle\varphi_q\rangle$ to the ε_q of formula (4.170); from the generalization of Eq. (4.34a), $\varepsilon_q = \bar{q}\varphi_q/\sqrt{2}$. Upon combining everything, we obtain the final result:

$$\gamma_q = -q^2\left[2\left(\frac{q^2}{\bar{q}^2}\right)\sum_k \theta^r_{q,k,-k}\bar{k}^{-4}k_y^2 k_x \frac{\partial\mathcal{N}_k}{\partial k_x}\right], \quad (4.178)$$

where the Cartesian components are taken *with respect to the direction of q* — e.g., $k_x \doteq \hat{q} \cdot k$.

---

[a] Thus use of the test-field model (Kraichnan, 1971; Bowman and Krommes, 1997; Krommes, 2002) is suggested.

If we go back through this derivation, we can trace the physical origin of each term in this expression. Basically, the result describes the mean-squared effect of random DW wave-number refraction due to advection by the CC's, constrained by conservation of the Casimir invariant:

$$
\underbrace{\gamma_q}_{\substack{\text{CC} \\ \text{energy} \\ \text{growth}}} = \underbrace{-1}_{\substack{\text{DW's} \\ \text{lose energy,} \\ \text{feeding the CC's}}} \times \underbrace{q^2}_{\substack{\text{size of the} \\ \text{CC advection} \\ \text{(squared)}}} \times \underbrace{(\tfrac{1}{2}\overline{q}^2)^{-2}}_{\substack{\text{proportionality} \\ \text{between} \\ \text{CC energy} \\ \text{and } |\varphi_q|^2}} \times \underbrace{q^2}_{\substack{\text{strength of the} \\ \text{refraction effect} \\ \text{(squared)}}}
$$

$$
\times \underbrace{\sum_k}_{\substack{\text{all DW} \\ \text{components} \\ \text{must be summed}}} \underbrace{\theta^r_{q,k,-k}}_{\substack{\text{characteristic} \\ \text{interaction time} \\ \text{between the} \\ \text{CC's and DW's}}} \times \underbrace{(\overline{k}^{-2})^2}_{\substack{\text{proportionality} \\ \text{between} \\ \mathcal{E} \text{ and } \mathcal{N} \\ \text{(squared)}}}
$$

$$
\times \underbrace{k_y^2}_{\substack{\text{advection} \\ \text{is due to} \\ \boldsymbol{E} \times \boldsymbol{B} \text{ velocity}}} \times \underbrace{k_x \frac{\partial}{\partial k_x}}_{\substack{\text{refraction} \\ \text{(squared)}}} \times \underbrace{\mathcal{N}_k}_{\substack{\text{the Casimir } \mathcal{N} \\ \text{is conserved under} \\ \text{modulation}}} . \tag{4.179}
$$

Formula (4.178) is considerably more complicated than expression (4.155) for the modulational instability growth rate; this is not surprising since one is analyzing a difficult self-consistent situation rather than a simple initial-value problem. Part of the apparent distinctions reside in the value of the triad interaction time, which is here determined by the nonlinear dampings η according to $\theta_{q,k,-k} = (\eta_q + 2\,\mathrm{Re}\,\eta_k)^{-1}$. To try to emulate the assumptions of the modulational-instability calculation, temporarily suppose that the drift waves have no nonlinear damping at all: $\eta_k \to 0$. Upon noting that η_q and γ_q are really the same object (except for a sign), so that $\theta_{q,k,-k} \sim \gamma_q^{-1}$, ignoring details of the derivative with respect to k_x (i.e., $k_x \partial_{k_x} \mathcal{N}_k \to \mathcal{N}_k$), and focusing specifically on ZF's ($\overline{q} = q$), one can solve the self-consistent equation $\gamma_q \sim \gamma_q^{-1} \ldots \mathcal{N}_k$ to find $\gamma_q \sim |\widehat{\boldsymbol{z}} \cdot \boldsymbol{q} \times \boldsymbol{k}||\varphi_k|$, which is the same basic scaling exhibited by the modulational instability. Although this is really required by dimensional consistency, the agreement provides a nice check on the details of the algebra.

I now point out that this calculation is intimately related to the eddy-viscosity calculations of Kraichnan (1976). Kraichan defined eddy viscosity μ_{eddy} in terms of the energy transfer into unresolved modes. When scale separation was assumed, he was able to perform an explicit calculation by asymptotic expansion of an isotropic Markovian closure (the test-field model); he found that $\mu_{\mathrm{eddy}} < 0$ in two dimensions. When that asymptotic expansion is extended to anisotropic situations and generalized Hasegawa–Mima dynamics, the result (4.178) is obtained precisely, as was shown by Krommes and Kim (2000).[a] The positivity of $\gamma_q = -q^2 \mu_{\text{eddy}}$ is consistent with negative μ_{eddy}. Thus, in essence, the heuristic derivation presented above provides a physical interpretation of the disparate-scale asymptotic expansion. Indeed, a close reading of Kraichnan (1976, Sec. 5) reveals that Kraichnan clearly understood the relevant physics, namely the change in wave number due to the random modulation. However, the nature of the appropriate (Casimir) invariant was perhaps not so clearly indicated.

Finally, I note that the long-wavelength zonal generation can be interpreted as due to Reynolds stresses due to the short-wavelength drift waves; see the discussion in Sec. 4.1.5.4. and especially footnote a, p. 142.

4.3.4.5. *Effects on the short scales*

So far we have examined the tendency for long-wavelength CC's to be excited by the nonlinear interactions of short-wavelength DW's. This implies an energy flow from the DW's to the CC's and a modification of the DW spectrum. Because the CC's randomly refract the short-scale wave packets, one anticipates a diffusive effect on the small-scale spectrum:

$$\frac{\partial \mathcal{N}_k}{\partial t} = \frac{\partial}{\partial \boldsymbol{k}} \cdot \mathsf{D}_k \cdot \frac{\partial \mathcal{N}_k}{\partial \boldsymbol{k}} + \cdots , \tag{4.180}$$

where the dots signify additional terms in the Fokker–Planck description.[b] The diffusion tensor is defined by $\mathsf{D}_k = \int_0^\infty dT \, \langle \dot{\boldsymbol{k}}(T) \dot{\boldsymbol{k}}(0) \rangle$, where T is the slow time in the random ray equation (4.165). Heuristically, one argues that the time integral is cut off by the natural triad interaction time. Upon squaring the right-hand side of Eq. (4.165), then averaging over the CC fluctuations, one readily finds $\mathsf{D}_k = \sum_q (\hat{q}\hat{q}) \theta_{q,k,-k} q^2 k_y^2 \langle \delta V_q^2 \rangle$, where $\delta \boldsymbol{V}_q \doteq i\hat{z} \times \boldsymbol{q} \delta \varphi_q$ is the CC velocity fluctuation. Now although the total $\mathcal{N}$ is conserved by the diffusion operator (4.180), energy is not. However, it is not difficult to verify that energy lost by the DW's due to the random refraction shows up properly in the CC's; that is, one can prove the energy conservation theorem

$$-\sum_k \frac{1}{k^2} \frac{\partial}{\partial \boldsymbol{k}} \cdot \left(\mathsf{D}_k \cdot \frac{\partial \mathcal{N}_k}{\partial \boldsymbol{k}} \right) = \sum_q \frac{1}{q^2} (2\gamma_q \mathcal{N}_q). \tag{4.181}$$

---

[a]The anisotropic expansion is tricky and tedious. Many other details of both conceptual and algebraic interest were also discussed by Krommes and Kim (2000).
[b]The standard form of a Fokker–Planck equation for PDF $f(z)$ is $\partial_t f + \partial_z(Vf) - \partial_z^2(Df) = 0$.

4.3.4.6. *Self-regulating systems of drift waves and zonal flows*

At this point, one can certainly appreciate the idea (Diamond et al., 1998) that DW's and ZF's form a self-regulating nonlinearly coupled system. The discussion in the previous section is based exclusively on the assumption of steady-state turbulence; statistical averaging procedures were employed. However, long-wavelength zonal flows can also arise from macroscopic plasma inhomogeneities, since radial electric fields must obey the radial component of the momentum equation. Any such (sheared) flows, once created, would be expected to affect the DW turbulence. A relevant review article is by Hahm (2002).

4.3.5. *Geodesic acoustic modes*

So far in this lecture, the discussion of zonal flows has been conducted in slab geometry. However, in the presence of toroidal effects a new mode of oscillation (Winsor et al., 1968) called the *geodesic acoustic mode* (GAM) appears. Unlike the standard ZF, the GAM has nonzero frequency[a] $\omega = \pm\sqrt{2}c_{\rm s}/R$, has both radial and poloidal variations, and is damped by parallel dissipation. Its importance is that it can couple to the standard ZF and provide a sink for ZF energy. Very detailed discussion of this mechanism has been given by Scott (2005).

In a general lecture of this nature, I cannot go deeply into specifically toroidal physics. However, it is worth indicating the basic physics of the GAM. Consider an ion fluid element moving poloidally under $\boldsymbol{E} \times \boldsymbol{B}$ flow. As this element moves into regions of stronger B, it is compressed since $\boldsymbol{\nabla} \cdot \boldsymbol{V}_E \approx -2\boldsymbol{\kappa} \cdot \boldsymbol{V}_E$, where the curvature vector $\boldsymbol{\kappa}$ is defined by Eq. (4.18); for a flow within the magnetic flux surface, this effect involves the *geodesic* (in the surface) component of the curvature. Now a fluid element at $\theta = \frac{1}{2}\pi$ moving to the left (into stronger B) by $\boldsymbol{E} \times \boldsymbol{B}$ would be associated with an electric field pointing downward. But the compression leads to an excess of ion density, which will then drift downward according to the magnetic drifts [see Eq. (4.63)], weakening the self-consistent electric field. Therefore, the process is not unstable; rather, an ion density perturbation will oscillate at a frequency built from the geometric mean of the characteristic frequencies (i) $2c_{\rm s}/R$, associated with geodesic curvature, and (ii) $c_{\rm s}/R$, associated with the magnetic drift. A recent paper discussing the linear and nonlinear physics of GAM's is by Itoh et al. (2005a).

The experimental measurement of ZF's and GAM's is of intense current interest. Recent observations are summarized by Itoh et al. (2006); see also

---

[a]The GAM frequency is thus higher than the standard ion acoustic frequency $\omega = \pm k_{\parallel}c_{\rm s} \approx c_{\rm s}/Rq$ by a factor of q, where q is the safety factor (inverse rotational transform).

the lectures by M. Shats (Chap. 5).

4.3.6. *Transport due to electron-temperature-gradient-driven modes*

There is an almost perfect analogy between the governing equations and physics of ITG and ETG modes that are obtained just by interchanging electron and ion species labels. If gyro-Bohm scaling were relevant, that would nominally imply that ETG transport would be small because $D_{\mathrm{gB},e}/D_{\mathrm{gB},i} \sim (m_e/m_i)^{1/2}$. However, there is an important difference having to do with the nature of the adiabatic response. For ITG, the electrons (possessing rapid parallel motion) are adiabatic except for convective cells, which are fluid in nature. Excitation of convective cells, especially zonal flows, tends to destroy radially extended streamers and to suppress the overall transport. For ETG, however, ion adiabatic response arises not because of parallel motion (the ions are very slow) but because at the perpendicular scales of interest ($k_\perp \rho_e \sim 1$) the ions are essentially unmagnetized. Adiabatic response holds for all modes. Zonal flows tend not to be formed, so streamers are not destroyed and electron transport can be considerably enhanced over the nominal prediction of gyro-Bohm scaling. Quantifying the correct level of ETG transport in realistic geometry is a subject of current research; unfortunately, space precludes a detailed discussion.

I have far from exhausted all that can be said about zonal flows in plasmas; further information can be found in the review articles by Diamond et al. (2005) and Itoh et al. (2006). We will also encounter them in the next lecture, where (along with streamers) they figure in the generation of coherent plasma "blobs" that are important near the plasma edge.

4.4. LECTURE 4 – Intermittency and Coherent Structures in Plasmas

In the first three lectures, I have (very incompletely) covered (i) the basic equations for confined plasmas in strong magnetic fields, (ii) renormalization techniques in the context of plasma physics, and (iii) some of the physics and analytical description of zonal flows. Much less understood are the significance and mathematical treatment of intermittency and coherent structures. In this lecture I provide an introduction to these challenging topics. I will mention

- the definition of intermittency;
- the naturally intermittent statistics of turbulent fluxes;

- the distinction between gradient-driven and flux-driven transport, the latter possibly giving rise to avalanche phenomena and self-organized criticality;
- and the intermittent formation of "blobs," as recently observed in the scrape-off regions of several devices as well as in computer simulations. The analytical theory of blob formation is not well developed and presents very interesting research challenges.

It has been asserted that fusion physics is now confronted with a new paradigm, involving fluctuations that are highly intermittent rather than essentially Gaussian. This follows increasing precision and number of fluctuation measurements, particularly near the edges of various toroidal devices. Time series of density or potential fluctuations obtained from probe measurements clearly display an intermittent aspect, recent time-dependent visualizations of 2D spatial fluctuations show the formation and propagation of coherent "blobs," *etc.* Clearly simple diffusion models are inadequate to describe such phenomena.

Although intermittent fluctuations are clearly observed, especially in edge regions, that in itself is not sufficient to warrant the abandonment of all that has come before. It could be that such fluctuations are the "tail on the dog," "dog" referring to the energy-containing fluctuations responsible for transport. For example, in the Hasegawa–Wakatani simulations of Hu et al. (1995) mentioned briefly in Sec. 4.2.5.2., the vorticity field was moderately intermittent, with a kurtosis (defined in the next section) in the range of 10–20, yet standard Markovian closure theory did a good job of predicting the turbulent particle flux. However, it is known that blobs (to be discussed in Sec. 4.4.4.) can carry as much as 50% or 60% of the total particle flux across the last closed flux surface, leading for example to difficult problems with wall erosion. Therefore further investigation is clearly warranted.

4.4.1.　*Definition of intermittency*

What is intermittency? Frisch (1995, Chap. 8) offers the definition that a signal is intermittent if its kurtosis increases without bound with the cutoff frequency of a high-pass filter. Kurtosis K is defined as $K \doteq F - 3$, where the *flatness factor* F is defined[a] for a fluctuating signal $\delta\psi$ as $F \doteq \langle\delta\psi^4\rangle/\langle\delta\psi^2\rangle^2$.[b] Non-Gaussian kurtosis is manifested by enhanced tails or "wings" of PDF's (thus the probability of large-amplitude events is

---

[a]Not everyone agrees on this definition; I follow Frisch (1995).

[b]A standard example of a signal with nonzero kurtosis can be constructed by considering a Gaussian time series that is on for only a fraction β of the time. The kurtosis of such a signal is $(\beta\langle\delta\psi^4\rangle)/(\beta\langle\delta\psi^2\rangle)^2 - 3 = 3\beta^{-1} - 3 = 3(1 - \beta)/\beta$. As $\beta \to 0$, the kurtosis increases without bound.

anomalously large). Note that nonlinearity precludes Gaussianity because the product of Gaussian random variables is not Gaussian. For example, the product of two centered Gaussian variables U with unit variance is the $n = 1$ special case of a χ^2 variable with n degrees of freedom. If $S_n \doteq \sum_{i=1}^n U_i^2$, then it is not difficult to show[a] that the PDF of $z \equiv S_n$ is

$$P_n(z) = 2^{-n/2} z^{n/2-1} e^{-z/2} / \Gamma(\tfrac{1}{2}n) \quad (z \geq 0). \tag{4.182}$$

The exponential decay (a straight line on a logarithmic ordinate, as compared to the inverted parabola of a Gaussian) is characteristic of intermittent distributions, which quite generally exhibit exponential or stretched exponential tails.

The non-Gaussianity of nonlinear dynamical evolution is most immediately manifested in a nonvanishing skewness or third-order cumulant (which is related to energy flow in k space). But especially for strong turbulence, the kurtosis can be nonnegligible as well, even if coherent structures are not playing a dominant role. The PDF (4.182) has a nonvanishing kurtosis (it is left as an exercise to calculate it), and we will see another, more relevant example in the next section.

Intermittency, then, can arise is various ways and at some level may be fairly ubiquitous. I will mention just a few of the simple possibilities.

4.4.2. *Flux and intermittency*

Both diagnostically and with computer simulations, it is possible to measure the PDF's of turbulent fluxes such as the radial particle transport. Suppose it turns out that such a flux is intermittent? Should that have been expected?

As we know, such fluxes involve the cross correlation between the fluctuations of turbulent radial velocity and transported quantity, e.g., $\Gamma \doteq \langle \delta V_x \delta n \rangle$. One insight is that even if each individual field is Gaussian, the product is not. For example, it is not hard to show that if one defines the random variable $\widetilde{\Gamma} \doteq \widetilde{x}\widetilde{y}$ and assumes that $\widetilde{x}$ and $\widetilde{y}$ are centered Gaussians with cross correlation coefficient ρ $(-1 \leq \rho \leq 1)$ and standard deviations σ_x and σ_y, then [see, for example, Carreras et al. (1996)]

$$P(\Gamma) = \frac{1}{\pi(1-\rho^2)^{1/2}} \exp\left[\left(\frac{\rho}{1-\rho^2}\right)\Gamma\right] K_0\left(\frac{|\Gamma|}{1-\rho^2}\right), \tag{4.183}$$

---

[a]One way of proceeding is to use characteristic functions, i.e., to work with the Fourier transform of the PDF. Thus, $P(z) = \langle \delta(z - \sum_{i=1}^n U_i^2) \rangle = (2\pi)^{-1} \int_{-\infty}^{\infty} dk \, \langle \exp(-ik \sum_{i=1}^n U_i^2) \rangle e^{ikz} = (2\pi)^{-1} \int_{-\infty}^{\infty} dk \, \langle \exp(-ikU^2) \rangle^n e^{ikz}$. The average can be easily calculated (it is essentially the normalization coefficient of a Gaussian with complex variance). The inverse Fourier transform provides a nice example of Hankel's integral representation of the inverse of the Gamma function.

where K_0 is the modified Bessel function of the second kind and x and y have been normalized to σ_x and σ_y, respectively. Note that

$$\lim_{\rho \to 1} P(\Gamma) = P_1(\Gamma), \qquad (4.184)$$

in agreement with the result (4.182) obtained above for a χ^2 variable with one degree of freedom. Also, for any ρ,

$$P(\Gamma) \to \frac{1}{(2\pi|\Gamma|)^{1/2}} e^{-|\Gamma|/(1\pm\rho)} \quad (\Gamma \to \pm\infty). \qquad (4.185)$$

The exponential tails exhibited in Eqs. (4.184) and (4.185) are quite characteristic of intermittent PDF's.

Thus, the relevant question is not "Is the flux intermittent?" but rather "*How* intermittent is the flux?", since Eq. (4.185) already predicts a nontrivial kurtosis. It is a somewhat tedious but straightforward exercise to find[a]

$$K(\rho) = 3 \left(\frac{3 + 14\rho^2 + 3\rho^4}{(1 + \rho^2)^2} \right) - 3 = 6 \left(\frac{1 + 6\rho^2 + \rho^4}{(1 + \rho^2)^2} \right), \qquad (4.186)$$

which increases monotonically from $K(0) = 6$ to $K(1) = 12$. These values set a baseline for "typical" kurtosis values for flux. If the measured $K(\rho)$ significantly exceeds the predictions of formula (4.186), clearly some interesting nonlinear physics is at work. But the fact that the prediction (4.186) is substantially nonzero also suggests that the flux is not the most sensitive indicator of intermittent dynamics; studying the non-Gaussian behavior of the primitive fields such as electrostatic potential may be more revealing.

4.4.3. *Flux-driven transport*

The designs of modern toroidal confinement devices are not necessarily kind to the analytical theorist. Although over the years much progress has been made based on the simple model of a magnetic equilibrium consisting of nested, circular surfaces of magnetic flux (with simple boundary conditions at the outermost surface, considered to be a conducting wall), in reality the magnetic geometry near the wall of the vacuum vessel is more complicated. To control the exhaust of charged particles that are transported outward, magnetic divertors are used in which open field lines exterior to a last closed flux surface lead to a divertor region where particles are exhausted by strong pumping. That is, there is a magnetic separatrix. Particles transported across that separatrix effectively experience a discontinuity in the physics. Generally the region immediately inside the separatrix is called

---

[a]As an exercise, verify that $K(1) = 12$ agrees with the kurtosis calculated from $P_1(\Gamma)$.

the *edge*, while the very-low-density region between the separatrix and the wall is called the *scrape-off layer* (SOL).

The basic models of gradient-driven transport, as exemplified by the various nonlinear-equation paradigms discussed in Lecture 1, have proven to be fruitful. However, a decomposition into mean and fluctuating parts is ill-motivated when both parts are comparable in size, as is the case for density fluctuations in the SOL. In such cases, it is better to work directly with the full field and let it evolve as it will. Sarazin and Ghendrih (1998) performed a detailed numerical study of interchange turbulence in an SOL model driven by a constant influx of particles. The resulting turbulence was found to be substantially intermittent and was interpreted in terms of *avalanche*-like dynamics, namely radial ballistic propagation of particles.

The notion that avalanche-type physics might apply to some regimes of tokamak discharges is due to Diamond and Hahm (1995), who argued for the relevance of the seminal papers on self-organized criticality (SOC) by Bak et al. (1987), Bak et al. (1988), and Hwa and Kardar (1992), who attempted to provide a dynamical basis for the $1/f^\alpha$ ($\alpha \approx 1$) noise frequently observed in experiments. As applied to plasmas, the basic idea is that turbulent fluxes typically turn on rapidly when the *local* profile gradient exceeds a critical or threshold value. The resulting strong transport can annihilate the gradient. The effect can propagate as an avalanche over a possibly macroscopic distance, producing self-similar and intermittent statistics as well as holding the profile close to the linearly marginal slope. Some aspects of the resulting SOC were reviewed by Jensen (1998).

Many of the dynamical models of SOC have been discrete cellular automata. In reality, however, one deals with continuous PDE's. Singh et al. (2005) studied a simple PDE model constructed with the above scenario in mind [and, in part, motivated by the analysis of Politzer (2000) of data from the D-IIID tokamak that appeared to be consistent with "avalanche-like phenomena"]. Specifically, they considered a coherently source-driven thermal diffusion equation with a two-state thermal conductivity that changes states (with hysteresis) over a nonzero time. For some ranges of drive, interesting random behavior emerged. Although the calculation is certainly not entirely first-principles, estimates of the several parameters of the model showed that it made reasonable predictions for experimentally relevant regimes of electron heat transport.

A full discussion of the recent work on plasma SOC models is not possible here; a specialized review is warranted. It should be noted, however, that although it is certainly true that various dynamical systems exhibit $1/f$ noise, that observation cannot by itself be interpreted as a proof of SOC. This was discussed by Krommes and Ottaviani (1999), who demonstrated a counterexample and attempted an alternate explanation in terms of the Eulerian statistics of the spectral balance equation.

4.4.4. *Blobs*

Although in ideal divertor operation particles transported outward across the separatrix would then be rapidly carried along the magnetic field lines to the divertor, it has been known for some time that the actual situation is more complicated; substantial amounts of plasma can propagate to the outer wall, possibly causing erosion problems. Krasheninnikov (2001) interpreted this rapid convective transport in the SOL in terms of coherent plasma filaments extended along the field lines. Viewed in the poloidal cross section, these filaments appear as *blobs* that can intermittently detach from the separatrix region, then propagate radially outward. Blob physics is now a major research activity.

Recently, significant progress has been made in visualizing fluctuations in the vicinity of the separatrix by means of a *gas puff imaging* (GPI) technique developed by S. Zweben and his collaborators at several institutions (Maqueda et al., 2003; Zweben et al., 2004). A puff of neutral gas is emitted into a region perpendicular to the local magnetic field in the vicinity of the separatrix. The neutral atoms are collisionally excited by interactions with the local plasma, and the resulting emission can be observed with a fast camera. That emission I is proportional to both the neutral density n_0 and a nontrivial function of local electron density and temperature: $I = n_0 f(n_e, T_e)$. If one makes the very rough assumption that temperature and density are advected together by the cross-field turbulence, one can imagine that density fluctuations are being measured. In any event, fluctuations can be resolved both spatially and temporally. A review of recent results has been given by Zweben (2006).

In the edge region, the GPI diagnostic reveals fluctuations that loosely resemble 2D turbulence. Movies,[a] however, also strikingly reveal the intermittent formation and subsequent outward propagation of *blobs* (regions of enhanced image intensity).

From the theoretical point of view, it is convenient to separate blob physics into two parts: generation and propagation. A fair amount is already known about the physics of the propagation of isolated blobs; much less is understood about their generation.

---

[a]Various videos of edge turbulence and further information on the GPI diagnostic can be found at `http://www.pppl.gov/~szweben/`.

4.4.4.1. *Model equations for the edge and scrape-off region*

The basic physics of blob propagation was identified by Krasheninnikov (2001) to be the plasma polarization due to the ∇B and curvature drifts. Roughly speaking, that process (dependent on the sign of charge) creates a vertical electric field, which then causes outward $\boldsymbol{E} \times \boldsymbol{B}$ motion. Accordingly, appropriate model equations must include curvature effects. Such equations can be derived by taking appropriate moments of the gyrokinetic equation (4.16). As an example of the result, consider the equations studied by Bisai et al. (2005a):

$$\frac{d\varpi}{dt} + \frac{g}{n}\frac{\partial n}{\partial y} - \mu\nabla^2_\perp \varpi = \langle n^{-1}\nabla_\| j_\| \rangle_\|, \qquad (4.187a)$$

$$\frac{dn_e}{dt} + g\left(\frac{\partial n_e}{\partial y} - n_e\frac{\partial \varphi}{\partial y}\right) - D\nabla^2_\perp n_e = \langle \nabla_\| j_{\|e} \rangle_\| + S_n. \qquad (4.187b)$$

Here $g \propto \rho_{\rm s}/R$ is a measure of the magnetic curvature, and $\langle \dots \rangle_\|$ signifies an average along the magnetic lines. Parallel current $j_\|$ must be treated quite differently inside and outside of the separatrix. Inside, one should recover the Hasegawa–Wakatani dissipative current response $\nabla_\| j_\| \propto D_\|(\delta\varphi - \delta n_e)$; the resulting system then provides a workable model of resistive ballooning turbulence (Singh et al., 2005). Outside, one must employ boundary conditions associated with the formation of a plasma sheath at the divertor plate. The details cannot be described here, but the result is standard. In order to smoothly connect the edge and SOL regions, Bisai et al. (2005a) used spatially varying turn-on functions $\widehat{\sigma}(x)$ and $\widehat{\chi}(x)$ and defined a *sheath conductivity* $\sigma(x)$ and a parallel diffusion rate $\alpha(x)$; the x dependence signifies a smooth but rapid transition across the separatrix. Then one finds

$$\langle \nabla_\| j_{\|e} \rangle = \alpha(x)(\delta\varphi - \delta n_e) + \sigma(x)n_e e^{\Lambda - \varphi}, \qquad (4.188a)$$

$$\langle n_e^{-1}\nabla_\| j_\| \rangle = \alpha(x)(\delta\varphi - \delta n_e) + \sigma(x)(1 - e^{\Lambda - \varphi}). \qquad (4.188b)$$

Here Λ is the floating potential.[a] This model is amenable to numerical simulation, as described by Bisai et al. (2005a), and show interesting turbulence and blob formation qualitatively not unlike that observed with the GPI diagnostic.

4.4.4.2. *Blob propagation*

Variants of the equations of Bisai *et al.* had been used earlier by D'Ippolito et al. (2002) to elucidate the basic physics of a blob moving in the SOL,

---

[a]For an understanding of the sheath-related terms, the paper of Sarazin and Ghendrih (1998) is very helpful.

assuming that it had somehow already been formed. One inquires into the possibility of steady-state, coherent solutions of Eqs. (4.187) propagating at constant velocity $\boldsymbol{u}$. The simplest balances considered by D'Ippolito *et al.* were between curvature drive (magnetic drift polarization) and parallel current flow to the sheath [the g term in Eq. (4.187a) and the σ term in Eq. (4.188b)], and between an ionizing density source $S_n = \xi n_e$ and parallel transport [the σ term in Eq. (4.188a)]. One finds that density blobs with Gaussian profiles of radius r_b propagate outward with velocity $u_x \propto g/\sigma r_b^2$; that is, smaller blobs propagate faster (because the internal polarization field is more intense). Many more details were reported by D'Ippolito et al. (2002) and D'Ippolito and Myra (2003); see also the more recent work of Bisai et al. (2005b).

4.4.4.3. *Blob generation*

At the time of writing (early 2006), relatively little is understood about the details of blob generation. Both simulations, such as those of Bisai et al. (2005a), and the GPI and probe diagnostics clearly reveal that the edge region is turbulent. Analysis of experimental data from GPI has led Myra (2005) to conclude that blobs are born near the radial zone where $|\nabla \ln P|$ is a maximum, which happens to reside inside of but near to the separatrix. This suggests a mechanism based on local turbulence rather than, for example, turbulence spreading from farther inside.

Bisai et al. (2005a) have suggested a plausible qualitative criterion for blob detachment. The scenario begins with the random formation of a radially directed *streamer*, characterized by its radial extent δ_x and poloidal extent δ_y. A radial electric field E_x gives rise to a sheared poloidal flow whose difference across the extent δ_x is $\Delta V_y \sim \delta_x[cE_x'(x)/B]$, where $E_x' \equiv \partial_x E_x$. It is then asserted that if in one eddy turnover time the relative poloidal displacement of the tips of the streamer exceeds the poloidal width δ_y, then the streamer will break and form a blob: $\Delta V_y \tau_{\mathrm{eddy}} > \delta_y$. One estimates $\tau_{\mathrm{eddy}}^{-1} \sim \delta_x^{-1}(cE_y/B)$; thus the final criterion for detachment is

$$E_x' \delta_x^2 / E_y \delta_y > 1. \tag{4.189}$$

Data from the simulations is in qualitative agreement with this criterion; most of the streamer breaking occurs at points where E_x changes sign (regions of maximum shear E_x').

Given that blobs do intermittently break off and propagate outward, one can at least pose the question of how to analytically calculate the rate of blob propagation across the separatrix. It is a formidable problem of current research to formulate that question precisely in terms of criteria such as Eq. (4.189). The difficulty is that while statistical closures provide some information about fluctuation scales, they typically predict effects only in mean square. For streamer breaking and blob formation, however, one needs to understand *microscopic* balances, transient fluctuations in a localized region, and, according to Eq. (4.189), at least the joint PDF of potential and wavevector. Such information is not supplied by the standard workable Markovian moment closures. More advanced approximations such as *mapping closure* (Krommes, 2002, and references therein) model entire PDF's, but are subtle and still require considerable development.

One methodology that has not been applied specifically to the blob problem but may give some feeling for the kind of formalism that could be useful involves the direct (approximate) evaluation of the MSR generating functional Z for a mixture of random turbulence and coherent structures. This has been recently attacked by Spineanu and Vlad (2005) for a model of randomly placed vortices of specific form (Meiss and Horton, 1982) interacting with turbulence. The basic idea is to formally decompose the potential into the sum of a vortex part V and a turbulence part T: $\varphi = \varphi^{\mathrm{V}} + \varphi^{\mathrm{T}}$. There are then four basic kinds of nonlinear interactions: V–V, V–T, T–V, and T–T. A basic vortex solution (V–V) is assumed to be known, but its shape is allowed to be plastically modified by interactions with the turbulence (V–T). Similarly, the standard turbulence problem in the absence of coherent structures is expressed by T–T, but the turbulence statistics are modified by interactions with randomly placed vortices (T–V). The modified vortex (V–V and V–T) and turbulence (T–V and T–T) interactions are considered separately; thus the generating functional is approximated as $Z \approx Z_V Z_T$. Further approximations, analogous to the semiclassical approximation in quantum mechanics, are made in order to evaluate each of Z_V and Z_T; then the two-point correlation function is derived by functional differentiation. The full calculation is rather involved. Although it is a valiant attempt, the approximations need to be further assessed; time will tell to what extent the methodology can be usefully applied to the various difficult problems of intermittency and coherent structures in plasmas.

4.4.5. *Summary and final remarks*

In these lectures I have tried to provide both some historical perspectives and a feeling for contemporary issues in the analytical treatment of plasma turbulence, particularly as inspired by magnetic confinement. I will organize the following brief summary in terms of fundamental equations, analytical methods, and physical phenomenology.

4.4.5.1. *Fundamental equations*

Truly fundamental are, of course, the Liouville and Klimontovich equations. But for low-frequency fluctuations in magnetized plasma, I stressed that the nonlinear gyrokinetic equation provides the basic analytical framework. That can be attacked numerically (regrettably, I was unable to discuss that here), but it is also the starting point for various nonlinear paradigms, including the original and generalized Hasegawa–Mima equation (the basic nonlinear paradigm for drift waves), the Hasegawa–Wakatani system (a paradigm for collisional drift waves in a slab), ITG models (relevant for turbulent ion heat transport) and ETG models (similarly relevant for electron transport), drift-resistive ballooning equations (appropriate for turbulence in the edge and scrape-off layer regions of a tokamak), *etc.*

4.4.5.2. *Analytical methods*

The quasilinear theory is surprisingly rich in both physical and analytical content, but in general it cannot saturate profile-driven fluid fluctuations. Weak-turbulence theory is an improvement and demonstrates the basic mode-coupling structure of statistical steady states; however, it is limited in physical applicability. Resonance-broadening theory was a valiant early attempt to treat strong, nonperturbative effects, and it still has useful qualitative points to teach us. The direct-interaction approximation gives a more systematic description of nonlinear mode coupling, but is extremely complicated for anisotropic, inhomogeneous plasmas. The DIA does provide one route to Markovian closure; however, the standard DIA-based EDQNM closure has problems with realizability in the presence of linear waves. A realizable modification, the RMC, can be practically successful. In the presence of disparate scale interactions, Markovian closures can be systematically expanded and used to justify and properly define certain adiabatic methods based on long-wavelength-modulated wave kinetic equations for the short-scale wave packets; the formalism is intimately related to statistical formulations of eddy viscosity. Hamiltonian formulation of $\boldsymbol{E} \times \boldsymbol{B}$ advection leads to the identification of Casimir invariants as the appropriate plasmon densities that evolve under the WKE.

The generating-functional/path-integral formalism of Martin, Siggia, and Rose provides an elegant and concise formulation of the renormalization problem. Although it is not a panacea, it has proven to be useful in analytical treatment of interacting zonal flows and drift waves, and has motivated serious attacks on the problem of coherent structures interacting with plasma turbulence.

Finally, systematic bifurcation theory (notably the construction of a local center manifold) has recently been employed to describe the transition to plasma turbulence in the face of the generation of zonal flows.

4.4.5.3. *Physical phenomenology*

I remarked on the onset of stochasticity both in unmagnetized plasmas (v-space diffusion) and magnetized ones (x-space diffusion). I also discussed the transition to forced, dissipative turbulence as modified by the generation of zonal flows; the latter are responsible for the so-called Dimits shift of the onset of collisionless ITG turbulence. The quasi-2D drift-wave turbulence has much in common with geostrophic turbulence in neutral fluids. In steady-state plasma turbulence, zonal flows (including geodesic acoustic modes) can interact with and regulate the transport-producing drift waves. Such systems can exhibit spectral bifurcations between modes of low and high confinement. The formation and possible destruction of streamers is also crucial to a proper understanding of turbulent plasma transport. The interaction of zonal flows and streamers has been suggested as a mechanism for the intermittent generation of coherent plasma blobs near the tokamak edge. Intermittency in plasmas has many facets, including the possibilities of avalanches and front propagation.

4.4.6. *The last word*

What I have probably not captured very clearly in these lectures are the enormous technical obstacles engendered by the complicated inhomogeneous containment geometry. The morass of complicated practical details in fusion research means that the difficulties of measuring the phenomena in the laboratory and of distilling tractable (clean) analytical problems and instructive paradigms are severe. Maybe one of you can help!

Acknowledgements

The lectures include results from research collaborations with J. Bowman, G. Hu, R. Kolesnikov, and M. Ottaviani. I enjoyed informative discussions on various of the included topics with T.-S. Hahm, G. Hammett, P. Holmes, A. Kaufman, P. Morrison, S. Zweben, and the participants in the Workshop on Turbulence and Coherent Structures in Fluids, Plasmas, and Granular Flows. Detailed suggestions and corrections on a draft of the manuscript were gratefully received from A. Brizard, T.-S. Hahm, W. W. Lee, T. Stoltzfus–Dueck, A. White, and S. Zweben. This work was supported by U. S. Dept. of Energy Contract No. DE-AC02-76-CHO-3073.

Appendix A. Notation

A.1. Abbreviations

BBGKY — Bogoliubov–Born–Green–Kirkwood–Yvon
c.p. — cyclic permutations
DIA — direct-interaction approximation
DNS — direct numerical simulation
EDQNM — eddy-damped quasinormal Markovian
ETG — electron-temperature-gradient-driven
FLR — finite-Larmor-radius
GAM — geodesic acoustic mode
GK — gyrokinetic
GPI — gas puff imaging
HM — Hasegawa–Mima
ITG — ion-temperature-gradient-driven

KAM — Kolmogorov–Arnold–Moser
LES — large-eddy simulation
LET — local energy transfer
MHD — magnetohydrodynamic
MSR — Martin–Siggia–Rose
NS — Navier–Stokes
NSTX — National Spherical Torus Experiment
PDE — partial differential equation
PDF — probability density function
QLT — quasilinear theory
RBT — resonance-broadening theory
RMC — Realizable Markovian Closure
SOC — self-organized criticality
TFM — test-field model
TFTR — Tokamak Fusion Test Reactor
WKE — wave kinetic equation
WTT — weak-turbulence theory

A.2. Basic physics symbols

$$— A, a, \alpha —$$

$\boldsymbol{A}(\boldsymbol{x}, t)$ — vector potential ($\boldsymbol{B} = \boldsymbol{\nabla} \times \boldsymbol{A}$)
$\tilde{a}$ — random acceleration
$\hat{\alpha}$ — in HM equation, projection operator into $k_{\parallel} \neq 0$ subspace; in HW equations, parallel coupling operator ($\hat{\alpha} \doteq -\omega_{ci}^{-1} D_{\parallel} \nabla_{\parallel}^2$);

$$— B, b, \beta —$$

$\boldsymbol{B}(\boldsymbol{x}, t)$ — magnetic field ($\boldsymbol{B} = \boldsymbol{\nabla} \times \boldsymbol{A}$)
$\hat{\boldsymbol{b}}$ — unit vector in direction of magnetic field ($\hat{\boldsymbol{b}} \doteq \boldsymbol{B}/|\boldsymbol{B}|$)
β — normalized plasma pressure [$\beta \doteq nT/(B^2/8\pi)$]

$$— C, c, \Gamma, \gamma —$$

$C_{\boldsymbol{k}}$ — Fourier transform of homogeneous correlation function
C_n — nth cumulant [$C_n(1, \ldots, n) \equiv \langle\langle \psi(1) \ldots \psi(n) \rangle\rangle$]
$\hat{C}$ — linearized collision operator
$C(\tau)$ — Lagrangian correlation function
$\mathsf{C}(1, 1')$ — Eulerian correlation matrix
$\mathcal{C}$ — Casimir invariant

c — speed of light

c_A — Alfvén velocity ($c_A^2 \doteq B^2/4\pi n_i m_i$)

c_s — sound speed [$c_s \doteq (ZT_e/m_i)^{\frac{1}{2}}$]

Γ — generic notation for a point in Γ space ($\Gamma \doteq \{X_i \mid i = 1,\ldots,N\}$)

$\Gamma_s(\boldsymbol{x},t)$ — flux of some quantity of species s

$\Gamma(\boldsymbol{k})$ — $\langle J_0^2(k_\perp v_\perp/\omega_{ci})\rangle_M = I_0(b)e^{-b}$

$\gamma(\boldsymbol{k}),\gamma_{\boldsymbol{k}}$ — linear growth rate [$\omega(\boldsymbol{k}) = \Omega(\boldsymbol{k}) + i\gamma(\boldsymbol{k})$]

$\gamma_{\boldsymbol{k}}^{(1)}$ — rate of wave-number evolution in weakly inhomogeneous medium ($\gamma_{\boldsymbol{k}}^{(1)} \doteq \boldsymbol{k} \cdot \nabla\Omega_{\boldsymbol{k}}$)

$\gamma_{\boldsymbol{q}}^{\mathrm{nl}}$ — long-wavelength nonlinear growth rate

$$- \ D,\ d,\ \Delta,\ \delta\ -$$

D — diffusion coefficient

D_B — Bohm diffusion coefficient ($D_B \doteq cT_e/eB$)

D_{gB} — gyro-Bohm diffusion coefficient [$D_{\mathrm{gB}} \doteq (\rho_s/L_n)D_B$]

$D_\parallel$ — diffusion coefficient in the parallel direction (along the magnetic field lines)

$D_\perp$ — diffusion coefficient in the perpendicular direction (across the magnetic field lines)

$\mathcal{D}$ — dissipation term in turbulent energy balance

$\mathcal{D}(\boldsymbol{k},\omega),\ \mathsf{D}(\boldsymbol{k},\omega)$ — dielectric function and tensor

$D[q]$ — volume element in functional integration

d — number of spatial dimensions

Δk — spectral width of wave packet

$\Delta t,\ \Delta v,\ \Delta x$ — elementary steps in random-walk process

$\Delta\Omega$ — frequency mismatch ($\Delta\Omega \doteq \Omega_{\boldsymbol{k}} + \Omega_{\boldsymbol{p}} + \Omega_{\boldsymbol{q}}$, with $\boldsymbol{k} + \boldsymbol{p} + \boldsymbol{q} = \boldsymbol{0}$)

δk — Fourier mode spacing in a box of size L ($\delta k \doteq 2\pi/L$)

$\delta_{\boldsymbol{k}}$ — nonadiabaticity factor in the $i\delta$ model [$\delta n_{\boldsymbol{k}}/n = (1 - i\delta_{\boldsymbol{k}})\delta\varphi_{\boldsymbol{k}}$]

$\delta_{k,k'}$ — Kronecker delta function [$\delta_{k,k'} \doteq 1\ \ (k' = k)$ or 0 otherwise; $\delta_k \equiv \delta_{k,0}$]

$\delta(x)$ — Dirac delta function. $\delta(x) = \delta_-(x) + \delta_+(x)$

$\delta_{\mp}(x)$ — one-sided delta functions [$\delta_{\mp}(x) \doteq \frac{1}{2}\delta(x) \pm (2\pi i)^{-1}\,\mathrm{P}(1/x)$]

$\delta f/\delta\eta(1)$ — functional derivative of $f[\eta]$ with respect to η at space-time point 1

$$- \ E,\ e,\ \epsilon,\ \varepsilon\ -$$

$E(k)$ — omnidirectional energy spectrum, normalized such that the total fluctuation energy is $E = \int_0^\infty dk\, E(k)$

$\boldsymbol{E}(\boldsymbol{x},t)$ — electric field

$\overline{E}_\parallel$ — effective, gyro-averaged $E_\parallel$ [$\overline{E}_\parallel \doteq J_0(k_\perp\rho)E_\parallel$]

$\mathcal{E}$ — fluctuation energy

e — base of the natural logarithms, approximately equal to 2.7 (Candlestick-maker, 1972)

e — electronic charge ($e = |e|$).

ϵ — a positive infinitesimal; the wave-number ratio $q/k \ll 1$; inverse aspect ratio r/a

$\epsilon_\perp$ — dielectric constant of the gyrokinetic vacuum ($\epsilon_\perp \approx \omega_{pi}^2/\omega_{ci}^2$)

ϵ_p — plasma discreteness parameter ($\epsilon_p \doteq 1/\overline{n}\lambda_D^3$)

$$— F, f —$$

F — flatness statistic ($F \doteq \langle \delta\psi^4 \rangle / \langle \delta\psi^2 \rangle^2$)
$F(\boldsymbol{x}, \mu, v_\parallel, t)$ — gyrokinetic PDF
$F_{\boldsymbol{k}}$ — covariance of incoherent noise
$f^{(\mathrm{ext})}(\boldsymbol{x}, t)$ — external random forcing in primitive amplitude equation
$\widetilde{f}_{\boldsymbol{k}}(\boldsymbol{x}, t)$ — incoherent noise in generalized Langevin equation
$f(X, t)$ — one-particle distribution function [$f(X, t) = \langle \widetilde{N}(X, t) \rangle$]
$f_M(\boldsymbol{v})$ — Maxwellian [$f_M(\boldsymbol{v}) \doteq (2\pi v_t^2)^{-\frac{3}{2}} \exp(-v^2/2v_t^2)$]
f_n — n-particle distribution function; $f_1 \equiv f$

$$— G, g, \eta —$$

G — Green's function
g — effective gravity, a measure of magnetic curvature
$g(1; 1')$, $g^{(0)}(1; 1')$ — particle propagators (renormalized and bare)
$\eta_{\boldsymbol{k}}$ — total damping (linear plus nonlinear) in Markovian closures ($\eta_{\boldsymbol{k}} = \eta_{\boldsymbol{k}}^{\mathrm{lin}} + \eta_{\boldsymbol{k}}^{\mathrm{nl}}$)
$\eta_{\boldsymbol{k}}^{\mathrm{nl}}$ — nonlinear damping in Markovian closures
$\eta(1)$, $\widehat{\eta}(1)$ — additive sources in field theory

$$— H, \theta —$$

$H(\Gamma)$ — N-particle Hamiltonian
$H(\tau)$ — Heaviside unit step function [$H(\tau) = 0$ ($\tau < 0$), $\frac{1}{2}$ ($\tau = 0$), or 1 ($\tau > 0$)]
θ — poloidal angle in (r, θ, ζ) coordinate system
$\theta_{\boldsymbol{k}}$ — wave phase
$\theta_{\boldsymbol{k}, \boldsymbol{p}, \boldsymbol{q}}$ — triad interaction time in EDQNM

$$— I, i —$$

I — identity matrix
$I_{\boldsymbol{k}}$ — fluctuation intensity (usually $I_{\boldsymbol{k}} \doteq \langle \delta\varphi_{\boldsymbol{k}}^2 \rangle$)
I_ν — modified Bessel function (first kind)
$\mathcal{I}$ — nonlinearly conserved invariant
i — $\sqrt{-1}$

$$— J, j —$$

$J_\nu(z)$ — Bessel function (first kind)
$\boldsymbol{j}(\boldsymbol{x}, t)$ — current density

$$— K, k, \kappa —$$

K — kurtosis statistic ($K \doteq F - 3$)
K_ν — modified Bessel function (second kind)
$\overline{k}$ — typical $k \doteq |\boldsymbol{k}|$ {in gyrokinetics, $\overline{k}^2 \doteq \tau[1 - \Gamma(\boldsymbol{k})]$}
$\boldsymbol{k}$ — Fourier-transform variable conjugate to $\boldsymbol{x}$ [$f(\boldsymbol{x}) \sim \exp(i\boldsymbol{k} \cdot \boldsymbol{x})$]
k_D — Debye wave number [$k_D^2 \doteq \sum_s k_{Ds}^2$, where $k_{Ds}^2 \doteq 4\pi(nq^2/T)_s$]
$\boldsymbol{\kappa}$ — magnetic curvature ($\boldsymbol{\kappa} \doteq \widehat{\boldsymbol{b}} \cdot \boldsymbol{\nabla}\widehat{\boldsymbol{b}}$)
κ_c — threshold for linear instability

κ_* — upper boundary of Dimits shift regime

κ_n, κ_T — inverses of density and temperature scale lengths ($\kappa_n \doteq L_n^{-1}$, $\kappa_T \doteq L_T^{-1}$)

$$ - \; L, \, l, \, \lambda \; - $$

L — box length for discrete Fourier transform; macroscopic system size

$\widehat{L}$, $\widehat{\mathsf{L}}$ — linear operator ($\partial_t \psi = \widehat{L}\psi$)

L_n, L_T — density and temperature scale lengths ($L_n^{-1} \doteq -\nabla \ln\langle n\rangle$, $L_T^{-1} \doteq -\nabla \ln\langle T\rangle$)

$\mathcal{L}$ — Lagrangian

ℓ — mixing length in turbulence; occasionally, arc length along $\boldsymbol{B}$

$\ln$, $\log$ — $\log_e$, $\log_{10}$

λ — linear eigenvalue (time dependence $\sim e^{\lambda t}$)

λ_D — Debye length ($\lambda_D \doteq k_D^{-1}$)

$$ - \; M, \, m, \, \mu \; - $$

$M_{\boldsymbol{k},\boldsymbol{p},\boldsymbol{q}}$ — symmetrized mode-coupling coefficient (symmetrical in $\boldsymbol{p}$ and $\boldsymbol{q}$)

m_e, m — electron mass

m_i, M — ion mass

μ — true magnetic moment

μ_0 — kinematic viscosity; lowest-order magnetic moment ($\mu_0 \approx v_\perp^2/2B$)

$$ - \; N, \, n, \, \nu \; - $$

N, N_s — total number of particles

$\widetilde{N}(X,t)$ — Klimontovich microdensity $[\widetilde{N}(X,t) \doteq \overline{n}_s^{-1} \sum_{i=1}^{N_s} \delta(X - \widetilde{X}_i(t))]$

$\mathcal{N}$ — wave action $[\mathcal{N}_k \doteq (\partial \operatorname{Re}\mathcal{D}/\partial\Omega_k)(\langle|\delta E_k|^2\rangle/8\pi)]$; statistically averaged Casimir invariant

$\overline{n}$, $\overline{n}_e$, $\overline{n}_i$ — mean number density ($\overline{n} \doteq N/V$)

$n_s(\boldsymbol{x},t)$, $n_s^G(\boldsymbol{x},t)$ — fluid density of particles (n) or gyrocenters (n^G)

ν — collision frequency

ν_d — resonance-broadening or diffusion frequency ($\nu_d \doteq \tau_d^{-1}$)

$\widehat{\nu}$ — dissipative linear operator

$$ - \; P, \, p, \, \pi, \, \varpi \; - $$

$P(\boldsymbol{x},t)$ — scalar pressure

$P(x)$ — probability density function

$P_N(\Gamma,t)$ — Liouville N-particle PDF

$\boldsymbol{p}$ — Fourier wave vector, as in $\boldsymbol{k} + \boldsymbol{p} + \boldsymbol{q} = \boldsymbol{0}$

π — approximately 3.14 (Sagan, 1985)

$\boldsymbol{\varpi}(\boldsymbol{x},t)$ — vorticity ($\boldsymbol{\varpi} \doteq \nabla \times \boldsymbol{u}$)

$\varpi(\boldsymbol{x},t)$ — z component of vorticity ($\varpi \doteq \nabla_\perp^2 \varphi$)

$$ - \; q \; - $$

q — generic charge, of unspecified species; canonical coordinate

q_e, q_i — $q_e \doteq -e$, $q_i \doteq Ze$

$\boldsymbol{q}$ — Fourier wave vector, as in $\boldsymbol{k} + \boldsymbol{p} + \boldsymbol{q} = \boldsymbol{0}$

$$— R,\ r,\ \rho —$$

R, $R^{(0)}$ — major radius of torus $(R = \mathcal{R}^{(0)} + r\cos\theta)$

$\boldsymbol{R}$ — gyrocenter position

Re, $\mathcal{R}$ — Reynolds number $(\mathrm{Re} \doteq UL/\mu_0)$

r — radial coordinate in (r,θ,ζ) coordinate system

$\rho(\boldsymbol{x},t)$ — charge density

ρ_e, ρ_i — gyroradii $[\rho_s \doteq v_{ts}/\omega_{cs}$ (s is in italics)$]$

ρ_s — sound radius $[\rho_\mathrm{s} \doteq c_\mathrm{s}/\omega_{ci}$ (s is in Roman)$]$

$\boldsymbol{\rho}$ — gyroradius vector

ρ^{pol} — polarization charge density

ρ_* — ratio of sound radius to system size $[\rho_* \doteq \rho_\mathrm{s}/a]$

$$— S,\ s,\ \Sigma,\ \sigma —$$

S — stochasticity parameter $(S \doteq$ island width/island separation); skewness statistic $(S \doteq \langle \delta\psi^3\rangle/\langle\delta\psi^2\rangle^{3/2})$

s — species label $(s = e,i)$

Σ — MSR mass-operator matrix; turbulent collision operator

$\sigma_{\boldsymbol{k}}$ — weighting factor that defines nonlinear invariant $(\mathcal{I} \doteq \sum_{\boldsymbol{k}} \sigma_{\boldsymbol{k}} C_{\boldsymbol{k}})$

$$— T,\ t,\ \tau —$$

T — periodicity length or total integration time in temporal Fourier transform; mean time $[T \doteq \frac{1}{2}(t + t')]$

T_s — temperature of species s

$T_s(\boldsymbol{x},t)$ — fluid temperature field

t, $\bar{t}$, t' — time

τ — time difference $(\tau \doteq t - t')$

$\boldsymbol{\tau}$ — Reynolds stress $(\boldsymbol{\tau} \doteq -\rho_m\langle\delta\boldsymbol{u}\,\delta\boldsymbol{u}\rangle)$

τ_{ac} — autocorrelation time $\{\tau_{\mathrm{ac}} \doteq [C(0)]^{-1} \int_0^\infty d\tau\, C(\tau)\}$

τ_d — turbulent diffusion time $(\tau_d \doteq \nu_d^{-1})$

$$— u —$$

$\boldsymbol{u}_s(\boldsymbol{x},t)$ — fluid velocity

$\boldsymbol{u}_{*s}(\boldsymbol{x},t)$ — diamagnetic flow velocity $[\boldsymbol{u}_{*s} \doteq (nq)_s^{-1}(c/B)\widehat{\boldsymbol{b}} \times \boldsymbol{\nabla}P_s]$

$$— V,\ v —$$

V — total volume of system (often taken to infinity in such a way that $\overline{n} \doteq N/V$ remains finite)

$\overline{V}$ — rms velocity

$\boldsymbol{V}_E$ — $\boldsymbol{E} \times \boldsymbol{B}$ velocity $(\boldsymbol{V}_E \doteq c\boldsymbol{E} \times \widehat{\boldsymbol{b}})$

V_{*s} — diamagnetic velocity $(V_{*s} \doteq -cT_s/q_sBL_n)$

$\boldsymbol{V}_{ps}$ — polarization drift velocity $[\boldsymbol{V}_{ps} \doteq \omega_{cs}^{-1}\partial_t(c\boldsymbol{E}_\perp/B)]$

$\boldsymbol{V}_{\boldsymbol{\nabla}B}$ — ∇B magnetic drift velocity $[\boldsymbol{V}_{\boldsymbol{\nabla}B} \doteq (\frac{1}{2}v_\perp^2/\omega_c)\widehat{\boldsymbol{b}} \times \boldsymbol{\nabla}\ln B]$

$\boldsymbol{V}_{\boldsymbol{\kappa}}$ — magnetic curvature drift velocity $[\boldsymbol{V}_{\boldsymbol{\kappa}} \doteq (v_\parallel^2/\omega_c)\widehat{\boldsymbol{b}} \times \boldsymbol{\kappa}]$

V_d — total magnetic drift velocity $(V_d \doteq \boldsymbol{V}_{\boldsymbol{\nabla}B} + v V_{\boldsymbol{\kappa}})$

v — particle velocity

$\boldsymbol{v}_{\mathrm{gr}}(\boldsymbol{k})$ — group velocity $[\boldsymbol{v}_{\mathrm{gr}} \doteq \partial\Omega(\boldsymbol{k})/\partial\boldsymbol{k}]$
$\boldsymbol{v}_{\mathrm{ph}}(\boldsymbol{k})$ — phase velocity $[\boldsymbol{v}_{\mathrm{ph}} \doteq [\Omega(\boldsymbol{k})/k]\hat{\boldsymbol{k}}]$
v_t, v_{te}, v_{ti} — thermal velocity $[v_{ts} \doteq (T_s/m_s)^{\frac{1}{2}}]$
v_{tr}, V_{tr} — microscopic (single-resonance) and macroscopic (all resonances) trapping velocity $[v_{\mathrm{tr},\boldsymbol{k}} \doteq (2q|\varphi_{\boldsymbol{k}}|/m)^{1/2}]$

$$- \, W, \, w \, -$$

$W(k)$, $W[\eta]$ — cumulant generating function and functional $(W \doteq \ln Z)$
$\mathcal{W}$ — Hasegawa–Mima invariant (potential enstrophy)
w — particle weight in the δf simulation method $(w \doteq \delta f/f_0)$
$\widetilde{w}(t)$ — Gaussian white noise

$$- \, x \, -$$

$\boldsymbol{x}$ — vector position in configuration space $[\boldsymbol{x} = (x, y, z)]$
x — Cartesian component of $\boldsymbol{x}$

$$- \, y \, -$$

$\boldsymbol{y}$ — vector position in configuration space
y — poloidal direction (perpendicular to both density gradient and magnetic field); Cartesian component of $\boldsymbol{x}$

$$- \, Z, \, z, \, \zeta \, -$$

Z — atomic number $(\overline{n}_e = Z\overline{n}_i)$
$Z(k)$, $Z[\eta]$ — characteristic (moment-generating) function and functional
z — toroidal direction; Cartesian component of $\boldsymbol{x}$
ζ — toroidal angle in (r, θ, ζ) coordinate system

$$- \, \phi, \, \varphi \, -$$

$\phi(\boldsymbol{x}, t)$ — dimensional electrostatic potential
$\varphi(\boldsymbol{x}, t)$ — dimensionless electrostatic potential: $\varphi \doteq e\phi/T_e$
$\overline{\varphi}$ — effective (gyro-averaged) potential $[\overline{\varphi}_{\boldsymbol{k}} \doteq J_0(k_\perp\rho)\varphi_{\boldsymbol{k}}]$

$$- \, \psi \, -$$

ψ — generic field

$$- \, \Omega, \, \omega \, -$$

$\Omega(\boldsymbol{k})$ — real normal-mode frequency
ω — Fourier variable conjugate to time t $[f(t) \sim \exp(-i\omega t)]$;
ω_c, ω_{ce}, ω_{ci} — gyrofrequencies $(\omega_{cs} \doteq q_s B/m_s c)$
ω_p, ω_{pe}, ω_{pi} — plasma frequency $[\omega_p^2 \doteq \sum_s \omega_{ps}^2,$ where $\omega_{ps}^2 \doteq 4\pi(\overline{n}q^2/T)_s]$
$\omega_*(k_y)$ — drift (diamagnetic) frequency $(\omega_* \doteq k_y V_*)$
ω_{tr}, Ω_{tr} — microscopic (single-resonance) and macroscopic (all resonances) trapping frequencies $(\omega_{\mathrm{tr}} \doteq k v_{\mathrm{tr}})$

A.3. Miscellaneous notation

$\widetilde{A}$ — Tilde signifies a random variable.

A^* — complex conjugate of A

A_k, $A(k)$ — The subscript and parentheses distinguish the discrete Fourier amplitude A_k from the continuum transform $A(k)$ when both are in use simultaneously.

$A_{ij}^\dagger$ — Hermitian conjugate $(A_{ij}^\dagger \doteq A_{ji}^*)$

$\overline{A(t)}$ — time average of A

$\overline{A}$ — space and ensemble average of A $[\overline{A} \doteq L^{-1} \int_0^L dx \, \langle A \rangle (x)]$

$\langle A \rangle$, δA — ensemble average and fluctuation of $\widetilde{A}$ $(\widetilde{A} = \langle A \rangle + \delta A$; the tilde is frequently omitted)

$\langle\langle x\, y\, z \rangle\rangle$ — cumulant of x, y, and z

$A[f]$ — A depends functionally on f.

$\{A, B\}$ — Poisson bracket of A and B

$[A, B]$ — Poisson bracket for $\boldsymbol{E} \times \boldsymbol{B}$ advection $([A, B] \doteq \widehat{\boldsymbol{z}} \cdot \boldsymbol{\nabla} A \times \boldsymbol{\nabla} B)$

$A_\pm(t)$ — one-sided functions $[A_\pm(t) \doteq H(\pm t) A(t)]$

A^T — transpose $[A_{ij}^T \doteq A_{ji}]$

$|A|$ — absolute value of A

$\underline{A}$, A' — long- and short-wavelength components of $A(\boldsymbol{x})$

$\widehat{\boldsymbol{k}}$ — unit vector $(\widehat{\boldsymbol{k}} \doteq \boldsymbol{k}/|\boldsymbol{k}|)$

$O(\epsilon)$ — asymptotically of order ϵ

$\mathrm{P}\!\int$ — principal value $\left[\mathrm{P}\!\int \doteq \lim_{\epsilon \to 0} \left(\int_{-\infty}^{\epsilon} + \int_{\epsilon}^{\infty}\right)\right]$

Re, Im — real and imaginary parts

$\doteq$ — definition

$\equiv$ — equivalent to $\left(\doteq \,\equiv\, \overset{\mathrm{def}}{=}\right)$

$\mathrm{sgn}(x)$ — sign of x

$\displaystyle\sum_{\Delta} - \sum_{\boldsymbol{p},\boldsymbol{q}} \delta_{\boldsymbol{k}+\boldsymbol{p}+\boldsymbol{q},\mathbf{0}}$

Bibliography

Amit, 2005: D. J. Amit, *Field Theory, the Renormalization Group, and Critical Phenomena: Graphs to Computers*, McGraw–Hill, New York, third edition, 2005.

Bak et al., 1987: P. Bak, C. Tang, and K. Wiesenfeld, Self-organized criticality: An explanation of $1/f$ noise, Phys. Rev. Lett., **59**:381, 1987.

Bak et al., 1988: P. Bak, C. Tang, and K. Wiesenfeld, Self-organized criticality, Phys. Rev. A, **38**:364, 1988.

Balescu, 1963: R. Balescu, *Statistical Mechanics of Charged Particles*, Interscience, London, 1963.

Ball et al., 2002: R. Ball, R. L. Dewar, and H. Sugama, Metamorphosis of plasma turbulence–shear-flow dynamics through a transcritical bifurcation, Phys. Rev. E, **66**:066408, 2002.

Ball, 2006: R. Ball, Distilled turbulence, in *Turbulence and Coherent Structures in Fluids, Plasmas and Nonlinear Mediums: Proceedings of the COSNET Workshop*, edited by J. Denier, World Scientific, Singapore, 2006.

Barenblatt, 1996: G. I. Barenblatt, *Scaling, Self-similarity, and Intermediate Asymptotics*, Cambridge University Press, Cambridge, 1996.

Benford and Thomson, 1972: G. Benford and J. J. Thomson, Probabilistic model of plasma turbulence, Phys. Fluids, **15**:1496, 1972.

Bisai et al., 2005a: N. Bisai, A. Das, S. Deshpande, R. Jha, P. Kaw, A. Sen, and R. Singh, Edge and scrape-off layer tokamak plasma turbulence simulation using two-field fluid model, Phys. Plasmas, **12**:072520, 2005.

Bisai et al., 2005b: N. Bisai, A. Das, S. Deshpande, R. Jha, P. Kaw, A. Sen, and R. Singh, Formation of a density blob and its dynamics in the edge and the scrape-off layer of a tokamak plasma, Phys. Plasmas, **12**:102525, 2005.

Biskamp and Zeiler, 1995: D. Biskamp and A. Zeiler, Nonlinear instability mechanism in 3D collisional drift-wave turbulence, Phys. Rev. Lett., **74**:706, 1995.

Biskamp et al., 1994: D. Biskamp, S. J. Camargo, and B. D. Scott, Spectral properties and statistics of resistive drift-wave turbulence, Phys. Lett. A, **186**:239, 1994.

Bohm, 1949: D. Bohm, Qualitative description of the arc plasma in a magnetic field, in *The Characteristics of Electrical Discharges in Magnetic Fields*, edited by A. Guthrie and R. K. Wakerling, chapter 1, p. 1, McGraw–Hill, New York, 1949.

Bourret, 1962: R. C. Bourret, Stochastically perturbed fields, with applications to wave propagation in random media, Nuovo Cimento, **26**:1, 1962.

Bowman and Krommes, 1997: J. C. Bowman and J. A. Krommes, The realizable Markovian closure and realizable test-field model. II: Application to anisotropic drift-wave turbulence, Phys. Plasmas, **4**:3895, 1997.

Bowman et al., 1993: J. C. Bowman, J. A. Krommes, and M. Ottaviani, The realizable Markovian closure. I. General theory, with application to three-wave dynamics, Phys. Fluids B, **5**:3558, 1993.

Braginskii, 1965: S. I. Braginskii, Transport processes in a plasma, in *Reviews of Plasma Physics, Vol. 1*, edited by M. N. Leontovich, pp. 205–311, Consultants Bureau, New York, 1965.

Brizard and Hahm, 2006: A. J. Brizard and T. S. Hahm, Foundations of nonlinear gyrokinetic theory, Rev. Mod. Phys., 2006, submitted.

Camargo et al., 1995: S. J. Camargo, D. Biskamp, and B. D. Scott, Resistive drift-wave turbulence, Phys. Plasmas, **2**:48, 1995.

Candlestickmaker, 1972: S. Candlestickmaker, On the imperturbability of elevator operators. LVII, Q. Jl R. astr. Soc., **13**:63, 1972.

Candy and Waltz, 2005: J. Candy and R. E. Waltz, Velocity-space resolution, entropy production and upwind dissipation in Eulerian gyrokinetic simulations, Phys. Plasmas, 2005, submitted.

Carnevale and Martin, 1982: G. F. Carnevale and P. C. Martin, Field theoretical techniques in statistical fluid dynamics: With application to nonlinear wave dynamics, Geophys. Astrophys. Fluid Dynamics, **20**:131, 1982.

Carreras et al., 1996: B. A. Carreras, C. Hidalgo, E. Sánchez, M. A. Pedrosa,

R. Balbín, I. García-Cortés, B. van Milligen, D. E. Newman, and V. E. Lynch, Fluctuation-induced flux at the plasma edge in toroidal devices, Phys. Plasmas, **3**:2664, 1996.

Cary and Kaufman, 1977: J. R. Cary and A. N. Kaufman, Ponderomotive force and linear susceptibility in Vlasov plasma, Phys. Rev. Lett., **39**:402, 1977.

Cary and Kaufman, 1981: J. R. Cary and A. N. Kaufman, Ponderomotive effects in collisionless plasma: A Lie transform approach, Phys. Fluids, **24**:1238, 1981.

Cary and Littlejohn, 1983: J. R. Cary and R. G. Littlejohn, Noncanonical Hamiltonian mechanics and its application to magnetic field line flow, Ann. Phys. (N.Y.), **151**:1, 1983.

Cary, 1981: J. R. Cary, Lie transform perturbation theory for Hamiltonian systems, Physics Rep., **79**:129, 1981.

Cercignani, 1998: C. Cercignani, *Ludwig Boltzmann: The man who trusted atoms*, Clarendon Press, Oxford, 1998.

Chandrasekhar, 1960: S. Chandrasekhar, *Plasma Physics*, U. of Chicago Press, Chicago, 1960.

Charney and Stern, 1962: J. G. Charney and M. E. Stern, On the stability of internal baroclinic jets in a rotating atmosphere, J. Atmos. Sci., **19**:159, 1962.

Ching, 1973: H. Ching, Large-amplitude stabilization of the drift instability, Phys. Fluids, **16**:130, 1973.

Chirikov, 1969: B. V. Chirikov, Research concerning the theory of non-linear resonance and stochasticity, Nuclear Physics Institute of the Siberian Section of the USSR Academy of Sciences Report 267, translated by A. T. Sanders (CERN Trans. 71–40), 1969.

Connor and Taylor, 1977: J. W. Connor and J. B. Taylor, Scaling laws for plasma confinement, Nucl. Fusion, **17**:1047, 1977.

Connor, 1988: J. W. Connor, Invariance principles and plasma confinement, Plasma Phys. Control. Fusion, **30**:619, 1988.

Cowley et al., 1991: S. C. Cowley, R. M. Kulsrud, and R. Sudan, Considerations of ion-temperature-gradient-driven turbulence, Phys. Fluids B, **3**:2767, 1991.

Davidson, 1972: R. C. Davidson, *Methods in Nonlinear Plasma Theory*, Academic Press, New York, 1972.

Davidson, 2001: R. C. Davidson, *Physics of Nonneutral Plasmas*, Imperial College Press, London, 2001.

de Dominicis and Martin, 1964a: C. de Dominicis and P. C. Martin, Stationary entropy principle and renormalization in normal and superfluid systems. I. Algebraic formulation, J. Math. Phys., **5**:14, 1964.

de Dominicis and Martin, 1964b: C. de Dominicis and P. C. Martin, Stationary entropy principle and renormalization in normal and superfluid systems. II. Diagrammatic formulation, J. Math. Phys., **5**:31, 1964.

Dewar and Abdullatif, 2006: R. L. Dewar and R. F. Abdullatif, Zonal flow generation by modulational instability, in *Turbulence and Coherent Structures in Fluids, Plasmas and Nonlinear Mediums: Proceedings of the COSNET Workshop*, edited by J. Denier, World Scientific, Singapore, 2006.

Dewar, 1973: R. L. Dewar, Oscillation center quasilinear theory, Phys. Fluids, **16**:1102, 1973.

Diacu and Holmes, 1996: F. Diacu and P. Holmes, *Celestial encounters: The origins of chaos and stability*, Princeton University Press, Princeton, NJ, 1996.

Diamond and Hahm, 1995: P. H. Diamond and T.-S. Hahm, On the dynamics of turbulent transport near marginal stability, Phys. Plasmas, **2**:3640, 1995.

Diamond et al., 1994: P. H. Diamond, Y.-M. Liang, B. A. Carreras, and P. W. Terry, Self-regulating shear flow turbulence: A paradigm for the L to H transition, Phys. Rev. Lett., **72**:2565, 1994.

Diamond et al., 1998: P. H. Diamond, M. N. Rosenbluth, F. L. Hinton, M. Malkov, J. Fleischer, and A. Smolyakov, Dynamics of zonal flows and self-regulating drift-wave turbulence, in *17th IAEA Fusion Energy Conference*, pp. 1421–1428, International Atomic Energy Agency, Vienna, 1998, IAEA–CN–69/TH3/1.

Diamond et al., 2005: P. H. Diamond, S.-I. Itoh, K. Itoh, and T. S. Hahm, Zonal flows in plasma—a review, Plasma Phys. Control. Fusion, **47**:R35, 2005.

Dimits et al., 2000: A. M. Dimits, G. Bateman, M. A. Beer, B. I. Cohen, W. Dorland, G. W. Hammett, C. Kim, J. E. Kinsey, M. Kotschenreuther, A. H. Kritz, L. L. Lao, J. Mandrekas, W. M. Nevins, S. E. Parker, A. J. Redd, D. E. Shumaker, R. Sydora, and J. Weiland, Comparisons and physics basis of tokamak transport models and turbulence simulations, Phys. Plasmas, **7**:969, 2000.

D'Ippolito and Myra, 2003: D. A. D'Ippolito and J. R. Myra, Blob stability and transport in the scrape-off-layer, Phys. Plasmas, **10**:4029, 2003.

D'Ippolito et al., 2002: D. A. D'Ippolito, J. R. Myra, and S. I. Krasheninnikov, Cross-field blob transport in tokamak scrape-off-layer plasmas, Phys. Plasmas, **9**:222, 2002.

Drake et al., 1988: J. F. Drake, P. N. Guzdar, and A. B. Hassam, Streamer formation in plasma with a temperature gradient, Phys. Rev. Lett., **61**:2205, 1988.

Drummond and Pines, 1962: W. E. Drummond and D. Pines, Non-linear stability of plasma oscillations, in *Proceedings of the Conference on Plasma Physics and Controlled Nuclear Fusion Research (Salzburg, 1961) [Nucl. Fusion Suppl. Pt. 3]*, pp. 1049–1057, International Atomic Energy Agency, Vienna, 1962.

Dubin and Krommes, 1982: D. H. E. Dubin and J. A. Krommes, Stochasticity, superadiabaticity, and the theory of adiabatic invariants and guiding center motion, in *Long Time Prediction in Dynamics*, edited by W. Horton, L. E. Reichl, and V. G. Szebehely, pp. 251–280, Wiley, New York, 1982.

Dubin et al., 1983: D. H. E. Dubin, J. A. Krommes, C. R. Oberman, and W. W. Lee, Nonlinear gyrokinetic equations, Phys. Fluids, **26**:3524, 1983.

Dupree and Tetreault, 1978: T. H. Dupree and D. J. Tetreault, Renormalized dielectric function for collisionless drift wave turbulence, Phys. Fluids, **21**:425, 1978.

Dupree, 1966: T. H. Dupree, A perturbation theory for strong plasma turbulence,

Phys. Fluids, **9**:1773, 1966.

Dupree, 1967: T. H. Dupree, Nonlinear theory of drift-wave turbulence and enhanced diffusion, Phys. Fluids, **10**:1049, 1967.

Dupree, 1972: T. H. Dupree, Theory of phase space density granulation in plasma, Phys. Fluids, **15**:334, 1972.

Dyson, 1949: F. J. Dyson, The radiation theories of Tomonaga, Schwinger, and Feynman, Phys. Rev., **75**:486, 1949.

Einstein, 1905: A. Einstein, Über die von der molekularkinetischen Theori der Wärme geforderte Bewegung von in ruhenden Flüssigkeiten suspendierten Teilchen, Ann. d. Phys., **17**:549, 1905, English translation in A. Einstein, *Investigations on the theory of the Brownian Movement*, edited by R. Fürth, translated by A. D. Cowper (Dover, New York, 1956), p. 1.

Einstein, 1908: A. Einstein, Elementare Theorie der Brownschen Bewegung, Zeit. für Elektrochemie, **14**:235, 1908, English translation (The elementary theory of the Brownian motion) in A. Einstein, *Investigations on the theory of the Brownian Movement*, edited by R. Fürth, translated by A. D. Cowper (Dover, New York, 1956), p. 68.

Fisch, 1987: N. J. Fisch, Theory of current drive in plasmas, Rev. Mod. Phys., **59**:175, 1987.

Frederiksen and Davies, 2000: J. S. Frederiksen and A. G. Davies, Dynamics and spectra of cumulant update closures for two-dimensional turbulence, Geophys. Astrophys. Fluid Dynamics, **92**:197, 2000.

Frederiksen et al., 1994: J. S. Frederiksen, A. G. Davies, and R. C. Bell, Closure theories with non-Gaussian restarts for truncated two-dimensional turbulence, Phys. Fluids, **6**:3153, 1994.

Frederiksen, 2006a: J. Frederiksen, 2006, private communication.

Frederiksen, 2006b: J. S. Frederiksen, Turbulence closures and subgrid-scale parameterizations, in *Turbulence and Coherent Structures in Fluids, Plasmas and Nonlinear Mediums: Proceedings of the COSNET Workshop*, edited by J. Denier, World Scientific, Singapore, 2006.

Frisch, 1995: U. Frisch, *Turbulence*, Cambridge University Press, Cambridge, 1995.

Guckenheimer and Holmes, 1983: J. Guckenheimer and P. Holmes, *Nonlinear Oscillations, Dynamical Systems, and Bifurcations of Vector Fields*, Springer, New York, 1983.

Gürcan et al., 2006: O. D. Gürcan, P. H. Diamond, and T. S. Hahm, Radial transport of fluctuation energy in drift-wave turbulence, Phys. Plasmas, 2006, submitted.

Guzdar et al., 2001: P. N. Guzdar, R. G. Kleva, A. Das, and P. K. Kaw, Zonal flow and zonal magnetic field generation by finite β drift waves: A theory for low to high transitions in tokamaks, Phys. Rev. Lett., **87**:015001, 2001.

Hahm et al., 1988: T. S. Hahm, W. W. Lee, and A. Brizard, Nonlinear gyrokinetic theory for finite-beta plasmas, Phys. Fluids, **31**:1940, 1988.

Hahm, 1988: T. S. Hahm, Nonlinear gyrokinetic equations for tokamak microturbulence, Phys. Fluids, **31**:2670, 1988.

Hahm, 2002: T. S. Hahm, Physics behind transport barrier theory and simulation, Plasma Phys. Control. Fusion, **44**:A87, 2002.

Hammett and Perkins, 1990: G. W. Hammett and F. W. Perkins, Fluid moment models for Landau damping with application to the ion-temperature-gradient instability, Phys. Rev. Lett., **64**:3019, 1990.

Hasegawa and Mima, 1978: A. Hasegawa and K. Mima, Pseudo-three-dimensional turbulence in magnetized nonuniform plasma, Phys. Fluids, **21**:87, 1978.

Hasegawa and Wakatani, 1983: A. Hasegawa and M. Wakatani, Plasma edge turbulence, Phys. Rev. Lett., **50**:682, 1983.

Helander and Sigmar, 2002: P. Helander and D. J. Sigmar, *Collisional Transport in Magnetized Plasmas*, Cambridge University Press, Cambridge, 2002.

Hinton and Horton, 1971: F. L. Hinton and C. W. Horton, Jr., Amplitude limitation of a collisional drift wave instability, Phys. Fluids, **14**:116, 1971.

Hirshman, 1980: S. P. Hirshman, Two-dimensional electrostatic $E \times B$ trapping, Phys. Fluids, **23**:562, 1980.

Holmes, 1990: P. Holmes, Poincaré, celestial mechanics, dynamical systems theory and "chaos", Phys. Rep., **193**:137, 1990.

Horton and Hasegawa, 1994: W. Horton and A. Hasegawa, Quasi-two-dimensional dynamics of plasmas and fluids, Chaos, **4**:227, 1994.

Horton et al., 1996: W. Horton, G. Hu, and G. Laval, Turbulent transport in mixed states of convective cells and sheared flows, Phys. Plasmas, **3**:2912, 1996.

Horton, 1986: W. Horton, Statistical properties and correlation functions for drift waves, Phys. Fluids, **29**:1491, 1986.

Horton, 1999: W. Horton, Drift waves and transport, Rev. Mod. Phys., **71**:735, 1999.

Hu and Krommes, 1994: G. Hu and J. A. Krommes, Generalized weighting scheme for δf particle-simulation method, Phys. Plasmas, **1**:863, 1994.

Hu et al., 1995: G. Hu, J. A. Krommes, and J. C. Bowman, Resistive drift-wave plasma turbulence and the realizable Markovian closure, Phys. Lett. A, **202**:117, 1995.

Hu et al., 1997: G. Hu, J. A. Krommes, and J. C. Bowman, Statistical theory of resistive drift-wave turbulence and transport, Phys. Plasmas, **4**:2116, 1997.

Hwa and Kardar, 1992: T. Hwa and M. Kardar, Avalanches, hydrodynamics, and discharge events in models of sandpiles, Phys. Rev. A, **45**:7002, 1992.

Ichimaru, 1973: S. Ichimaru, *Basic Principles of Plasma Physics—A Statistical Approach*, W. A. Benjamin, New York, 1973.

Ichimaru, 1992a: S. Ichimaru, *Statistical Plasma Physics. Volume I: Basic Principles*, Addison–Wesley, Reading, Massachusetts, 1992.

Ichimaru, 1992b: S. Ichimaru, *Statistical Plasma Physics. Volume II: Condensed Plasmas*, Addison–Wesley, Reading, Massachusetts, 1992.

Isichenko et al., 1992: M. B. Isichenko, W. Horton, D. E. Kim, E. G. Heo, and D.-I. Choi, Stochastic diffusion and Kolmogorov entropy in regular and random Hamiltonians, Phys. Fluids B, **4**:3973, 1992.

Itoh et al., 2005a: K. Itoh, K. Hallatschek, and S.-I. Itoh, Excitation of geodesic acoustic mode in toroidal plasmas, Plasma Phys. Control. Fusion, **47**:451,

2005.

Itoh et al., 2005b: K. Itoh, Y. Nagashima, S.-I. Itoh, P. H. Diamond, A. Fujisawa, M. Yagi, and A. Fukuyama, On the bicoherence analysis of plasma turbulence, Phys. Plasmas, **12**:102301, 2005.

Itoh et al., 2006: K. Itoh, S.-I. Itoh, P. H. Diamond, T. S. Hahm, A. Fujisawa, G. R. Tynan, M. Yagi, and Y. Nagashima, Physics of zonal flows, Phys. Plasmas, 2006, in press.

Jenko et al., 2000: F. Jenko, W. Dorland, M. Kotschenreuther, and B. N. Rogers, Electron temperature gradient driven turbulence, Phys. Plasmas, **7**:1904, 2000.

Jensen, 1998: H. J. Jensen, *Self-Organized Criticality*, Cambridge University Press, Cambridge, UK, 1998.

Joyce and Montgomery, 1973: G. Joyce and D. Montgomery, Negative temperature states for the two-dimensional guiding-centre plasma, J. Plasma Phys., **10**:107, 1973.

Kadomtsev, 1965: B. B. Kadomtsev, *Plasma Turbulence*, Academic Press, New York, 1965, Translated by L. C. Ronson from the 1964 Russian edition *Problems in Plasma Theory*, edited by M. A. Leontovich. Translation edited by M. C. Rusbridge.

Kadomtsev, 1975: B. B. Kadomtsev, Tokamaks and dimensional analysis, Sov. J. Plasma Phys., **1**:295, 1975, [Fiz. Plazmy **1**, 531 (1975)].

Karney, 1978: C. F. F. Karney, Stochastic ion heating by a lower hybrid wave, Phys. Fluids, **21**:1584, 1978.

Kaufman, 1972: A. N. Kaufman, Reformulation of quasi-linear theory, J. Plasma Phys., **8**:1, 1972.

Klimontovich, 1967: Y. L. Klimontovich, *The Statistical Theory of Non-equilibrium Processes in a Plasma*, M.I.T. Press, Cambridge, Massachusetts, 1967, Edited by D. ter Harr, translated by H. S. H. Massey and O. M. Blunn.

Klinger et al., 1997: T. Klinger, A. Latten, A. Piel, G. Bonhomme, T. Pierre, and T. D. de Wit, Route to drift wave chaos and turbulence in a bounded low-β plasma experiment, Phys. Rev. Lett., **79**:3913, 1997.

Kolesnikov and Krommes, 2005a: R. A. Kolesnikov and J. A. Krommes, Bifurcation theory of the transition to collisionless ion-temperature-gradient-driven plasma turbulence, Phys. Plasmas, **12**:122302, 2005.

Kolesnikov and Krommes, 2005b: R. A. Kolesnikov and J. A. Krommes, The transition to collisionless ion-temperature-gradient-driven plasma turbulence: A dynamical systems approach, Phys. Rev. Lett., **94**:235002, 2005.

Koniges et al., 1991: A. E. Koniges, J. A. Crotinger, W. P. Dannevik, G. F. Carnevale, P. H. Diamond, and F. Y. Gang, Equilibrium spectra and implications for a two-field turbulence model, Phys. Fluids B, **3**:1297, 1991.

Kotschenreuther, 1991: M. Kotschenreuther, Particle simulations with greatly reduced noise, in *Proceedings of the 14th International Conference on the Numerical Simulation of Plasmas*, Arlington, Virginia, 1991, Office of Naval Research, Paper PT20.

Kraichnan and Montgomery, 1980: R. H. Kraichnan and D. Montgomery, Two-dimensional turbulence, Rep. Prog. Phys., **43**:547, 1980.

Kraichnan, 1959: R. H. Kraichnan, The structure of isotropic turbulence at very high Reynolds numbers, J. Fluid Mech., **5**:497, 1959.

Kraichnan, 1961: R. H. Kraichnan, Dynamics of nonlinear stochastic systems, J. Math. Phys., **2**:124, 1961, Erratum: J. Math. Phys. **3**, 205 (1962).

Kraichnan, 1964: R. H. Kraichnan, Kolmogorov's hypotheses and Eulerian turbulence theory, Phys. Fluids, **7**:1723, 1964.

Kraichnan, 1967: R. H. Kraichnan, Inertial ranges in two-dimensional turbulence, Phys. Fluids, **10**:1417, 1967.

Kraichnan, 1971: R. H. Kraichnan, An almost-Markovian Galilean-invariant turbulence model, J. Fluid Mech., **47**:513, 1971.

Kraichnan, 1976: R. H. Kraichnan, Eddy viscosity in two and three dimensions, J. Atmos. Sci., **33**:1521, 1976.

Kraichnan, 1980: R. H. Kraichnan, Realizability inequalities and closed moment equations, in *Nonlinear Dynamics*, edited by R. H. G. Helleman, p. 37, New York Academy of Sciences, New York, 1980.

Kraichnan, 1985: R. H. Kraichnan, Decimated amplitude equations in turbulence dynamics, in *Theoretical Approaches to Turbulence*, edited by D. L. Dwoyer, M. Y. Hussaini, and R. G. Voight, p. 91, Springer, New York, 1985.

Krasheninnikov, 2001: S. I. Krasheninnikov, On scrape off layer plasma transport, Phys. Lett. A, **283**:368, 2001.

Krommes and Hu, 1994: J. A. Krommes and G. Hu, The role of dissipation in simulations of homogeneous plasma turbulence, and resolution of the entropy paradox, Phys. Plasmas, **1**:3211, 1994.

Krommes and Kim, 2000: J. A. Krommes and C.-B. Kim, Interactions of disparate scales in drift-wave turbulence, Phys. Rev. E, **62**:8508, 2000.

Krommes and Kolesnikov, 2004: J. A. Krommes and R. A. Kolesnikov, Hamiltonian description of zonal-flow generation, Phys. Plasmas, **11**:L29, 2004.

Krommes and Ottaviani, 1999: J. A. Krommes and M. Ottaviani, Long-time tails do not necessarily imply self-organized criticality or the breakdown of the standard transport paradigm, Phys. Plasmas, **6**:3731, 1999.

Krommes et al., 1986: J. A. Krommes, W. W. Lee, and C. Oberman, Equilibrium fluctuation energy of gyrokinetic plasma, Phys. Fluids, **29**:2421, 1986.

Krommes, 1993: J. A. Krommes, Dielectric response and thermal fluctuations in gyrokinetic plasma, Phys. Fluids B, **5**:1066, 1993.

Krommes, 1997: J. A. Krommes, The clump lifetime revisited: Exact calculation of the second-order structure function for a model of forced, dissipative turbulence, Phys. Plasmas, **4**:655, 1997.

Krommes, 2002: J. A. Krommes, Fundamental statistical theories of plasma turbulence in magnetic fields, Phys. Rep., **360**:1, 2002.

Krommes, 2004a: J. A. Krommes, Comments on "Dynamics of zonal flow saturation in strong collisionless drift wave turbulence" [Phys. Plasmas **9**, 4530 (2002)], Phys. Plasmas, **11**:1744, 2004.

Krommes, 2004b: J. A. Krommes, Statistical plasma physics in a strong magnetic field: Paradigms and problems, in *Mathematical and Physical Theory of Turbulence: Proceedings of the International Turbulence Workshop*, edited by J. Cannon and B. Shivamoggi, New York, 2004, Marcel Dekker, In press.

Krommes, 2006: J. A. Krommes, The transition to ion-temperature-gradient-driven plasma turbulence, in *Turbulence and Coherent Structures in Fluids, Plasmas and Nonlinear Mediums: Proceedings of the COSNET Workshop*, edited by J. Denier, World Scientific, Singapore, 2006.

Kruskal, 1965: M. Kruskal, Asymptotology, in *Plasma Physics*, p. 373, International Atomic Energy Agency, Vienna, 1965.

Kubo, 1959: R. Kubo, Some aspects of the statistical-mechanical theory of irreversible processes, in *Lectures in Theoretical Physics*, edited by W. E. Brittin and L. G. Dunham, volume I, p. 181, Interscience, New York, 1959.

Kubo, 1962a: R. Kubo, Generalized cumulant expansion method, J. Phys. Soc. Japan, **17**:1100, 1962.

Kubo, 1962b: R. Kubo, Stochastic theory of line shape and relaxation, in *Fluctuation, Relaxation, and Resonance in Magnetic Systems*, edited by D. ter Harr, p. 23, Oliver and Boyd, Edinburgh, 1962.

Langevin, 1908: P. Langevin, Sur la théorie du mouvement brownien, C. R. Acad. Sci., **146**:530, 1908.

Lee and Tang, 1988: W. W. Lee and W. M. Tang, Gyrokinetic particle simulations of ion temperature gradient drift instabilities, Phys. Fluids, **31**:612, 1988.

Lee, 1983: W. W. Lee, Gyrokinetic approach in particle simulation, Phys. Fluids, **26**:556, 1983.

Lichtenberg and Lieberman, 1992: A. J. Lichtenberg and M. A. Lieberman, *Regular and Chaotic Dynamics*, Springer, New York, second edition, 1992.

Lin et al., 1998: Z. Lin, T. S. Hahm, W. W. Lee, W. M. Tang, and R. B. White, Turbulent transport reduction by zonal flows: Massively parallel simulations, Science, **281**:1835, 1998.

Lithwick and Goldreich, 2003: Y. Lithwick and P. Goldreich, Imbalanced weak magnetohydrodynamic turbulence, Ap. J., **582**:1220, 2003.

LoDestro et al., 1991: L. L. LoDestro, B. I. Cohen, R. H. Cohen, A. M. Dimits, Y. Matsuda, W. M. Nevins, W. A. Newcomb, T. J. Williams, A. E. Koniges, W. P. Dannevik, J. A. Crotinger, R. D. Sydora, J. M. Dawson, S. Ma, V. K. Decyk, W. W. Lee, T. S. Hahm, H. Naitou, and T. Kamimura, Comparison of simulations and theory of low-frequency plasma turbulence, in *Plasma Physics and Controlled Nuclear Fusion Research, 1990*, volume II, p. 31, International Atomic Energy Agency, Vienna, 1991.

Manneville, 1990: P. Manneville, *Dissipative Structures and Weak Turbulence*, Academic Press, Boston, 1990.

Maqueda et al., 2003: R. J. Maqueda, G. A. Wurden, D. P. Stotler, S. J. Zweben, B. LaBombard, J. L. Terry, J. L. Lowrance, V. J. Mastrocola, G. F. Renda, D. A. D'Ippolito, J. R. Myra, and N. Nishino, Gas puff imaging of edge turbulence, Rev. Sci. Instrum., **74**:2020, 2003.

Martin et al., 1973: P. C. Martin, E. D. Siggia, and H. A. Rose, Statistical dynamics of classical systems, Phys. Rev. A, **8**:423, 1973.

Martin, 1976: P. C. Martin, 1976, Private communication.

Martin, 1979: P. C. Martin, Schwinger and statistical physics: A spin-off success story and some challenging sequels, Physica, **96A**:70, 1979.

McComb, 1990: W. D. McComb, *The Physics of Fluid Turbulence*, Clarendon Press, Oxford, 1990.

McComb, 2004: W. D. McComb, *Renormalization Methods: A Guide for Beginners*, Oxford University Press, Oxford, 2004.

Mehra and Milton, 2000: J. Mehra and K. A. Milton, *Climbing the Mountain— The Scientific Biography of Julian Schwinger*, Oxford University Press, Oxford, 2000.

Meiss and Horton, 1982: J. D. Meiss and W. Horton, Fluctuation spectra of a drift wave soliton gas, Phys. Fluids, **25**:1838, 1982.

Miyamoto, 1978: K. Miyamoto, Recent stellarator research, Nucl. Fusion, **18**:243, 1978.

Miyamoto, 2005: K. Miyamoto, *Plasma Physics and Controlled Nuclear Fusion*, Springer, Berlin, 2005.

Montgomery, 1977: D. Montgomery, Implications of Navier–Stokes turbulence theory for plasma turbulence, Proc. Indian Acad. Sci., **8A**:87, 1977.

Morrison and Greene, 1980: P. J. Morrison and J. M. Greene, Noncanonical Hamiltonian density formulation of hydrodynamics and ideal magnetohydrodynamics, Phys. Rev. Lett., **45**:790, 1980.

Morrison, 1998: P. J. Morrison, Hamiltonian description of the ideal fluid, Rev. Mod. Phys., **70**:467, 1998.

Moyer et al., 2001: R. A. Moyer, G. R. Tynan, C. Holland, and M. J. Burin, Increased nonlinear coupling between turbulence and low-frequency fluctuations at the L–H transition, Phys. Rev. Lett., **87**:135001, 2001.

Myra, 2005: J. R. Myra, Blobbirth and transport in NSTX: GPI data analysis and theory, Bull. Am. Phys. Soc., **50**:323, 2005.

Nevins et al., 2005: W. M. Nevins, G. W. Hammett, A. M. Dimits, W. Dorland, and D. E. Shumaker, Discrete particle noise in particle-in-celll simulations of plasma microturbulence, Phys. Plasmas, **12**:122305, 2005.

Newell and Whitehead, 1969: A. C. Newell and J. A. Whitehead, Finite bandwidth, finite amplitude convection, J. Fluid Mech., **38**:279, 1969.

O'Neil, 1995: T. M. O'Neil, Plasma with a single sign of charge (an overview), Phys. Scripta, **T59**:341, 1995.

Orszag and Kraichnan, 1967: S. A. Orszag and R. H. Kraichnan, Model equations for strong turbulence in a Vlasov plasma, Phys. Fluids, **10**:1720, 1967.

Orszag, 1977: S. A. Orszag, Lectures on the statistical theory of turbulence, in *Fluid Dynamics*, edited by R. Balian and J.-L. Peube, pp. 235–374, Gordon and Breach, New York, 1977.

Ottaviani et al., 1997: M. Ottaviani, M. Beer, S. Cowley, W. Horton, and J. Krommes, Unanswered questions in ion-temperature gradient driven turbulence, Phys. Rep., **283**:121, 1997.

Parker and Lee, 1993: S. E. Parker and W. W. Lee, A fully nonlinear characteristic method for gyrokinetic simulation, Phys. Fluids B, **5**:77, 1993.

Politzer, 2000: P. A. Politzer, Observation of avalanchelike phenomena in a magnetically confined plasma, Phys. Rev. Lett., **84**:1192, 2000.

Qin et al., 2000: H. Qin, W. M. Tang, and W. W. Lee, Gyrocenter-gauge kinetic theory, Phys. Plasmas, **7**:4433, 2000.

Reynolds, 1895: O. Reynolds, On the dynamical theory of incompressible viscous fluids and the determination of the criterion, Phil. Trans. Roy. Soc. London, **186**:123, 1895.

Roberson and Gentle, 1971: C. Roberson and K. W. Gentle, Experimental test of the quasilinear theory of the gentle bump instability, Phys. Fluids, **14**:2462, 1971.

Roberson et al., 1971: C. Roberson, K. W. Gentle, and P. Nielsen, Experimental test of quasilinear theory, Phys. Rev. Lett., **26**:226, 1971.

Rogers et al., 2000: B. N. Rogers, W. Dorland, and M. Kotschenreuther, Generation and stability of zonal flows in ion-temperature-gradient mode turbulence, Phys. Rev. Lett., **85**:5336, 2000.

Rose, 1985: H. A. Rose, An efficient non-Markovian theory of non-equilibrium dynamics, Physica D, **14**:216, 1985.

Rosenbluth, 1965: M. N. Rosenbluth, Microinstabilities, in *Plasma Physics*, pp. 485–513, International Atomic Energy Agency, 1965.

Ruelle and Takens, 1971: D. Ruelle and F. Takens, On the nature of turbulence, Commun. Math. Phys., **20**:167, 1971.

Sagan, 1985: C. Sagan, *Contact*, Simon & Schuster, New York, 1985.

Sagdeev and Galeev, 1969: R. Z. Sagdeev and A. A. Galeev, *Nonlinear Plasma Theory*, W. A. Benjamin, New York, 1969.

Sarazin and Ghendrih, 1998: Y. Sarazin and P. Ghendrih, Intermittent particle transport in two-dimensional edge turbulence, Phys. Plasmas, **5**:4214, 1998.

Schwinger, 1951a: J. Schwinger, On the Green's functions of quantized fields. I., Proc. Natl. Acad. Sci., **37**:452, 1951.

Schwinger, 1951b: J. Schwinger, On the Green's functions of quantized fields. II., Proc. Natl. Acad. Sci., **37**:455, 1951.

Schwinger, 1951c: J. Schwinger, The theory of quantized fields. I., Phys. Rev., **82**:914, 1951.

Scott, 2005: B. D. Scott, Energetics of the interaction between electromagnetic $e \times b$ turbulence and zonal flows, New J. Phys., **7**:92, 2005.

Similon and Sudan, 1990: P. L. Similon and R. N. Sudan, Plasma turbulence, Annu. Rev. Fluid Mech., **22**:317, 1990.

Singh et al., 2005: R. Singh, V. Tangri, P. Kaw, and P. N. Guzdar, Coupled drift-wave-zonal flow model of turbulent transport in the tokamak edge, Phys. Plasmas, **12**:092307, 2005.

Smith and Kaufman, 1975: G. R. Smith and A. N. Kaufman, Stochastic acceleration by a single wave in a magnetic field, Phys. Rev. Lett., **34**:1613, 1975.

Smolyakov and Diamond, 1999: A. I. Smolyakov and P. H. Diamond, Generalized action invariants for drift waves-zonal flow systems, Phys. Plasmas, **6**:4410, 1999.

Spineanu and Vlad, 2005: F. Spineanu and M. Vlad, Statistical properties of an ensemble of vortices interacting with a turbulent field, Phys. Plasmas, **12**:112303, 2005.

Stix, 1992: T. H. Stix, *Waves in Plasmas*, Am. Inst. of Phys., New York, 1992.

Strauss, 1976: H. R. Strauss, Nonlinear, three-dimensional magnetohydrodynamics of noncircular tokamaks, Phys. Fluids, **19**:134, 1976.

Sugama et al., 2003: H. Sugama, T.-H. Watanabe, and W. Horton, Comparison between kinetic and fluid simulations of slab ion temperature gradient driven turbulence, Phys. Plasmas, **10**:726, 2003.

Taylor and McNamara, 1971: J. B. Taylor and B. McNamara, Plasma diffusion in two dimensions, Phys. Fluids, **14**:1492, 1971.

Taylor and Thompson, 1973: J. B. Taylor and W. B. Thompson, Fluctuations in guiding center plasma in two dimensions, Phys. Fluids, **16**:111, 1973.

Taylor, 1915: G. I. Taylor, Eddy motion in the atmosphere, Philos. Trans. R. Soc. London, Ser. A, **215**:1, 1915.

Taylor, 1921: G. I. Taylor, Diffusion by continuous movements, Proc. London Math. Soc., Ser. 2, **20**:196, 1921, Reprinted in *Turbulence: Classic Papers on Statistical Theory*, edited by S. K. Friedlander and L. Topper (Interscience, New York, 1961), p. 1.

Taylor, 1974: J. B. Taylor, Dielectric function and diffusion of a guiding-center plasma, Phys. Rev. Lett., **32**:199, 1974.

Tennekes and Lumley, 1972: H. Tennekes and J. L. Lumley, *A First Course in Turbulence*, MIT Press, Cambridge, MA, 1972.

Terry and Horton, 1982: P. Terry and W. Horton, Stochasticity and the random phase approximation for three electron drift waves, Phys. Fluids, **25**:491, 1982.

Tetreault, 1976: D. J. Tetreault, *Renormalization of the wave particle resonance in turbulent plasma*, PhD thesis, Massachusetts Institute of Technology, 1976.

Treve, 1978: Y. M. Treve, Theory of chaotic motion with application to controlled fusion research, in *Topics in Nonlinear Dynamics: A Tribute to Sir Edward Bullard*, edited by S. Jorna, p. 147, New York, 1978, AIP.

Uhlenbeck and Ornstein, 1930: G. E. Uhlenbeck and L. S. Ornstein, On the theory of the Brownian motion, Phys. Rev., **36**:823, 1930, Reprinted in *Selected Papers on Noise and Stochastic Processes*, edited by N. Wax (Dover, New York, 1954), p. 93.

Vahala, 1974: G. Vahala, Transport properties of the three-dimensional guiding-centre plasma, J. Plasma Phys., **11**:159, 1974.

van Kampen, 1976: N. G. van Kampen, Stochastic differential equations, Phys. Rep., **24**:171, 1976.

Vedenov et al., 1962: A. Vedenov, E. Velikhov, and R. Sagdeev, The quasi-linear theory of plasma oscillations, in *Proceedings of the Conference on Plasma Physics and Controlled Nuclear Fusion Research (Salzburg, 1961) [Nucl. Fusion Suppl. Pt. 2]*, pp. 465–475, International Atomic Energy Agency, Vienna, 1962, Translated in U.S.A.E.C. Division of Technical Information document AEC–tr–5589 (1963), pp. 204–237.

Wagner et al., 1982: F. Wagner, G. Becker, K. Behringer, D. Campbell, A. Eberhagen, W. Engelhardt, G. Fussmann, O. Gehre, J. Gernhardt, G. v. Gierke, G. Haas, M. Huang, F. Karger, M. Keilhacker, O. Klüber, M. Kornherr, K. Lackner, G. Lisitano, G. G. Lister, H. M. Mayer, D. Meisel, E. R. Müller, H. Murmann, H. Niedermeyer, W. Poschenrieder, H. Rapp, H. Röhr, F. Schneider, G. Siller, E. Speth, A. Stäbler, K. H. Steuer, G. Venus, O. Vollmer, and Z. Yü, Regime of improved confinement and high beta in neutral-beam-heated divertor discharges of the ASDEX tokamak, Phys. Rev. Lett., **49**:1408, 1982.

Wakatani and Hasegawa, 1984: M. Wakatani and A. Hasegawa, A collisional drift wave description of plasma edge turbulence, Phys. Fluids, **27**:611, 1984.

Wang and Uhlenbeck, 1945: M. C. Wang and G. E. Uhlenbeck, On the theory of the Brownian motion II, Rev. Mod. Phys., **17**:323, 1945, Reprinted in *Selected Papers on Noise and Stochastic Processes*, edited by N. Wax (Dover, New York, 1954), p. 113.

Weiland, 2000: J. Weiland, *Collective Modes in Inhomogeneous Plasma*, IOP Publishing, Bristol, UK, 2000.

Weinstock, 1968: J. Weinstock, Turbulent diffusion, particle orbits, and field fluctuations in a plasma in a magnetic field, Phys. Fluids, **11**:1977, 1968.

Wesson, 2004: J. Wesson, *Tokamaks*, Clarendon Press, Oxford, 3rd edition, 2004.

White et al., 2006: A. E. White, S. J. Zweben, M. J. Burin, T. A. Carter, T. S. Hahm, J. Krommes, and R. J. Maqueda, Bispectral analysis of L–H transitions on NSTX using the GPI diagnostic, Phys. Plasmas, 2006, submitted.

White, 2001: R. B. White, *The Theory of Toroidally Confined Plasmas*, Imperial College Press, London, 2001.

Whitham, 1974: G. B. Whitham, *Linear and Nonlinear Waves*, Wiley, New York, 1974.

Winsor et al., 1968: N. Winsor, J. L. Johnson, and J. M. Dawson, Geodesic acoustic waves in hydromagnetic systems, Phys. Fluids, **11**:2448, 1968.

Yoshizawa et al., 2001: A. Yoshizawa, S.-I. Itoh, K. Itoh, and N. Yokoi, Turbulence theories and modelling of fluids and plasmas, Plasma Phys. Control. Fusion, **43**:R1, 2001.

Zakharov, 1984: V. E. Zakharov, Statistical descriptions and plasma physics, in *Handbook of Plasma Physics*, edited by A. A. Galeev and R. N. Sudan, volume 2, chapter 5.1, p. 3, North–Holland, Amsterdam, 1984.

Zaslavskii and Chirikov, 1972: G. M. Zaslavskii and B. V. Chirikov, Stochastic instability of non-linear oscillations, Usp. Fiz. Nauk, **105**:3, 1972, [Sov. Phys. Usp. **14**:549 (1972)].

Zaslavskii and Sagdeev, 1967: G. M. Zaslavskii and R. Z. Sagdeev, Limits of statistical description of a nonlinear wave field, Zh. Eksp. Teor. Fiz., **52**:1081, 1967, [Sov. Phys. JETP **25**:718 (1967)].

Zinn-Justin, 1996: J. Zinn-Justin, *Quantum Field Theory and Critical Phenomena*, Oxford University Press, Oxford, third edition, 1996.

Zweben et al., 2004: S. J. Zweben, R. J. Maqueda, D. P. Stotler, A. Keesee, J. Boedo, C. E. Bush, S. M. Kaye, B. LeBlanc, J. L. Lowrance, V. J. Mastrocola, R. Maingi, N. Nishino, G. Renda, D. W. Swain, J. B. Wilgen, and the NSTX Team, High-speed imaging of edge turbulence in NSTX, Nucl. Fusion, **44**:134, 2004.

Zweben, 2006: S. J. Zweben, APS05 talk, Phys. Plasmas, 2006, in press.

Chapter 5

Experimental Studies of Plasma Turbulence

Michael Shats and Hua Xia

The Australian National University, Canberra ACT 0200, Australia
E-mail: Michael.Shats@anu.edu.au, Hua.Xia@anu.edu.au

This chapter is based on two lectures given by M. Shats at the Summer School describing studies of turbulence in toroidal plasma confinement experiments. Some plasma diagnostics relevant for turbulence studies are reviewed. The data analysis techniques and methods are described in the context of the turbulence studies performed in the low-temperature plasma of the H-1 toroidal heliac, with particular emphasis on the analysis of spectral transfer in turbulent spectra. Experimental results on self-organization of the two-dimensional fluid turbulence are presented to illustrate some similarity with processes in quasi-two-dimensional plasma turbulence. Experimental signatures of zonal flows in plasma are illustrated.

Contents

233

5.1. Introduction

Our interest in experimental studies of turbulence in magnetized plasma
has been driven by the need to understand anomalously high loss of particles and energy across confining magnetic field in toroidal magnetic confinement experiments. Anomalous transport in plasma placed in magnetic
field has been attributed to microscopic plasma turbulence as early as 1949
by Bohm[1] who suggested that oscillating electric fields due to plasma instabilities can substantially increase diffusion. Due to massive theoretical
effort, large number of linear instabilities have been proposed as candidates
for driving fluctuations in the plasma density and electrostatic potential
(for review see e.g.[2,3]).

Experimental studies of the low-frequency (typically below 1 MHz) turbulent fluctuations in the high-temperature plasma in tokamaks initially
focused on characterization of the small-scale density turbulence. Measurements of the frequency and wave number spectra of the density fluctuations
in the high-temperature plasma regions were performed using scattering of
the microwave[4,5] and laser[6,7] radiation. Experiments performed between
1976 and 1979 have revealed several important characteristics of the plasma
turbulence, and have triggered theoretical work beyond linear theory of various instabilities.

One of them was the observation of the broad frequency and the wave
number spectra at both the inner (high temperature) plasma regions and
at the periphery.[3] Spectra of developed plasma turbulence do not show any
obvious features which correspond to an underlying linear instability and
typically have maxima in the frequency (and wave number, k) range which
is much lower than that of the expected linear (drift) instability. Some
of these observations have been partially understood within the frame of
simple nonlinear models, such as the Hasegawa-Mima (H-M) model.[8] The
analysis of the spectral evolution in the H-M model has shown the possibility of the dual cascade *via* the three-wave interactions and the transfer
of energy and potential enstrophy in k-space[9] (see also Chapter 4.4. by
J. Krommes). This process was found to be similar to the energy and
enstrophy cascades in two-dimensional (2D) Navier-Stokes turbulence (see

Chapter 1 by G. Falkovich). The inverse energy cascade in the plasma turbulence, similarly to the 2D fluid turbulence, tends to condense at low k. As a result, a potential structure is formed whose poloidal component of the wave vector is small, $k_\theta \approx 0$, while its radial component k_r is finite. Such structures are referred to as zonal flows.[9]

Zonal flows have been studied theoretically and in numerical simulations (for review see[10]). Recently several experimental reports on observations of zonal flows[11–17] have confirmed basic theoretical predictions and demonstrated the universality of zonal flows. Zonal flows may play many important roles in magnetically confined plasma, such as the regulation of the drift-wave turbulence,[18] formation of the transport barriers[19] and others.

Detailed experimental studies of the zonal flow - turbulence interaction have become possible due to the remarkable progress in diagnostics for turbulence studies, such as the heavy-ion beam probe, the Doppler reflectometry, beam emission spectroscopy, and others. More traditional fluctuation diagnostics, such as the Langmuir probe arrays in combination with modern signal analysis techniques also contributed to improved understanding of the plasma turbulence.

The main goal of these lectures given at the Summer School by M. Shats is to introduce some aspects of turbulence in toroidal magnetized plasma from experimental point of view. Though it is impossible in two lectures to even mention all plasma diagnostics relevant to the turbulence studies, we will select several diagnostics, which in our view, have significantly contributed to the progress in understanding plasma turbulence in recent years. This selection is highly subjective, nevertheless it gives a feeling about the direction in which experimental plasma turbulence studies are moving.

In Section 2, we describe some experimental techniques for studying turbulence in magnetized plasma, such as the Langmuir probes, collective scattering of electromagnetic waves by the density fluctuations, reflectometry, the Doppler reflectometry, beam emission spectroscopy, and the heavy-ion beam probe technique. In Section 3 spectral analysis techniques are described, in particular higher-order spectral analysis and spectral transfer. Section 4 illustrates application of the spectral transfer analysis to experimental results from the H-1 heliac. Experimental evidence of the inverse energy cascade is presented. In Section 5 we describe a model experiment in two-dimensional fluid turbulence to compare physics of the inverse energy cascade and generation of large coherent structures in fluids with generation of zonal flows in plasma turbulence. Experimentally identifiable signatures of zonal flows in plasma are also illustrated.

5.2. Experimental techniques and diagnostic tools in plasma turbulence

Main difficulties in experimental studies of turbulence are related to:

(1) usual problems in understanding turbulence in any medium;
(2) limited accessibility due to the high temperature and high vacuum conditions, intense fluxes of particles and heat from the plasma;
(3) difficulties in interpretation of measurements which are typically limited to just a few spatial locations;
(4) difficulty of imaging and visualization of turbulent fields in fully ionized plasma;
(5) the multi-field nature of the plasma turbulence: simultaneous presence of fluctuations in the density, temperature, electric fields, magnetic fields, etc.

Plasma turbulence has been studied using a variety of tools and techniques. One can adopt several ways of classifying the plasma fluctuation diagnostics. For example, they can be classified according to the measured plasma parameters, such as the fluctuations in the electron density, electrostatic potential, electron temperature, magnetic field, etc. Turbulence diagnostics can also be viewed according to the physical principles used for the measurements, such as the refractive index measurements, scattering of the electromagnetic radiation, measurements of the plasma particle fluxes and others.

In this section we do not overview plasma turbulence diagnostics. There are several good review papers, which cover diagnostics used in plasma turbulence studies, such as.[20,21] For the state-of-the-art methods in plasma diagnostics, including the turbulence measurements, one should consult Proceedings of the Topical Conferences on High-Temperature Plasma Diagnostics held bi-annually. The latest Proceedings have been published in *Reviews of Scientific Instruments*, Volume 75, Issue 10, 2004. Here we briefly describe selected experimental techniques which have been particularly successful in characterizing key plasma turbulence phenomena in the last 15-20 years.

5.2.1. *Langmuir probes*

Langmuir probes[22,23] are probably the most basic tools for the plasma turbulence studies. They are relatively easy to design, and they have excellent spatial resolution which is determined by the size of the probe tip and by the accuracy of its positioning within the plasma. However they have a somewhat limited applicability in toroidal plasma experiments since they cannot withstand very high fluxes of particles and energy. As a result, their applications are limited to either the peripheral plasma in large high-temperature experiments, or to the low-temperature plasma in the smaller-scale laboratory experiments, such as for example, the H-1 heliac.

Langmuir probes provide measurements of the electron density, n_e, electrostatic potential, ϕ, the electron temperature, T_e and their fluctuations, $\tilde{n}_e$, $\tilde{\phi}$, and $\tilde{T}_e$. When several probe tips are mounted on the same probe shaft, Langmuir probes can be used to measure the fluctuation wave numbers and various components of the turbulent electric fields.

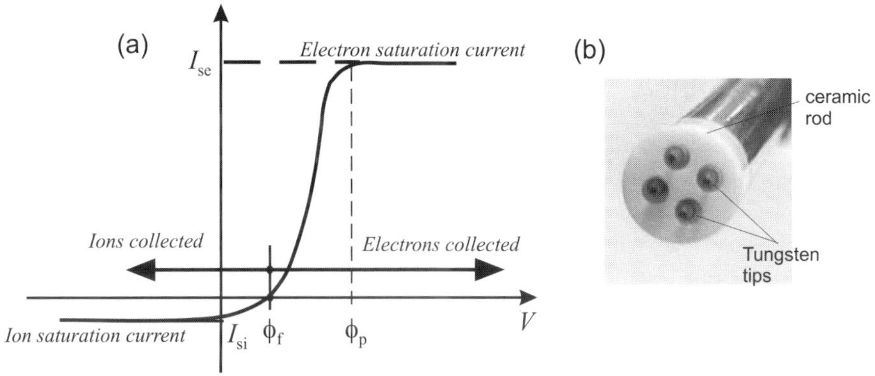

Fig. 5.1.   (a) Current-voltage characteristic of the Langmuir probe in plasma. (b) Design of the 4-pin Langmuir probe

The current-voltage characteristic of a Langmuir probe in the plasma is shown in Fig. 5.1.(a). The potential at the probe corresponding to the zero current is referred to as the *floating potential*, ϕ_f. A Langmuir probe will draw an electron current, I, when biased to a voltage $V > \phi_f$. When the probe is biased at $V < \phi_f$, it collects ions. At sufficiently high negative bias, the current to the probe saturates at the level of the *ion saturation current*:

$$I_{si} = e^{-1/2} q A_p n_e c_s, \qquad (5.1)$$

where

$$c_s = \sqrt{\frac{q(ZT_e + T_i)}{m_i}}. \qquad (5.2)$$

is the ion acoustic velocity, T_e and T_i are the electron and ion temperatures respectively, A_p is the probe collecting area, m_i is the ion mass, and Z is the ion charge state.

At sufficiently high positive bias voltage, the electron probe current saturates at the level of the *electron saturation current*,

$$I_{se} = \frac{1}{4} q A_p n_e V_{te},$$
(5.3)

where

$$V_{te} = \sqrt{\frac{8qT_e}{\pi m_e}}.$$
(5.4)

is the electron thermal velocity. The positive bias potential ϕ_p, corresponding the the electron saturation current, is referred to as the *plasma potential*. For $V < \phi_p$, the probe current can be expressed as[22]

$$I = I_{se} \exp\left(\frac{V - \phi_p}{T_e}\right) - I_{si}.$$
(5.5)

The electron temperature T_e can be determined from the slope of a semi-log plot of I *versus* V. This slope is equal to e/kT_e in the electron-retarding region ($V < \phi_p$).

Thus several basic plasma parameters and their fluctuations can be deduced from the probe $I - V$ characteristic if the bias voltage is swept sufficiently fast. In some cases, the sweep frequency may limit the time resolution of the probe measurements.

An alternative solution may be in using a probe which has several tips biased at different potentials. Such a probe design is illustrated in Fig. 5.1.(b). This an extension of the principle of the *triple probe*.[24] Triple probes allow instantaneous values of the electron temperature T_e, as well as the electron density to be determined. The electron temperature can be derived by continuously sampling two points on the characteristic, the floating potential, ϕ_f, and a positive potential, ϕ_+, corresponding to a current which is equal to I_{si} but is oppositely directed. In this case the electron temperature can be determined as[24]

$$T_e = \frac{(\phi_+ - \phi_f)}{\ln 2}.$$
(5.6)

The plasma potential can also be determined from the triple probe as[23]

$$\phi_p = \phi_f + \alpha T_e,$$
(5.7)

where

$$\alpha = -\frac{1}{2} \ln\left(2\pi \frac{m_e}{m_i}\left(1 + \frac{T_i}{T_e}\right)(1 - \delta)^{-2}\right),$$
(5.8)

and δ is the secondary electron emission coefficient.

The ability of the triple probe to simultaneously measure fluctuations in the floating potential and in the electron temperature is particularly valuable in the turbulence studies. As will be shown below, fluctuations in the plasma potential is one of the key parameters which determine the turbulence-driven transport. The plasma potential can be deduced from the triple probe data using Eq. (5.7). Also, the time-resolved measurements of T_e allow time-varying electron density n_e to be derived from the measurements of the ion saturation current I_{si}, Eq. (5.1).

5.2.2. *Characterization of turbulent transport using probes*

Langmuir probes can be arranged in arrays to characterize plasma turbulence. In this subsection we consider measurements of the turbulence-driven fluxes using Langmuir probes. We limit our discussion to the measurements of the particle fluxes due to electrostatic fluctuations. The particle flux measurements are crucial for understanding the roles of turbulence in the plasma confinement and require simultaneous characterization of the fluctuations in the radial velocity of particles and the density fluctuations. The Langmuir probe array is the only diagnostic capable of performing such measurements with required spatial and temporal resolution.

In case of the electrostatic turbulence (e.g., drift-wave turbulence, see Chapter 4 by J. Krommes) electrons and ions fluctuate in the radial direction due to the $E \times B$ drift in the fluctuating poloidal electric field, $\widetilde{E}_\theta$,

$$\widetilde{v}_{rad} = \frac{\widetilde{E}_\theta}{B} = \frac{k_\theta \widetilde{\phi}}{B}. \tag{5.9}$$

The fluctuation-driven flux is then

$$\widetilde{\Gamma}_{fl} = \frac{\widetilde{n}\widetilde{E}_\theta}{B} = \frac{k_\theta}{B}\left(\widetilde{n}\widetilde{\phi}\right). \tag{5.10}$$

The time-average fluctuation-driven particle flux can also be defined in the frequency domain as:[25]

$$\Gamma_{fl} = \frac{2}{B} \int_0^\infty d\omega\, [P_{nn}P_{EE}]^{1/2} |\gamma_{nE}| \cos{[\alpha_{nE}]}, \tag{5.11}$$

where P_{nn} and P_{EE} are the spectral power densities of the fluctuations in the electron density and poloidal electric field. The coherence $0 \leq |\gamma_{nE}| \leq 1$ is defined *via* the cross- and auto-power spectra of the fluctuations as

$$\gamma_{nE}(\omega_k) = \left(\frac{[Re\,(P_{nE}(\omega_k))]^2 + [Im\,(P_{nE}(\omega_k))]^2}{P_{nn}(\omega_k)P_{EE}(\omega_k)}\right)^{1/2}, \tag{5.12}$$

while α_{nE} is the phase shift between $\tilde{n}$ and $\tilde{E}_\theta$:

$$\alpha_{nE}(\omega_k) = \arctan\left(\frac{Im\left(P_{nE}(\omega_k)\right)}{Re\left(P_{nE}(\omega_k)\right)}\right). \qquad (5.13)$$

Here Im and Re denote the imaginary and the real parts of the cross-spectra.

Eq. (5.11) shows that the particle flux can be reduced either by suppressing the turbulence (P_{nn} and P_{EE} reduction), or by decorrelating density and E_θ fluctuations ($\gamma_{nE} \to 0$), or by changing the relative phase α_{nE} between them. Depending on the phase shift $\alpha_{\tilde{n}\tilde{E}}$ between $\tilde{n}$ and $\tilde{E}_\theta$, the time-average flux can be zero ($\alpha_{\tilde{n}\tilde{E}} = \pi/2$), positive (radially outward), or negative (radially inward). Below we will illustrate that all these parameters affect turbulent particle fluxes in magnetized plasma.

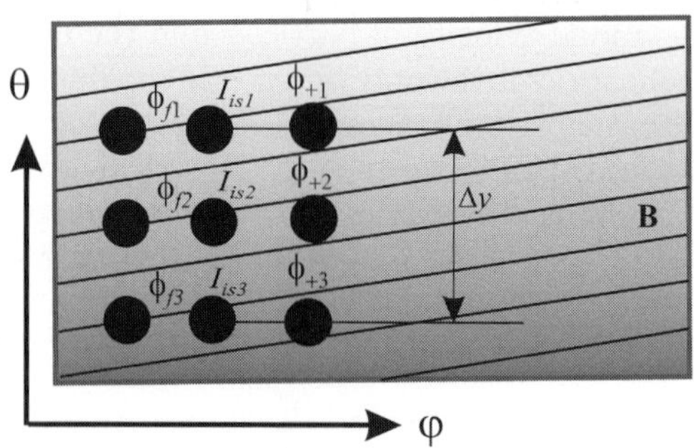

Fig. 5.2. Example of the probe array suitable for the characterization of the turbulence-driven particle flux.

Figure 5.2. shows the geometry of the Langmuir probe array capable of measuring the fluctuation-driven flux. Three triple probes, $(I_{is1}, \phi_{f1}, \phi_{+1})$, $(I_{is2}, \phi_{f2}, \phi_{+2})$, and $(I_{is3}, \phi_{f3}, \phi_{+3})$ measure the electron temperature and the plasma potentials at three poloidally shifted locations in the plasma. Since fluctuations in toroidal plasma are strongly elongated in toroidal direction (toroidal wave numbers, k_φ, are much smaller than either poloidal k_θ, or radial k_r wave numbers), a small relative shift of the triple probe tips in toroidal direction does not affect phases of the fluctuations. Probes 1 and 3 give ϕ_{p1} and ϕ_{p3} using Eq. (5.7), such that the fluctuations in the poloidal electric field can be computed as $\tilde{E}_\theta = (\tilde{\phi}_{p1} - \tilde{\phi}_{p2})/(\Delta y)$. This $\tilde{E}_\theta$ is multiplied with the fluctuations in the electron density to obtain $\tilde{\Gamma}_{fl}$ using Eq. (5.10) or Eq. (5.11). $\tilde{n}$ is deduced from the ion saturation current of the probe 2 using Eq. (5.1) and T_{e2}.

In practice, in many experiments floating potentials are often used instead of the plasma potentials. This can only be justified when $\widetilde{\phi}_f$ and $\widetilde{T}_e$ are in phase, which needs to be proven experimentally. If this is the case, the probe array can be greatly simplified.

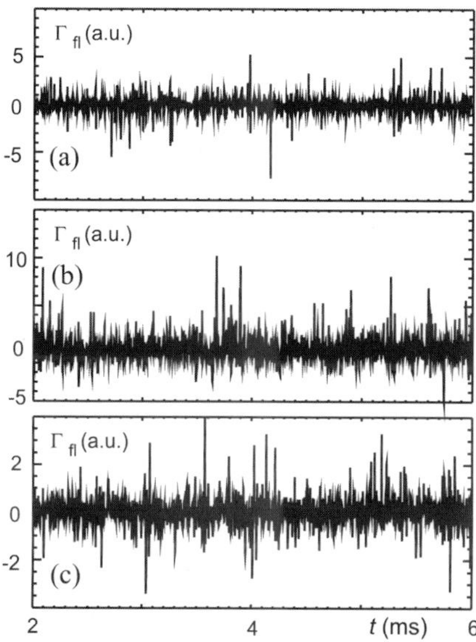

Fig. 5.3. Time-resolved fluctuation-driven flux, $\Gamma_{fl} = \widetilde{n}\, \widetilde{E}_\theta \, / \, B$

An example of the experimentally measured time-resolved fluctuation-driven flux is shown in Fig. 5.3.. It is seen that Γ_{fl} has a bursty structure, such that it is the statistics of positive and negative bursts which determines the direction of the time-average flux. Three plots in Fig. 5.3. correspond to the inward (a), outward (b), and zero-average flux (c).

So far we have assumed that the particle fluxes driven by the turbulent fluctuations are the same for electrons and ions, in other words, $\Gamma_{fl}^{e} = \Gamma_{fl}^{i}$. There are however several physics effects which can break this balance. It has been mentioned in the Introduction that turbulence can drive plasma flows, such as, for example, zonal flow. The Reynolds stress is an example of such a mechanism. Turbulence induced Reynolds stress drives the plasma flow which can be associated with the radial current:[26]

$$ J_r = \frac{m_i n_e}{eB} \frac{\partial}{\partial r} \langle \widetilde{v}_{ri} \widetilde{v}_{\theta i} \rangle . \tag{5.14} $$

Another example is the flow driven due to the finite-Larmor-radius

(FLR) effect.[27,28] In any case such flow generation would be manifested as the fluctuation-driven radial current

$$J_r = e \left(\left\langle \tilde{n}_i \tilde{V}_{ri} \right\rangle - \left\langle \tilde{n}_e \tilde{V}_{re} \right\rangle \right), \tag{5.15}$$

which can change the radial electric field and generate plasma flow in either poloidal or toroidal direction.

The first experimental evidence of the fluctuation-driven radial electric current has been found in the H-1 heliac.[29] Fluctuations in the ion radial velocity were measured using the so-called Mach (or paddle) probe. Such probes are used to provide information about plasma flow velocities.[30] A Mach probe typically consists of two identical collectors separated by an insulator. Both collectors are negatively biased into the ion saturation current. According to Eq. (5.1) the ion saturation current is dependent on the velocity at which the ions stream towards the probe. Therefore, if the plasma drifts with some velocity perpendicular to the axis of the probe, the two probe tips will collect ions arriving with different velocities and therefore measure different currents.

If the ion gyroradius is larger than the probe size, then such a probe is referred to as "unmagnetized". A revision of the Bohm theory suitable for the unmagnetized Mach probe has been presented in.[31] Since the Mach probe is unmagnetized, it can be oriented to be sensitive to the ion radial velocity and its fluctuations $\tilde{V}_{ri}$. This provides an independent estimate for the ion fluctuation-driven flux $\Gamma_i = \left\langle \tilde{n}_i \tilde{V}_{ri} \right\rangle$. The fluctuation-induced flux for the electrons was assumed a result solely of $\tilde{E} \times B$, where the main contribution comes from the poloidal $\tilde{E}$ component ($\tilde{E}_\theta$) and the toroidal B component (B_t). It was found that the fluctuations in the ion radial velocity are significantly lower than those for electrons. As a result the fluctuation-driven fluxes are different for electrons and ions, which leads to the production of a radial current.[29]

5.2.3. *Collective (Bragg) scattering of electromagnetic waves by density fluctuations*

Experimental results obtained using *collective scattering diagnostic* have greatly influenced studies of turbulence in 1970s and 1980s. The most important result is that the wave number spectra of the plasma turbulence are broad and have maximum at rather low wave numbers. Collective, or Bragg scattering diagnostic is capable of directly measuring the wave number spectra of the density fluctuations in plasma.

The process of scattering of electromagnetic waves in plasma can be thought of as follows. The incident electromagnetic wave impinges on plasma particles. Particles are accelerated in the wave. Accelerated particles emit electromagnetic radiation in all directions. This emitted radiation is the scattered wave.

Whether the wave is scattered by electrons participating in collective motion, or by the unshielded electrons, is determined by the scattering parameter α:

$$\alpha = \frac{1}{k\lambda_D},\tag{5.16}$$

where λ_D is the Debye length and k is the wave number of the plasma density fluctuations. When $\alpha \geq 1$, the main contribution to the scattered wave comes from oscillations of wavelength longer than the Debye length. This is called the *collective domain.* In the process of scattering the incident electromagnetic wave $(\boldsymbol{k_0}, \omega_0)$ interacts with the plasma wave $(\boldsymbol{k}, \omega)$ such that the scattered wave $(\boldsymbol{k_s}, \omega_s)$ is generated. The momentum and energy are conserved in each act of scattering:

$$\boldsymbol{k} = \boldsymbol{k_s} - \boldsymbol{k_0}, \quad \omega = \omega_s - \omega_0,\tag{5.17}$$

The wave vector diagram illustrating this is shown in Fig. 5.4.. The angle, θ between the wave vectors of the incident and the scattered waves is called the scattering angle. By selecting different scattering angles, scattering by different wave lengths in the plasma can be studied. The relation between the scattering angle and the fluctuation wave vector k is given by the Bragg rule:

$$k = 2k_0 \sin \theta/2,\tag{5.18}$$

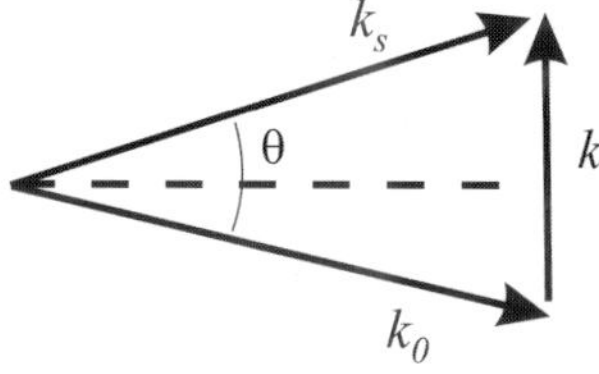

Fig. 5.4.   Wave vectors of the waves participating in Bragg scattering

The Bragg rule follows from the fact that the wave vector triangle in Fig. 5.4. is isosceles because

$$\omega_0 \approx \omega_s >> \omega, \quad k_s = \frac{\omega_s}{c} = \frac{\omega + \omega_0}{c} \approx k_0 = \frac{\omega_0}{c}.\tag{5.19}$$

An example of the microwave scattering geometry is shown in Fig. 5.5. Detailed description of this diagnostic can be found in.[32] The microwave beam at the wavelength of $\lambda_0 \approx 2$ mm is focused into the plasma using the horn-mirror antenna. This radiation, scattered at four different angles in the plasma, is collected using four similar receiving antennas. It should be noted that the size of the scattering volume is determined by the intersection of the radiation patterns of the incident and the receiving antennas. This volume is larger for the smaller scattering angles (longer wave length of the plasma fluctuations). The larger the scattering angle, the better the spatial resolution of the diagnostic.

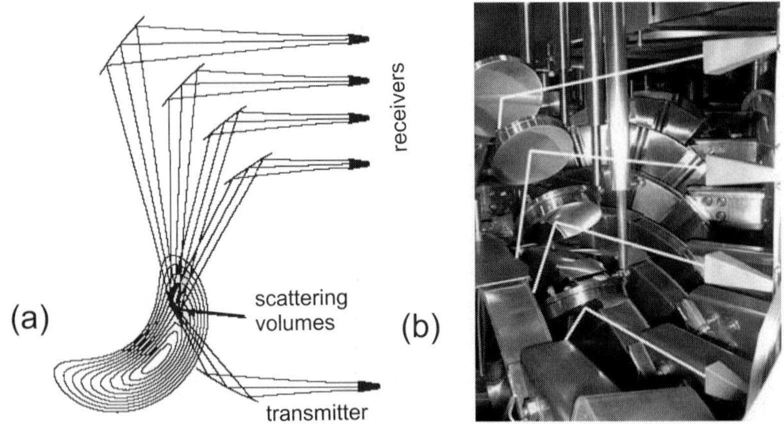

Fig. 5.5. (a) Schematic of the microwave scattering diagnostic in the H-1 heliac. (b) Photograph of the receiving microwave mirror-horn antennas installed inside the H-1 vacuum tank.

As seen from Eq. (5.18), for a given wave number of the density fluctuation, the scattering angle becomes very small if lasers are used instead of the microwave sources. For example, to detect fluctuations having $\lambda = 2$ mm one needs to collect the microwave beam at $\lambda_0 = 2$ mm scattered at $\theta = 60$ degrees. If one uses a CO_2 laser at $\lambda_0 = 10.6$ μm, the scattering angle should be $\theta \approx 0.3$ degrees, or 2.66 mrad. For such small scattering angles the size of the scattering volume extends beyond the diameter of the plasma cross-section, such that the spatial resolution becomes poor. In this case other techniques, such as the crossed laser beams are used. For details see, for example.[3]

Scattered radiation is then analyzed, and if the heterodyne detection scheme is used,[32] the direction of the propagating density waves can be determined in addition to their amplitude and the frequency spectra.

5.2.4. *Reflectometry in fluctuation studies*

Reflectometry has been widely used for measuring density fluctuations in the high temperature plasma. Measurements rely on the existence in the plasma of the cut-off layer, at which the refractive index for the probing electromagnetic wave is zero. For example, for the ordinary electromagnetic wave in plasma (a wave whose electric field vector $\boldsymbol{E}$ is parallel to the local vector of the magnetic field $\boldsymbol{B}$)

$$N^2 = 1 - \frac{\omega_{pe}^2}{\omega^2} = 0. \tag{5.20}$$

Here ω_{pe} is the electron plasma frequency, and ω is the frequency of the probing electromagnetic wave. The critical density, at which the wave is reflected at the cut-off layer is given by

$$n_c = \frac{m_e \epsilon_0 \omega^2}{e^2}. \tag{5.21}$$

The phase difference between the incident and reflected waves is sensitive to the position of the cut-off layer. Since in the presence of turbulence the layer of the critical density fluctuates, measurements of the phase and of the amplitude of the reflected wave can, in principle, give information about local fluctuations at the cut-off layer. A great advantage of the reflectometers is a relatively easy access to the plasma since both the incident and the reflected beams are transmitted through the same vacuum window. Reflectometers allow radial scans of the cut-off layer by sweeping the microwave frequency. They possess good spatial resolution since the cut-off layer is very thin.

The schematic of a reflectometer is shown in Fig. 5.6.. This is an example of the heterodyne detection scheme in which the reflected wave at the frequency $f_0 + \Delta f$ is mixed with the wave of the local oscillator (LO) to produce a signal at the intermediate frequency, $f_0 - f_{LO} + \Delta f$, which is then processed in the quadratic (IQ) detector to produce two low-frequency output signals proportional to the sine and cosine of the phase. The reflected wave is characterized by its amplitude, A, and the phase, φ.

If the reflected wave is coherent, strong phase fluctuations are usually measured, while the fluctuations in the amplitude are weak. However, in many experimental situations the reflected wave is incoherent, such that its phase is randomly distributed around zero. In this case the interpretation of the reflectometer data is not straightforward and requires additional modelling.

Main reasons for difficulties in the interpretation of the reflectometer signals are as follows:

- "Rigid" motion of the cutoff surface changes the path length x (see Fig. 5.6.) which affects the phase φ of the reflected wave;

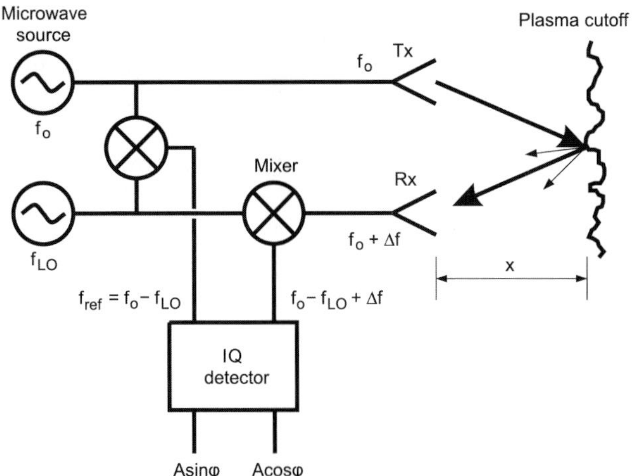

Fig. 5.6.   Schematic of the microwave reflectometer. Figure courtesy G.D. Conway.

- Bragg back scattering, or the scattering at $\theta = 180$ degrees contributes to reflected wave. In this case fluctuations in the cut-off layer, whose wave number satisfy $k = 2k_0$ (see Eq. 5.18), "pollute" the reflected signal.
- The interference in reflected microwaves may be caused by "roughness" of the reflection surface when $k \perp \nabla n_e$ or $k \perp B$.

For details on the problems with the interpretation of the reflectometer data and ways of solving it see references[33–36] which describe various models used in the interpretation of measurements ("random phase screen", "distorted mirror" etc.).

5.2.5.   *Doppler reflectometry*

The Doppler reflectometer can be considered as a hybrid between a reflectometer and collective scattering diagnostic.[17,37] The schematic of the Doppler reflectometer is shown in Fig. 5.7.. The experimental setup consists of the microwave reflectometer with antennas poloidally tilted to deliberately misalign the angle θ between the incident beam and the normal to the plasma cut-off layer. The diagnostic is sensitive to the perpendicular density fluctuation having a wave number determined by the Bragg rule (Eq. (5.18)), $k_\perp = 2k_0 \sin \theta/2$.

Poloidal motion of the density turbulence at the cutoff layer induces a Doppler frequency shift f_D in the reflected signal. This frequency shift

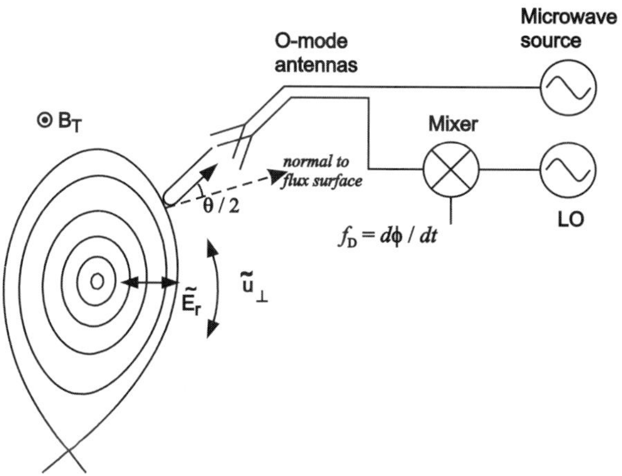

Fig. 5.7.   Schematic of the Doppler reflectometer. Figure courtesy G.D. Conway.

is proportional to the perpendicular rotation velocity of turbulence, $u_\perp = V_{E \times B} + V_{phase}$ (V_{phase} is the phase velocity of turbulent fluctuation in the plasma frame of reference):

$$f_D = \frac{u_\perp k_\perp}{2\pi} = \frac{u_\perp 2 \sin \theta/2}{\lambda_0}. \tag{5.22}$$

By changing the tilt angle θ and by measuring the received power, a perpendicular wave number spectrum $S(k_\perp)$ can be obtained.

In many experimental situations the Doppler shift due to the $E \times B$ drift dominates over the phase velocity in the plasma frame, $V_{E \times B} >> V_{phase}$. As a result, the frequency shift, f_D, will be proportional to the radial electric field, E_r. Radial electric field fluctuations $\tilde{E}_r$ will appear in f_D and can be detected as a spectrally broadened feature around f_D in the scattered wave spectrum. This technique has been successfully used to detect low-frequency oscillations in E_r due to the presence of the geodesic acoustic mode.[17]

5.2.6.   *Optical imaging of turbulent fluctuations*

The spectral line radiation emitted by excited neutrals and ions contains useful information about plasma parameters, such as the electron temperature and density. However, in the high-temperature interior of the fusion-relevant plasma, atoms are fully ionized, such that only the impurity radiation can be measured for the diagnostic purposes.

When the neutral beam is injected into the plasma, the beam atoms become collisionally excited and radiate. This radiation due to the interaction of the beam particles with electrons and ions is used to derive local density and its fluctuations in the diagnostic method referred to as the *beam emission spectroscopy* (BES).[38,39]

Figure 5.8. shows experimental setup of the BES diagnostic on the DIII-D tokamak. A beam of deuterium atoms having energy of $E = 75$ keV is launched tangentially into the plasma. Such a high energy of the beam atoms leads to a Doppler shift in the wave length of the radiated emission. This can be used to distinguish the radiation emitted by the beam particles from the neutral emission originated at the plasma edge. The light is collected along several chords perpendicular to the neutral beam. In this example a matrix of 4×4 optical fibres is imaged into the plasma such that the light in each of the optical channels comes from a small volume in the plasma determined by the intersection between the neutral beam and the fibre image in the plasma. Spatial resolution also depends on the radiative lifetime, τ, of a chosen excited state. This should be small (10^{-9} - 10^{-8} s) because spatial resolution is proportional to the square root of the beam energy (velocity) and the lifetime: $l \propto \sqrt{E}\tau$.

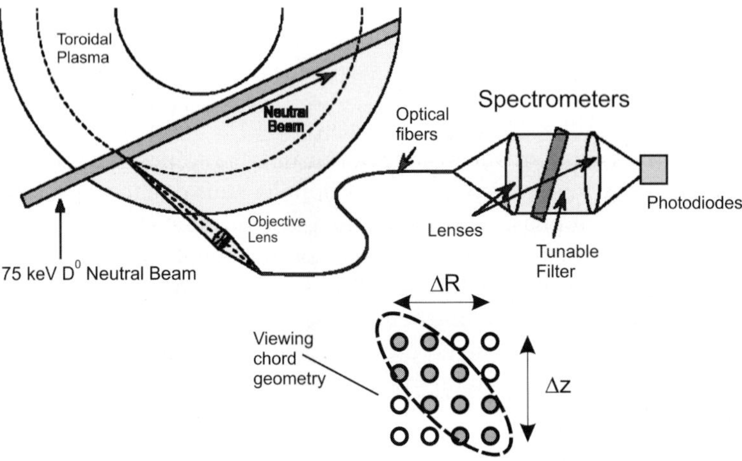

Fig. 5.8. Setup of the beam emission spectroscopy on the DIII-D tokamak. Figure courtesy G. McKee

The relative level of the local density fluctuations in each of the intersection volumes is proportional to the relative level of the fluctuations in the intensity of the light emission I:

$$\frac{\tilde{n}}{n} = K(T_e, n_e)\frac{\tilde{I}}{I}, \tag{5.23}$$

where the proportionality constant K is determined by the atomic physics relevant to the excitation of a given spectral line and depends on the electron temperature and density.

Though the line emissivity is proportional to the electron density, fluctuating velocity field can also be obtained from the measured fluctuating density. In this case two-dimensional cross-correlation analysis of the $\tilde{n}$-field is used. This method has been successful in identifying radially sheared zonal flows in the DIII-D tokamak.[13]

5.2.7. *Heavy ion beam probe*

Measurements of the electrostatic potential fluctuations is one of the most challenging problems in experimental plasma turbulence studies. With the exception of the Langmuir probes, whose operational range is limited to the cold edge plasma, the *heavy ion beam probe* (HIBP) is the only diagnostic capable of providing information about the electrostatic potential from the plasma core.[40,41]

The principle of the diagnostic is as follows. A beam of heavy ions (e.g. gold, caesium, or thallium) of very high energy (hundreds of keV) is launched into the magnetized plasma. High energy and large mass of ions are needed to increase the ion gyroradius, $\rho_i = (m_i v_{i\perp})/qB$ (where m_i is the ion mass, $v_{i\perp}$ is the component of the ion velocity perpendicular to the magnetic field B, and q is the ion charge). In this case ρ_i exceeds the diameter of the plasma column and ions will not be confined by the magnetic field. As the incident ion beam propagates through the plasma, the probe ions are ionized through the electron impact collisions. As a result, their charge increases and the trajectories of these secondary ions deviate from the trajectory of the primary ion beam as shown in Fig. 5.9.. A small fraction of the primary beam ions enters the detector. The energy of the secondary ions originated in a small sample volume (determined by the intersection of trajectories of primary and secondary ions) is then analyzed. Their energy exceeds the energy of the primary ions by the amount equal to the electric potential in the sample volume. The intensity of the secondary beam reflects the electron density in sample volume. The intensity S of the secondary beam is given by:[16]

$$S = \sigma n_e(r)(\delta r)e^{-\int \sigma n_e dl} I_0, \tag{5.24}$$

where I_0 is the injected beam current, σ is the ionization cross-section by electron impact, δl is the length of the sample volume determined by the width of the detector aperture. The integral extends along the beam trajectory. The position of the sample volume can be swept through plasma by changing the direction and energy of the primary beam.

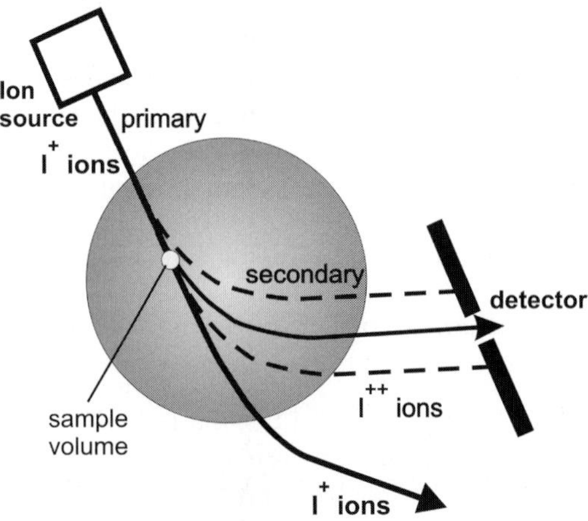

Fig. 5.9.   Schematic of the heavy ion beam diagnostic.

Fluctuations in the observed energy and intensity of the secondary ion beam are proportional to fluctuations in electrostatic potential and electron density respectively. The HIBP diagnostic has become a powerful tool capable of providing valuable information on the low-frequency potential fluctuations from the inner plasma regions in tokamaks and stellarators (see, for example,[15,16]).

5.3.   Spectral analysis techniques

In this section spectral analysis techniques used in experimental turbulence studies are overviewed. Turbulence, which to large extent determines the plasma behaviour, is characterized by broad wave number spectra whose maxima are observed at the longest measured scales.[3] Since unstable waves, which generate turbulence are initially unstable only in a limited spectral range, mechanisms of the nonlinear wave-wave interaction are needed to explain observations.

For example, three-wave interactions may lead to the energy cascade which spreads spectral energy over the spectrum. In two-dimensional (2D) turbulence[42] (see Chapter 1) and in magnetized plasma[9] spectral energy is transferred to lower wave numbers (larger scales) in the process of the so-called inverse energy cascade. If the energy dissipation at large scales is low, spectral energy can condense in large coherent structures, such as, for example, vortex structures (for review see[43]) and zonal flows.[44]

Signatures of nonlinear interactions in broad spectra can be revealed by analysing the higher order moments of the turbulence spectra. The presence of the three-wave interactions can be detected using a *bispectrum*. Four-wave interactions can be revealed by means of a *trispectrum*.

Bispectra, which measure the amount of the phase correlation between three spectral components, have been used in plasma research for a long time.[45–47] Other higher order spectral characteristics, such as the bicoherence (normalized bispectrum), trispectrum,[48,49] *etc.*, have been developed in recent years.

The higher order spectral analysis does not determine however the direction of the energy transfer. The nature and the direction of the spectral transfer have direct impact on the way in which the instability-driven turbulence is saturated, on the magnitude and shape of the spectrum, and ultimately on the nature of the particle and energy transport produced by turbulence.

A method of computing the power transfer function (PTF) was developed and applied to the fluid and plasma turbulence in 1980s. The method allows quantitative estimates of the nonlinear coupling coefficients and the energy cascades from experimentally measured turbulent signals to be made.[50,51] In the PTF technique, linear and quadratic transfer functions are estimated from the measured fluctuation signals $x(s)$ and $y(s)$ in either temporal, or in spatial domain. Spectral transfer is described by the wave coupling equation which is appropriate in a single-field turbulence. The PTF method is based on the quantitative description of nonlinear interactions between different scales using statistically averaged estimation of the power spectra, bispectra and other higher order moments.

The technique was first applied to experimental data in the nonlinear stages of a transition flow of a wake behind a thin flat plate.[52] Later this technique was applied to the turbulence measured at the edge plasma of the Texas Experimental Tokamak.[51]

A modified version of the PTF technique was proposed in.[53] In the modified method, non-ideal spectra which do not participate in the three-wave interactions are taken into account. The method was tested using simulated 1D turbulence data generated using the Hasegawa-Mima model and 2D data using the Terry-Horton model. Both of these are the single-field models. The modified technique was able to accurately reproduce the input characteristics (the linear growth rate and nonlinear energy transfer) of the simulated data.

In this section we will overview the higher order spectral analysis (HOSA) techniques, and then will describe the PTF technique. The amplitude correlation method is another technique suitable for studying energy transfer between different spectral regions. The amplitude correlation technique complements the PTF analysis in situations where the coherent phase interactions dominates over the random-phase interactions.

5.3.1. *Higher-order spectral analysis*

The presence of three-wave interactions in turbulence can be detected by means of a bispectrum.[45] The auto-bispectrum of a signal is defined as:

$$B(f_1, f_2) = \langle X_f X_{f_1}^* X_{f_2}^* \rangle, \qquad f = f_1 + f_2, \tag{5.25}$$

where X_f is a Fourier transform of the signal under investigation and $*$ denotes the complex conjugate.

Auto-bispectra measure the statistical relationship between spectral components at the frequencies f_1, f_2 and $f = f_1 + f_2$.

If the Fourier transform of the signal is $X_f = A_f e^{\phi_f}$, the auto-bispectrum of the signal can then be expressed as:

$$B(f_1, f_2) = \langle A_{f_1} A_{f_2} A_f e^{(\phi_f - \phi_{f_1} - \phi_{f_2})} \rangle, \qquad f = f_1 + f_2 . \tag{5.26}$$

If waves at f_1, f_2 and f have statistically independent random phases (like in a Gaussian signal), the resulting biphase $\phi = \phi_f - \phi_{f_1} - \phi_{f_2}$ of the polar representation (Eq. (5.26)) will be random and the averaged value of the bispectrum will be zero. If, however, a coherent phase relationship exists due to the nonlinear coupling between these waves, the auto-bispectra (averaged over many realizations) will have a finite value.

A similar definition can be given to a bispectrum between two signals, $x(t)$ whose Fourier transform is X_f and $y(t)$ whose Fourier transfer is Y_f. A cross-bispectrum is a useful characteristic of three-wave coupling effects between two turbulent spectra.

$$B(f_1, f_2) = \langle Y_f X_{f_1}^* X_{f_2}^* \rangle \qquad f = f_1 + f_2 . \tag{5.27}$$

A nonzero bispectrum is indicative of either (1) strong three-wave interactions or (2) weak interactions between spectral components having large amplitudes, as can be seen from the definition of the bispectrum, Eq. (5.26). To avoid this ambiguity, the *bicoherence*, which is the bispectrum normalized by the amplitude of the interacting waves can be used to accentuate the strength of the three-wave interactions:

$$Bic^2(f_1, f_2) = \frac{|\langle Y_f X_{f_1}^* X_{f_2}^* \rangle|^2}{\langle Y_f Y_f^* \rangle \langle X_{f_1} X_{f_1}^* \rangle \langle X_{f_2} X_{f_2}^* \rangle}, \quad f = f_1 + f_2 \tag{5.28}$$

The value of the bicoherence varies between 0 and 1, similarly to the usual 1st order coherency given by Eq. (5.12).

5.3.2. *Wave coupling equation*

The wave coupling equation which describes the time evolution of the spectral components in the turbulence spectra lies in the heart of the spectral transfer analysis. This equation has already been introduced in Chapter 1 (Eq. (5)) as a kinetic wave equation and in Chapter 4 (Eq. (96)) during the discussion of the weak-turbulence theory.

The wave coupling equation can be expressed as follows:

$$\frac{\partial \phi(k,t)}{\partial t} = (\gamma_k + i\bar{\omega}_k)\phi(k,t) + \frac{1}{2} \sum_{\substack{k_1,k_2, \\ k=k_1+k_2}} \Lambda_k^Q(k_1,k_2)\phi(k_1,t)\phi(k_2,t), \quad (5.29)$$

where $\psi(x,t)$ is the fluctuation field, $\phi(k,t)$ is the spatial Fourier spectrum of the fluctuation field $\psi(x,t) = \sum_k \phi(k,t)e^{ikx}$.

The wave coupling equation describes the rate of change of the spectral components due to linear and nonlinear effects, namely, due to the mode growth at the rate γ_k, its dispersion $\bar{\omega}_k$, and due to the wave-wave coupling. The coupling coefficient $\Lambda_k^Q(k_1,k_2)$ represents the strength of the wave coupling. A wave (k,ω) thus decays into two waves (k_1,ω_1) and (k_2,ω_2) or two waves merge into one.

This equation describes the wave coupling in the weak-turbulence theory in the random-phase approximation, but a similar equation can be derived from a more specific models, such as the Hasegawa-Mima model (see Section 1.4 of Chapter 4). Equation (5.29) can also be constructed on purely phenomenological grounds using a black-box approach as will be discussed in the next subsection. It should also be noted that to analyze the spectral transfer in plasma turbulence using the wave kinetic equation one needs to justify the validity of the single field description of turbulence. Such description is not always valid, however it is possible in several important models (e.g. in the Hasegawa-Mima model) and in some experiments, as will be discussed below.

The Hasegawa-Mima equation is the basis of the simple drift wave model,[8,54]

$$\frac{\partial}{\partial t}(\nabla^2 \phi - \phi) - [(\nabla \phi \times \hat{z}) \cdot \nabla]\left[\nabla^2 \phi - \ln\left(\frac{n_0}{\omega_{ci}}\right)\right] = 0. \quad (5.30)$$

If we expand $\phi(x,t)$ in a spatial Fourier series, as

$$\phi(x,t) = \frac{1}{2}\sum_k (\phi_k(t)e^{ik\cdot x} + c.c), \quad (5.31)$$

where k is $k_\perp$, Eq. (5.30) is reduced to

$$\frac{\partial \phi(k,t)}{\partial t} + i\omega_k^* \phi_k(t) = \frac{1}{2} \sum_{\substack{k_1,k_2 \\ k=k_1+k_2}} \Lambda_k(k_1,k_2)\phi(k_1,t)\phi(k_2,t). \qquad (5.32)$$

Here, the matrix element Λ_{k_1,k_2} is given by

$$\Lambda_{k_1,k_2} = \frac{1}{1+k^2}(k_1 \times k_2) \cdot \hat{z}[k_2^2 - k_1^2]. \qquad (5.33)$$

ω_k^* is the normalized (by ω_{ci}) drift wave frequency given by

$$\omega_{ci}^* = \frac{-k_\theta T_e \partial(\ln n_0)/\partial r}{eB_0(1+k^2)\omega_{ci}}. \qquad (5.34)$$

Equation (5.32) is the Hasegawa-Mima equation in the Fourier space. It contains the mode coupling of different modes of fluctuations and the linear dispersion of the modes, and it has exactly the same form as the wave coupling equation, Eq. (5.29).

As explained in Chapter 4 (section 1.4.1), when the background density gradient and the adiabatic electron response are neglected, the Hasegawa-Mima equation, Eq. (5.30), closely resembles the equation for the stream function ψ, which can be derived from the 2D Euler equation for the vorticity:

$$\frac{\partial}{\partial t}\nabla^2\psi - [(\nabla\psi \times \hat{z}) \cdot \nabla]\nabla^2\psi = 0. \qquad (5.35)$$

The Fourier-transform of the 2D Euler equation for the streamfunction takes a form which is very similar to the wave coupling equation, Eq. (5.29).[8]

5.3.3. *Computation of the power transfer function*

Below we follow the description of the PTF technique given in.[50,51]

The spectrum $\phi(k,t)$ in Eq. (5.29) can be represented by its amplitude and phase. The amplitude is slowly varying in time compared with the phase changes $\phi(k,t) = |\phi(k,t)|e^{i\Theta(k,t)}$.

The spectrum change in time $\dfrac{\partial \phi(k,t)}{\partial t}$, can then be estimated using a differential approach:

$$\frac{\partial \phi(k,t)}{\partial t} = \lim_{\tau \to 0}\left(\frac{||\phi(k,t+\tau)| - |\phi(k,t)||}{\tau}\frac{1}{|\phi(k,t)|} + i\frac{\Theta(k,t+\tau) - \Theta(k,t)}{\tau}\right)\phi(k,t).$$

$$(5.36)$$

Substituting Eq. (5.36) into Eq. (5.29) and solving for $\phi(k, t+\tau)$ (where τ is very small), we obtain:

$$\phi(k, t+\tau) = \frac{\Lambda_k^L \tau + 1 - i[\Theta(k, t+\tau) - \Theta(k, t)]}{e^{-i[\Theta(k,t+\tau)-\Theta(k,t)]}} \phi(k, t)$$

$$+ \frac{1}{2} \sum_{\substack{k_1, k_2, \\ k=k_1+k_2}} \frac{\Lambda_k^Q(k_1, k_2)\tau}{e^{-i[\Theta(k,t+\tau)-\Theta(k,t)]}} \times \phi(k_1, t)\phi(k_2, t), \tag{5.37}$$

where $\Lambda_k^L = \gamma_k + i\varpi_k$.

The spectrum at time $t + \tau$, $\phi(k, t + \tau)$ is thus defined by the spectrum $\phi(k, t)$ at t through linear coefficient Λ_k^L and quadratic coefficient $\Lambda_k^Q(k_1, k_2)$. To simplify, the following definitions are used:

$$X_k = \phi(k, t), \qquad Y_k = \phi(k, t+\tau),$$

$$L_k = \frac{\Lambda_k^L \tau + 1 - i[\Theta(k, t+\tau) - \Theta(k, t)]}{e^{-i[\Theta(k,t+\tau)-\Theta(k,t)]}},\tag{5.38}$$

$$Q_k^{k_1, k_2} = \frac{\Lambda_k^Q(k_1, k_2)\tau}{e^{-i[\Theta(k,t+\tau)-\Theta(k,t)]}},$$

where $k = k_1 + k_2$. Equation (5.37) can now be written as:

$$Y_k = L_k X_k + \frac{1}{2} \sum_{\substack{k_1, k_2, \\ k=k_1+k_2}} Q_k^{k_1, k_2} X_{k_1} X_{k_2} \tag{5.39}$$

The wave coupling equation Eq. (5.29) is thus related to a nonlinear system (described by Eq. (5.39)) in which the output Y_k is composed of linear and quadratic nonlinear responses to the input signal X_k.

Equation (5.39) is the simplest form of an equation which can be related to wave-wave coupling, assuming that the four-wave coupling and the higher order processes are much weaker than the three-wave coupling. The technique to derive coefficients L_k and $Q_k^{k_1, k_2}$ of Eq. (5.39) from the measured fluctuation series will be described in the next subsection. These coefficients serve as fundamental quantities for the estimation of the growth rate, the wave-wave coupling coefficients, and eventually of the energy transfer in the spectrum.

The phase shift at wave number k between t and $t+\tau$ can be estimated from the cross-power spectrum,

$$e^{-i[\Theta(k,t+\tau)-\Theta(k,t)]} = \frac{\langle Y_k X_k^* \rangle}{|\langle Y_k X_k^* \rangle|}. \tag{5.40}$$

The coupling coefficient $\Lambda_k^Q(k_1, k_2)$ in Eq. (5.29) can be derived from the wave coupling coefficients L_k and $Q_k^{k_1, k_2}$ as,

$$\Lambda_k^Q(k_1, k_2) = Q_k^{k_1, k_2} \frac{\langle Y_k X_k^* \rangle}{|\langle Y_k X_k^* \rangle|} \frac{1}{\tau}, \qquad k = k_1 + k_2 . \tag{5.41}$$

Multiplying the wave coupling equation (5.29) by $\phi^*(k,t)$, we can write a wave kinetic equation for the spectral power $P_k = \langle \phi_k \phi_k^* \rangle$ in terms of coupling coefficients Λ_k^L and $\Lambda_k^Q(k_1, k_2)$.

Since $\frac{\partial}{\partial t}[\phi(k,t)\phi^*(k,t)] = \frac{\partial \phi(k,t)}{\partial t}\phi^*(k,t) + \frac{\partial \phi^*(k,t)}{\partial t}\phi(k,t)$, the wave kinetic equation can be written as:

$$\frac{\partial P_k}{\partial t} \approx 2\gamma_k P_k + \sum_{\substack{k_1,k_2, \\ k=k_1+k_2}} T_k(k_1, k_2) \tag{5.42}$$

The power transfer function $T_k(k_1, k_2)$ quantifies the spectral power exchanged between different waves in the spectrum due to the three-wave coupling. It is related to the quadratic coupling coefficient, as

$$T_k(k_1, k_2) = Re\left[\Lambda_k^Q(k_1, k_2)\langle \phi_k^* \phi_{k_1} \phi_{k_2} \rangle\right]. \tag{5.43}$$

The energy stored in the electrostatic fluctuations ϕ_k can be expressed as $W_k = \left(1 + k_\perp^2\right)|\phi_k|^2$.[9] The nonlinear energy transfer function can be defined as:

$$W_{NL}^k = \left(1 + k_\perp^2\right) \sum_{\substack{k_1,k_2, \\ k=k_1+k_2}} T_k(k_1, k_2), \tag{5.44}$$

The nonlinear energy transfer function (NETF), W_{NL}^k in Eq. (5.44), and the linear growth rate, γ_k in Eq. (5.42), are the main spectral quantities used in the experimental analysis of the spectral transfer.

5.3.3.1. *Derivation of coupling coefficients (Ritz method)*

The method of computing the coupling coefficients which characterize a nonlinear system described by a single input and a single output, has been proposed by Ritz *et al.*[50] The output of such a black-box system, Y_p, (in either spatial or in temporal domain) contains linear and quadratic responses to the input signal, X_p, in the form:

$$Y_p = L_p X_p + \sum_{\substack{p_1 > p_2 \\ p=p_1+p_2}} Q_p^{p_1,p_2} X_{p_1} X_{p_2} + \epsilon_p \tag{5.45}$$

where ϵ_p is added to represent the noise in the signal.

Coefficients L_p and $Q_p(p_1, p_2)$ quantify linear and quadratic responses. The subscript p in Equation (5.45) represents either the wave number k, or the frequency f, depending on a system.

We assume that the measured signals are stationary and that they can be divided into many statistically similar segments, the realizations. To derive the coupling coefficients, Eq. (5.45) is multiplied by the complex conjugate of X_p. Then by ensemble averaging over many realizations, denoted as $<\ >$, one obtains:

$$\langle Y_p X_p^* \rangle = L_p \langle X_p X_p^* \rangle + \sum_{\substack{p_1 > p_2, \\ p = p_1 + p_2}} Q_p^{p_1, p_2} \langle X_p^* X_{p_1} X_{p_2} \rangle. \qquad (5.46)$$

Here $\langle Y_p X_p^* \rangle$ is the cross-power spectrum of the fluctuations, $\langle X_p X_p^* \rangle$ is the auto-power spectrum. $\langle X_p^* X_{p_1} X_{p_2} \rangle$ is the auto-bispectrum. Note that in Eq. (5.46), the cross-power spectrum term $\langle \epsilon_p X_p^* \rangle$ is ignored since the cross-power spectrum (and any higher order spectrum, such as $\langle \epsilon_p X_{p_1}^* X_{p_2}^* \rangle$ which will be encountered in the derivation of Eq. (5.47)) averages to zero between the signals and noise.

By multiplying Eq. (5.45) with $X_{p_1}^{'*} X_{p_2}^{'*}$ and by ensemble averaging, we obtain a second equation which contains linear and quadratic transfer functions,

$$\left\langle Y_p X_{p_1}^{'*} X_{p_2}^{'*} \right\rangle = L_p \left\langle X_p X_{p_1}^{'*} X_{p_2}^{'*} \right\rangle + \sum_{\substack{p_1 > p_2, \\ p = p_1 + p_2}} Q_p^{p_1, p_2} \left\langle X_{p_1} X_{p_2} X_{p_1}^{'*} X_{p_2}^{'*} \right\rangle,$$

$$(5.47)$$

where $p = p_1 + p_2 = p_1' + p_2'$. Here $\langle Y_p^* X_{p_1} X_{p_2} \rangle$ is the cross-bispectrum and $\left\langle X_{p_1} X_{p_2} X_{p_1}^{'*} X_{p_2}^{'*} \right\rangle$ is the fourth-order moment.

Eq. (5.47) can be simplified by approximating the fourth-order moments $\left\langle X_{p_1} X_{p_2} X_{p_1}^{'*} X_{p_2}^{'*} \right\rangle$ by the square of the second-order moments $\left\langle |X_{p_1} X_{p_2}|^2 \right\rangle$ (by neglecting terms with $(p_1', p_2') \neq (p_1, p_2)$).[50] This approximation is based on the random-phase assumption, similarly to the weak turbulence theory.

Under this approximation, Eq. (5.47) is reduced to:

$$\langle Y_p X_{p_1} X_{p_2} \rangle = L_p \langle X_p X_{p_1} X_{p_2} \rangle + Q_p^{p_1, p_2} \left\langle |X_{p_1} X_{p_2}|^2 \right\rangle \qquad (5.48)$$

The determination of the coupling coefficients L_p and $Q_p^{p_1, p_2}$ is usually not straightforward: a set of dependent equations (5.46) and (5.48) need to be solved iteratively.

An example where the coupling coefficients can be easily determined is a Gaussian input signal. For a Gaussian signal, the auto-bispectrum goes to zero, $\langle X_p X_{p_1} X_{p_2} \rangle = 0$ such that the coupling coefficients $Q_p^{p_1, p_2}$ and L_p are simply determined from Eq. (5.48).

However, many systems such as turbulent fluids and plasmas, do not allow such a restrictive assumption about the input signal. Generally, the input should not be considered Gaussian because of the nonlinear history of the fluctuations.

5.3.3.2.　*Derivation of coupling coefficients: modified Ritz method*

Applications of the technique described above sometimes yield unphysically large transfer coefficients for the measured fluctuation data.[50] This problem may arise because the method does not account for the non-ideal fluctuations. For example, Eq. (5.45) contains only linear response and the three-wave interactions. Turbulence may contain spectral components which do not participate in the wave coupling described by Eq. (5.45). The higher-order nonlinear coupling (e.g., fourth-order, or higher), systematic errors, *etc.*, may also need to be taken into consideration. To address this problem, a modified method has been proposed.[53]

In the modified method, the measured spectra (X_p, Y_p) are represented as the sum of an ideal spectrum (β_p, α_p), which is driven by linear and quadratic processes, and a non-ideal spectrum (X_p^{ni}, Y_p^{ni}) whose components are not involved in the linear and the three-wave coupling processes:

$$X_p = \beta_p + X_p^{ni}, \qquad Y_p = \alpha_p + Y_p^{ni}. \tag{5.49}$$

The non-ideal spectrum (X_p^{ni}, Y_p^{ni}) is assumed to be completely uncorrelated with the ideal fluctuation spectrum (β_p, α_p), which is reasonable for the noise or any spectrum not described by Eq. (5.45).

Using Eq. (5.49), equation (5.39) can be rewritten in the form

$$Y_p - Y_p^{ni} = L_p(X_p - X_p^{ni}) + \sum_{\substack{p_1 \geq p_2, \\ p=p_1+p_2}} Q_p^{p_1,p_2}(X_{p_1} - X_{p_1}^{ni}) \times (X_{p_2} - X_{p_2}^{ni}).$$

$$\tag{5.50}$$

The same procedure as in the original method is applied to Eq. (5.50). First, Eq. (5.50) is multiplied by X_p^* and $X_{p_1'}^* X_{p_2'}^*$ respectively. Then, the ensemble averaging over many statistically similar realizations is performed. In the two equations obtained, the terms with cross terms containing the non-ideal spectrum can be removed due to the zero-correlation assumption. The exceptions are the auto-power spectra $\langle \beta_p \beta_p^* \rangle$, $\langle \alpha_p \alpha_p^* \rangle$. As a result, the following set of equations is obtained,[53]

$$\langle Y_p X_p^* \rangle = L_p \langle \beta_p \beta_p^* \rangle + \sum_{\substack{p_1 \geq p_2, \\ p=p_1+p_2}} Q_p^{p_1,p_2} \langle p_{p_1} X_{p_2} X_p^* \rangle,$$

$$\langle \alpha_p \alpha_p^* \rangle = L_p \langle X_p Y_p^* \rangle + \sum_{\substack{p_1 \geq p_2, \\ p=p_1+p_2}} Q_p^{p_1,p_2} \langle X_{p_1} X_{p_2} Y_p^* \rangle,$$

$$\langle Y_p X_{p_1}^* X_{p_2}^* \rangle = L_p \langle X_p X_{p_1}^* X_{p_2}^* \rangle$$

$$+ \sum_{\substack{p_1 \geq p_2, \\ p=p_1+p_2=p_1'+p_2'}} Q_k^{p_1',p_2'} \langle X_{p_1'} X_{p_2'} X_{p_1}^* X_{p_2}^* \rangle,$$

$$\tag{5.51}$$

An additional relationship is required to complete the set of equations needed for the derivation of the four unknown variables, $L_p, Q_p^{p_1,p_2}, \langle \beta_p \beta_p^* \rangle, \langle \alpha_p \alpha_p^* \rangle$. The power spectrum is considered stationary for fully developed turbulence:

$$\langle \beta_p \beta_p^* \rangle = \langle \alpha_p \alpha_p^* \rangle \tag{5.52}$$

The fourth-order moment $\langle X_{p_1'} X_{p_2'} X_{p_1^*} X_{p_2^*} \rangle$ is either retained, or it is substituted using the approximation $\left\langle X_{p_1} X_{p_2} X_{p_1'}^* X_{p_2'}^* \right\rangle = \left\langle |X_{p_1} X_{p_2}|^2 \right\rangle$ as discussed above.

5.3.4. *Amplitude correlation technique*

The amplitude correlation is a spectral transfer analysis technique which in many aspects complements the PTF method. The amplitude correlation was first applied to the measurement of nonlinear interactions between drift waves and low-frequency flute-like modes.[55] The nonlinear interactions in which two drift waves interact with the low-frequency flute-like mode were identified as the mechanism responsible for the drift-wave saturation. Later, the method was applied to study the 'frequency doubling' interaction drift waves. The justification of the method was given in reference.[56]

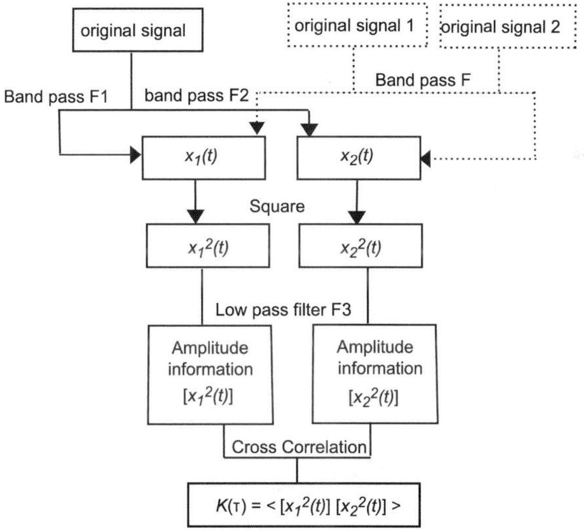

Fig. 5.10.   The flow chart of the amplitude correlation technique

The basic idea of the amplitude correlation is to obtain a new signal from a fluctuation signal. This new signal represents the time-envelope of the original signal in a particular frequency band. This envelope signal can then be cross-correlated with other envelope signals derived in a similar way. The resulting maximum correlation between the two signals and the time delay of the maximum correlation are indicative of the degree of coherent coupling and of the energy flow direction between the two frequency bands.

The procedure is illustrated schematically in Fig. 5.10.. From the original fluctuation signal $x(t)$, two time series, $x_1(t)$ and $x_2(t)$, are obtained by applying two band-pass filters F_1 and F_2 centered on frequencies f_1 and f_2 respectively to the signal. These two time series, $x_1(t)$ and $x_2(t)$, are then squared and passed through a low-pass filter F_3 to obtain the slow varying amplitude components denoted as $[x_i^2(t)], i = 1, 2$. Then the cross-correlation function (CCF) between these signals is computed as

$$K(\tau) = \left\langle \left[x_1^2(t) \right] \left[x_2^2(t + \tau) \right] \right\rangle \tag{5.53}$$

where the angle brackets $<>$ denote ensemble average.

If the original signals x_1 and x_2 contain frequencies $\omega_{1,2} \pm \Delta\omega$, the squared signals would contain high-frequency bands centered around $2\omega_{1,2}$ and a low-frequency band extending from zero to $2\Delta\omega$. It is this low-frequency band that contains the amplitude information and the purpose of the filter F_3 is to remove the upper band. It also serves to remove the mean values of $x_{1,2}^2$.

$K(\tau)$ can be used for intuitive interpretations. For example, in a wave propagation experiment using the amplitude correlation, applying identical filters F_1 and F_2 to signals from probes separated in space, a direct measurement of the group velocity of waves in the selected frequency band can be obtained.[56] One of the most important usages of the amplitude correlation method is determination of the energy flow in a turbulent spectrum.

The direction of the energy flow from one frequency domain to the other can be determined from the sign of the time delay (τ_{lag}) of the peak value of the cross-correlation function $K(\tau)$. The two domains are defined by the two bandpass filter F_1 and F_2. Positive time lag means the first signal $x_1(t)$ leads the second signal $x_2(t)$ in phase. As a result, one can speculate that the frequency domain around F_1 $(x_1(t))$ could be the energy source of that of F_2 $(x_2(t))$. Similarly, a negative time lag τ suggests that the region around $x_2(t)$ is the energy supplier for $x_1(t)$. The amplitude correlation method can, in principle, help to estimate the growth/damping rate of the driven mode and the nonlinear energy throughput rate.

In the applications of the amplitude correlation technique, the two frequency bands under consideration do not necessarily need to come from the same fluctuation signal. Nonlinear interaction between different fluctuation fields can also be detected through this method.

An important issue in interpreting the time lag is that the time delay,

τ_{lag}, is indicative of the energy flow between the two frequency bands only when the two frequency bands are strongly correlated.

5.4. Experimental evidence of the inverse energy cascade in plasma

In this section we illustrate how the above methods of the spectral transfer analysis are applied to experimental data. The examples are based on the results from the H-1 toroidal heliac.[57,58]

5.4.1. *Applicability of the nonlinear spectral transfer model*

The wave coupling equation, Eq. (5.29), describes turbulence in which a single-field description is valid and the three-wave interactions are permitted and dominant. A distinct feature of plasma as a continuous medium is that the responses of the electrons and ions are not identical, hence they induce collective electromagnetic fields. The dynamical equations for a plasma in the fluid limit are usually constructed using a two-fluid picture.

A single-field description of the plasma turbulence needs to be justified on a case-to-case basis. Consider the electrostatic wave turbulence, when the magnetic field fluctuations can be neglected. In a stable drift wave, electrons relax, along the magnetic field, to acquire Boltzmann distribution in the wave potential:

$$\tilde{n}_e = n_0 \exp(e\tilde{\phi}/T). \tag{5.54}$$

When the drift wave becomes unstable, the electron response is perturbed. This perturbation is characterized by the so-called non-adiabatic electron response, δn_e. If the normalized level of the potential fluctuations is small, $(e\tilde{\phi}/T) << 1$, this can be written as

$$\tilde{n}_e = n_0(e\tilde{\phi}/T) + \delta n_e. \tag{5.55}$$

In the unstable wave $\delta n_e \neq 0$ and the n_e and ϕ fluctuations are out of phase. This phase shift can be detected in experiment.

For a single-field description to be valid, the n_e and ϕ fluctuations should be in phase. A well-known example is the Hasegawa-Mima model (see Section 1.4 of Chapter 4) where $\delta n_e = 0$. This also means that the fluctuation-driven particle flux is zero. As follows from Equation (5.11),

$$\Gamma_{fl} = \frac{2}{B} \int\limits_0^\infty d\omega \, [P_{nn}P_{EE}]^{1/2} |\gamma_{nE}| \cos [\alpha_{nE}] = 0,$$

since $\tilde{n}_e$ and $\tilde{E}_\theta$ have a $\pi/2$ phase shift ($\tilde{E}_\theta = -\nabla_\theta \tilde{\phi}$) and $\cos (\alpha_{nE}) = 0$.

While confirming the phase difference between $\tilde{n}_e$ and $\tilde{\phi}$, it should be kept in mind, that it is fluctuations in the *plasma* potential, rather than in the *floating* potential which should be analyzed. In other words, fluctuations in the electron temperature should be included (see Eq. (5.7)).

Fig. 5.11. (c) shows the phase shift between fluctuations $\tilde{\phi}_p$ and $\tilde{\phi}_f$ obtained from the experimental data in H-1. The phase shift between the two fluctuating quantities is close to zero in the broad spectral range from $f \approx 0$ to 60 kHz. For the spectral energy transfer analysis, the phase information is much more important than the amplitude, since most of the spectral quantities used in the analysis (e.g. auto- and cross-bicoherence) are normalized.

In the given example from the H-1 heliac, $\tilde{\phi}_f$ and $\tilde{\phi}_p$ are in phase and the analysis can be simplified: $\tilde{\phi}_f$ is used in the PTF analysis.

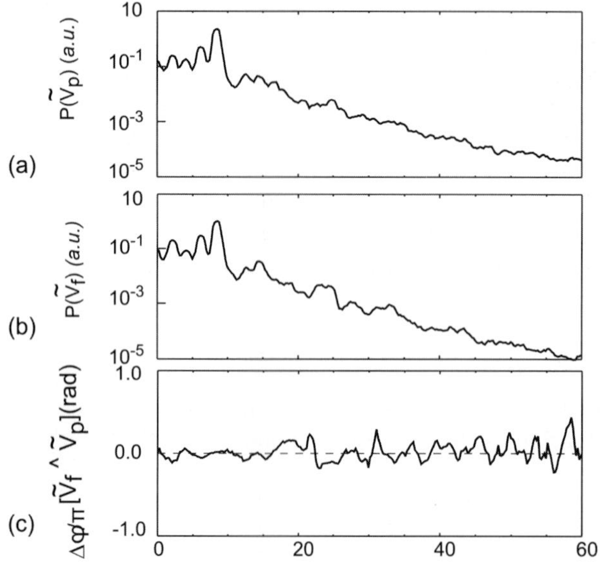

Fig. 5.11. Power spectra of (a) fluctuations in the plasma potential $\tilde{\phi}_p$, and (b) fluctuations in the plasma floating potential $\tilde{\phi}_f$. (c) Spectra of the phase shift between $\tilde{\phi}_p$, and $\tilde{\phi}_f$.

The phase difference between fluctuations in the poloidal electric field, $\tilde{E}_\theta$, and the density, $\tilde{n}_e$ is shown in Fig. 5.12.. This phase shift is close to $\pi/2$ over the most of the spectrum suggesting that the electron response is adiabatic. Thus, the applicability of the single-field description in the described experiments is justified.

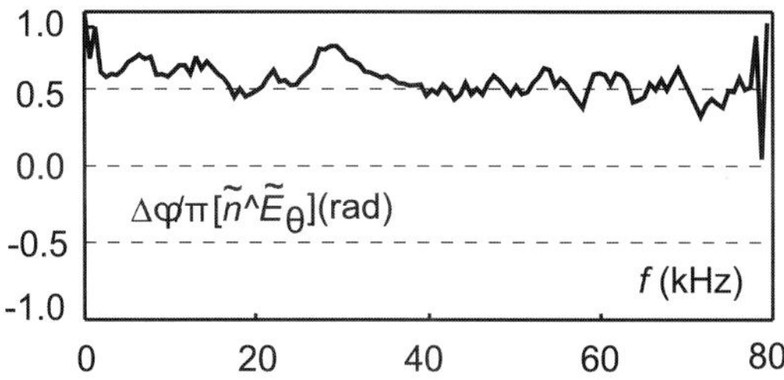

Fig. 5.12.   Spectrum of the phase shift between fluctuations in the electron density, $\tilde{n}_e$, and poloidal electric field, $\tilde{E}_\theta$.

Another point which should be made here is that in experiments, spectra are measured in the frequency (f) domain, while the wave kinetic equation is defined in the wave number (k) domain. The PTF analysis of experimental data would only be valid if a linear $k - f$ relationship is confirmed experimentally. Below we illustrate how this can be done in experiment.

In the laboratory frame of reference, frequencies of the fluctuations are Doppler shifted due to the presence of the $E \times B$ drift in practically all toroidal plasmas: $\omega_{lab} = \omega_{plasma} + k_\theta V_{E \times B}$. As explained in Section 2.5, in most cases, $E \times B$ drift dominates over the phase velocity in the plasma frame. Thus, the fluctuation frequencies in the lab frame are proportional to the poloidal wave numbers of the fluctuations. Since in the broadband turbulence the wave number spectra are isotropic, $k_\theta \approx k_{r,,}$[3] one can assume that $k \approx \sqrt{2}k_\theta \propto \omega$. The $E \times B$ Doppler shift plays in such cases a role of the wave number spectrograph.

5.4.2.   *Results on the spectral transfer analysis*

Figure 5.13. shows the power spectrum of the $\tilde{\phi}_f$ fluctuations in the H-1 plasma.[57] Spectral power decreases with frequency in the range $f = (0-80)$ kHz. In the frequency range of $f < 20$ kHz several coherent modes are observed.

Before applying the power transfer analysis technique to the fluctuation data, the linear $k - f$ relationship needs to be tested, as discussed above.

Wave numbers of fluctuations are measured using two poloidally separated probes, as $k_\theta = \Delta\phi/\Delta y$, where ϕ is the phase shift and Δy is the distance between the probes. This wave number k_θ is shown as a function of frequency, $k_\theta(f)$, in Fig. 5.13. (b). Though the $k_\theta(f)$ plot has large ripple, a linear trend, represented using a black line, is clearly observed.

The fluctuation phase velocity in the poloidal direction, $V = \omega_{lab}/k_\theta$, derived from this linear fit agrees within 10% with the measured $\mathbf{E} \times \mathbf{B}$

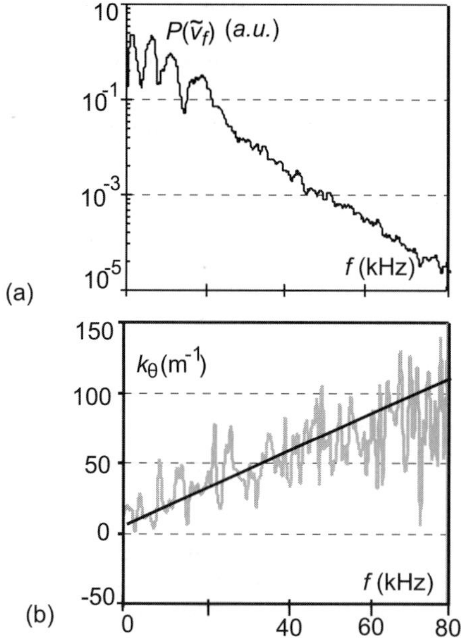

Fig. 5.13. (a): Power spectrum of the fluctuations in the floating potentials, $\tilde{V}_f$, (b) measured poloidal wave number spectrum $k_\theta(f)$ (grey line) with the linear fit (black line)

drift velocity in this radial region, confirming that $k_\theta V_{E \times B} >> \omega_{plasma}$.

Thus, the spectral power transfer can be studied in the frequency domain. The three-wave interactions satisfying matching rules for the wave numbers, $k = k_1 + k_2$, also obey the frequency selection rule, $f = f_1 + f_2$.

It should be noted, that while estimating the temporal evolution of the turbulence spectra, one needs to take into account that turbulence drifts in the lab frame. As discussed in Section 3.3, the change in the turbulence spectrum is estimated using the differential approach represented by Eq. (5.36). During the time interval, τ, turbulence will drift in poloidal direction by $\Delta y = \tau V_{E \times B}$. As a result, the turbulence evolution should be studied using two probes separated poloidally. The time delay, τ, in the Equation (5.36) should be estimated using the distance between the probes, Δy, and $V_{E \times B}$.

The nonlinear energy transfer function (NETF), W_{NL}^k, and the linear growth rate, γ_k, derived from Eq. (5.42) in Section 5.3.3. are shown in Fig. 5.14. W_{NL}^k is negative in the broadband spectral region of $f = (20 -$

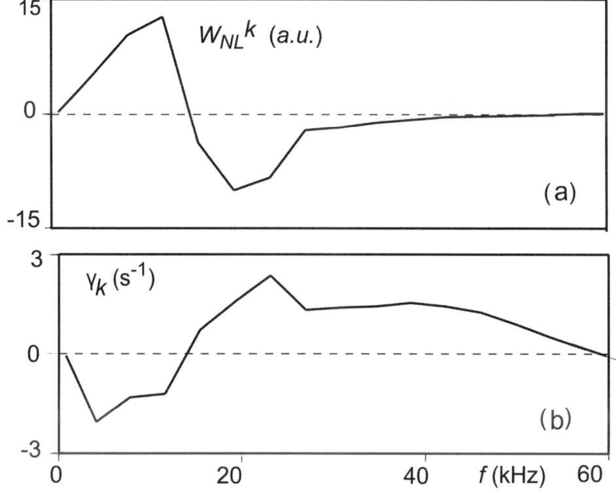

Fig. 5.14. (a) The nonlinear energy transfer function W_{NL}^k; (b) linear growth rate γ_k derived from Eq. (5.42). The frequency resolution is $\Delta f \approx 4$ kHz.

50) kHz suggesting that waves in this range on-average lose energy, whereas the lower frequency spectral range ($f < 20$ kHz) gains spectral energy due to the three-wave interactions. The linear growth rate shown in Fig. 5.14. (c) has positive maximum at $f \approx 25$ kHz. This spectral range is the range where initially unstable waves develop.

The NETF shown in Fig. 5.14. illustrates the *inverse energy cascade* in the broadband turbulence. Its computation required substantial statistical averaging. The turbulence signals digitized at the rate of 1 MHz during 80 ms of the plasma discharge are divided into 460 overlapping segments, such that the spectral moments needed for the PTF computation are then averaged over these segments. Such averaging is needed to correctly estimate spectral transfer *via* the random-phase wave interactions.[58]

The inverse energy cascade is the mechanism of spreading spectral energy from the instability range into a broad range of the wave numbers and frequencies. The PTF method[50,53] has been successfully used to demonstrate the existence of the inverse energy cascade in toroidal plasma.[57,58] However, the NETF in Fig. 5.14.(a) does not show the fine structure, which would correspond to coherent spectral features seen in the spectrum of Fig. 5.13.(b). This may be suggestive of the non-cascade origin of these features.

5.5. Quasi-two-dimensional turbulence in fluids and plasma and generation of zonal flows

The inverse energy cascade in 2D fluid turbulence[42] is the flow of spectral energy from smaller to larger scales, leading to the $E_k \propto k^{-5/3}$ scaling of the energy spectrum in this spectral range (see Chapter 1, Section 4, by G. Falkovich). R. Kraichnan has predicted that the spectral energy may pile up at the largest scale allowed by the system size and noted a similarity between the condensation of the turbulent energy and the Bose-Einstein condensation of the 2D quantum gas.[42] The condensate formation in 2D fluids has been confirmed in experiments[59,60] and in numerical simulations,[61,62] for review see.[63] Below we illustrate how turbulence condenses in the 2D fluid experiment.

In plasma, quasi-2D turbulence can also be generated *via* 3-wave interactions.[8] The structure of the Hasegawa-Mima equation, which describes spectral evolution of the drift-wave turbulence, is identical to the Charney equation describing the evolution of nonlinear Rossby waves in planetary atmosphere.[9,64] These models, similarly to the models of the 2D fluid turbulence described by the 2D Navier-Stokes equation, have two conserved quantities, energy and enstrophy. As a consequence, there are two inertial ranges which correspond to (a) the inverse cascade of energy, and (b) forward cascade of enstrophy. Similarly to the fluid dynamics in 2D, spectral energy in plasma can condense in the largest scale.[9,65] In particular, such condensation may lead to the formation of zonal flows and other coherent structures which is a form of the plasma self-organization.[66] We will illustrate generation of zonal flow in plasma experiment and will discuss experimental signatures of such flow.

5.5.1. *Spectral condensation of 2D turbulence*

One of the first convincing experimental evidence of the inverse energy cascade in 2D turbulence was presented by J. Sommeria in 1987.[59] In this experiment, turbulence was generated in a thin layer of mercury in a cell. The fluid was placed in the vertical magnetic field. 36 biased electrodes of varying polarity generated electric currents in a layer which, by interacting with the vertical magnetic field generated 36 planar vortices in a cell. By varying the current and the depth of the mercury layer, the forcing and the linear damping could be finely controlled. Sommeria observed the $k^{-5/3}$ scaling due to the inverse cascade in the energy inertial range (though in a rather narrow k-range) and also reported the observation of the largest vortex limited by the cell size at low linear dissipation due to the process of the spectral condensation.

Detailed measurements of the spectral energy scaling were presented by Paret and Tabeling[60,67] in 1997-98 in experiments in the stratified layers of electrolyte. In these experiments 2D turbulent flows were studied by generating $J \times B$-driven vortices in thin layers of fluid. These experiments have confirmed the existence of the inverse energy cascade in the quasi-2D turbulence, the $k^{-5/3}$ scaling, as predicted by Kraichnan,[42] and also have confirmed the generation of the spectral condensate at low damping.

The most recent experiment in which the spectral condensation of turbulence was reproduced, has been aimed at the comparison between fluid and plasma turbulence.[68] Below we summarize some of these results.

Figure 5.15. shows experimental setup used in references[60] and reproduced in.[68] Turbulence is generated *via* the interaction between $J \times B$-driven vortices whose sizes and the distances between their centres (positions of the magnetic dipoles) determine the scale at which energy is injected into the system. This scale is characterized by the wave number k_i. The flow on the free surface of the light fluid is visualized by placing small latex particles on the surface and by recording their trajectories using video camera. The particle image velocimetry (see Chapter 7 by J. Soria) allows the velocity and the vorticity fields to be reconstructed from the consecutive images of the particles in the flow.

After the current is turned on, and if the vortex interaction energy is sufficiently high, the inverse energy cascade leads to the aggregation of the spectral energy at larger and larger scales. The maximum of the energy spectrum thus moves from k_i toward the lower-k range of the wave numbers. In the absence of the energy dissipation, a wave number corresponding to the maximum spectral energy, k_m, can not be constant in time. However, if there is damping for large scales, for example *via* linear damping μ, k_m stabilizes at

$$k_m \approx \left(\frac{\mu^3}{\epsilon}\right)^{1/2}, \tag{5.56}$$

where ϵ is the energy dissipation rate (see Chapter 1).

Another characteristic scale in 2D turbulence is determined by the size of the system, k_s, such as the size of the fluid cell. If the system size is larger than the resistive scale, or, $k_s < k_m$, one should observe the stationary spectrum which has a maximum at k_m. If however the resistive wave number [Eq. (5.56)] is larger than the cell size, $k_s > k_m$, spectral condensation of turbulence becomes possible when spectral energy accumulates at the system size. This is illustrated schematically in Fig. 5.16..

The evolution of the turbulent 2D flow in a cell is shown in Fig. 5.17. for the case when the damping to the bottom of the fluid cell is reduced due to the fluid stratification, such that the spectral condensation of turbulence

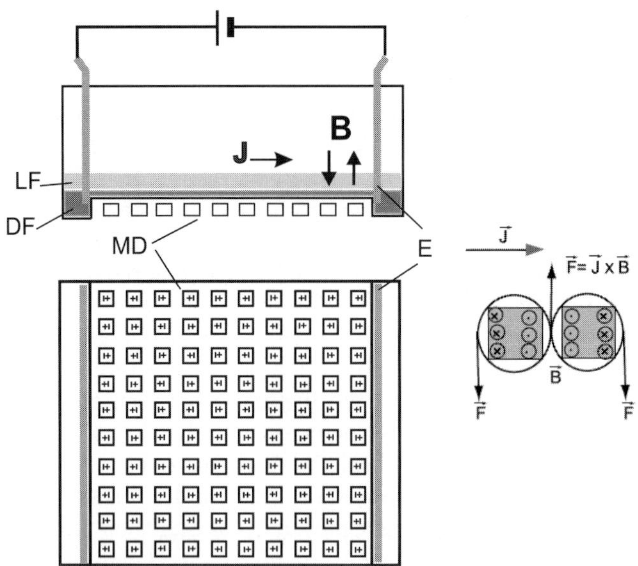

Fig. 5.15.　Schematic of the electromagnetically-driven 2D turbulence in electrolyte. A thin layer of a light fluid (LF) rests on the layer of denser fluid (DF). A matrix of 10×10 magnetic dipoles (MD) is placed under the bottom of the fluid cell. Current J flowing between two electrodes (E) generates 100 vortices due to the $J \times B$ force, as illustrated on the right.

is possible. Initially only externally forced vortices are seen [Fig. 5.17.(a)]. The inverse energy cascade destroys these vortices and leads to the generation of a broadband spectrum of eddies seen in Fig. 5.17.(b). At some stage, a large vortex is formed, which then persists in steady-state as shown in Fig. 5.17.(c).

The corresponding evolution of the wave number spectra, derived from the velocity field of the trace particles, is illustrated in Fig. 5.18.. At the early stage of the flow evolution, $t = t_1$, spectrum shows a peak at the forcing scale, $k = k_i$. Later, at $t = t_2$, this peak is washed out and the maximum of the spectral energy shifts to lower k. Eventually spectral energy is accumulated at k_s which represents the scale of the large vortex shown in Fig. 5.17.(c).

It should be noted, that the increase in spectral energy in the condensate wave number range at $t = t_3$ coincides with the reduction in spectral energy in the broadband turbulence, including the forcing range k_i.

After the formation of the largest vortex, the condensate persists in steady-state. The energy necessary to overcome damping at this largest scale can only come from the energy source at k_i. The fact that spectral

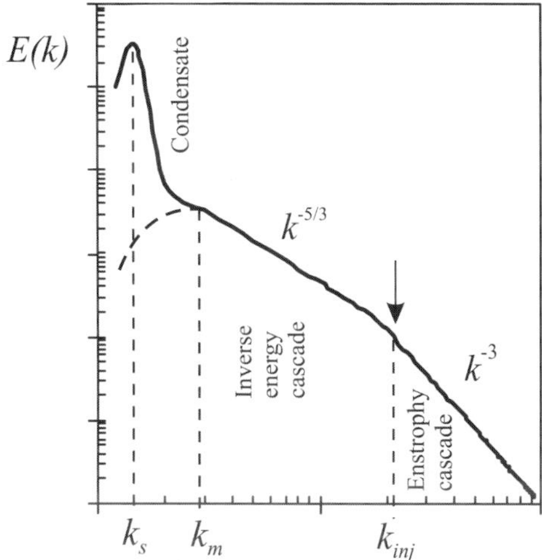

Fig. 5.16.   Schematic of the 2D turbulence spectrum.

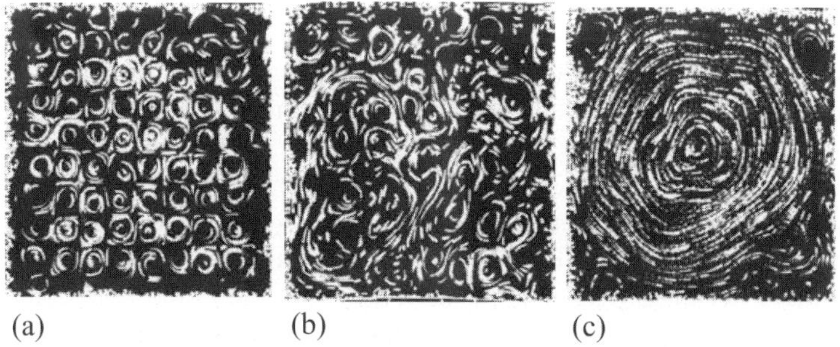

(a)   (b)   (c)

Fig. 5.17.   Trajectories of the tracer particles averaged over 12 frames of recorded video during spectral condensation of turbulence. (a) The initial stage, $t = 3$ seconds after switching on the current; (b) the inverse energy cascade stage, $t = 16$ s; (c) steady-state spectral condensate stage, $t = 55$ s.

energy in the spectral range $k_s < k$ is reduced in the presence of the condensate may be indicative of the stronger spectral coupling between spectral regions of k_i and k_s, such that the spectral energy is delivered from k_i to k_s directly, rather than through a multi-step process of the energy cascade.

5.5.2.   *Zonal flows in plasma turbulence*

In the above example we have seen that strong anisotropic flow may develop as a result of accumulation of the turbulent energy of the 2D turbulence

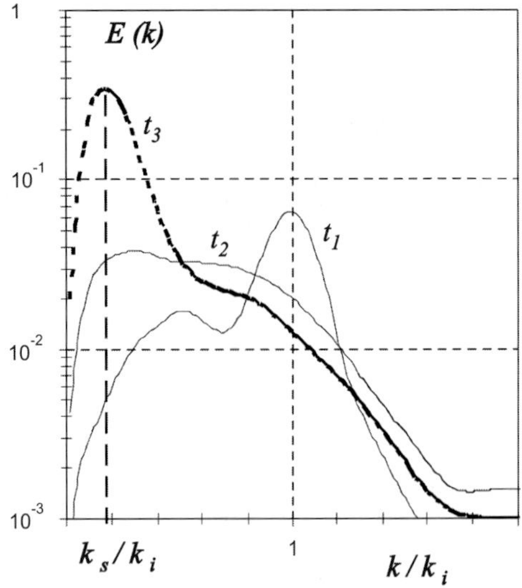

Fig. 5.18.   Wave number spectra of the velocity field during the development of spectral condensate in 2D fluid. $t_3 > t_2 > t_1$.

at low wave numbers. Understanding physics of such an interplay between stationary flows and turbulent fields is also important in plasma turbulence studies (for review see[69]). In addition to flows driven by the momentum input (or loss), turbulence-driven flows play important roles in the plasma dynamics. A complex interplay between turbulence and flows in magnetically confined plasma has been a focus of the turbulence related studies in the last two decades. This topic is of great practical importance, due to its relevance to such phenomena as improvement of the plasma confinement and formation of the plasma barriers. We will only briefly discuss this topic in these lectures.

First, it is important to clarify which flow is the most important in this context, since in plasma, unlike in neutral fluids, a number of flows appear (electron and ion diamagnetic drifts, $E \times B$ drift, etc.). Out of them, the $E \times B$ flow has unique status in plasma physics and it also has a central role in the plasma-flow interaction physics.

It has been shown theoretically[70] in the equations governing electrostatic turbulence, the only convective term in the equations is the $E \times B$ drift. The $E \times B$ flow is the sole advectant of fluctuations in density, temperature, and flow. As explained in,[69] this statement is a result of empirical observations and careful theoretical calculations. It does not simply follow from the fact that the $E \times B$ drift velocity is the same for all particles regardless of their charge or mass, simply because this is not universally correct, for example in complex magnetic geometry. A good illustration of the difference in the $E \times B$ flow and in the ion mass flow velocities is given in.[71] What is important for us here is the fact that it is the $E \times B$ flow, rather than bulk plasma flow, which is relevant in the turbulence-flow interaction.

In toroidal plasma the $E \times B$ flow is determined by the radial component of the electric field. Radial electric field in the plasma can be estimated using the radial force balance equation (related to the ion momentum balance). This equation can be expressed as[69]

$$E_r = \frac{1}{q_i n_i} \frac{\partial}{\partial r} p_i + \frac{m_i}{q_i} \frac{\partial}{\partial r} \langle \tilde{u}_{\theta i} \tilde{u}_{ri} \rangle - u_{\theta i} B_\phi + u_{\phi i} B_\theta, \qquad (5.57)$$

where q_i is the ion charge, n_i is the ion density, m_i is the ion mass and u_i is the ion velocity. Subscripts θ and ϕ indicate poloidal and toroidal components of the flow velocities and of the magnetic field. The second term on the right-hand-side is the Reynolds stress. Reynolds stress in the fluid turbulence is discussed by J. Jiménez in Chapter 6 (Section 3). Reynolds stress in the plasma[26] (for review see[69]) provides a mechanism of generating stationary flows by turbulence.

Some of these flows have already been mentioned. These are zonal flows, or poloidally and toroidally symmetric potential structures driven by the plasma fluctuations. Reynolds stress is not the only theoretical mechanism which can drive zonal flows (see Chapter 4, Lecture 3). Poloidal and toroidal symmetry means that the poloidal and toroidal components of the wave number are zero: $k_\theta = k_\phi = 0$, while their radial wave number, k_r, remains finite. Such a structure is shown schematically in Fig. 5.19.. Arrows indicate the direction of the flow in poloidal plasma cross-section. Once again, the *flow* here is the $E \times B$ flow, which does not coincide with the ion flow whose velocity $u_{\theta i}$ appears in the radial force balance, Equation (5.57).

Theories of zonal flows in plasma are discussed in Lecture 3 by J. Krommes (see Chapter 4). For detailed review on the theory of zonal flows see.[10] Experimental studies of zonal flows are described in references.[11–17] The most recent collection of experimental results on zonal flows has been published in the special issue of the Plasma Physics and Controlled Fusion journal (Volume 48, Number 4, April 2006).

It is impossible to overview experimental results on the zonal flows in toroidal plasma in this section, however, it seems appropriate to illustrate how zonal flows are identified in experiments. Sometimes, geodesic acoustic modes, or GAMs are also referred to as the higher frequency branch of zonal flows (see Section 3.5 of Chapter 4). These modes appear in toroidal plasma, and are observed as the finite-frequency coherent potential structures. They will not be discussed here.

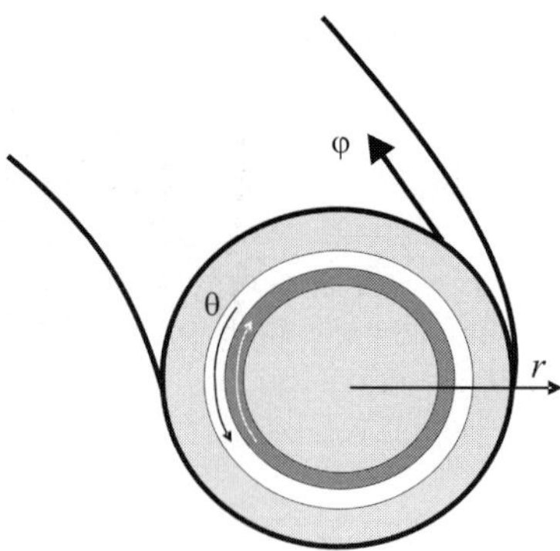

Fig. 5.19.   Schematic of the zonal flow in toroidal plasma.

First we discuss experimental evidence of the stationary ($f_{ZF} = 0$) zonal flows. There are two theoretically expected mechanisms of the zonal flow generation in plasma, which seem to be supported by observations. The first one is the generation of zonal flow *via* the inverse energy cascade in broadband turbulence of drift waves (see[10]). This mechanism is somewhat similar, but not identical, to the inverse energy cascade in 2D fluid turbulence described in the previous subsection. The second mechanism of the zonal flow generation is modulational instability whose theory is described in Section 3.1 of Chapter 4 by J. Krommes.

In the high-temperature plasma, the first observation of the stationary zonal flow whose frequency was very close to zero, $f_{ZF} \approx 0$, was reported in the CHS torsatron[15] by Fujisawa *et al.* Two heavy ion beam probe diagnostics (see Section 2.7) were set at two toroidal locations, such that the toroidal mode number $n = 0$ of the observed $f \approx 0$ potential structure could be confirmed experimentally.

Similar spectral feature has been also observed in the low-temperature plasma in the H-1 heliac using Langmuir probes.[72]

Spectrally broadened low-frequency band around $f_{ZF} = 0$ has signatures of zonal flow. Poloidal and toroidal mode numbers have been estimated from poloidally (1 and 2) and toroidally separated (2 and 3 probes). It is usually difficult to align toroidally separated probes to exactly the same poloidal position. As a result, a phase shift between probes 1 and 3 will occur due to the uncertainty in the poloidal separation between the probes, Δy_{13}:

$$\Delta\varphi_{13}(f) = k_{\parallel}(f)\Delta L_{\parallel} + k_{\theta}(f)\Delta y_{13}, \qquad (5.58)$$

where $\Delta L_{\parallel}$ and y_{13} are toroidal and poloidal separation between probes 1 and 3 respectively, and $k_{\parallel}(f)$ is known from the phase difference between

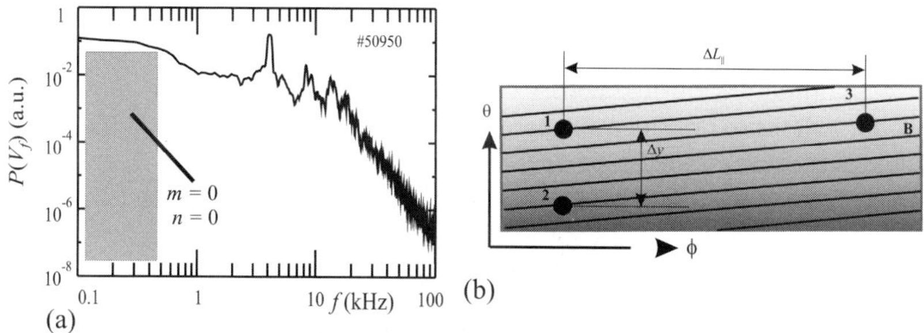

Fig. 5.20. (a) Frequency spectrum of electrostatic potential. Shaded $f < = 0.5$ kHz frequency band corresponds to $m = n = 0$ zonal flow. (b) Probe setup used for identification of poloidal and toroidal mode numbers.

probes 1 and 2. In case of zonal flow, $m = 0$, and the second term on the right-hand side becomes zero (since $k_\theta = 0$), such that the poloidal uncertainty Δy_{13} becomes unimportant and the toroidal wave number can be reliably estimated by measuring $\Delta\varphi_{13}$.

Accumulation of spectral energy in the spectrally broadened zero-frequency zonal flow illustrated in Fig. 5.20.(a) possibly occurs *via* the inverse energy cascade (see Section 4.2). It has been suggested in reference,[68] that the processes of spectral condensation in 2D fluid turbulence (section 5.1) may be in several aspects similar to the generation of zonal flows in turbulent plasma. Experimental data[68] generally agree with the theoretically proposed idea that the spectral energy is gradually accumulated in larger scales by cascading from the unstable spectral range (injection scale in 2D fluid), until the largest structure, or zonal flow develops. At this stage, the spectral transfer may be changing from spectrally local (wave numbers participating in the energy cascade are comparable) to non-local, such that the energy can be delivered into zonal flow directly from the unstable range scales $k_{ur} >> k_{ZF}$. Such a possibility has been proposed in[73] and has been reviewed in.[66]

Generation of zonal flow *via* the modulational instability (Chapter 4, Section 3.1) is another possible scenario. The first experimental evidence of the zonal flow development which is consistent with this has been reported in.[74] In this case, zonal flow develops in the improved confinement mode (H-mode), in which the level of the broadband turbulence was substantially reduced. The generation of zonal flow was correlated with the development of the secondary instability in the plasma. This instability is driven by the E_r shear and is similar to the Kelvin-Helmholtz instability. Initially coherent oscillations appear at $f = 15$ kHz. As the instability develops and the fluctuation spectrum becomes broader, a low frequency E_r spectral feature develops, as illustrated in Fig. 5.21.. Zonal flow is the spectral feature at $f \approx 1$ kHz seen in the wavelet plot of the E_r fluctuations of Fig. 5.21.(a). Corresponding 1 kHz feature in the fluctuations of E_{pol} (Fig. 5.21.(b)) is considerably weaker, since for zonal flows $k_r >> k_\theta \approx 0$. In this example,

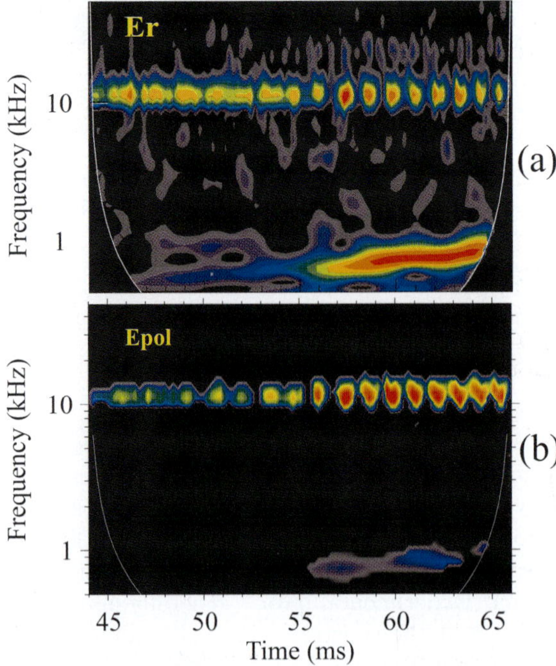

Fig. 5.21.   Wavelet plots of the time evolution of the fluctuations in (a) radial and (b) poloidal electric fields. For details see.[74]

zonal flow has low but finite frequency, as is expected from the theory of the modulational instability (see paper by R. L. Dewar and R. F. Abdullatif, in the *Workshop Proceedings*).

5.6.   Conclusion

In this chapter we gave a brief introduction to modern methods and results in experimental studies of turbulence in the plasma confined by toroidal magnetic field. Considerable progress has been made due to novel sophisticated diagnostics and analysis techniques, but also due to closer interaction between plasma theory, numerical simulations and experiments. This is particularly true with regard to studies of large turbulence-driven structures in plasma, such as zonal flows, geodesic acoustic modes, streamers etc. We have not even attempted to review all experimental results related to these topics in two lectures.

Several fundamental questions however remain unresolved. Among them is the role of turbulent fluxes in the net particle and energy transport. Answers to this question remain controversial. Though it is common to believe that turbulence generates fluxes which can affect net fluxes, it is clear that in some situations the role of turbulence is limited to the generation of strongly sheared $E \times B$ flows in the plasma. In this case the improvement in particle confinement may be achieved due to, for example particle orbit squeezing. To some extent this agrees with recent theoretical ideas about intimate interrelation between neoclassical and turbulent fluxes in plasma.[75]

Understanding plasma turbulence can be greatly enhanced through interdisciplinary comparative studies of plasma and fluid turbulence. In this chapter we have illustrated one of the first attempts in this direction, which we find appropriate for this cross-disciplinary collection of lectures on turbulence.

The authors thank G. Conway for his valuable comments.

Bibliography

1. D. Bohm. In A. Guthrie and R. K. Wakerling, editors, *The Characteristics of Electrical Discharges in Magnetic Fields*, volume 1 of *National Nuclear Energy Series*, page 77. McGraw-Hill Book Company, Inc., New York, 1949.
2. A. B. Mikhailovskii. *Instabilities in a confined plasma*. Institute of Physics Pub., Bristol, 1998.
3. P. C. Liewer. Measurements of microturbulence in tokamaks and comparisons with theories of turbulence and anomalous transport. *Nuclear Fusion*, 25, 543–621, 1985.
4. E. Mazzucato. Small-scale density fluctuations in the adiabatic toroidal compressor. *Physical Review Letters*, 36, 792–4, 1976.
5. E. Mazzucato. Low-frequency microinstabilities in the plt tokamak. *Physics of Fluids*, 21, 1063–9, 1978.
6. C. M. Surko and R. E. Slusher. Study of the density fluctuations in the adiabatic toroidal compressor scattering tokamak using co/sub 2/ laser. *Physical Review Letters*, 37, 1747–50, 1976.
7. R. E. Slusher and C. M. Surko. Study of density fluctuations in the alcator tokamak using co/sub 2/ laser scattering. *Physical Review Letters*, 40, 400–403, 1978.
8. A. Hasegawa and K. Mima. Pseudo-three-dimensional turbulence in magnetized nonuniform plasma. *Physics of Fluids*, 21, 87–92, 1978.
9. A. Hasegawa, C. G. Maclennan, and Y. Kodama. Nonlinear behavior and

turbulence spectra of drift waves and rossby waves. *Physics of Fluids*, 22, 2122–9, 1979.

10. P. H. Diamond, S. I. Itoh, K. Itoh, and T. S. Hahm. Zonal flows in plasma - a review. *Plasma Physics and Controlled Fusion*, 47, R35–161, 2005.

11. M. G. Shats and W. M. Solomon. Experimental evidence of self-regulation of fluctuations by time-varying flows. *Physical Review Letters*, 88, 045001/1–4, 2002.

12. M. Jakubowski, R. J. Fonck, and G. R. McKee. Observation of coherent sheared turbulence flows in the DIII-D tokamak. *Physical Review Letters*, 89, 265003/1–4, 2002.

13. G. R. McKee, R. J. Fonck, M. Jakubowski, K. H. Burrell, K. Hallatschek, R. A. Moyer, W. Nevins, D. L. Rudakov, and X. Xu. Observation and characterization of radially sheared zonal flows in DIII-D. *Plasma Physics and Controlled Fusion*, 45, A477–85, 2003.

14. S. Coda, M. Porkolab, and K. H. Burrell. Signature of turbulent zonal flows observed in the DIII-D tokamak. *Physical Review Letters*, 86, 4835–8, 2001.

15. A. Fujisawa, K. Itoh, H. Iguchi, K. Matsuoka, S. Okamura, A. Shimizu, T. Minami, Y. Yoshimura, K. Nagaoka, C. Takahashi, M. Kojima, H. Nakano, S. Ohsima, S. Nishimura, M. Isobe, C. Suzuki, T. Akiyama, K. Ida, K. Toi, S. I. Itoh, and P. H. Diamond. Identification of zonal flows in a toroidal plasma. *Physical Review Letters*, 93, 165002/1–4, 2004.

16. Y. Hamada, A. Nishizawa, T. Ido, T. Watari, M. Kojima, Y. Kawasumi, K. Narihara, and K. Toi. Zonal flows in the geodesic acoustic mode frequency range in the JIPP T-IIU tokamak plasmas. *Nuclear Fusion*, 45, 81–8, 2005.

17. G. D. Conway, B. Scott, J. Schirmer, M. Reich, and A. Kendl. Direct measurement of zonal flows and geodesic acoustic mode oscillations in ASDEX upgrade using doppler reflectometry. *Plasma Physics and Controlled Fusion*, 47, 1165–85, 2005.

18. P. H. Diamond, M. N. Rosenbluth, F.L. Hinton, M. Malkov, J. Fleischer, and A. Smolyakov. Dynamics of zonal flows and self-regulating drift-wave turbulence. In *17th IAEA Fusion Energy Conference*, volume IAEA-CN-69, page TH3/1, Yokohama, Japan, 1998. International Atomics Energy Agency, Vienna, 1998.

19. H. Punzmann and M. G. Shats. Formation and structure of transport barriers during confinement transitions in toroidal plasma. *Physical Review Letters*, 93, 125003/1–4, 2004.

20. K. W. Gentle. Diagnostics for magnetically confined high-temperature plasmas. *Reviews of Modern Physics*, 67, 809–36, 1995.

21. N. Bretz. Diagnostic instrumentation for microturbulence in tokamaks. *Review of Scientific Instruments*, 68, 2927–64, 1997.

22. N. Hershkowitz. *Plasma Diagnostics*, volume 1 of *Plasma–materials interactions*. Academic Press, Boston, 1989.

23. P. C. Stangeby and G. M. McCracken. Plasma boundary phenomena in tokamaks. *Nuclear Fusion*, 30, 1225–379, 1990.

24. S. L. Chen and T. Sekiguchi. Instantaneous direct display system of plasma parameters by means of triple probe. *Journal of Applied Physics*, 36, 2363–75,

1965.

25. E. J. Powers. Spectral techniques for experimental investigation of plasma diffusion due to polychromatic fluctuations. *Nuclear Fusion*, 14, 749–52, 1974.

26. P. H. Diamond and Y. B. Kim. Theory of mean poloidal flow generation by turbulence. *Physics of Fluids B (Plasma Physics)*, 3, 1626–33, 1991.

27. R. E. Waltz. Magnetic fluctuations, ambipolarity, charge filamentation, and plasma rotation in tokamaks. *Physics of Fluids*, 25, 1269–78, 1982.

28. T. E. Stringer. Neoclassical transport in the presence of fluctuations. *Nuclear Fusion*, 32, 1421–32, 1992.

29. W. M. Solomon and M. G. Shats. Nonambipolarity of fluctuation-driven fluxes and its effect on the radial electric field. *Physical Review Letters*, 87, 73–6, 2001.

30. I. H. Hutchinson. *Principles of plasma diagnostics*. Cambridge University Press, Cambridge, New York, 1987.

31. W. M. Solomon and M. G. Shats. Fluctuation studies using combined mach/triple probe. *Review of Scientific Instruments*, 72, 449–52, 2001.

32. W. M. Solomon, M. G. Shats, D. Korneev, and K. Nagasaki. Collective microwave scattering diagnostic on the H-1 heliac. *Review of Scientific Instruments*, 72, 352–4, 2001.

33. R. Nazikian, G. J. Kramer, and E. Valeo. A tutorial on the basic principles of microwave reflectometry applied to fluctuation measurements in fusion plasmas. *Physics of Plasmas*, 8, 1840–55, 2001.

34. B. B. Afeyan, A. E. Chou, and B. I. Cohen. The scattering phase shift due to bragg resonance in one-dimensional fluctuation reflectometry. *Plasma Physics and Controlled Fusion*, 37, 315–27, 1995.

35. G. D. Conway, L. Schott, and A. Hirose. Plasma density fluctuation measurements from coherent and incoherent microwave reflection. *Plasma Physics and Controlled Fusion*, 38, 451–66, 1996.

36. G. D. Conway, L. Schott, and A. Hirose. Comparison of reflectometer fluctuation measurements from experiment and two-dimensional numerical simulation. *Review of Scientific Instruments*, 67, 3861–70, 1996.

37. M. Hirsch and E. Holzhauer. Doppler reflectometry with optimized temporal resolution for the measurement of turbulence and its propagation velocity. *Plasma Physics and Controlled Fusion*, 46, 593–609, 2004.

38. G. McKee, R. Ashley, R. Durst, R. Fonck, M. Jakubowski, K. Tritz, K. Burrell, C. Greenfield, and J. Robinson. The beam emission spectroscopy diagnostic on the DIII-D tokamak. *Review of Scientific Instruments*, 70, 913–16, 1999.

39. G. R. McKee, C. Fenzi, R. J. Fonck, and M. Jakubowski. Turbulence imaging and applications using beam emission spectroscopy on DIII-D (invited). *Review of Scientific Instruments*, 74, 2014–19, 2003.

40. T. P. Crowley. Rensselaer heavy ion beam probe diagnostic methods and techniques. *IEEE Transactions on Plasma Science*, 22, 291–309, 1994.

41. A. Fujisawa, H. Iguchi, S. Lee, T. P. Crowley, Y. Hamada, S. Hidekuma, and E. Kojima. Active trajectory control for a heavy ion beam probe on the compact helical system. *Review of Scientific Instruments*, 67, 3099–107,

1996.

42. R.H. Kraichnan. Inertial ranges in two-dimensional turbulence. *Physics of fluids*, 10(7), 1417, 1967.

43. W Horton and Y.H. Ychikawa. *Chaos and Structures in Nonlinear Plasmas.* World Scientific, Singapore, 1996.

44. A. Hasegawa and M. Wakatani. Self-organization of electrostatic turbulence in a cylindrical plasma. *Physical Review Letters*, 59, 1581–4, 1987.

45. Y. C. Kim and E. J. Powers. Digital bispectral analysis of self-excited fluctuation spectra. *Physics of Fluids*, 21, 1452–3, 1978.

46. Y. C. Kim and E. J. Powers. Digital bispectral analysis and its applications to nonlinear wave interactions. *IEEE Transactions on Plasma Science*, PS-7(2), 120–31, 1979.

47. Y. C. Kim, J. M. Beall, E. J. Powers, and R. W. Miksad. Bispectrum and nonlinear wave coupling. *Physics of Fluids*, 32(2), 258–63, 1980.

48. V. Kravtchenko-Berejnoi, F. Lefeuvre, D. Krasnossel'skikh, and D. Lagoutte. On the use of tricoherent analysis to detect non-linear wave-wave interactions. *Signal processing*, 42, 291–309, 1995.

49. W.B. Collis, P.R. White, and J.K. Hoammond. Higher-order spectra: the bispectrum and trispectrum. *Mechanical systems and signal processing*, 12(3), 375–94, 1998.

50. C. P. Ritz and E. J. Powers. Estimation of nonlinear transfer functions for fully developed turbulence. *Physica D*, 20D(2-3), 320–34, 1986.

51. C. P. Ritz, E. J. Powers, and R. D. Bengtson. Experimental measurement of three-wave coupling and energy cascading. *Physics of Fluids B-Plasma Physics*, 1(1), 153–63, 1989.

52. C. P. Ritz, E. J. Powers, R. W. Miksad, and R. S. Solis. Nonlinear spectral dynamics of a transitioning flow. *Physics of Fluids*, 31(12), 3577–88, 1988.

53. J. S. Kim, R. D. Durst, R. J. Fonck, E. Fernandez, A. Ware, and P. W. Terry. Technique for the experimental estimation of nonlinear energy transfer in fully developed turbulence. *Physics of Plasmas*, 3(11), 3998–4009, 1996.

54. A. Hasegawa and K. Mima. Stationary spectrum of strong turbulence in magnetized nonuniform plasma. *Physical Review Letters*, 39(4), 205–8, 1977.

55. F. J. Crossley, P. Uddholm, P. Duncan, M. Khalid, and M. G. Rusbridge. Experimental study of drift-wave saturation in quadrupole geometry. *Plasma Physics Controlled Fusion*, 34(2), 235–62, 1992.

56. P.J. Duncan and M.G. Rusbridge. The 'amplitude correlation' method for the study of nonlinear interactions of plasma waves. *Plasma Physics Controlled Fusion*, 35, 825–35, 1993.

57. H. Xia and M. G. Shats. Inverse energy cascade correlated with turbulent-structure generation in toroidal plasma. *Physical Review Letters*, 91, 155001/1–4, 2003.

58. H. Xia and M. G. Shats. Spectral energy transfer and generation of turbulent structures in toroidal plasma. *Physics of Plasmas*, 11, 561–71, 2004.

59. J. Sommeria. Experimental study of the two-dimensional inverse energy cascade in a square box. *Journal of Fluid Mechanics*, 170, 139–68, 1986.

60. J. Paret and P. Tabeling. Intermittency in the two-dimensional inverse cas-

cade of energy: Experimental observations. *Physics of Fluids*, 10(12), 3126–36, 1998.

61. M. Hossain, W. H. Matthaeus, and D. Montgomery. Long-time states of inverse cascades in the presence of a maximum length scale. *Journal of Plasma Physics*, 30, 479–93, 1983.

62. L.M Smith and V. Yakhot. Bose condensation and small-scale structure generation in a random forced driven 2d turbulence. *Physical Review Letters*, 71, 352–355, 1993.

63. P. Tabeling. Two-dimensional turbulence: a physicist approach. *Physics Reports*, 362(1), 1–62, 2002.

64. D. Fyfe and D. Montgomery. Possible inverse cascade behavior for drift-wave turbulence. *Physics of fluids*, 22(2), 246–8, 1979.

65. A. Hasegawa and M. Wakatani. Plasma edge turbulence. *Physical Review Letters*, 50(9), 682–6, 1983.

66. W Horton and A Hasegawa. Quasi-two-dimensional dynamics of plasmas and fluids. *Chaos*, 4(2), 227–251, 1994.

67. J. Paret and P. Tabeling. Experimental observation of the two-dimensional inverse energy cascade. *Physical Review Letters*, 79(21), 4162–5, 1997.

68. M. G. Shats, H. Xia, and H. Punzmann. Spectral condensation of turbulence in plasmas and fluids and its role in low-to-high phase transitions in toroidal plasma. *Physical Review E*, 71, 046409, 2005.

69. P. W. Terry. Suppression of turbulence and transport by sheared flow. *Reviews of Modern Physics*, 72, 109–65, 2000.

70. Y. B. Kim, P. H. Diamond, H. Biglari, and J. D. Callen. Theory of neoclassical ion temperature-gradient-driven turbulence. *Physics of Fluids B (Plasma Physics)*, 3, 384–94, 1991.

71. K. H. Burrell. Effects of $\mathbf{E} \times \mathbf{B}$ velocity shear and magnetic shear on turbulence and transport in magnetic confinement devices. *Physics of Plasmas*, 4, 1499–518, 1997.

72. M. G. Shats, H. Xia, and M. Yokoyama. Mean $\mathbf{E} \times \mathbf{B}$ flows and gam-like oscillations in the H-1 heliac. *Plasma Physics and Controlled Fusion*, 48, S17–29, 2006.

73. A.M. Balk, V.E. Zakharov, and S.V. Nazarenko. Nonlocal turbulence of drift waves. *Journal of Experimental and Theoretical Physics*, 71(2), 249–60, 1990.

74. M. G. Shats and W. M. Solomon. Zonal flow generation in the improved confinement mode plasma and its role in confinement bifurcations. *New Journal of Physics*, 4, 30.1–30.14, 2002.

75. K. C. Shaing. On the relation between neoclassical transport and turbulent transport. *Physics of Plasmas*, 12, 82508–1–2, 2005.

Chapter 6

The Numerical Computation of Turbulence

Javier Jiménez

School of Aeronautics
Universidad Politécnica, 28040 Madrid SPAIN
E-mail: jimenez@torroja.dmt.upm.es

This chapter deals primarily with the numerical computation of three-dimensional Navier–Stokes incompressible turbulence. As such, it leaves a lot of ground uncovered. It will be made clear early in the chapter that turbulence is not restricted to such flows, or even only to fluids, but this is the most common case in industrial applications, and most engineering computational schemes have been developed for it. Even when compressibility is important for the flow as a whole, such as in transonic flight, it does not affect turbulence strongly, and most of what is said here can be used with minor modifications. More extreme departures, such as hypersonics, plasmas, or the strong stratification of many geophysical flows, require changes that would make much of the material in this chapter irrelevant, except perhaps for the direct numerical simulations discussed in §6.5.. It would make little sense, for example, to use current RANS or LES models for two-dimensional turbulence.

This chapter has two parts. Sections §6.1. to §6.3. are a quick review of classical turbulence theory, with the goal of establishing the tools to be used later, and specially of defining the relevant turbulent length scales. The second part, formed by sections §6.4. to §6.7., describe the three levels of detail at which turbulence is nowadays computed. It is difficult in such a limited space to do more than to familiarize the reader with the requirements of each technique, and with the results that can be expected from them. The final result should at least be to allow him or her to determine which method to use in a particular situation.

A subject such as this cannot be discussed without some reference to numerical analysis, which is the tool of simulations. This is specially true in the case of turbulence, where the flow field is not smooth, and therefore not easy to represent numerically. Any reader seriously interested in computing turbulence should, of course, familiarize her or himself as much as possible with numerical techniques.

Contents

6.1.   The self-similar energy cascade

The central experimental observation on turbulent flows[1,2] is that they dissipate energy at a rate that is independent of the fluid viscosity ν, even if it is a rigorous consequence of the Navier–Stokes equations that viscosity is the only process that can dissipate energy, and that the dissipation per unit mass is

$$\varepsilon_v = \nu V^{-1} \int_V |\nabla \boldsymbol{u}|^2 \, \mathrm{d}^3 \boldsymbol{x}. \tag{6.1}$$

How this comes about was a minor mystery of physics until it was realized that energy and dissipation reside in turbulent flows at different scales, separated by a self-similar eddy cascade that carries one into the other.

This explanation was first proposed by Richardson,[3] who imagined that large turbulent eddies, generated by the external forces, become unstable and break into smaller ones that in turn break again, until they became small enough to be damped by viscosity (Figure 6.1.).

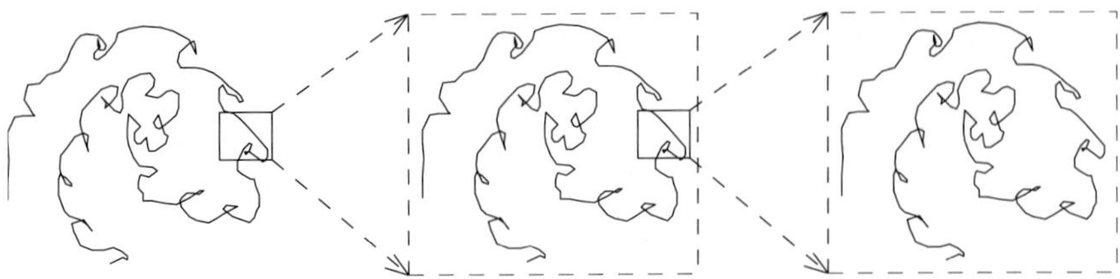

Fig. 6.1.   A self-similar cascade of eddies in a turbulent flow.

The next step was taken by Kolmogorov,[4] who introduced the concept of an energy cascade, and computed the scaling exponent of the self-similar 'fractal' cascade process. He argued that the role of the break-up is to transfer energy from the larger eddies, where it can not be dissipated, to the smaller ones where it can. Let us label an 'eddy' by a velocity difference u_ℓ across a distance ℓ. We can for example use the r.m.s. velocity difference $u_\ell = \langle (u(\boldsymbol{x}+\boldsymbol{\ell}) - u(\boldsymbol{x}))^2 \rangle^{1/2}$, where the average $\langle \rangle$ is taken over some statistically homogeneous set of points $\boldsymbol{x}$. The precise definition has to take into account the direction of the velocity component being considered with respect to the separation vector ℓ (see Ref.[5]), but the argument depends only on scaling similarity, which is essentially the same as dimensional analysis, and is independent of the details of how the velocities are measured.

There are two time scales for the evolution of the eddy. In the 'inertial' time $T_\ell = \ell/u_\ell$ the eddy deforms so much that we should have, for example, to divide it into two individual smaller pieces if we wished to continue studying it. In the 'viscous' time $T_\nu = \ell^2/\nu$ viscosity smoothes the velocity differences. Because of equation (6.1), the energy of the eddy can only be dissipated in the viscous time scale. The ratio $T_\nu/T_\ell = \ell u_\ell/\nu$ is the Reynolds number Re_ℓ. If it is large, $T_\nu \gg T_\ell$, the eddy deforms before losing its energy. This is the situation in the first steps of the Kolmogorov cascade. Energy is injected into the largest eddies, and it is transferred from one scale to the next without dissipation. But the eddies become smaller with every cascade step and so do their Reynolds numbers. Eventually, the energy reaches an eddy size for which $Re_\ell \approx 1$, and viscosity cannot be neglected. The hypothesis of Kolmogorov is that energy is predominantly transferred from larger to smaller eddies and that the process is self-similar and uniformly distributed in space. It then follows that the flux of energy is independent of the eddy size at which it is measured.

Those assumptions are enough to determine the scaling exponent for the velocity fluctuations. The kinetic energy per unit mass of the eddies of size ℓ is proportional to u_ℓ^2, and it is transferred to the next cascade stage in the deformation time T_ℓ. The rate of energy transfer is proportional to

$$\varepsilon \sim u_\ell^2/T_\ell = u_\ell^3/\ell. \tag{6.2}$$

For equilibrium flows, this quantity is the same as the rate of energy dissipation in (6.1), since the energy cannot reach the viscous scales unless it is transported there by the inviscid cascade. Equation (6.2) then leads directly to a power law for the velocity increments,

$$u_\ell = C_1(\varepsilon\ell)^{1/3}, \tag{6.3}$$

where $C_1 \approx 1.4$ is a universal empirical constant. This relation also allows us to estimate the order of magnitude of the inner 'cutoff' scale for the self-similar range, which is reached when Re_ℓ becomes of order unity,

$$u_\ell \ell/\nu \approx 1, \qquad \ell \approx \eta = (\nu^3/\varepsilon)^{1/4}. \tag{6.4}$$

The velocity difference across these 'Kolmogorov' eddies is

$$u_K = (\varepsilon\eta)^{1/3} = (\nu\varepsilon)^{1/4}. \tag{6.5}$$

Equation (6.3) implies that the velocity 'gradient' $u_\ell/\ell \sim \ell^{-2/3}$ increases as ℓ decreases. As long as $\ell > \eta$, there is no limit that can be interpreted as a derivative. Only for distances shorter that the Kolmogorov scale can the flow be considered smooth, and it is only then that the classical tools of analysis, or of numerical analysis, can be used. As the viscosity is made to vanish, the Kolmogorov scale tends to zero, and the velocity becomes singular.

The functional forms of the previous equations can be derived more easily if we accept the experimental fact that the energy dissipation rate ε is independent of viscosity. Consider for example the velocity difference u_ℓ across a distance ℓ that is small enough for the overall dimensions of the system to be irrelevant, but large enough for viscosity to be unimportant. The details of the large eddies cannot influence the behaviour at scale ℓ, and the only magnitudes 'known' to the eddy are ε and ℓ. The only dimensionless group is $\varepsilon\ell/u_\ell^3$, which cannot depend on anything, and therefore has to be a universal constant. This leads directly to (6.3). Furthermore, if we want to estimate the distance where viscosity becomes important, and we assume that it can only depend on ε and on ν, the only dimensionally correct combination is (6.4).

In general we can only say that, for eddies that are small enough for the effect of the large-scale forcing to be negligible,

$$u_\ell/u_K = F(\ell/\eta), \tag{6.6}$$

where F is some function that tends to $(\ell/\eta)^{1/3}$ when $\ell/\eta \gg 1$. Equation (6.3) cannot however hold for arbitrarily large separations, since u_ℓ would grow indefinitely for larger eddies, and it would eventually become larger than the available kinetic energy $K \equiv u'^2/2$ per unit mass. The upper limit for the validity of (6.3) is then given by $u_\ell \approx u'$, or

$$\ell \approx L_\varepsilon = \frac{u'^3}{\varepsilon}, \tag{6.7}$$

which defines the 'integral' scale L_ε of the turbulence. It follows from the previous arguments that the kinetic energy of the flow resides at scales $O(L_\varepsilon)$, while the energy is dissipated primarily at scales $O(\eta)$.

It is only for $\eta \ll \ell \ll L_\varepsilon$ that the self-similar relation (6.3) holds, and it is only in that range that the flow is a 'fractal'. Because the viscous forces are not important, those scales are usually called 'inertial'. The extent of the inertial range can be derived from the previous relations as

$$\frac{L_\varepsilon}{\eta} = \left(\frac{u' L_\varepsilon}{\nu}\right)^{3/4} = Re_L^{3/4}, \tag{6.8}$$

and can be expressed as a power of the turbulent Reynolds number Re_L. It is customary in isotropic turbulence to write this relation in terms of a different *microscale* Reynolds number, defined for historical reasons as

$$Re_\lambda = (15 Re_L)^{1/2}. \tag{6.9}$$

6.1.1. *Spectra*

The velocity distribution among eddies of difference sizes can also be expressed in terms of spectra. Briefly, the wavenumber associated to a length scale ℓ is $\kappa = 2\pi/\ell$, and the velocity components u_j are expressed as Fourier series or integrals. For a single spatial coordinate we can for example define

$$u(x) = \int_{-\infty}^{\infty} \widehat{u}(\kappa) \exp(i\kappa x)\, d\kappa. \tag{6.10}$$

For functions of the three spatial coordinates, (6.10) is trivially generalized by substituting κ by a three-dimensional wave-vector with components κ_n, and the argument of the exponential by the scalar product, $i\kappa_n x_n$. The Fourier representation is only useful for spatially homogeneous fields, since the exponentials on which it is based are invariant to translations. If a flow is statistically homogeneous in some directions but not in others, it makes sense to use mixed expansions which are only Fourier-like over some coordinates.

The importance of Fourier expansions in signal representation is due to the simple way in which they express the energy in a given range of scales. The definition (6.10) is not the most useful one for turbulent quantities, which are not usually square-integrable over infinite intervals, but it can be modified to cover that case (see §2 of Ref.[5]). The orthogonality of the Fourier basis functions then allows us to write Parseval's theorem,

$$\langle u(x)\, v(x)^* \rangle = \int_{-\infty}^{\infty} \widehat{u}(\kappa)\, \widehat{v}^*(\kappa)\, d\kappa, \tag{6.11}$$

linking the mean value of products in physical space to integrals of the Fourier coefficients. If u is real and $u = v$, the mean-square velocity can be expressed either as the mean value over physical space of the square of the velocity, or as the sum of the squares of the absolute magnitudes of the Fourier transform. Note that this is just a generalization of Pythagoras' theorem for right-angled triangles.

It is then natural to interpret $|\widehat{u}(\kappa)|^2$ as an energy spectrum, proportional to the energy associated with the wavenumber κ. For real variables the energy spectrum is symmetric with respect to $\kappa = 0$, and is traditionally defined only for $\kappa \geq 0$. The proportionality constant in its definition is adjusted so that its integral is the mean-square value of the velocity. For the one-dimensional spectrum of u_1 along x_1,

$$\langle u_1^2 \rangle = \int_0^\infty E_{11}(\kappa_1)\,\mathrm{d}\kappa_1. \tag{6.12}$$

Because u_ℓ, as defined above, is roughly the contribution to the energy of the eddies up to size ℓ, the spectrum and the second order 'structure function' u_ℓ^2 contain identical information. In particular (6.3) is equivalent to

$$E_{11}(\kappa_1) = C_2 \varepsilon^{2/3} \kappa_1^{-5/3}, \quad \text{with} \quad C_2/C_1^2 \approx 0.248854. \tag{6.13}$$

As in the case of u_ℓ, it is easy to show that (6.13) is the only dimensionally consistent form for the spectrum if the only available variables are ε and κ.

Power-law spectra, and structure functions for which u_ℓ is not proportional to ℓ, indicate non-smoothness. Neither of them can extend to all length scales. If all the derivatives of a function are continuous, its spectrum decays at high wavenumbers faster than any power of κ. For wavelengths much shorter than the Kolmogorov dissipation scale, $\kappa\eta \gg 1$, viscosity smooths the velocity, and the spectrum decays at least exponentially.

Equations (6.3) and (6.13) are two of the few real analytic predictions of turbulence theory. They were proposed by Kolmogorov[4] before adequate data were available, and only later tested experimentally. Although they are known to be only lowest-order approximations,[6] they fit experiments exceedingly well (see Fig. 6.2.).

Because spectra contain statistical information about the relative behaviour of the velocity at two different points, they can be manipulated to obtain statistical one-point moments for the derivatives. For example, since differentiating (6.10) with respect to x brings down a factor $\mathrm{i}\kappa$ to each Fourier component, the power spectrum of the velocity derivative $\partial_1 u_1 = \partial u_1/\partial x_1$ is proportional to $\kappa_1^2 E_{11}(\kappa_1)$. In fact we can define for statistically isotropic flows a one-dimensional dissipation spectrum such that

$$\varepsilon_v = \nu \int_0^\infty G_{11}(\kappa)\,\mathrm{d}\kappa_1, \tag{6.14}$$

and express it in terms of the energy spectrum as $G_{11} = 15\kappa_1^2 E_{11}(\kappa_1)$. The numerical coefficient comes from the effect of the different components of the velocity gradient tensor, under the assumption of isotropy.[5]

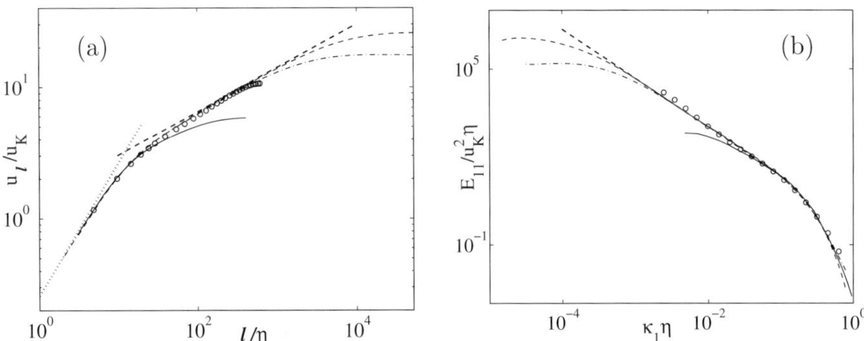

Fig. 6.2. (a) R.m.s. longitudinal velocity differences, and (b) one-dimensional spectra of the longitudinal velocities for several turbulent flows, displayed in Kolmogorov scaling. The dashed straight lines are (6.3) and (6.13) with $C_1 = 1.4$. The dotted line to the left of (a) is $u_\ell \sim \ell$, which characterizes differentiable flows. Longer lines correspond to higher Reynolds numbers.

Note that, while $E_{11}(\kappa) \sim \kappa_1^{-5/3}$ decays with increasing wavenumbers, $G_{11} \sim \kappa_1^{1/3}$ increases, confirming our previous conclusion that energy resides in the larger scales, and dissipation in the smaller ones.

The reason for spectral power laws such as (6.13) is that the inviscid flow equations are invariant to multiplicative scalings, and that the important relations are those between a given length or velocity scale and their multiples, rather than between scales which differ by a fixed amount. It is for this reason that spectra and structure functions are often plotted in logarithmic coordinates, such as in figure 6.2.. The logarithm converts scaling factors to increments along the axes, and self-similar power laws are displayed as straight lines.

In doing so, the spectrum loses one of its useful graphic properties, which is to represent energies by integrals or by areas. To remedy that, it is sometimes useful to use semilogarithmic plots of the premultiplied spectrum, $\kappa_1 E_{11}(\kappa_1)$ versus $\log \kappa_1$. The extra factor in front of the spectrum compensates for the differential of the logarithm, and the integral property is restored,

$$\langle u_1^2 \rangle = \int_0^\infty \kappa_1 E_{11}(\kappa_1)\, \mathrm{d}(\log \kappa_1). \qquad (6.15)$$

The areas underneath premultiplied spectra corresponds to energies, and spectral peaks show where the energy is concentrated.

An illustration is given in figure 6.3., which displays premultiplied spectra of the velocity u_1 and of the velocity gradient $\partial_1 u_1$ for a variety of Reynolds numbers and flow conditions. As expected from the previous

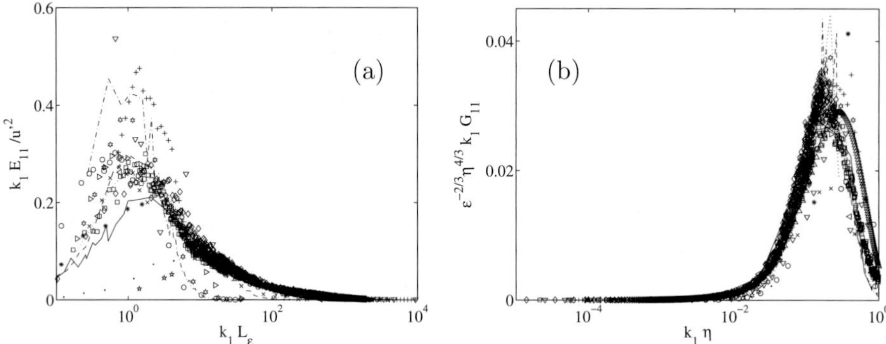

Fig. 6.3. Premultiplied one-dimensional spectra for several turbulent flows at different Reynolds numbers. (a) Streamwise velocity. The wavenumber is normalized with the integral scale. (b) Velocity derivative, normalized with the Kolmogorov scale.

discussion, the former collapse around the integral length L_ε, while the latter collapse in Kolmogorov units. Note that the gradients, which are responsible for the dissipation, peak around $\kappa_1 \eta \approx 0.1$, which corresponds to $\ell/\eta \approx 50$. In figure 6.2. the structure function also becomes regular at $\ell/\eta \approx 10$, rather than at η itself. Scaling arguments only provide orders of magnitudes, and it is not always necessary to resolve turbulence down to the level of η to study it. In particular, a resolution of several Kolmogorov scales is often enough in numerical simulations, depending on the numerical scheme.

While the spectra of energy and of dissipation collapse in their respective scalings, it is impossible to collapse both at the same time for different Reynolds numbers. We saw in (6.8) that the main effect of the Reynolds number is precisely to increase the ratio L_ε/η, but it is important to emphasize that the total energy transfer rate ε is determined by large-scale quantities, such as u' and L_ε in (6.7), which are fixed by such factors as the velocity and the size of the body creating the turbulence. Decreasing the viscosity, and therefore increasing Re_L, broadens the inertial range through which the energy has to be transferred, but it does not change ε.

Note that, if we consider each smooth viscous eddy as an object, (6.8) implies that the number of independent degrees of freedom in a volume L_ε^3 is

$$N_T \sim Re_L^{9/4} \sim Re_\lambda^{9/2}. \tag{6.16}$$

Fully turbulent flows require Re_λ to be larger than approximately 100; the instabilities needed for the cascade do not develop below $Re_\lambda \approx 30$. The highest Reynolds numbers, $Re_\lambda \approx 10^4$, have been measured in the atmospheric boundary layer, although indications of higher ones have been found in astrophysics. Typical industrial flows have $Re_\lambda \approx 100 - 1,000$. The Reynolds number in the wake of a running person is $Re_\lambda \approx 500$ and, in the boundary layer of a large commercial aircraft, $Re_\lambda \approx 3,000$. These are all large numbers, showing that most 'practical' flows are turbulent. They imply that the number of degrees of freedom in turbulent flows is typically $N_T = 10^9 - 10^{20}$, and tends to infinite in the high-Reynolds number limit.

Systems in which the number of degrees of freedom is finite are in some sense trivially solvable, since it is possible to write a computer program to simulate them as exactly as required. If the number is very large, that may be impractical, and a theoretical estimate might be more useful, but computers improve rapidly, and presently impractical simulations may become reasonable in the future. Problems like turbulence, in which the number of degrees of freedom can be made astronomical large by changing the parameters, cannot be fully solved in this way, and it is for them that genuine theories are required. The numerical techniques described later in this chapter can be used to analyze turbulent flows at finite Reynolds numbers, but they can at most suggest the asymptotic behaviour of fully-developed turbulence in the limit $Re_L \gg 1$.

It is clear from figure 6.3. that the quality of the collapse of the two spectral peaks is not the same. Even if it is usually easier to measure large scales than small ones, it is the latter that collapse best. The reason is that large eddies are close to the different energy injection mechanisms of the different flows, and that they therefore change from one flow to another. It is only after the eddies are small enough to be separated from the injection scales by several instability steps, $\kappa L_\varepsilon \gg 1$, that we can hope for all turbulent flows to be more or less similar.

6.2. Other fractal processes in physics

The previous discussion would be interesting even if it only applied to turbulence in fluids, since most macroscopic flows in the natural sciences are turbulent, but the appearance of a non-analytic range of scales is not an exclusive property of the Navier–Stokes equations. Consider the following interpretation of transition to turbulence. Assume that we inject energy into a fluid by stirring it at velocity U with a paddle of size L. The force on the paddle is $F = O(\rho U^2 L^2)$, and the energy transferred to the fluid by unit time and mass is $\varepsilon = FU/\rho L^3 = O(U^3/L)$. If we now assume that the flow is smooth, the velocity gradients are $O(U/L)$, and the energy that the viscous forces can dissipate is, according to equation (6.1), $\varepsilon_\nu = O(\nu U^2/L^2)$. The ratio between the two, $\varepsilon/\varepsilon_\nu = UL/\nu$, is the Reynolds number and, if it is large, viscosity cannot directly get rid on the energy that is being injected. The transition to turbulence is the response of the system to this impossibility by creating gradients that allow viscosity to do its job.

There are many other examples of fractalization in strongly forced systems. Perhaps the most familiar is the fracture of solids. When we inject an energy E into a brittle solid of size L, for example by hitting it, most of the energy goes into elastic deformation waves, and is eventually dissipated into heat. As in the case of viscosity, elastic deformation has its limits and, if enough energy is injected, molecular bonds break, and the body fractures. Breaking bonds creates new surface area, and the energy absorbed in that way is $\sigma \Delta A$, where σ is the surface tension and ΔA is the area that has been created. If the body breaks into n pieces its volume is conserved, and the diameter of each piece is $L_n = Ln^{-1/3}$. The total surface of the fragments is $nL_n^2 = L^2 n^{1/3}$. Equating the surface energy to the forcing we can estimate that $n \approx (E/\sigma L^2)^3$. The size of the final pieces is then $L_n = L^3 \sigma / E$.

In the limit of infinite energy the body is reduced to dust, and its area becomes infinite. The process is similar to that in turbulent flows. Weak forcing can be dissipated by smooth deformations but, if the forcing is increased beyond the point in which this ceases to be possible, the system responds by fractalizing, and by creating non-analytic mechanisms to accommodate the higher input. In this sense flows become turbulent because, beyond a certain energy input, they 'break'.

Fractalization is not necessarily destructive. Living beings need oxygen to generate energy and, in the absence of other mechanism, obtain it by diffusion through their surfaces. As they become larger, their oxygen requirements increase approximately as their volume, while the diffusion flux, which is proportional to the area and to the concentration gradient, increases only as their linear dimensions. It can be shown that this would limit the size of living beings to at most a few millimetres. Larger ones, like ourselves, have developed fractal exchange mechanisms, like vascular circulation, branching lungs and the leaves of trees, to handle the diffusion of gases. It may be said that the reason why fractals, geometric or otherwise, are prevalent in nature is the need to handle solicitations that cannot be managed smoothly.

6.3. Inhomogeneity and anisotropy

We have implicitly assumed up to now that the flow is statistically homogeneous and, by only considering one-dimensional quantities, that all spatial directions are equivalent. In fact, few turbulent flows are strictly isotropic or homogeneous, and most technologically important flows are neither. In isotropic turbulence the kinetic energies $\langle u_i^2 \rangle$ of the three velocity components are identical, and the cross products $\langle u_i u_j \rangle$ vanish. In other situations, those cross products play an important role in turbulence theory. Consider a turbulent velocity field, and separate at each point the mean value and the fluctuation,

$$u_j(\boldsymbol{x},\, t) = U_j(\boldsymbol{x}) + \widetilde{u}_j(\boldsymbol{x},\, t), \tag{6.17}$$

so that $\langle \widetilde{u}_j \rangle = 0$. We can average the Navier-Stokes equations to obtain the evolution of the mean velocities,

$$\partial_t U_i + \partial_j U_i U_j + \rho^{-1}\partial_i P = \partial_j \left[\nu \partial_j U_i - \langle \widetilde{u}_i \widetilde{u}_j \rangle \right], \tag{6.18}$$

$$\partial_j U_j = 0. \tag{6.19}$$

Repeated indices will from now on imply summation. In the first of these equations the unknown symmetric tensor

$$\tau_{ij} = -\langle \widetilde{u}_i \widetilde{u}_j \rangle, \tag{6.20}$$

represents the momentum flux due to the turbulent fluctuations, and plays a role comparable to the viscous stresses. It is typically much larger than the latter. If we assume from their definitions that $\tau_{ij} = O(u'^2)$ and that $\partial_j U_i = O(u'/L_\varepsilon)$, it follows that

$$\frac{|\tau_{ij}|}{|\nu \partial_j U_i|} = O(Re_L). \tag{6.21}$$

Estimating those 'Reynolds' stresses is the main problem in the practical computation of turbulent flows.

6.3.1. *The energy equation*

The role of the Reynolds stresses is seen most clearly in shear-driven flows. It was in these flows that turbulence was first recognized, and they have continued to be the subject of technological attention because they appear in many problems of practical interest. They are characterised by the presence of a mean velocity gradient that provides a continuous source of turbulent kinetic energy.

Consider the 'Reynolds' decomposition (6.17). By separating the energy equation for the complete flow into mean and fluctuating components, it is possible to write an evolution equation for the kinetic energy of the turbulent fluctuations, $k = \langle \widetilde{u}_j \widetilde{u}_j \rangle/2$,

$$(\partial_t + U_j \partial_j)k + \partial_j \phi_j = \Pi - \varepsilon_v. \tag{6.22}$$

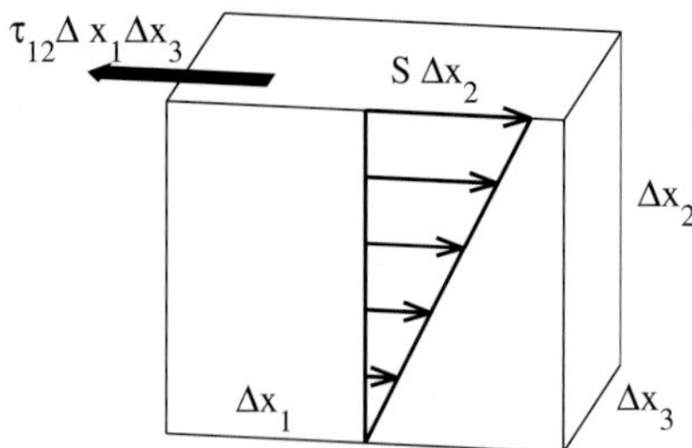

Fig. 6.4.   The production of turbulent energy in equation (6.23) can be understood by considering a fluid parallelepiped in a simple shear. If we fix our frame of reference to the bottom of the fluid element, the Reynolds stress $\mathrm{d}F = \tau_{12}\Delta x_1 \Delta x_3$ acts on the top surface, which is moving with a velocity $u_1 = S\Delta x_2$. The resulting work $u_1 \mathrm{d}F$, is used to power the turbulent cascade.

The left-hand side of (6.22) contains the total derivative of the energy and the divergence of an energy flux. In the right hand side we find the energy dissipation (6.1), and a new term,

$$\Pi = \tau_{ij}\partial_j U_i, \tag{6.23}$$

that feeds energy into the fluctuations, and that can be interpreted as the work of the mean flow against the Reynolds stresses (see figure 6.4.). It is this production term that is responsible for maintaining inhomogeneous shear turbulence.

It is possible to define one-dimensional 'cospectra' between different velocity components as

$$E_{ij}(\kappa_1) \sim \widehat{u}_i\widehat{u}_j^* + \widehat{u}_j\widehat{u}_i^*, \quad \text{such that} \quad \langle u_i u_j\rangle = \int_0^\infty E_{ij}\,\mathrm{d}\kappa_1. \tag{6.24}$$

They describe how the Reynolds stresses are distributed among the different length scales, and their general form can be estimated using dimensional arguments similar to those used for E_{11}. Consider, for example, the anisotropy induced by a simple shear $S = \partial U_1/\partial x_2$. From symmetry considerations, all the off-diagonal Reynolds stresses vanish except for $\tau_{12} = \tau_{21}$. The latter also vanishes for $S = 0$, and changes sign with S. If the shear is not too strong we may therefore assume that the cospectrum E_{12} is linear in S and, in the self-similar inertial range, we can also assume that the only other relevant parameters are κ and ε. The only possible combination is, on dimensional grounds,

$$E_{12}(\kappa_1) = -C_0 S \varepsilon^{1/3} \kappa_1^{-7/3}. \tag{6.25}$$

The experimental value of C_0 is about 0.15.

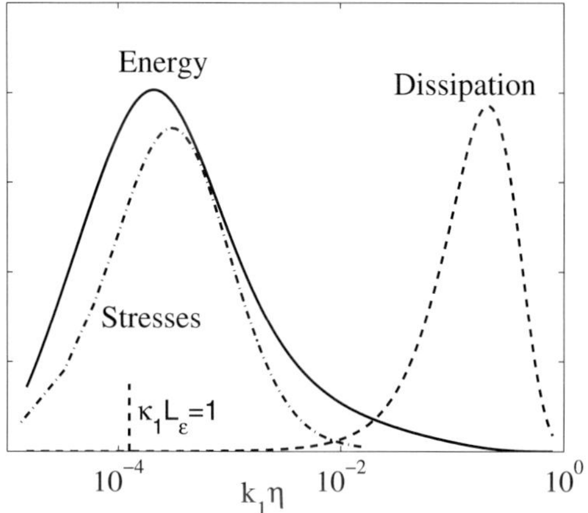

Fig. 6.5. One-dimensional premultiplied spectra of: ———— , $\kappa_1 E_{11}(\kappa_1)$. ———— , $\kappa_1 G_{11}(\kappa_1)$; —·— , $\kappa_1 E_{12}(\kappa_1)$. Arbitrary units. The flow is a boundary layer. The vertical dashed line marks the location of the integral scale.

This implies that the Reynolds shear stresses are large-scale quantities, and that their cospectra decay at high wavenumbers faster than the energy spectrum (6.13). The 'Corrsin' scale ℓ_c is defined by the wavenumber for which $|E_{12}| \approx E_{11}$, and marks the limit between the larger anisotropic scales that are influenced by the shear, and the smaller ones that are not. It follows from (6.13) and (6.25) that

$$\kappa_c = \ell_c^{-1} = (S^3/\varepsilon)^{1/2}. \tag{6.26}$$

If we further assume that the flow is approximately in equilibrium, so that $\Pi = \tau_{12}S \approx \varepsilon$, we can eliminate S from (6.26) to obtain $\ell_c \approx \tau_{12}^{3/2}/\varepsilon$. It follows from (6.20) that $\tau_{12} = O(u'^2)$, and we finally obtain that $\ell_c \approx L_\varepsilon$. The isotropic, homogeneous, universal theory is therefore a good model for eddies smaller than some fixed fraction – of the order of $1/10$ – of the integral scale, but larger eddies require a knowledge of the driving mechanism.

The general classification of turbulent scales in shear flow is therefore as in figure 6.5.. The large-scale peak near L_ε contains the energy, as well as the Reynolds stresses that generate it from the mean shear. The small scales near η contain the dissipation. In the intermediate inertial range neither of them is important, but there is an spectral energy flux, not represented in the figure, that takes the energy from one peak to the other.

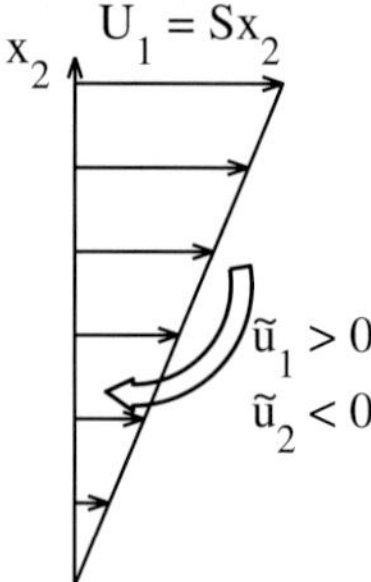

Fig. 6.6.   The Reynolds stress due to momentum exchange in a simple shear is usually of the same sign as the shear, and the corresponding eddy viscosity is positive.

6.3.2.   *The eddy viscosity approximation*

The assumption that the cospectrum is proportional to the mean shear seems to imply, by integration, that the Reynolds stresses should also be so. This is not strictly correct, because the linearity assumption is only valid for those eddies, smaller than the Corrsin scale, for whom the effect of the shear is weak. It is less well founded for the larger scales that dominate the Reynolds stresses. Nevertheless this linear approximation was one of the first to be tried in computing turbulent flows, and is still the basis of most of the engineering computational methods. It is expressed in the tensor form as

$$\tau_{ij} - \frac{1}{3}(\tau_{ii})\delta_{ij} = 2\nu_\varepsilon S_{ij}, \quad \text{where} \quad S_{ij} = \frac{1}{2}(\partial_i U_j + \partial_j U_i), \tag{6.27}$$

which preserves both the symmetry of the stress tensor and the incompressibility of the velocity. The *eddy viscosity* ν_ε is in general unknown, and it is not expected to be constant throughout the flow. It is nevertheless usually positive, which is important for the stability of the flow. Consider in fact a situation such as the one in figure 6.6., where $S > 0$, and consider a fluid particle moving downwards across the shear, $\tilde{u}_2 < 0$. Whenever it dumps its momentum in a lower layer, its velocity would in general be higher than that of the mean stream, because it comes from a faster layer, so that $\tilde{u}_1 > 0$. The result is that in the mean $\tau_{12} = -\langle \tilde{u}_1 \tilde{u}_2 \rangle > 0$. It is easy to see that the same is true for upwards-moving particles, and that signs change with the sign of the shear.

We will see shortly some ways of estimating ν_ε. First it is important to remark that, even if its magnitude could be adjusted in an optimum way, the assumption (6.27) is very strong, implying both the linear dependence of two tensors, and that they are coaxial. None of those assumptions are justified, and there are many examples in which they do not work, even to the point that the Reynolds stresses and the turbulent diffusion fluxes move 'countergradient', implying negative eddy viscosities. But there are also many examples in which the results are good, including some in which there are few theoretical reasons to expect it. Eddy viscosity, at least in situations close to equilibrium, is probably the first approximation that should be tested in estimating the behaviour of a turbulent flow. But it should not be used uncritically, and it is only quantitatively useful when there are experimental observations for a similar range of parameters that can be used to guide the choice of the empirical constants that are typically present in such models.

We may note at this stage that, even if turbulent flows dissipate energy more efficiently than laminar ones, turbulent dissipation is still a comparatively weak phenomenon. The basic inertial relation (6.3) can be recast as a relation between the kinetic energy of the transverse velocity fluctuations and a local time scale,

$$\frac{u_\ell^2}{\varepsilon} \approx C_1^3 \frac{\ell}{u_\ell} \approx 10 \frac{\ell}{u_\ell}, \tag{6.28}$$

where the coefficient in the rightmost part of (6.28) comes from three-dimensional considerations. The left-hand side of this equation is the time that it would take for the dissipation to damp an eddy, and it is proportional, in its right-hand side, to the eddy turnover time. This proportionality agrees with our assumptions about the inertial range, but the coefficient is large, and the result is that an eddy can be sheared by a relatively large amount, $10\,\ell$, before the subgrid dissipation has time to damp it.

Energy considerations such as these can be used to derive a rough estimate for the eddy viscosity. If we use the energy production (6.23) as a surrogate for the dissipation ε, we can express it as

$$\varepsilon \approx \Pi = \tau_{ij} S_{ij} = 2\nu_\varepsilon S_{ij} S_{ij}. \tag{6.29}$$

If we then estimate the rate of strain as $S = u'/L_\varepsilon$, (6.28) becomes

$$\nu_\varepsilon \approx 0.1\, L_\varepsilon u'. \tag{6.30}$$

which can be written as

$$u' L_\varepsilon / \nu_\varepsilon \approx 10. \tag{6.31}$$

This is the basis for a rule of thumb, sometimes used in rough engineering estimates, that the effective Reynolds number of *any* turbulent flow, based on its eddy viscosity, is $\Delta U L/\nu_\varepsilon \approx 10 - 50$. The present derivation shows that the rule is a lowest-order approximation, still coarser and even less justified than the eddy viscosity assumption, but nevertheless useful for *preliminary* evaluations.

The expression in (6.30) for the eddy viscosity as the product of a velocity and a length scale could have been anticipated on dimensional grounds. The energy argument identifies the two scales as u' and L_ε, which are consistent with our previous conclusion that the Reynolds stresses are large-scale quantities. Other approximate formulas express ν_ε as the product of an energy and a time scale (such as u'^2 and L_ε/u'). Which combination is chosen in a given case depends on which flow variables are considered to be best known. We will see several examples in the second part of this chapter.

Note that, for the same reasons as in (6.21), the ratio between ν_ε and the molecular viscosity is the (molecular) Reynolds number, and the molecular viscosity is always negligible. This is the main property of turbulence from the applications point of view, and extends to the diffusion of scalar contaminants. One example might help to bring it into focus. The time L^2/κ that the molecular diffusivity κ needs to mix a drop of milk into a cup of coffee is about one day, while we are all familiar with how much faster is mixing with even moderate levels of stirring.

6.4.   Computing turbulence

Although turbulence is a complex phenomenon, it is governed by the deterministic Navier-Stokes equations. Given initial and boundary conditions, there should be no difficulty in computing numerically any particular turbulent flow, but we shall see that this is only true in an abstract sense. The number of degrees of freedom is usually so large that the cost of direct computations makes that approach impractical. It is also not always necessary to know everything about a particular flow realization. It is often more useful to consider it as a random sample from a set of equivalent flows, of which we only need statistical information. A large part of the art of turbulence computation is how to *approximate* some aspects of the flow using limited resources, rather than how to compute it in detail.

Computing does not automatically bring understanding, but there are fundamental reasons why being able to compute something is important. In the first place, quantifying a theory is a prerequisite for making predictions. The main difference between science and philosophy is that, while the latter is generally content with giving a view of the World which is intellectually coherent and personally satisfying, the former insists on testable theories. The core of the scientific method is to develop theories, derive predictions, and *test* them. The last step requires estimating what happens when we manipulate the system, and it is most convincing when it is done quantitatively. Until then, a theory cannot be accepted as science.

In the first part of this chapter we have outlined the currently accepted 'philosophical' model of three-dimensional Navier-Stokes turbulence, which is coherent and qualitatively explains the observations. This is a necessary first step, but we cannot be content until we are able to compute what would happen if we changed the flow in ways that are different from those of the original experiments.

This has traditionally been a problem in complex systems. Classical mechanics initially concerned itself with analytic solutions. The mathematics of the motion of a rigid body are within the reach of pencil and paper calculations. We can find closed-form solutions for the motion of a cannonball, and use them to predict the effect of changing the shooting angle. We can use the confidence gained in that way to send a spacecraft to Mars, but even that example soon gets complicated. Without including the perturbations due to the Moon and Jupiter we would miss Mars by an impractical margin. Very simple systems, formed by elements which are perfectly understood, can be too complicated for analytic solutions. The rules of chess are few and clear, but the game is interesting precisely because we cannot deduce analytically from them an optimal strategy.

Such complex systems used to be considered games, or at most engineering questions to be treated with empirical rules of thumb. They were not considered proper subjects for science, precisely because no quantitative predictions could be done.

Turbulent flows used to be in that category, and a lot of effort was spent looking for theoretical results that could be cast in closed form. Even today very few are known. This failure frustrated some of those working in the field, and convinced them that there was something special, and even 'mysterious', about turbulence that had to be cracked before the problem could be considered understood.

Although analytic work will always have an important place in scientific study, the availability of ever more powerful computers has changed the classical view of what it means to understand something. Complex systems can now be simulated numerically, essentially exactly, and their theories can be tested even when no analytic results are known. It has always been possible to test turbulence theories by checking experimentally whether the effect of changing a parameter agrees with a given model. Most of the concepts of classical turbulent theory were confirmed in that way. But computers allow us to do more. We can now change the constitutive equations, and quantitatively compute the consequences, and in that way check whether our understanding of a particular term is correct or not.

This of course does not mean that there is no role left for turbulence theory. We have already mentioned that, although numerical experiments are possible, many are impractical. The latter include most cases of industrial interest, and simple theories are crucial in providing a bridge between what we can, and what we want to compute.

This leads us to the technological reasons for computing turbulence. Turbulent flows are extraordinarily common, and differ qualitatively from laminar ones. We would not be able to fly aeroplanes, or to build efficient power plants, without understanding to a good approximation what turbulence does. Turbulence theories, embodied in the form of models, are the core of the approximate computational methods needed by industry, and better theories sooner of later lead to better models.

In the rest of these notes we will briefly survey the different methods currently used for computing turbulence, starting with 'exact' ones, which are for the moment mainly interesting for testing scientific theories, and moving towards the approximations which have to be made in industrial applications.

6.5.   Direct numerical simulations

We have seen above that one of the defining characteristics of high-Reynolds number turbulence is that the kinetic energy and the Reynolds stresses are associated with length scales that are much larger than those responsible for the energy dissipation, and that the two sets of scales are linked by a quasi-equilibrium inertial range that is isotropic and universal. The different strategies for computing turbulent flows differ in which scales they compute explicitly and which ones they model.

At one end of the spectrum of computational methods are direct numerical simulations (DNS), which explicitly compute everything up to, and including, the energy dissipating scales. The velocity field is smooth at those sizes, and standard numerical analysis can be used. For a low-resolution numerical scheme the smallest relevant details of the functions that are being computed should have no less than five or six grid points and, since we have seen that the turbulent velocities are only smooth for distances shorter than approximately 10η, the grid spacing of a simulation using those schemes should not be much larger than the Kolmogorov scale. Higher-resolution numerical schemes allow somewhat coarser grids, but all numerical differentiation formulas fail at distances shorter than two grid points, implying that the flow field should at least be smooth at those scales. When the grid spacing is finer than those minimum requirements the accuracy of good numerical schemes increases rapidly, and the results of the simulations differ from the analytic ones by negligible amounts.

Note on the other hand that it would make no sense to use discrete approximations with grid spacings in the inertial range because, as already noted in §6.1., the numerical estimate $\Delta u/\Delta x$ for the derivative does not converge to a constant limit until $\Delta x \approx \eta$, at which point we are back into the sub-Kolmogorov DNS range.

With these precautions, the quality of DNS is only limited by how much we are willing to spend in resolution, domain size, and running time to collect statistics. When they can be obtained, the results of DNS are indistinguishable from laboratory experiments, and the scatter among careful simulations is generally smaller than among comparable experiments.

Unfortunately their price is high. Since the ratio between the integral and the dissipation lengths is $O(Re_L^{3/4})$, the number of grid points needed to simulate a cube whose size is of the order of the integral scale is

$$N_g \sim (L_\varepsilon/\eta)^3 = O(Re_L^{9/4}) = O(Re_\lambda^{9/2}), \tag{6.32}$$

that determines the amount of computer storage that should be used. Direct simulations are the numerical equivalent of experiments. The flow fields are unsteady, as in real flows, and the compilation of statistics requires that they should be run at least for a few turnover times, L_ε/u', while considerations of numerical accuracy limit the time step to be shorter than the time it takes for a fluid particle to cross one grid element, $\Delta x/u'$. The number of time steps needed is then of the order of $N_T \sim L_\varepsilon/\Delta x = O(Re_L^{3/4})$, and the total number of operations is

$$N_g N_T = O(Re_L^3) = O(Re_\lambda^6). \tag{6.33}$$

Even for low Reynolds numbers, $Re_\lambda \approx 100$, this means 10^{12} operations, and memory requirements of 10^9 variables. Other considerations, such as the need for several variables per grid point and of several operations per variable, add one or two orders of magnitude.

Modern tabletop computers process 10^9 floating-point operations per second (one Gigaflop/s), and one hundred Teraflop/s (10^{14}) are possible in demonstration machines. Petaflop/s (10^{15}) commercial machines are expected by the year 2010. Computer memories, measured in bytes, have traditionally been of the same order as the computing speed in flops. Simple direct simulations, with grids of several million points, run now in a few hours on commercial computers, and in a few seconds on prototypes. They are expected to run in milliseconds in twenty years. These are reasonable times, and DNS is a very useful tool for studying simple flows at low or moderate Reynolds numbers.

Flows that have been computed directly include isotropic turbulence up to $Re_\lambda \approx 1200$, turbulent pipes and channels up to $Re_\tau \approx 2000$, and mixing and boundary layers up to similar Reynolds numbers. Recent reviews of what is possible are.[7,8] Figure 6.7. summarizes the recent historical evolution of computer speed, and how it has been used for the direct simulation of a particular class of turbulent flows.

 Javier Jiménez

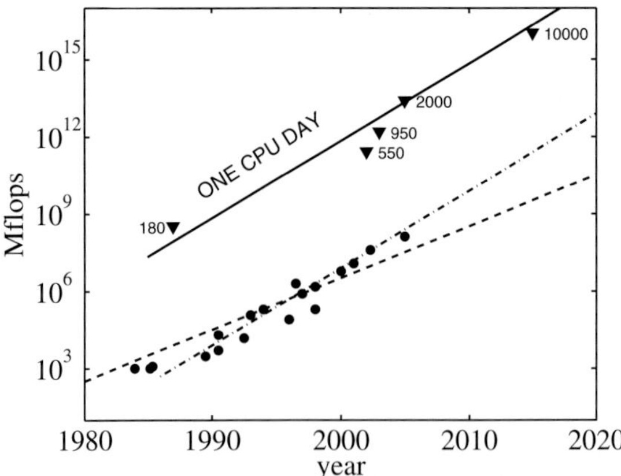

Fig. 6.7. Computational cost of selected simulations of turbulent channel flows. The circles are the computational speed (in flops) of the largest computers available in a given year. The two discontinuous straight lines correspond to the single-processor doubling of speed every 18 months, resulting in a factor of 100 per decade, and to the faster recent historical rate for parallel machines of 500 per decade. The triangles are the total cost of the simulations, in flops $\times$ s, labelled by their Re_τ, which is roughly equivalent to Re_L. The solid line is one day of dedicated time in a machine following the historical trend for parallel computers.[7] The trend suggests that a direct simulation of a channel at $Re_\tau = 10000$ could be run by the year 2015.

Simulations have several advantages over laboratory experiments. An obvious one is that, once a flow has been simulated, it is completely accessible to observation, including three-dimensional views and variables which are difficult to obtain in any other way. Even more important is the possibility of simulating 'imaginary' flows, using equations and boundary conditions which differ from the real ones in almost arbitrary aspects, and allowing us to check partial processes or hypothetical mechanisms, or to test proposed control strategies. Direct simulations have already made important contributions to turbulence research, and will undoubtedly be used increasingly as improved hardware and algorithms extend their capabilities. Many of the results that are now generally accepted, specially on the structure of turbulence, have either been derived or supported by direct numerical simulations.

They are however likely to remain limited in scope for some time. All the examples cited above refer to simple idealized geometries, and geometrical complexity has to be traded for Reynolds number. The exponent in (6.33) is high, and a small increase in Reynolds number implies a large one in operation count. Industrial turbulent flows have Reynolds numbers in the range of $Re_\lambda \approx 10^3$, and atmospheric turbulence, of 10^4. The speed of large

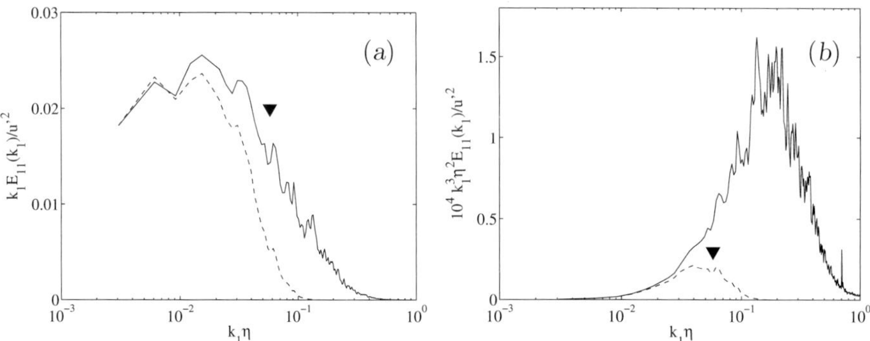

Fig. 6.8. Premultiplied spectra of a typical turbulent velocity signal (solid lines), and of its filtered version (dashed). Gaussian filter (6.36) with $\sigma = 17\eta$. (a) Velocity. (b) Velocity gradients. The arrowhead is $k_1\sigma = 1$. Note that the subgrid component has much less energy than the filtered velocity, but contains most of the gradients.

parallel computers has historically increased by a factor of 500 per decade (figure 6.7.). Assuming that the same trend continues to hold, it will take 20 years until direct numerical simulations of complete *simple* industrial flows are possible, and 50 years before we can tackle geophysical ones.

6.6. Large-eddy simulations

Fortunately, other strategies are available. The isotropic inertial range is more or less universal, and can be parameterized only by the energy transfer rate. If we can estimate that rate, and use it to model the effect of the inertial range, we should be able to avoid computing not only the dissipation scales, but all those which are approximately isotropic and in equilibrium. This is the principle of large eddy simulations (LES). The large scales are computed directly, and the dissipation scales, and most of the inertial cascade, are substituted by a 'subgrid' model.

Large eddy simulations are implemented in terms of filtered variables,

$$\overline{u}_i = \int Q(\boldsymbol{x} - \boldsymbol{y})u_i(\boldsymbol{y})\mathrm{d}\boldsymbol{y}, \qquad \text{etc.} \tag{6.34}$$

where $Q(\boldsymbol{x})$ is a smoothing kernel. Common examples are the 'box' filter

$$Q(x) = \begin{cases} 1/\sigma & \text{if} \quad |x| < \sigma/2, \\ 0 & \text{otherwise,} \end{cases} \tag{6.35}$$

and the Gaussian one,

$$Q(x) = \frac{1}{\sigma\sqrt{2\pi}}\mathrm{e}^{-x^2/2\sigma^2}. \tag{6.36}$$

These are essentially local averages, and the result is a local version of the Reynolds decomposition of the flow into a mean value and fluctuations. In Fourier space, smoothing filters damp the spectrum for wavenumbers $\kappa \gtrsim 1/\sigma$, while leaving lower wavenumbers relatively unaffected. Since we have seen that the higher wavenumbers are responsible in turbulent flows for the gradients and for the dissipation, while the energy resides in the lower ones, what the filtering operation does is to separate the turbulent velocity into a 'large-eddy' component containing the energy and the Reynolds stresses, and a 'sub-grid' component containing the dissipation (see figure 6.8.).

Another effect of smoothing is to permit the differentiation of the filtered field, which would not have been possible otherwise in the inertial range of scales. Assume that the filter commutes with differentiation, which is strictly true only when the filter width is uniform, but which is nevertheless a good approximation in most cases. Applying it to the incompressible Navier–Stokes equations, we obtain

$$\partial_t \overline{u}_i + \partial_j(\overline{u}_i \overline{u}_j) + \rho^{-1}\partial_i p^* = \partial_j(\nu \partial_j \overline{u}_i + \overline{\tau}_{ij}),$$
$$\partial_i \overline{u}_i = 0, \tag{6.37}$$

which are similar to the Reynolds averaged equations (6.18)–(6.19) with a modified subgrid stress tensor. The latter has been separated into a trace-free component

$$\overline{\tau}_{ij} = T_{ij} - (T_{kk}/3)\delta_{ij}, \quad \text{where} \quad T_{ij} = \overline{u}_i \overline{u}_j - \overline{u_i u_j}, \tag{6.38}$$

and an isotropic one that has been absorbed into a modified pressure

$$p^* = \overline{p} - \rho\, T_{kk}/3. \tag{6.39}$$

If we think of the filter as a local average, and of the velocity as separated into its smoothed part and a subgrid fluctuation, the subgrid stresses are the momentum fluxes due to the fluctuations across the 'walls' of the filter, while the correction to the pressure is the subgrid kinetic energy. The latter needs no explicit modelling. The role of pressure in incompressible flows is to insure continuity, and its only physical constraint is to be a scalar. Continuity automatically gives the correct effective pressure, and the correction term in (6.39) is due to the difference between enforcing continuity for the true velocity field, or for the filtered one. A separate model for the subgrid kinetic energy is only required if the real pressure is needed for some reason, such as in aeroacoustics or in liquids involving cavitation.

The divergence of the subgrid shear stresses, on the other hand, always has to be modelled. The most popular model[9] is an extension of the eddy-viscosity idea. The subgrid stresses are assumed to be proportional to the filtered rate-of-strain tensor,

$$\overline{\tau}_{ij} = 2\nu_\varepsilon \overline{S}_{ij}. \tag{6.40}$$

The subgrid eddy viscosity is written as

$$\nu_\varepsilon = C_S \Delta^2 |\overline{S}|, \tag{6.41}$$

where Δ is a measure of the width of the filter, $\overline{S}_{ij}$ is the rate-of-strain tensor computed with the filtered velocity $\overline{u}$, and $|\overline{S}|^2 = \overline{S}_{ij}\overline{S}_{ij}$. In this equation Δ has been used as the length scale for the eddy viscosity, and $|\overline{S}|\Delta$ as the velocity scale. As with other eddy-viscosity formulations, there is little reason to believe that (6.40) is true, and in particular that the tensors $\overline{\tau}_{ij}$ and $\overline{S}_{ij}$ are parallel. In fact, checks on real flows show that both tensors are only weakly correlated, in spite of which (6.40) works well in many situations. The reason seems to be that all that the flow needs is a mechanism to dissipate the correct amount of energy at the end of the cascade.

If an equation for the kinetic energy of the filtered velocities is derived from (6.37), the product $-\Pi_S = -\overline{\tau}_{ij}\overline{S}_{ij}$ appears in the right-hand side as an energy sink. If the Reynolds number is large, and if the filter is far enough from the Kolmogorov scale, the large-scale viscous dissipation $\nu|\nabla\overline{u}|^2$ is negligible, and Π_S acts as a substitute for ε. The energy is transferred to the sub-grid scales, where viscosity would normally get rid of it, and the dissipation is done by the model.

This can be used to estimate the Smagorinsky constant C_S. The dissipation is written as

$$\varepsilon = \Pi_S = 2C_S\Delta^2|\overline{S}|^3, \tag{6.42}$$

and $|\overline{S}|$ is expressed in terms of the energy spectrum. The norm $|\overline{S}|^2$ can be written as a weighted integral of the spectrum over a neighbourhood of the origin in Fourier space and, because the spectrum of the velocity gradients increases with increasing wavenumbers, is dominated by scales of $O(\Delta)$ (see figure 6.8.). If the filter is narrow enough for those scales to be isotropic, that allows us to compute $|S|$ as a function of the filter width Δ, and of the inertial-range spectrum, which only depends on ε. When that estimate is plugged back into (6.42), the result is an equation for C_S which depends only on the filter shape. For the particular case in which the filtering operation is local averaging over a cubic box of side Δ, $C_S \approx 0.03$. In practice the filter is seldom cubical but, because we are essentially integrating a spherically symmetric spectrum over a volume in Fourier space, anisotropic filters can be used approximately with the same C_S as long as we take $\Delta^3 = \Delta_1\Delta_2\Delta_3$.

In shear flows, which is where LES is really needed, the corrections due to anisotropy are fairly large, and more sophisticated strategies have been developed. The reader should consult recent reviews in Refs.[10,11] for details, as well as for successful applications. Basically LES can be used for simulations of quasi-industrial flows with non-trivial geometries. Examples that have been computed include wakes behind simple obstacles and separating flows, such as diffusers and the trailing edge of wings. There are technical problems involved in the boundary conditions near walls, but most flows in which walls are not important, such as jets, mixing layers and the injectors for combustion systems, can be simulated, essentially independent of their Reynolds number. More complicated geometries in which the Reynolds number is not too high, such as blade cascades in turbines and compressors, are also beginning to be possible.

That subgrid models work well in spite of all their known shortcomings is actually a demonstration of the idea, which we have implicitly used up to now, of a predominantly one-directional cascade. Causality is from large to small scales, and how the energy is dissipated in the latter does not influence the former, as long as the amount is correct.

It is interesting to note that, in spite of this one-directionality of the flow of causality, the cascade itself is not really one-directional. The subgrid transfer Π_S provides a definition for the point-wise energy transfer, which can be measured by explicitly filtering fully resolved simulations. The result is that the cascade is only direct (large to small) in about 60% of the points, and inverse in the rest.[12] The exact percentage depends on the filter shape. The overall direct energy transfer is the difference between two large opposing fluxes. This is the reason for the main limitation of eddy viscosity models, including Smagorinsky's, which is that, as long as $\nu_\varepsilon > 0$ the cascade is direct everywhere. Locally negative eddy viscosities are generally forbidden by numerical considerations, and the result is that eddy viscosity models need too little stress to produce a given dissipation.[13] It is therefore generally impossible to obtain with them the right dissipation and the right stresses at the same time, and LES using these models only works if the filter is chosen narrow enough that just the dissipation has to be modelled. The stresses have to be carried by the resolved field.

This still makes LES a practical alternative for the computation of free-shear flows, even at very high Reynolds numbers. We saw above that the stresses are carried by scales which are at most one order of magnitude smaller than L_ε. Including them requires only that $\Delta \approx L_\varepsilon/10$, so that a large-eddy simulation requires at most a few thousand grid points per cubic integral scale. While this is a large number, it is independent of the Reynolds number, and can therefore be applied equally well to research flows and to industrial ones. The expected accuracy of LES, specially far from walls, is at present of the order of 5-10%.

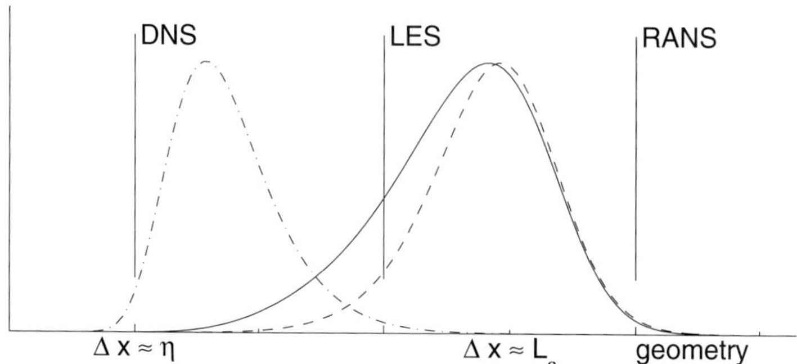

Fig. 6.9. Sketch of the resolution requirements for the different simulation schemes for turbulent flows. The three curves are premultiplied spectra for, ———— , turbulent kinetic energy; – – – – , shear stress; — · — , energy dissipation. Note that the abscissae, $\Delta x \sim \kappa^{-1}$, are plotted in the opposite direction to those in figure 6.5.

6.7. Reynolds-averaged Navier–Stokes simulations

The oldest and least general scheme for computing turbulent flows is to solve directly the Reynolds-averaged Navier–Stokes (RANS) equations (6.18)–(6.19). These equations are not 'closed', because of the presence of the unknown Reynolds stresses. It is possible to write evolution equation for those terms, but they contain triple products, which cannot be expressed in term of simpler quantities, etc. This process leads to an infinite hierarchy of equations of progressively higher moments, which has to be closed at some point with a model. We have used several times in these notes 'one-point' closures, in the form of eddy viscosities that we have estimated in *ad-hoc* ways. Those choices are generally different for free-shear flows than for those in the neighbourhood of walls, although it is often possible to give a-posteriori physical reasons for the differences. This is a general problem with RANS. Since they try to model *all* the turbulent scales, including the non-universal energy-containing ones, RANS models are not universal, and have to be adapted to the different cases.

The development of RANS models, either in the form of eddy viscosity formulas, or in the more sophisticated one of evolution equations for the Reynolds stresses, is an industry in itself, and has been quite successful in providing approximations for flows of practical interest. Good introductions are.[14,15]

A typical example is the popular k–ε method. It solves two 'evolution' equations: one for the fluctuating kinetic energy k, which has relatively few arbitrary assumptions, and another one for the dissipation ε, which is written in a *ad-hoc* way using the same structure as in the first one. The diffusion fluxes in both equations are assumed to be proportional to the gradients of the respective variables. The kinetic energy provides a velocity scale, and the dissipation a length scale. The eddy viscosity is then chosen proportional to $\nu_\varepsilon \sim k^2/\varepsilon$, which is the product of an estimated large-scale velocity, $k^{1/2}$, and of an approximation to the integral length scale $u'^3/\varepsilon \approx k^{3/2}/\varepsilon$. There are various empirical coefficients that are adjusted to give the right results for canonical flows, such as decaying turbulence, the logarithmic velocity profile, etc. Despite the crudeness of this procedure, which means that the expected accuracy is never much higher than 30-50%, this model has been fairly successful in computing the mean velocity profiles of industrial flows. Most aeroplanes, cars, or air-conditioners have been computed, at best, using a k–ε model.

Part of the reason is that the way of adjusting the empirical coefficients is designed to give the right results for flows which are not too different from the ones used for calibration. It is an accepted procedure in RANS to use different sets of coefficients for different flows. In this sense RANS models are sophisticated interpolation tables, but the best ones are based on sound physical reasoning, paying attention to such requirements as conservation properties in the non-dissipative limit, tensor invariance, etc., which go a long way towards guaranteeing the correct physics. Even if RANS is mostly seen today as an industrial device, it has a distinguished history in turbulence research, and many of the intuitive concepts of time and length scales have been honed by their application to it.

The resolution requirements for RANS are not different, in principle, from LES, but they are quite cheaper in practice. Since all the turbulence fluctuations are included in the model, there is usually no need to consider the flow as being unsteady, and it is enough to look for equilibrium solutions. It is also not necessary to choose the grid based on the integral turbulent scales, and the resolution is solely controlled by the geometry of the flow. The geometry also controls the integral scales, so that the requirements for RANS and LES are proportional to one another, independent of the Reynolds number. But the proportionality constant may be $O(10-100)$ in each direction, for a total saving of several orders of magnitude. In exchange for those savings, RANS methods are *ad-hoc*, work poorly in non-equilibrium situations, and cannot be trusted if calibration experiments are not available.

The resolution requirements for the three simulation methods are summarized in figure 6.9., relative to the turbulence spectrum. DNS has to resolve the dissipation peak, LES the energy and stress peaks, while RANS only has to resolve the geometry.

Acknowledgments

The preparation of this notes was supported in part by the CICYT contract DPI2003-03434.

Bibliography

1. G.H.L. Hagen. Über die Bewegung des Wassers in engen cylindrischen Röhren. *Poggendorfs Ann. Physik Chemie*, 16, 1839.
2. H. Darcy. Recherches expérimentales rélatives au mouvement de l'eau dans les tuyeaux. *Mém. Savants Etrang. Acad. Sci. Paris*, 17:1–268, 1854.
3. L.F. Richardson. *Weather prediction by numerical process*. Cambridge U. Press, 1922.
4. A.N. Kolmogorov. The local structure of turbulence in incompressible viscous fluids a very large Reynolds numbers. *Dokl. Akad. Nauk. SSSR*, 30:301–305, 1941. Reprinted in *Proc. R. Soc. London.* A **434**, 9–13 (1991).
5. G.K. Batchelor. *The theory of homogeneous turbulence*. Cambridge U. Press, 1953.
6. U. Frisch. *Turbulence. The legacy of A.N. Kolmogorov*. Cambridge U. Press, 1995.
7. J. Jiménez. Computing high-Reynolds number flows: Will simulations ever substitute experiments? *J. of Turbulence*, 22, 2003.
8. P. Moin and K. Mahesh. Direct numerical simulation: A tool in turbulence research. *Ann. Rev. Fluid Mech.*, 30:539–578, 1998.
9. J. Smagorinsky. General circulation experiments with the primitive equations. *Mon. Weather Rev.*, 91(3):99–16, 1963.
10. M. Lesieur and O. Metais. New trends in large-eddy simulations of turbulence. *Ann. Rev. Fluid Mech.*, 28:45–82, 1996.
11. P. Moin. Progress in large eddy simulation of turbulent flows. *AIAA Paper*, 97–0749, 1997.
12. U. Piomelli, W.H. Cabot, P. Moin, and S. Lee. Subgrid-scale backscatter in turbulent and transitional flows. *Phys. Fluids*, A 3:1766–1771, 1991.
13. J. Jiménez and R.D. Moser. LES: where are we and what can we expect? *AIAA J.*, 38:605–612, 2000.
14. B.E. Launder and D.B. Spalding. *Mathematical models of turbulence*. Academic, 1972.
15. D.C. Wilcox. *Turbulence modelling for CFD*. DCW Industries, 1993.

Chapter 7

Particle Image Velocimetry - Application to Turbulence Studies

Julio Soria

Laboratory for Turbulence Research in Aerospace & Combustion (LTRAC), Department of Mechanical Engineering, Monash University P.O. Box 31, VIC 3800, Australia
E-mail: julio.soria@eng.monash.edu.au

In this chapter one of the most widely used techniques in experimental studies of fluid turbulence, particle image velocimetry (PIV), is reviewed. Principles and main characteristics of the PIV method are complemented by examples of its implementation for studies of turbulent jet.

Contents

7.1.   Introduction to laser-based velocimetry and PIV

Laser-based fluid velocity measurement techniques can be broadly divided into point measurement techniques and field measurement techniques. An example of the former is laser Doppler anemometry (LDA)[1] and of the latter laser speckle velocimetry (LSV).[2] Particle image velocimetry (PIV), sometimes referred to particle image displacement velocimetry (PIDV), and laser speckle velocimetry (LSV) can be used to measure instantaneous in-plane velocity fields by recording the location of the images of the markers at multiple times. These techniques are different from the conventional particle tracking methods that suffer from the complication of identifying and following individual particles in the flow field. PIV and LSV do not require the tracking of individual particles as the in-plane velocity components in the flow are determined using a finite interrogation region or analysis window in an image of multi exposed or single exposed tracer particles. As with most optical field and point velocity measurement techniques, PIV and LSV use as their cornerstone the fundamental definition of velocity to arrive at an estimate of the local velocity $\boldsymbol{u}$, i.e.

$$\boldsymbol{u}(\boldsymbol{x}, t) = \lim_{\Delta t \to 0} \frac{\Delta \boldsymbol{x}(\boldsymbol{x}, t)}{\Delta t} \qquad (7.1)$$

where $\Delta \boldsymbol{x}$ is the displacement of the marker particles, located at $\boldsymbol{x}$ at time t, over a short time period Δt. The principle behind the implementation of Eq. (7.1) in PIV is simple: the flow is seeded with small tracer particles which faithfully follow the flow; the flow is illuminated by a powerful light source in the form of a thin light sheet; the particle positions are recorded by imaging the scattered light from the tracer particles in the light sheet onto a recording medium (i.e. photographic film or an electronic photo-array detector). The in-plane displacement and hence the in-plane velocity is then estimated from sequential images of particle positions. The most significant attraction of PIV for fluid velocity measurements is its ability to simultaneously measure, as a minimum, the in-plane velocity of a relatively large region of the flow. Historically, laser speckle velocimetry preceded PIV in its application to fluid flow measurements. LSV has its foundations in solid mechanics, where coherent light scattered from solid surfaces forms an interference pattern known as a specklegram. A manual analysis of a double-exposed speckle patterns to determine displacement is not possible because the human brain cannot untangle the superimposed random-looking speckle fields. Analysis of such superimposed fields became possible with the development of a fast and efficient statistical analysis method known as Young's fringe method of interrogation.[3] In the Young's fringe method an interrogation spot on a double-exposed specklegram is illuminated by coherent light such as from a laser beam. The speckle field from each exposure diffracts the light wave from the coherent interrogation beam, which interfere to form a Young's fringe pattern (see Fig. 7.1. (a)). The orientation of the fringes is perpendicular to the direction of the displacement and the spacing is inversely proportional to the magnitude of the displacement.

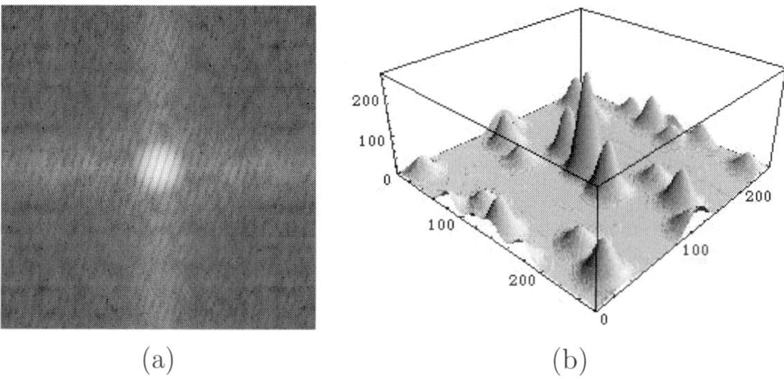

(a)   (b)

Fig. 7.1. (a) Typical Young's fringe pattern, (b) corresponding auto-correlation function.

Barker & Fourney[4] applied this technique in low Reynolds number Poiseuille flow, while Simpkins & Dudderar[5] used LSV to study transient Benard convection. Meynart[6] applied LSV to velocity measurements in axisymmetric vortex roll-up and pairing events of an air jet. Laser speckle velocimetry (LSV) requires a very high seeding density in order to form speckle patterns. However in most practical cases low flow tracer seeding is the prevailing situation. The result of low seeding density is that individual particles are imaged rather than the interference pattern generated by neighbouring particle images. This fact was first pointed out by Adrian.[7] Therefore, most of the fluid experiment measurements conducted using multi-exposed images are PIV measurements rather than LSV measurements.

Kompenhans & Reithmuth[8] and Lourenco & Krothapalli[9] looked at turbulent wake flows from circular cylinder using PIV measurements. Vogel & Lauterborn[10] studied the flow around cavitation bubbles by measuring the flow field using PIV. Adrian[11] showed how PIV flow field measurements could be used to study the flow field of water jets, while Shepherd *et al.*[12] showed how PIV can be used to measure the in-plane velocity field of a laminar circular cylinder wake flow without and with transverse forced oscillations. Velocity field data obtained using PIV was used by Wu *et al.*[13] to study the feedback effect of sound and velocity perturbations on the flow around two airfoils in tandem. Liu *et al.*[14] have investigated the spatial turbulent structure in a channel flow at Re = 2872 by using auto-correlation PIV analysis yielding more than 10,000 velocity points per image. PIV was used by Grant *et al.*[15] to measure the mean flow and turbulence characteristics in the separated flow behind a rearward facing step. Arroyo & Saviron[16] have studied the spatial features of Rayleigh-Benard convection in a small box using spatial velocity PIV measurements.

PIV has also been recently applied to studies of flows of a more applied nature where great care has been exercised in the application of PIV methodology to arrive at accurate velocity field measurements. For example Shih *et al.*[17] used PIV velocity field measurements to study the unsteady flow past an airfoil pitching at a constant rate. Shekarriz *et al.*[18] studied the development of the junction and tip vortex behind a sail attached to an axisymmetric underwater body by using PIV measurements to map the instantaneous velocity field around this geometry. Dong *et al.*[19,20] implemented quantitative flow visualisations within the volute of a centrifugal pump by using PIV measurements. Shekarriz *et al.*[21] investigated the near-field behaviour of a tip vortex trailing behind a low aspect ratio wing attached to an axisymmetric body. There are a number of review articles on PIV, which have been written over the twenty years that PIV has been available and the interested reader is referred to them for numerous examples of the application of PIV to fluid mechanics.[2,22–29] The journal issue containing the paper by Adrian,[29] contains invited papers to this journal that demonstrate the most recent advances and application in PIV, Stereo-PIV, Digital Image Holography and other related experimental techniques to PIV.

PIV measurements and its variations can be divided into two separate stages: (i) a recording stage and (ii) an analysis stage. Each stage can be implemented separately, for example the recording of multi-exposed images can be performed using photographic film or using an electronic photosensitive array; the analysis can take the form of analog optical processing or digital optical processing. Various combinations of these two stages are possible. Each stage can be optimised separately, although, it must be noted that there is an interrelationship between both stages which will be pointed out in subsequent sections where appropriate.

The specific PIV technique that is the topic of this chapter is one whereby single exposed images are acquired in digital form using a CCD array. These single exposed images are subsequently analysed using cross-correlation digital analysis of small appropriately sampled interrogation windows extracted from a pair of single exposed images.[30–32] Therefore, this velocity field measurement technique is referred to as cross-correlation digital PIV (CCDPIV). This technique can be implemented to measure: (1) 2 velocity components in a 2 dimensional plane (2C-2D), (2) 3 velocity components in a 2 dimensional plane (3C-2D) and (3) 3 components in a 3 dimensional volume (3C-3D) with the complexity of the experimental setup and the post-processing of the experimental data increasing as well as the measuring capability.

7.2.　Principles and characteristics of PIV

It is useful to provide a brief summary of the principles and characteristics of PIV before embarking on the mathematical theory that underpins the cross-correlation digital PIV of single exposed image pairs. A more extensive description of the principles and characteristics of PIV can be found in Raffel *et al.*[33]

Some of the more pertinent characteristics of PIV are that PIV is based on imaging (intensity or holographically) tracer particles in the fluid and deducing the instantaneous velocity of the tracer particles and thus, the instantaneous velocity of the fluid. The imaging of the particles is typically via Mie scattering using a laser as the light source. In principle a laser is not required to undertake a PIV measurements, however, a laser has very beneficial illumination characteristics that are exploited in PIV, such as: (i) the laser beam diverges little (*i.e.* laser beams go in straight lines with little spread), (ii) by using shape forming optics the laser beam can be shaped into a well-defined thin light sheet and (iii) lasers, such as Nd:YAG, produce very short, high intensity laser pulses, *e.g.* 5 ns pulses with 500 mJ energy, which allow the effective freezing of the tracer particle motion of large areas of the flow. Because PIV is an optical technique it works non-intrusively. This allows the application of PIV to flows which may be disturbed or topologically altered by the presence of probes. However, like laser Doppler velocimetry, the PIV technique measures the velocity of a fluid element indirectly by means of the measurement of the velocity of tracer particles within the flow. In most applications, these tracer particles have been added to the flow before the commencement of the experiment and have well defined characteristics. PIV allows the recording of images of "large" parts of flow fields and to extract the velocity information from these images. This feature is unique to the PIV technique. All other fluid velocity measurement techniques only allow the measurement at a single point, albeit in most cases with high temporal resolution. With PIV the temporal resolution (*i.e.* frame rate of recording PIV images) is limited only by technical restrictions, primarily sensor technology and bandwidth. The instantaneous PIV image capture and reasonably high spatial resolution (dependent on sensor technology) in PIV allows the detection of spatial structures even in unsteady flow fields.

314   *Julio Soria*

The need to employ tracer particles for the inference of the local flow velocity requires a careful analysis in each experiment to ascertain whether the particles will faithfully follow the motion of the fluid elements. In general small, neutrally buoyant particles will follow the flow better. Equation (7.2), which is based on Stokes flow assumption, can be used to calculate the response or relaxation time of a particle to a step change in the velocity. This equation is easily derived by solving the equation of motion.

$$\tau_p = \frac{d_p^2}{18\,\nu_f}\left(\frac{\rho_p}{\rho_f}\right) \tag{7.2}$$

Figure 7.2. shows the relaxation time constants for typical tracer particles used in water and air PIV experiments as a function of the tracer particle diameter. In air experiments the oil droplets are typically 1 μm in diameter, which indicates that these particles have a typical response time to velocity changes of the air of the order of 1 μs. In water experiments the hollow glass spheres have a typical diameter of 10 μm, which indicates that these particles have a typical response time to velocity changes of the water of the order of 7 μs.

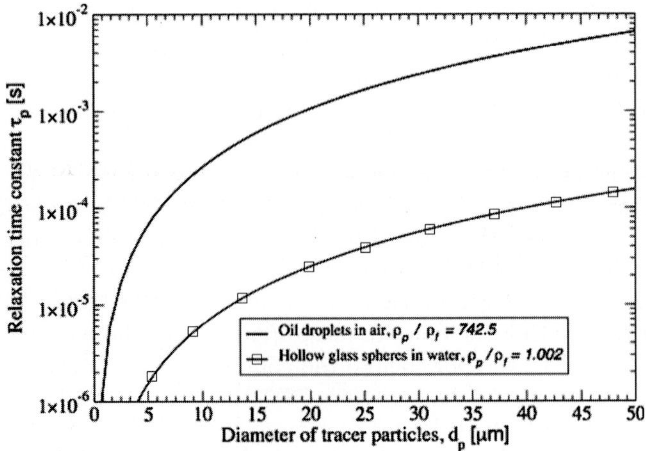

Fig. 7.2.   Relaxation time constant for oil droplets in air and hollow glass spheres in water as a function of the tracer particle diameter.

The tracer particle seeding density of the flow is an importantfactor in a PIV experiment as low particle seeding can result in a failure to measure the velocity PIV or a large quantity of invalid vectors. For PIV a medium image density is optimum with homogeneously dispersed tracer particle seeding. Typically 5 - 10 particle images per interrogation area will assure more than 95% valid CCDPIV evaluations.[34,35]

The illumination in PIV as previously pointed out is via lasers. High power Nd:YAG lasers are typically required for illumination in order that the light scattered by the tiny tracer particles (*e.g.* $1 - 5 \ \mu m$ in diameter) will expose the recording medium. In liquid flows larger particles (*e.g.* 10 $- 30 \ \mu m$ in diameter) can usually be employed that scatter much more light and lasers of considerably lower peak power can be used. However, nowadays the double cavity pulsed Nd:YAG laser is the laser of choice for both gas and liquid PIV measurements, The pulsed Nd:YAG laser has the advantage of providing a very short illumination light pulse, typically 6 *ns* that freezes the motion of the particles during the pulse exposure, without blurring of the images and streaks. As it is the two laser pulses that exposed the single exposed image pairs, the time delay between these two pulses provides the required time for the displacement of the tracer particles. The time delay between the illumination pulses must be long enough so that the displacement between the images of the tracer particles can be determined using cross-correlation PIV analysis with sufficient precision and short enough to: (i) avoid particles with out-of-plane velocity component leaving the light sheet between illumination pulse pairs and (ii) minimise the interrogation area and hence, maximise the spatial resolution. Note, that the latter condition can be relaxed with advanced analysis techniques, *e.g.* MCCDPIV.[36]

The size of the interrogation window plays a major role in CCDPIV and PIV in general. The size of the interrogation window at the CCDPIV analysis stage must be small enough so that velocity gradients have no significant effect on the result. This issue is analysed in detail in subsequent sections of this chapter, however at this introductory stage Fig. 7.3., which shows the separated flow over a blunt leading edge at a Reynolds number of 1000, serves as an illustration of the effect of the size of the interrogation window on the measurement of complex vortical flows. The interrogation windows range from 128 px down to 16 px in Fig. 7.3..

The size of the IW coupled with the separation between the velocity vectors Δ also determines: (i) the number of statistically independent velocity vectors and (ii) the maximum spatial resolution of the measured velocity field that can be obtained at a given optical spatial resolution of the sensor employed for recording.

An important stage in the PIV analysis stage is the validation of the measured velocity data. This stage in the PIV analysis process is outside the scope of this chapter and the interested reader is referred to Raffel *et al.*[33] for details and a thorough discussion of this topic.

To conclude this introductory section Fig. 7.4. shows a schematic of

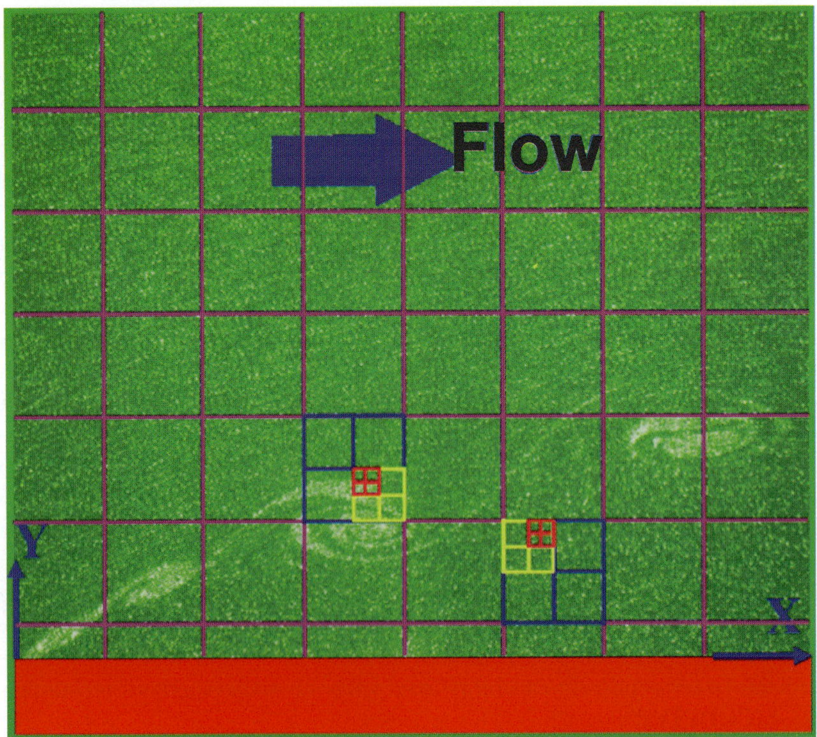

Fig. 7.3.   Flow over a blunt leading edge at Re = 1000. Different size IW windows are indicated to illustrated their integration effect and relationship to the size of the vortical structures in the flow. The size of the IW ranges from 128 px down to 16 px.

the full PIV experimental procedure with pertinent equipment and the different stages ranging from the experiment to the cross-correlation digital PIV analysis of single exposed image pairs to the post-processing of the measured velocity fields.

7.3.   Mathematical analysis of cross-correlation of single exposed image pairs

This section develops the underlying mathematical analysis framework that is employed in the image processing of single exposed images to deduce the local in-plane fluid velocity. The basic principle of PIV analysis of single exposed images is the local cross-correlation of images recorded within a small interrogation window, IW, of tracer particles. If the recorded image

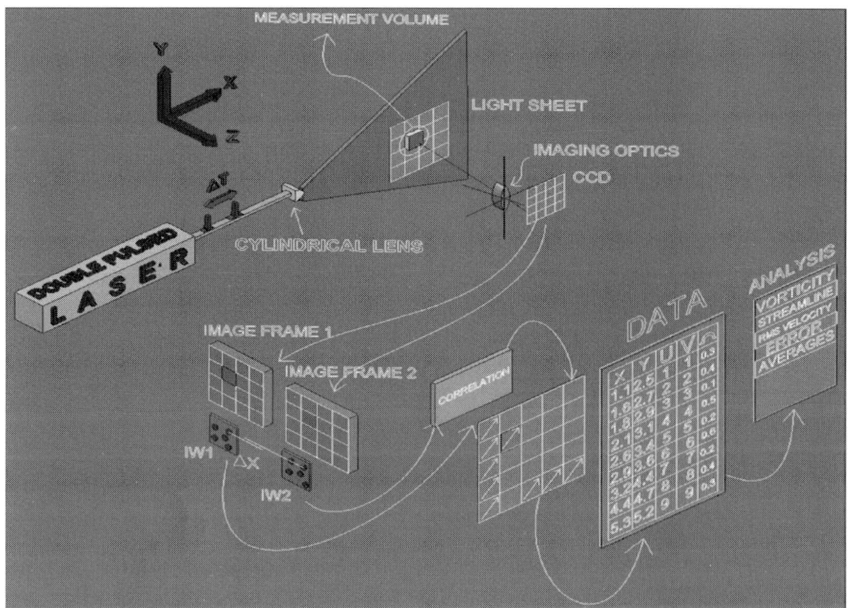

Fig. 7.4.   Illustration of the single exposed PIV Image Pair acquisition, CCDPIV analysis process and post-processing of the measured velocity data.

intensity distribution of the ith particle located in the image plane at (x_i, y_i) can be represented by $I_i(x - x_i, y - y_i)$, then, denoting the interrogation window by $\mathcal{R}$, which contains N particle images, the image intensity within $\mathcal{R}$ that describes the image intensity distribution at $t = t_1$ is given by

$$\mathcal{I}_1(x, y) = \sum_{i=1}^{N} I_i(x - x_i, y - y_i), \ (x_i, y_i) \in \mathcal{R} \tag{7.3}$$

Figure 7.5. shows a typical intensity distribution for one particle and for a number of particles within IW. Note that for all analysis in this chapter associated with the domain of the interrogation window, the origin of the (x, y) local coordinate system associated with said interrogation window will be assumed to be in its geometric centre.

Consider the case where each particle whose image intensity is located within $\mathcal{R}$ undergoes a motion, such that the resulting displacement in the image plane during a time interval Δt, as illustrated in Fig. 7.6., is given by

$$\Delta x_i = u_i \Delta t$$

$$\Delta y_i = v_i \Delta t \tag{7.4}$$

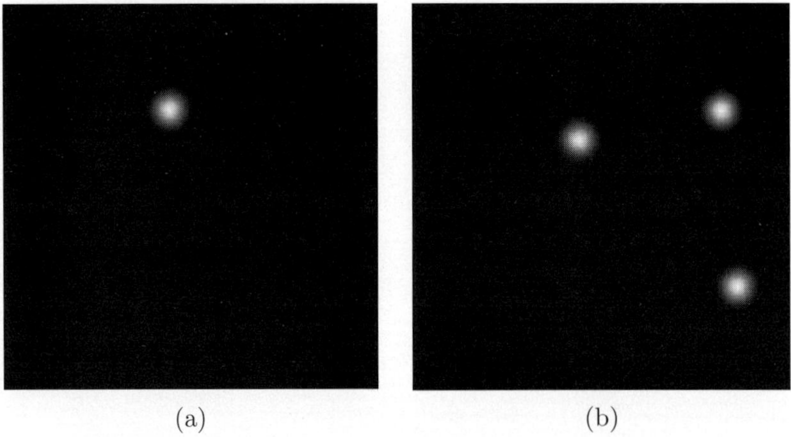

(a)   (b)

Fig. 7.5. (a) Intensity distribution of one particle represented by $I_i(x - x_i, y - y_i)$, and (b) intensity distribution of three particles represented by Eq. (7.3).

where the subscript i refers to the ith particle and its location and $(u_i(x, y), v_i(x, y))$ is its velocity.

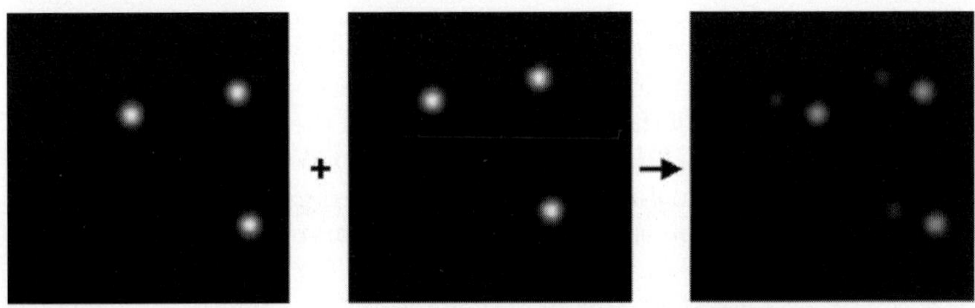

Fig. 7.6. This figure illustrates the displacement of tracer particles undertaken between time interval Δt, *i.e.* at $t = t_1$ (bright) and $t = t_1 + \Delta t = t_2$ (faint).

It will be assumed that during the motion the N particles remain within $\mathcal{R}$ - something that in practice is not assured. Then at $t = t_2 = t_1 + \Delta t$ the image intensity in the image domain $\mathcal{R}$ is given by

$$\mathcal{I}_2(x, y) = \sum_{i=1}^{N} I_i(x - (x_i + \Delta x_i), y - (y_i + \Delta y_i)), \qquad (7.5)$$

where $(x_i + \Delta x_i, y_i + \Delta y_i) \in \mathcal{R}$. The cross-correlation function between $\mathcal{I}_1(x, y)$ and $\mathcal{I}_2(x, y)$ is given by

$$R_{12}(\eta, \xi) = \int_{\mathcal{R}} \mathcal{I}_1(x, y) \mathcal{I}_2(x + \eta, y + \xi) \, dx \, dy. \qquad (7.6)$$

Using the Correlation Theorem, the cross-spectrum $G_{12}(f_x, f_y)$ is given by

$$G_{12}(f_x, f_y) = \mathcal{F}[R_{12}(\eta, \xi)]$$
$$= \mathcal{F}[\mathcal{I}_1(x, y)]\mathcal{F}[\mathcal{I}_2(x, y)]^* \tag{7.7}$$

where $\mathcal{F}[g]$ denotes the Fourier transform of g and $*$ denoted the complex conjugate. Conversely, the cross-spectrum is related to the cross-correlation function through the inverse Fourier transform

$$R_{12}(\eta, \xi) = \mathcal{F}^{-1}[G_{12}(f_x, f_y)] \tag{7.8}$$

Defining,

$$\hat{I}_i(f_x, f_y) = \mathcal{F}[I_i(x, y)] \tag{7.9}$$

and using the Shift Theorem, Eq. (7.9), results in the Fourier transform of Eq. (7.3) being given by

$$\mathcal{F}[\mathcal{I}_1(x, y)] = \sum_{i=1}^{N} \mathcal{F}[I_i(x - x_i, y - y_i)]$$
$$= \sum_{i=1}^{N} \hat{I}_i(f_x, f_y) \exp[-i2\pi(f_x x_i + f_y y_i)] \tag{7.10}$$

and similarly the Fourier transform of Eq. (7.5) is given by

$$\mathcal{F}[\mathcal{I}_2(x, y)] = \sum_{i=1}^{N} \mathcal{F}[I_i(x - (x_i + \Delta x_i), y - (y_i + \Delta y_i))]$$
$$= \sum_{i=1}^{N} \hat{I}_i(f_x, f_y) \exp[-i2\pi(f_x(x_i + \Delta x_i) + f_y(y_i + \Delta y_i))] \tag{7.11}$$

The process indicated by Eqs. (7.10) and (7.11) is illustrated in Fig. 7.7. for an image with three tracer particles. The origin of the (f_x, f_y) coordinates is located at the centre of the images, which represent the magnitude of the Fourier transform of $\mathcal{I}_1(x, y)$ and $\mathcal{I}_2(x, y)$.

The cross-spectrum $G_{12}(f_x, f_y)$ defined by Eq. (7.7) can now be written down using Eqs.(7.10) and (7.11) as

$$G_{12}(f_x, f_y) = \left(\sum_{i=1}^{N} \hat{I}_i(f_x, f_y) \times \exp[-i2\pi(f_x x_i + f_y y_i)]\right)$$
$$\times \left(\sum_{j=1}^{N} \hat{I}_j^*(f_x, f_y) \times \exp[i2\pi(f_x(x_j + \Delta x) + f_y(y_j + \Delta y))]\right) \tag{7.12}$$

Figure 7.8. illustrates the magnitude of the cross-spectrum given by Eq. (7.12) for the example of the images given in Fig. 7.7. where the origin of the (f_x, f_y) coordinates is located at the centre of the image.

The product of the two sums given by Eq. (7.12) contributes two types of terms to $G_{12}(f_x, f_y)$:

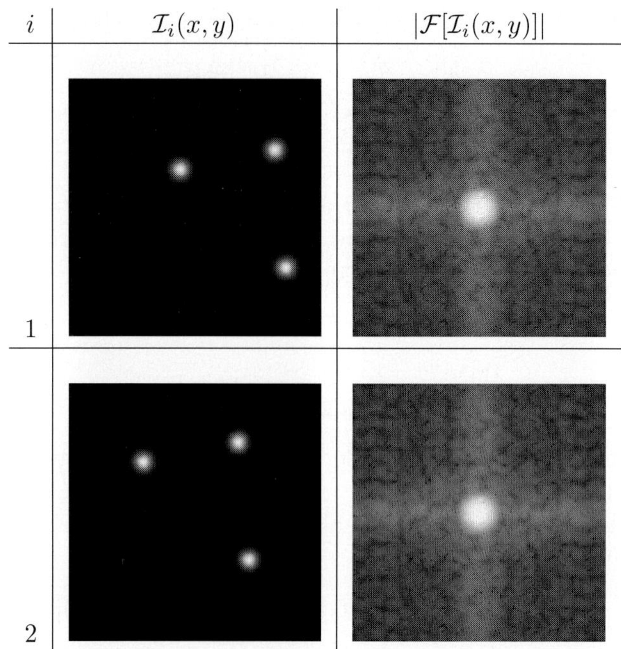

Fig. 7.7. Two images and corresponding magnitude of the Fourier transform of the images. The origin of the coordinates (x, y) of the particle images is at the lower left hand corner, the origin of the frequency coordinates (f_x, f_y) of the Fourier transform of the images is located at the centre of those images.

(1) there are correlated contributions corresponding to $i = j$:

- these terms contribute to the cross-spectrum via the following terms: $\left[\hat{I}_i(f_x, f_y)\hat{I}_i^*(f_x, f_y) \exp[i2\pi(f_x\Delta x_i + f_y\Delta y_i)]; \right]$
- this term is the particle pair contribution to the cross-spectrum due to the displacement of the particles, which is usually assumed to be homogeneous, but is typically not due to the velocity gradients in most practical flows, particularly turbulent flows;

(2) there are uncorrelated contributions corresponding to $i \neq j$:

- these
  terms contribute to the cross-spectrum via the following terms: $\left[\hat{I}_i(f_x, f_y)\hat{I}_j^*(f_x, f_y) \exp[i2\pi\{f_x(x_j - x_i + \Delta x_j) + f_y(y_j - y_i + \Delta y_j)\}] \right]$
- this type of term is the correlation due to each particle with every other particle in the second recording that is NOT its pair;
- *i.e.* it effectively represents effectively uncorrelated noise.

Fig. 7.8. The magnitude of the cross-spectrum between the two images shown in Fig. 7.7. - the origin of the frequency coordinates (f_x, f_y) is located at the centre of this image.

Defining the auto-spectrum and cross-spectrum of two individual particle images as

$$G_{ii}(f_x, f_y) = \hat{I}_i(f_x, f_y)\hat{I}_i^*(f_x, f_y) \text{ and} \tag{7.13}$$

$$G_{ij}(f_x, f_y) = \hat{I}_i(f_x, f_y)\hat{I}_j^*(f_x, f_y) \tag{7.14}$$

respectively. Then the first of these relations is interpreted as the cross-spectrum of the ith particle image with itself, whilst the second of these relations is interpreted as the cross-spectrum between the ith and the jth particle. Thus, the cross-spectrum represented by Eq. (7.12) between image 1 and image 2 can be written as:

$$G_{12}(f_x, f_y) = \sum_{i=1}^{N} G_{ii}(f_x, f_y) \times \exp[i2\pi(f_x\Delta x_i + f_y\Delta y_i)]$$

$$+ \sum_{\substack{i=1, j=1 \\ i \neq j}}^{N} G_{ij}(f_x, f_y)$$

$$\times \exp[i2\pi\left\{f_x(x_j - x_i + \Delta x_j) + f_y(y_j - y_i + \Delta y_j)\right\}] \tag{7.15}$$

The first term in Eq. 7.15 contributes to the cross-correlation due to the coherent motion of the seed particles, whilst the second term contributes incoherent noise to the cross-correlation.

Using the Correlation Theorem in the form of Eq. (7.8) yields the following relationship for the cross-correlation function between image 1

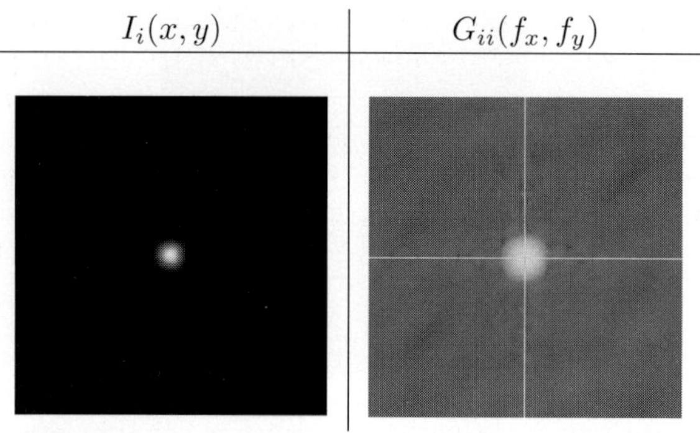

$I_i(x,y)$ $G_{ii}(f_x,f_y)$

Fig. 7.9. Single particle image and its corresponding auto-spectrum - the origin of the frequency coordinates (f_x, f_y) is located at the centre of this image.

and 2:

$$R_{12}(\eta,\xi) = \sum_{i=1}^{N} \mathcal{F}^{-1}\left[G_{ii}(f_x,f_y) \times \exp(i2\pi\{f_x\Delta x_i + f_y\Delta y_i\})\right]$$

$$+ \sum_{\substack{i=1,j=1 \\ i\neq j}}^{N} \mathcal{F}^{-1}\left[G_{ij}(f_x,f_y)\right.$$

$$\left. \times \exp(i2\pi\{f_x(x_j - x_i + \Delta x_j) + f_y(y_j - y_i + \Delta y_j)\})\right]$$

$$(7.16)$$

In order to more clearly visualise the meaning of the terms in Eq. 7.16, it is useful to define the following correlation functions:

$$R_{ii}(\eta,\xi) \equiv \mathcal{F}^{-1}[G_{ii}(f_x,f_y)] = \int_{\mathcal{R}} I_i(x,y)I_i(x+\eta,y+\xi)\, dx\, dy \qquad (7.17)$$

$$R_{ij}(\eta,\xi) \equiv \mathcal{F}^{-1}[G_{ij}(f_x,f_y)] = \int_{\mathcal{R}} I_i(x,y)I_j(x+\eta,y+\xi)\, dx\, dy. \qquad (7.18)$$

Equation (7.17) is interpreted as the correlation of the ith particle image with itself when both are located at the same location in the image plane, whereas Eq. (7.18) represents the correlation of the ith particle image with j^{th} particle image when both are located at the same location in the image plane. Using the defining Eqs. (7.17), (7.18) and the Shift Theorem, the cross-correlation function between $\mathcal{I}_1(x,y)$ and $\mathcal{I}_2(x,y)$ given by Eq. (7.16), becomes

$$R_{12}(\eta,\xi) = \sum_{i=1}^{N} R_{ii}(\eta - \Delta x_i, \xi - \Delta y_i)$$

$$+ \sum_{\substack{i=1,j=1 \\ i\neq j}}^{N} R_{ij}(\eta - (x_j - x_i + \Delta x_j), \xi - (y_j - y_i + \Delta y_j))$$

$$(7.19)$$

Equation 7.19 states that the cross-correlation function between the two images is made up of the linear superposition of two parts: (i) the auto-correlation functions of each individual particle image with the auto-correlation function origin (*i.e.* its peak location) shifted to the position in (η, ξ) space which corresponds to the displacement of that particle and (ii) all possible cross-correlation functions of each particle in the first image with each particle in the second image except itself, with the peak located at a position in (η, ξ) space that is dependent on the position of the two particles involved in the cross-correlation and the displacement of the particle identified in the second image. The first part, made up of the superposition of the auto-correlation functions, contributes to the average displacement of all the particles within the IW. The second part, made up of cross-correlation functions, contributes to the noise and introduces experimental uncertainty in the measurement. This effect is illustrated in Fig. 7.10..

$\mathcal{I}_1(x,y) + \mathcal{I}_2(x,y)$	$|G_{12}(f_x, f_y)|$	$R_{12}(\eta, \xi)$

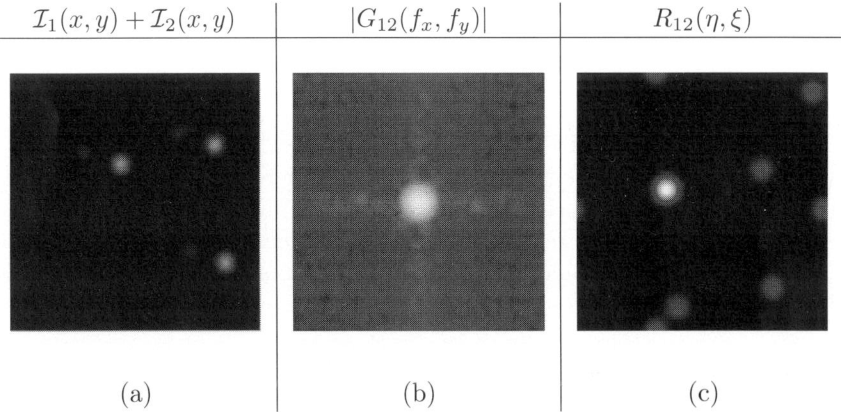

(a)	(b)	(c)

Fig. 7.10. Illustration of the cross-correlation process: (a) a superimposed image of image one (bright particles) and two (faint particles), (b) the magnitude of the cross-spectrum function and (c) the cross-correlation function. The origins of the independent variables for the cross-spectrum and the cross-correlation is in the centre of each image.

Equation 7.19 is quite general and can be used to study: (a) the effect of velocity gradients, (b) the effect of particle size, (c) the effect of particle image intensity and its variation across the image, as well as other effects on the cross-correlation function between two interrogation windows and hence, the ability to identify the correct peak in the cross-correlation function and therefore, the correct average displacement of the particles within the IW.

For the case of homogeneous motion within the IW, *i.e.*

$$\Delta x_i = \Delta x$$
$$\Delta y_i = \Delta y,$$

(7.20)

the cross-correlation function expression given by Eq. 7.19 simplifies to

$$R_{12}(\eta, \xi) = \sum_{i=1}^{N} R_{ii}(\eta - \Delta x, \xi - \Delta y)$$

$$+ \sum_{\substack{i=1,j=1 \\ i \neq j}}^{N} R_{ij}(\eta - (x_j - x_i + \Delta x), \xi - (y_j - y_i + \Delta y)).$$

(7.21)

This equation clearly demonstrates that the individual auto-correlation functions R_{ii} of each particle image with their origins shifted to $(\Delta x, \Delta y)$ in the $\eta - \xi$ plane add constructively to form the cross-correlation peak of R_{12} that indicates the location of the average displacement of all tracer particles in the IW. Whereas, the individual cross-correlation functions R_{ij} of each particle image i in the first exposed IW with another particle image j in the second exposed IW contributes uncorrelated noise to R_{12}.

7.3.1. *Cross-correlation analysis using Gaussian particle image models*

The tracer particles in PIV experiments are typically quite small such that when illuminated by the laser they can be considered as point light sources and they have an intensity distribution in the image plane given by the Airy function, which is reasonably well approximated by an axisymmetric Gaussian distribution function. This allows the imaged particles within the IW to be modelled by an intensity $I_i(x, y)$ given by

$$I_i(x, y) = I_{0_i}\, e^{\frac{2\left(-x^2 - y^2\right)}{\sigma_i^2}}$$

(7.22)

The σ_i value is related to the particle image size, which is also often referred to as d_p, the particle diameter. The term diameter is incorrectly used in this context, as $d_p = \sigma$ represents the location from the peak intensity of the particle where the intensity has decayed to 0.135 of the peak intensity and represents a quantity more correctly associated with the radius of a particle. The Fourier transform of Eq. 7.22 is

$$\hat{I}_i(k_x, k_y) = \frac{I_{0_i}}{4}\, e^{-\frac{1}{8}\left(k_x^2 + k_y^2\right)\sigma_i^2}\sigma_i^2$$

(7.23)

where $k_x = 2\pi f_x$ and $k_y = 2\pi f_y$ represent the wavenumber in the $x-$ and $y-directions$ respectively. For the case of a Gaussian particle image model Eq. 7.14 and Eq. 7.17 becomes

$$G_{ii}(k_x, k_y) = \hat{I}_i(k_x, k_y)\hat{I}_i^*(k_x, k_y) = \frac{I_{0_i}^2}{16}\, e^{\frac{1}{4}\left(-k_x^2 - k_y^2\right)\sigma_i^2}\sigma_i^4$$

(7.24)

$$R_{ii}(\eta, \xi) = \frac{I_{0_i}^2 \, \sigma_i^2}{8} \, e^{-\frac{\eta^2 + \xi^2}{\sigma_i^2}} \tag{7.25}$$

respectively. Figure 7.11. illustrates that for Gaussian intensity models the size of the cross-correlation peak will always be larger than the size of the image intensity. For each particle intensity its corresponding correlation contribution to Eq. 7.19 is an auto-correlation given by Eq. 7.25 which is a factor of $\sqrt{2}$ larger than its effective image radius given by $\sigma_i = d_{p_i}$. Similar relationships are easily derived for $G_{ij}(k_x, k_y)$ and $R_{ij}(\eta, \xi)$ which, upon substitution in Eq. 7.19, becomes

$$R_{12}(\eta, \xi) = \sum_{i=1}^{N} \frac{I_{0_i} \, \sigma_i^2}{8} \, e^{-\frac{(\eta - \Delta x_i)^2 + (\xi - \Delta y_i)^2}{\sigma_i^2}}$$

$$+ \sum_{\substack{i=1, j=1 \\ i \neq j}}^{N} \frac{I_{0_i} \, I_{0_j} \, \sigma_i^2 \sigma_j^2}{4\left(\sigma_i^2 + \sigma_j^2\right)} \, e^{-\frac{2\left((\eta - (x_j - x_i + \Delta x_j))^2 + (\xi - (y_j - y_i + \Delta y_j))^2\right)}{\sigma_i^2 + \sigma_j^2}}$$

$$\tag{7.26}$$

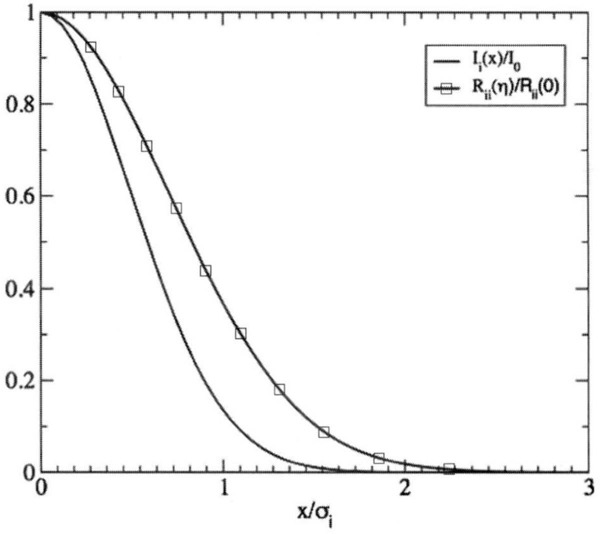

Fig. 7.11. This figure illustrates the relationship between the size of an Gaussian image distribution and its corresponding auto-correlation in one dimension.

If the motion of the tracer particles represented by Gaussian intensity models is further assumed to be homogeneous within the IW, given by Eq.

7.20, then Eq. 7.26 simplifies to:

$$R_{12}(\eta, \xi) = \sum_{i=1}^{N} \frac{I_{0_i}\,\sigma_i^2}{8}\, e^{-\frac{(\eta-\Delta x)^2+(\xi-\Delta y)^2}{\sigma_i^2}}$$

$$+\,\mathbf{NOISE} \tag{7.27}$$

For clarity the second sum in Eq. 7.26, which represents the uncorrelated noise contribution, is referred to as NOISE in Eq. 7.27. Equation 7.27 clearly shows the constructive superposition of all the Gaussian-like autocorrelations corresponding to each particle image within the IW which make up the peak in the cross-correlation of the two single exposed images. This peak in the cross-correlation function between two single exposed images will always be larger than the size of the particle images. This is the case even if the particle images are not Gaussian-like.

7.4.   Spatial resolution

Spatial resolution plays an undervalued role in PIV that if not taken appropriately into account can have disastrous effects on the accuracy of PIV measurements. Before considering two of the major effects of spatial resolution in detail it is useful to consider the spectral response of PIV by assuming that a PIV measurement is a continuous signal. This permits an analysis to be carried out by computing the spectral response due to averaging procedure over the IW. Thus, providing information of how the size of the IW relative to the spatial variation of the velocity field via the characteristic length scale in the form of the wavelength affects the velocity measurement. The results of this analysis is shown in Fig. 7.12.. It indicates that if the IW has a size $l_x = 0.2\,\lambda_x$, where λ_x is the wavelength of the spatial variation of the velocity in the $x - direction$, the measured velocity is underestimated by 6%. Additional numerical results are provided in the table provided in Fig. 7.12..

One of the major results of lack of spatial resolution is the destruction of the cross-correlation peak, which locates the average displacement of the tracer particles within the IW in the analysis of single exposed PIV image pairs. This has the consequence of the complete failure of cross-correlation analysis to determine the average velocity within an IW, even for constant velocity gradients within the IW. This effect is analysed in the following subsection.

The other major effect of spatial resolution is its effect on the determination of the velocity gradient tensor components within acceptable precision even though velocity vectors can be determined. This effect is analysed in the second subsection.

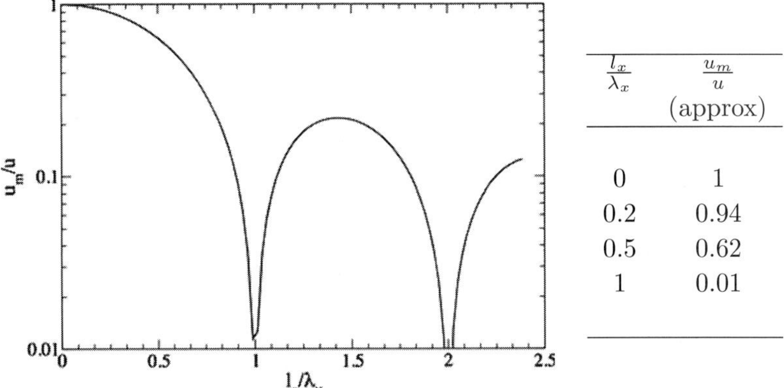

Fig. 7.12. Spectral response of PIV. u_m is the measured velocity while u represents the actual velocity, l_x is the size of the *IW* and the length over which the velocity is integrated, while λ_x represents the spatial wavelength of the velocity signal.

7.4.1. *The effect of velocity gradients on cross-correlation analysis of single exposed image pairs*

The generality of Eq. 7.19 permits its use to analyze the effect of velocity gradients on the cross-correlation function between two co-located interrogation windows extracted from single exposed image pairs and the ability to determine the location of the global cross-correlation maximum and hence, the average velocity within the IW. It is appropriate to begin with an illustration of the effect of a constant shear strain rate on the cross-correlation function shown in Fig. 7.13.. This data shows the effect of shear strain field only with varying values of s_{12} such that within the IW the value of $\frac{|\Delta u|}{|u|}$ ranges from 0 to 0.51.

The illustration in Fig. 7.13. clearly shows the disastrous effect that a velocity shear gradient across the IW has on the fidelity of the cross-correlation as the magnitude of the velocity gradient increases. The effect of velocity gradients on the determination of the local velocity within an IW using cross-correlation PIV analysis of single exposed image pair can be studied analytically in more detail using the two-dimensional model of a linear fluid velocity field within the IW given by Eq. (7.28).

$$u(x, y) = u_0 + a_{11}\, x + a_{12}\, y$$

$$v(x, y) = v_0 + a_{21}\, x + a_{22}\, y$$

$$(7.28)$$

where a_{ij} is the velocity gradient tensor, which is assumed constant within the IW and can be decomposed into: (i) s_{ij}, the symmetric rate-of-strain

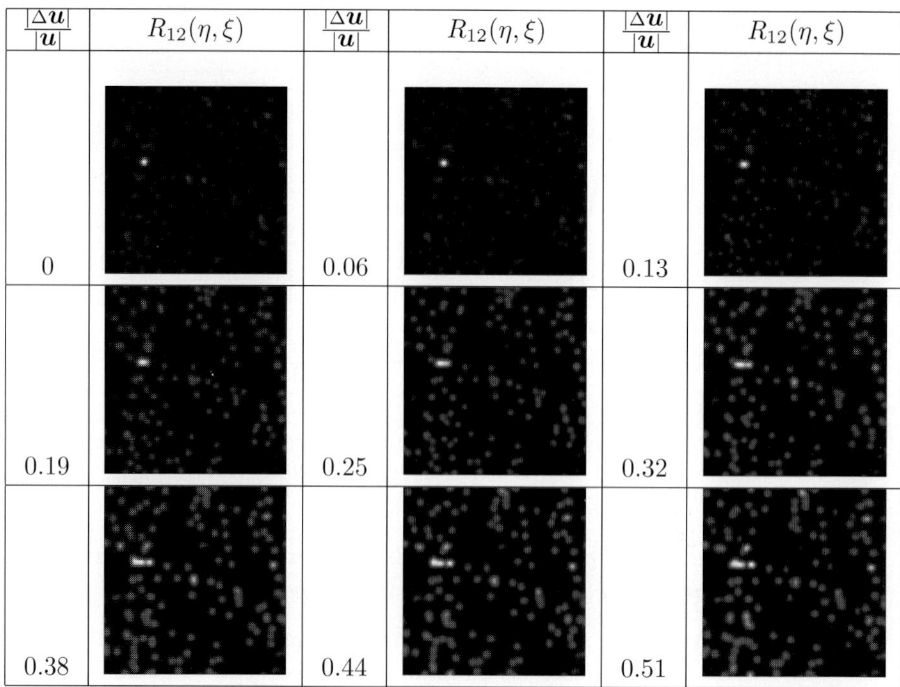

Fig. 7.13. Cross-correlation functions for the same mean displacement of $\Delta x_0 = -0.3$, $\Delta y_0 = 0.1$ with superimposed constant strain-rate s_{12} of increasing strength such that $\frac{|\Delta u|}{|u|}$ varies from 0 to 0.51. The origin of $(\Delta x, \Delta y)$ coordinate system is located in the centre of each cross-correlation function.

tensor and (ii) w_{ij}, the skew-symmetric rate-of-rotation tensor, such that

$$a_{ij} = s_{ij} + w_{ij}. \tag{7.29}$$

Note that as previously stated, the origin of the local (x, y) coordinate system attached to the IW is located at the centre of the IW. Therefore, within the IW using Eqs. (7.28) and (7.29) the fluid displacement between the two exposures during the time delay Δt is given by

$$\Delta x(x, y) = \Delta x_0 + a_{11}\, x\, \Delta t + a_{12}\, y\, \Delta t$$

$$\tag{7.30}$$

$$\Delta y(x, y) = \Delta y_0 + a_{21}\, x\, \Delta t + a_{22}\, y\, \Delta t$$

where $(\Delta x_0, \Delta y_0) = (u_0\, \Delta t, v_0\, \Delta t)$ represents the fluid displacement at the centroid of the interrogation window, *i.e.* the fluid displacement at the origin of the (x, y) coordinate system. Thus, the displacement of the image of the ith tracer particle located at (x_i, y_i) can be written as

$$\Delta x_i = \Delta x(x_i, y_i) = \Delta x_0 + a_{11}\, x_i\, \Delta t + a_{12}\, y_i\, \Delta t$$

$$\Delta y_i = \Delta y(x_i, y_i) = \Delta y_0 + a_{21}\, x_i\, \Delta t + a_{22}\, y_i\, \Delta t. \tag{7.31}$$

The incompressibility restriction is now imposed for convenience as it allows the development of analytical relationships that expose the effect of the velocity gradients on the cross-correlation PIV analysis more clearly. Due to incompressibility $s_{22} = -s_{11}$ and the tensor properties of s_{ij} and w_{ij}, the velocity gradient tensor for a purely two-dimensional flow field is given by

$$\begin{aligned}
a_{11} &= s_{11} \\
a_{12} &= s_{12} + w_{12} \\
a_{21} &= s_{12} - w_{12} \\
a_{22} &= -s_{11}.
\end{aligned} \tag{7.32}$$

Note that in general 2C-2D PIV measures the projected in-plane velocity field of a 3C-3D velocity field and that this fact introduces additional errors that can be analysed in the framework presented by Eq. (7.19), but are not considered here. Furthermore, the remainder of the analysis is only concerned with the effect of velocity gradients within the interrogation window on the coherent motion represented in $R_{12}(\eta, \xi)$, *i.e.* the first sum of Eq. (7.19) - this is were the damage is done by the velocity gradients. It is understood that in addition to this detrimental effect on $R_{12}(\eta, \xi)$ there is uncorrelated noise, given via the second sum of Eq. (7.19), that also reduces the signal-to-noise ratio of the cross-correlation PIV analysis further.

It is convenient to use polar coordinates to represent the location of the ith tracer particle image within the IW relative to the centroid of the IW, *i.e.*

$$\begin{aligned}
x_i &= r_i \, \cos(\theta_i) \\
y_i &= r_i \, \sin(\theta_i)
\end{aligned} \tag{7.33}$$

where r_i is the radial distance of the tracer particle image from the centroid of the IW and θ_i is the angular position measured counter-clockwise from the horizontal $x - axis$. Equation (7.19) indicates that the signal contribution of the ith particle image is given by $R_{ii}(\eta - \Delta x_i, \xi - \Delta y_i)$ - the auto-correlation function of the ith particle with its maximum located at:

$$\begin{aligned}
\eta_i &= \Delta x_i \\
\xi_i &= \Delta y_i.
\end{aligned} \tag{7.34}$$

Using Eqs. (7.31) and (7.33) in Eq. (7.34) yields the location of the maximum of $R_{ii}(\eta - \Delta x_i, \xi - \Delta y_i)$ as:

$$\begin{aligned}
\eta_i &= \Delta x_0 + a_{11}\, r_i\, \cos(\theta_i)\, \Delta t + a_{12}\, r_i\, \sin(\theta_i)\, \Delta t \\
\xi_i &= \Delta y_0 + a_{21}\, r_i\, \cos(\theta_i)\, \Delta t + a_{22}\, r_i\, \sin(\theta_i)\, \Delta t.
\end{aligned} \tag{7.35}$$

 Julio Soria

From Eqs. (7.19) and (7.35) it is clearly deducible that the signal component of $R_{12}(\eta, \xi)$, which is given by the superposition of the individual auto-correlation functions of each tracer particle within the IW, $R_{ii}(\eta - \Delta x_i, \xi - \Delta y_i)$, is dependent on the magnitude of the components of a_{ij} and the position of the tracer particle given by (r_i, θ_i). More specifically with regards to a_{ij} from Eq. (7.32) it is dependent on the magnitudes of s_{11}, s_{12} and w_{12}. The following specific characteristics can be deduced:

(1) **Average Displacement:** the average displacement within the IW of the linear velocity field Eq. (7.28) is given by $(u_0 \, \Delta t, v_0 \, \Delta t)$. This is the quantity that cross-correlation PIV analysis of single exposed image pairs aims to measure. If there are no velocity gradients present in the IW, then all particles contribute to the signal part of Eq. (7.19) as previously demonstrated by Eq. (7.21). If there are velocity gradients present in the IW, the average displacement is only contributed to exactly by the tracer particle that is located at the centroid of the IW during the first exposure of the single exposed PIV image pair, because $r_i = 0$ and the maximum of $R_{ii}(\eta - \Delta x_i, \xi - \Delta y_i)$ of this tracer particle image contributing to the signal part of Eq. (7.19) is located at

$$\begin{aligned}
\eta_0 &= \Delta x_0 = u_0 \, \Delta t \\
\xi_0 &= \Delta y_0 = v_0 \, \Delta t.
\end{aligned} \tag{7.36}$$

This is truely independent of the strength of the velocity gradients.

(2) **General Location of the $R_{ii}(\eta - \Delta x_i, \xi - \Delta y_i)$ Maximum:** from Eq. (7.35) the location of the maximum of $R_{ii}(\eta - \Delta x_i, \xi - \Delta y_i)$ when $r_i \neq 0$ is in general given by

$$\begin{aligned}
\eta_i &= \Delta x_0 + s_{11} \, r_i \cos(\theta_i) \, \Delta t + (s_{12} + w_{12}) \, r_i \sin(\theta_i) \, \Delta t \\
\xi_i &= \Delta y_0 + (s_{12} - w_{12}) \, r_i \cos(\theta_i) \, \Delta t - s_{11} \, r_i \sin(\theta_i) \, \Delta t.
\end{aligned} \tag{7.37}$$

This equation shows that when there are velocity gradients present within the IW, the location of the maximum of $R_{ii}(\eta - \Delta x_i, \xi - \Delta y_i)$ due to a tracer particle image that is located during the first exposure at (r_i, θ_i) is distributed around the location of (η_0, ξ_0). The position of this maximum relative to (η_0, ξ_0) is given by:

$$\begin{aligned}
\Delta \eta_i &= s_{11} \, r_i \cos(\theta_i) \, \Delta t + (s_{12} + w_{12}) \, r_i \sin(\theta_i) \, \Delta t \\
\Delta \xi_i &= (s_{12} - w_{12}) \, r_i \cos(\theta_i) \, \Delta t - s_{11} \, r_i \sin(\theta_i) \, \Delta t.
\end{aligned} \tag{7.38}$$

The distance of the maximum of $R_{ii}(\eta - \Delta x_i, \xi - \Delta y_i)$ from (η_0, ξ_0) is given by:

$$\rho_i = \Delta t \, r_i \sqrt{s_{11}^2 + s_{12}^2 + w_{12}^2 + 2 \, w_{12} \left(s_{11} \sin(2\theta_i) - s_{12} \cos(2\theta_i) \right)} \tag{7.39}$$

Defining β_i as the counter-clockwise angle from the $\eta - axis$ made by a line joining (η_0, ξ_0) to (η_i, ξ_i) the angular orientation of the location of the maximum of $R_{ii}(\eta - \Delta x_i, \xi - \Delta y_i)$ from (η_0, ξ_0) is given by:

$$\tan \beta_i = -\frac{s_{11} \sin(\theta_i) + (-s_{12} + w_{12}) \cos(\theta_i)}{s_{11} \cos(\theta_i) + (s_{12} + w_{12}) \sin(\theta_i)}. \tag{7.40}$$

Equation (7.39) shows that the distance of the maximum of $R_{ii}(\eta - \Delta x_i, \xi - \Delta y_i)$ from (η_0, ξ_0) is directly proportional to Δt and r_i and dependent in a more complicated fashion on the velocity gradient field and the relative orientation of the tracer particle image to the velocity gradient field.

(3) **Location of the $R_{ii}(\eta - \Delta x_i, \xi - \Delta y_i)$ Maximum In Irrotational Flow:**
in irrotational flow $w_{ij} = 0$ and Eq. (7.39) simplifies to

$$\rho_i = \Delta t\, r_i \sqrt{s_{11}^2 + s_{12}^2} = \Delta t\, r_i \alpha \tag{7.41}$$

where α is the stretching principal strain rate which corresponds to the largest eigenvalue of the rate-of-strain tensor. Note that the contracting principal strain rate is given by $-\alpha$. This implies that the distance of the maximum of $R_{ii}(\eta - \Delta x_i, \xi - \Delta y_i)$ from (η_0, ξ_0) is also directly proportional to the strength of the irrotational velocity gradient field via α. The orientation of the location of the maximum of $R_{ii}(\eta - \Delta x_i, \xi - \Delta y_i)$ from (η_0, ξ_0) is given by

$$\tan \beta_i = -\frac{s_{11} \sin(\theta_i) - s_{12} \cos(\theta_i)}{s_{11} \cos(\theta_i) + s_{12} \sin(\theta_i)}. \tag{7.42}$$

(4) **Location of the $R_{ii}(\eta - \Delta x_i, \xi - \Delta y_i)$ Maximum In Purely Rotational Flow:**
in purely rotational flow $s_{ij} = 0$ and Eq. (7.39) simplifies to

$$\rho_i = \Delta t\, r_i |w_{12}| = \frac{\Delta t\, r_i\, \omega}{2} \tag{7.43}$$

where ω is the vorticity of the rotational flow. This implies that the distance of the maximum of $R_{ii}(\eta - \Delta x_i, \xi - \Delta y_i)$ from (η_0, ξ_0) is also directly proportional to the strength of the purely rotational flow via ω. The orientation of the location of the maximum of $R_{ii}(\eta - \Delta x_i, \xi - \Delta y_i)$ from (η_0, ξ_0) is given by

$$\tan \beta_i = -\cot(\theta_i) \tag{7.44}$$

This analysis shows that the maximum of $R_{12}(\eta, \xi)$ in Eq. (7.19) is deteriorated by the velocity gradients in the IW via displacements of the individual maxima of the $R_{ii}(\eta - \Delta x_i, \xi - \Delta y_i)$ in the first sum of Eq. (7.19). The displacement of the maximum of the $R_{ii}(\eta - \Delta x_i, \xi - \Delta y_i)$ depends directly on the distance that the ith tracer particle image is from the centre of the IW, the exposure time, the strength of the irrotational and rotational fields and orientation the ith tracer particle image. Another implicit dependence is the size of the tracer particle image.

It is conceivable that in a worst case scenario the location of the average velocity is in fact a crater surrounded by the individual maxima of the $R_{ii}(\eta - \Delta x_i, \xi - \Delta y_i)$ due to the tracer particles that are not located at the centre of the IW. This is a possibility because as the distance from the centre of the IW increase so does the surface area where tracer particles are possible and hence, and therefore this scenario is likely. Naturally this depends also to a large extend on the number of tracer particle images within the IW. Equation (7.39) provides the relative distance of location of the maximum of $R_{ii}(\eta - \Delta x_i, \xi - \Delta y_i)$ to $(\Delta x_0, \Delta y_0)$ to the velocity magnitude.

$$\frac{\rho_i}{\sqrt{\eta_0^2 + \xi_0^2}} = \frac{r_i}{\sqrt{u_0^2 + v_0^2}} \sqrt{s_{11}^2 + s_{12}^2 + w_{12}^2 + 2\,w_{12}\left(s_{11}\sin\left(2\theta_i\right) - s_{12}\cos\left(2\theta_i\right)\right)}$$

$$(7.45)$$

The effects that have been analytically deduced above are illustrated in Fig. 7.14.. The results presented in Fig. 7.14. represent only the first sum of Eq. (7.19), computed using 20 randomly position tracer particle images of normalised size given by $\sigma = 0.09375$. This particle size is equivalent to a particle with a $\sigma = 1.5\,px$ when a square IW is 16 px in size. The size of the IW is normalised ranging from -0.5 to 0.5 in the independent variables (x, y) and (η, ξ). The gray contourplot of $R_{12}(\eta, \xi)$ is normalised by the value of $R_{12}(\Delta x_0, \Delta y_0)$ without any velocity gradients - the maximum normalised $R_{12}(\eta, \xi)$ is 1. A normalised average displacement of $(\Delta x_0 = 0.133, \Delta y_0 = 0.133)$ was used in this illustration, although without loss of generality the effects shown in Fig 7.14. could have been demonstrated with an average displacement of 0 of the tracer particles between the two single exposed images. The velocity gradient field has both rate-of-strain components and rate-of-rotation component. The former was varied, while the latter was kept constant for the cases considered. The conclusions do not change because of this. The range of values for the rate-of-strain components are given in Fig. 7.14.. Table 7.4.1. gives the corresponding values for the magnitude of the principal stretching strain rate α, the vorticity ω and the normalised $R_{12}(\Delta x_0, \Delta y_0)$ for the cases given in Fig. 7.14. (a) - (h). The values of the rate-of-strain components, rate-of-rotation component, α, and ω represent consistently normalised values.

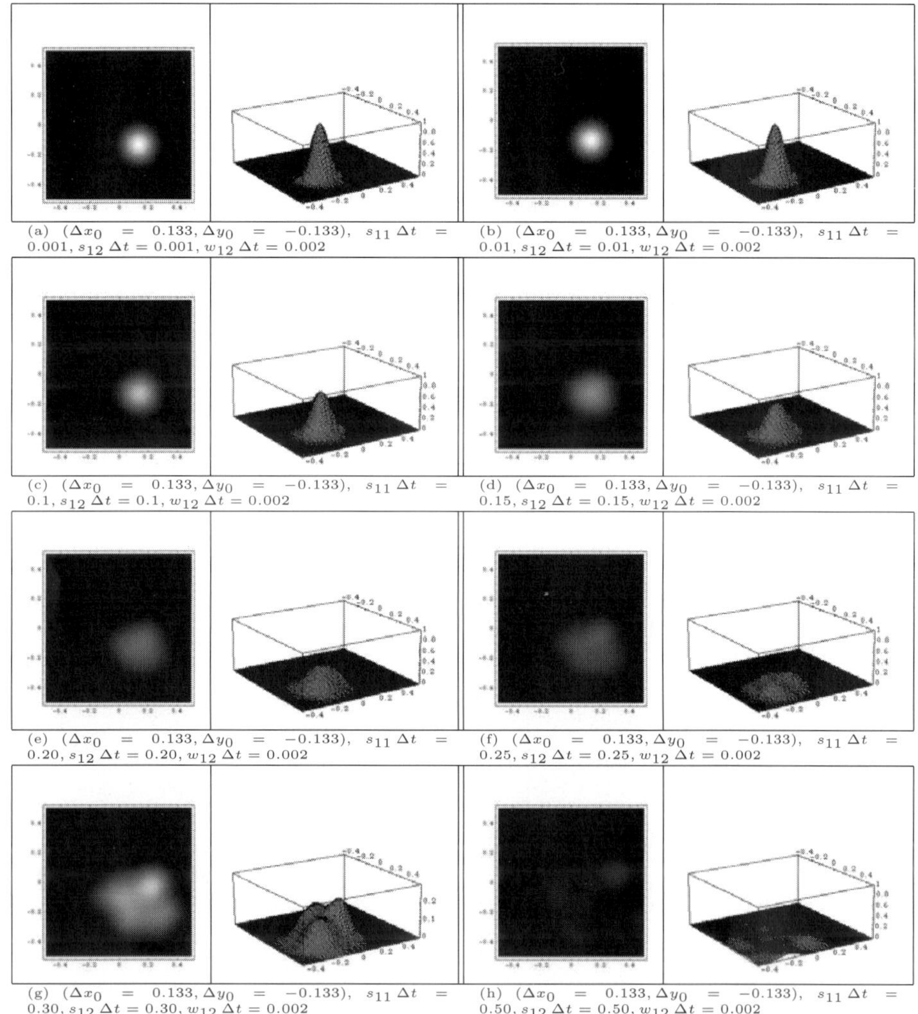

(a) $(\Delta x_0 = 0.133, \Delta y_0 = -0.133)$, $s_{11}\Delta t = 0.001$, $s_{12}\Delta t = 0.001$, $w_{12}\Delta t = 0.002$

(b) $(\Delta x_0 = 0.133, \Delta y_0 = -0.133)$, $s_{11}\Delta t = 0.01$, $s_{12}\Delta t = 0.01$, $w_{12}\Delta t = 0.002$

(c) $(\Delta x_0 = 0.133, \Delta y_0 = -0.133)$, $s_{11}\Delta t = 0.1$, $s_{12}\Delta t = 0.1$, $w_{12}\Delta t = 0.002$

(d) $(\Delta x_0 = 0.133, \Delta y_0 = -0.133)$, $s_{11}\Delta t = 0.15$, $s_{12}\Delta t = 0.15$, $w_{12}\Delta t = 0.002$

(e) $(\Delta x_0 = 0.133, \Delta y_0 = -0.133)$, $s_{11}\Delta t = 0.20$, $s_{12}\Delta t = 0.20$, $w_{12}\Delta t = 0.002$

(f) $(\Delta x_0 = 0.133, \Delta y_0 = -0.133)$, $s_{11}\Delta t = 0.25$, $s_{12}\Delta t = 0.25$, $w_{12}\Delta t = 0.002$

(g) $(\Delta x_0 = 0.133, \Delta y_0 = -0.133)$, $s_{11}\Delta t = 0.30$, $s_{12}\Delta t = 0.30$, $w_{12}\Delta t = 0.002$

(h) $(\Delta x_0 = 0.133, \Delta y_0 = -0.133)$, $s_{11}\Delta t = 0.50$, $s_{12}\Delta t = 0.50$, $w_{12}\Delta t = 0.002$

Fig. 7.14.   Cross-correlation function with contribution from the *signal* part of Eq. (7.19) only. Twenty randomly positioned particles all with the same $\sigma = 0.09375$ were used to calculate the *signal* contribution to the cross-correlation function.

Figure 7.14. clearly shows that for small velocity gradients (a) - (b) there is very little degradation in the observability of the cross-correlation function maximum that identifies the location of the average displacement of the tracer particles within the IW, *i.e.* . for normalised strain rates less than $0.014/\Delta t$ the normalised cross-correlation function maximum has a value greater than 0.99. However, for normalised strain rates larger than $0.014/\Delta t$, the normalised cross-correlation function maximum deteriorates fast and when $\alpha\,\Delta t = 0.078$, the normalised cross-correlation function maximum has dropped to a value of 0.91. The cross-correlation function around its maximum begins to flatten and the cross-correlation peak broadens as the strain rates increase. such that when $\alpha\,\Delta t = 0.424264$, the normalised cross-correlation function value at $(\Delta x_0, \Delta y_0)$ has a value of 0.17 and is surrounded by other maxima was deduced by the preceding analysis and clearly demonstrated in Fig. 7.14. (g). Note that the cross-correlation functional scale in Fig. 7.14. (g) ranges up to 0.3 not 1.0.

Case	$\alpha\,\Delta t$	$\omega\,\Delta t$	Normalised $R_{12}(\Delta x_0, \Delta y_0)$
(a)	0.00141421	0.001	0.999887
(b)	0.0141421	0.001	0.996609
(c)	0.141421	0.001	0.741763
(d)	0.212132	0.001	0.535809
(e)	0.282843	0.001	0.36528
(f)	0.353553	0.001	0.245905
(g)	0.424264	0.001	0.169404
(h)	0.707107	0.001	0.062025

The conclusion that can be drawn from this analysis is that in order to keep the cross-correlation peak above a normalised value of 0.9 the velocity gradient field must not result in relative displacements across the IW that are larger than 7.8% of the IW size, and if the relative displacements across the IW due to velocity gradients can be kept to less than 1.4% of the IW size than the normalised cross-correlation will have a value of not less than 0.99. Of course this analysis has ignored in the uncorrelated noise component in Eq. (7.19), which may reduce these normalised cross-correlation values.

7.4.2. *Effective wavenumber relationships: accuracy of spatial numerical differentiation of PIV measurements*

One of the major post-processing and analysis of measured PIV velocity fields is the measurement of derived quantities such as velocity gradients and vorticity. The following analysis will derive and discuss the requirement for high spatial resolution in PIV measurements when the data post-processing of the PIV measurements involves spatial differentiation. The following analysis is not only applicable to instantaneous PIV velocity measurements, but is more general and its results are applicable to: (i) the calculation of phase-averaged velocity gradients and vorticity from hot-wire anemometry or laser Doppler anemometry measurements; (ii) the calculation of mean velocity gradients and vorticity from hot-wire anemometry or laser Doppler anemometry measurements and (iii) the calculation of spatial gradients of higher order statistics from hot-wire anemometry or laser Doppler anemometry measurements.

Consider that the PIV velocity data is available on a regular grid in a x-y Cartesian coordinate system – this is usually the case for PIV measurements and the other velocity measurements as shown in Fig. 7.15..

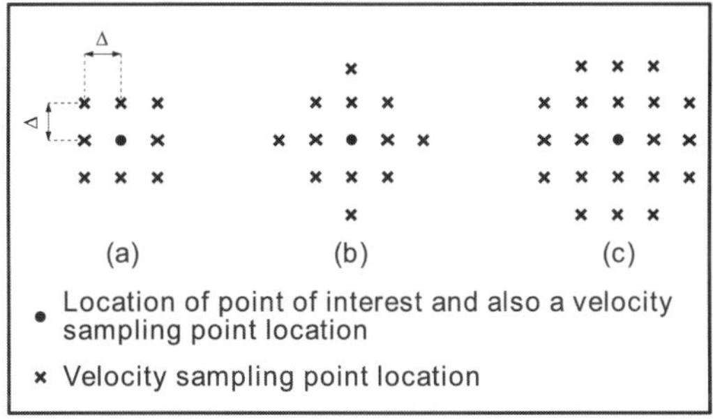

Fig. 7.15. Structured grid spacing showing spatial discretisation and location where PIV data is available.

There are a number of methods available for the spatial differentiation of experimental velocity data:[37] (i) Centered finite-differencing (FD), (ii) Adaptive Gaussian filtering with centered finite differencing (AGW$_x$–FD), where "$_x$" denotes number of sampled grid points used in the operation and (iii) χ^2 local fit of a second order polynomial to the velocity components with analytic differentiation – χ^2_x.

To analyse the wavenumber characteristics (*i.e.* the filter characteristics) of each of these differentiation techniques, let

$$f(x) = e^{i\,k_x x} \tag{7.46}$$

then exact differentiation yields

$$f'(x) = ik_x\, e^{i\,k_x x} = ik_x\, f(x). \tag{7.47}$$

For any numerical differentiation scheme an effective wavenumber can be defined

$$f'(x) = ik_{x\,eff}\, f(x). \tag{7.48}$$

A measure of the accuracy of the numerical differentiation scheme is obtained by comparing k_x with $k_{x\,eff}$ for each scheme. A similar process can be implemented for the more general:

$$f(x,y) = e^{i\,k_x x}\, e^{i\,k_y y}. \tag{7.49}$$

Defining a non-dimensional wavenumber:

$$k'_i \equiv k_i\, \Delta/\pi \tag{7.50}$$

where Δ = sampling spacing between the velocity data, which is assumed to be the same in the x and y directions. The non-dimensional effective wavenumber can be analytically derived for each numerical differentiation scheme, *e.g.* for the χ_9^2 and FD numerical differentiation schemes, which are identical:

$$k'_{x\,eff} = \frac{\sin(\pi\, k'_x)}{\pi}. \tag{7.51}$$

The non-dimensional effective wavenumber for higher order methods are more complex and involve cross-talk between the waves in the x and y directions, *e.g.* for the χ_{21}^2 numerical differentiation scheme:

$$k'_{x\,eff} = \frac{1}{65\,\pi}\left(4\sin(\pi\,(\,k'_x + k'_y\,)) - 5\sin(\pi\,(\,k'_x + 2\,k'_y\,))\right. \tag{7.52}$$

$$-4\sin(\pi\,(\,-k'_x + k'_y\,)) + 5\sin(\pi\,(\,-k'_x + 2\,k'_y\,)) \tag{7.53}$$

$$-8\sin(\pi\,(\,-2\,k'_x + k'_y\,)) + 8\sin(\pi\,(\,2\,k'_x + k'_y\,)) \tag{7.54}$$

$$+14\sin(\,2\,\pi\,k'_x\,) + 7\sin(\pi\,k'_x\,)) \tag{7.55}$$

Figure 7.16. shows the spectral response characteristics of a number of different differentiation schemes. Figure 7.16. shows that for all numerical differentiation schemes except for the χ_9^2 and FD schemes, which are identical, significant attenuation of information and/or noise occurs if $k'_x > 0.1$. Therefore, in order to minimise the attenuation of high wavenumber components during spatial differentiation it is necessary for

$$\Delta \leq \frac{\lambda}{20} \tag{7.56}$$

where λ = wavelength or a "characteristic length scale".

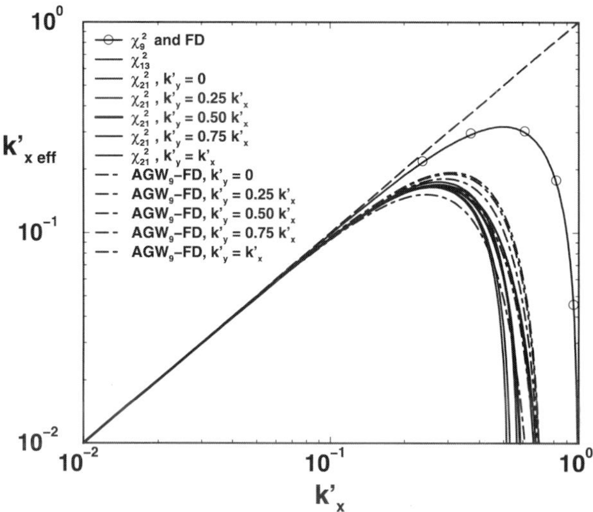

Fig. 7.16.   Non-dimensional effective wavenumber versus non-dimensional wavenumber for a number of different differentiation schemes. Differentiation is in the $x - direction$.

The effect of lack of spatial resolution is illustrated by considering experimental PIV measurements and vorticity measurements of a vortex ring. the experimental details of the vortex ring apparatus, the PIV experimental setup and PIV analysis details can be found in.[37–40] Figure 7.17. clearly demonstrates the effect of spatial resolution both on the velocity and more specifically on the out-of-plane vorticity fields. Low spatial resolution as the preceding analysis demonstrated results in smoothing of the data and loss of detailed features. In this case the vortex ring is not axisymmetric, but low spatial resolution would us believe from the vorticity that it is axisymmetric, whereas high spatial resolution clearly shows that it is not. These results are further emphasised in Fig. 7.18., which show measured vorticity profiles of the vortex ring in two orthogonal directions for a range of different sampling spacings of the velocity. These results show that low spatial resolution results in large bias error in the form of underestimation of the vorticity. The random uncertainty of the vorticity, which can also be computed from the uncertainty of the velocity using the methodology introduced by,[37] shows a smaller error than the bias error due to low spatial resolution.

  Julio Soria

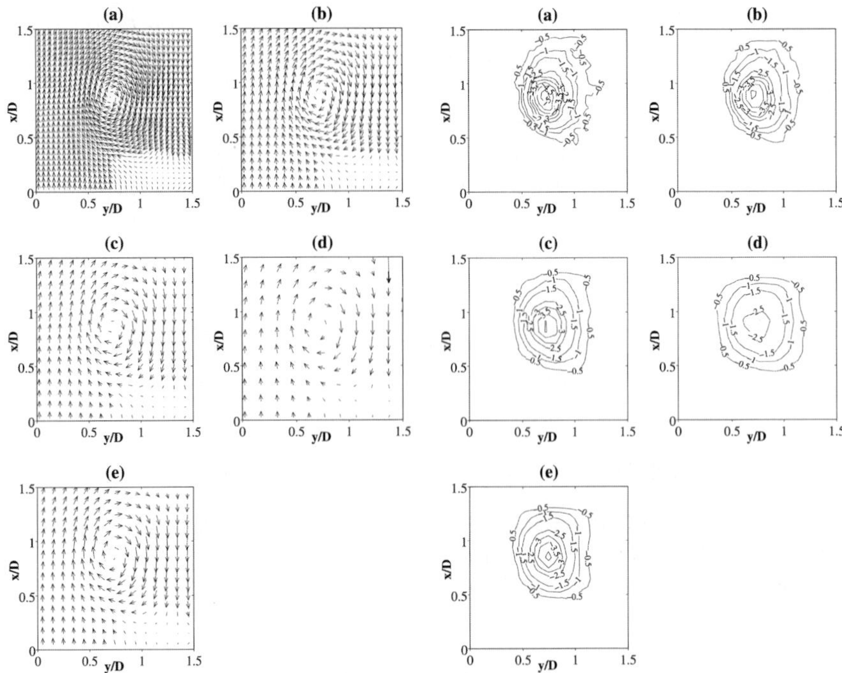

Fig. 7.17. Velocity and out-of-plane vorticity measurements using PIV of a vortex ring using different sized IW and sampling spacing between the velocity vectors.

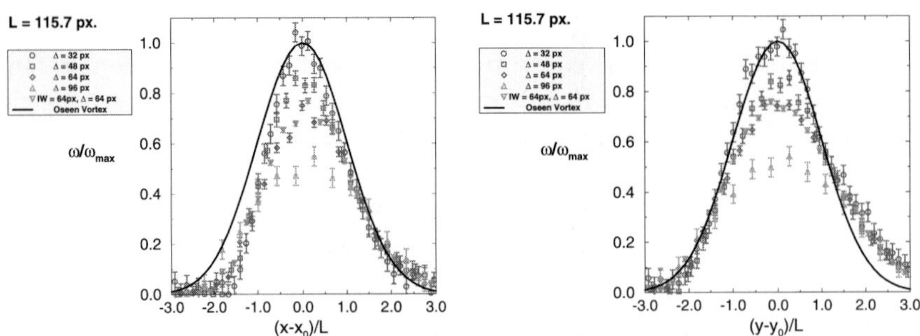

Fig. 7.18. Vorticity profiles in two orthogonal directions of the vortex ring.

7.5. Example of PIV - turbulent round jet

This section illustrates some of the issues in the light of the previous analysis and discussion that need to be taken into account when using PIV to measure turbulent flows through the example of PIV measurements of a turbulent jet. The same issues arise in other turbulent flows (*e.g.* other turbulent free shear flows, turbulent wall-bounded flows, and turbulent rotational flows) and these can be considered using the framework developed in this section as well as the previous sections.

The measurements were undertaken using 2C-2D multigrid cross-correlation digital PIV (MCCDPIV)[36,39–42] in the near region up to $x/d < 9$ of a round jet at a Reynolds number based on diameter (D_0) and average velocity $(\overline{U}_0)$ at the orifice of 4000 as an example of PIV in a turbulent flow. The experiments were conducted using water as the working fluid.

7.5.1. *Round jet geometry, coordinates and parameters*

The geometry and coordinate system for the free round jet flow is depicted in Figure 7.19.. The primary parameter governing the free round jet flow is the Reynolds number Re defined as

$$Re = \frac{\overline{U}_0 \ D_0}{\nu} \tag{7.57}$$

where $\overline{U}_0$ is the mean velocity of the jet at the orifice, D_0 is the diameter of the orifice and ν is the kinematic viscosity of the fluid. The experimental data presented in this manuscript pertains to a free round jet at $Re = 4000$.

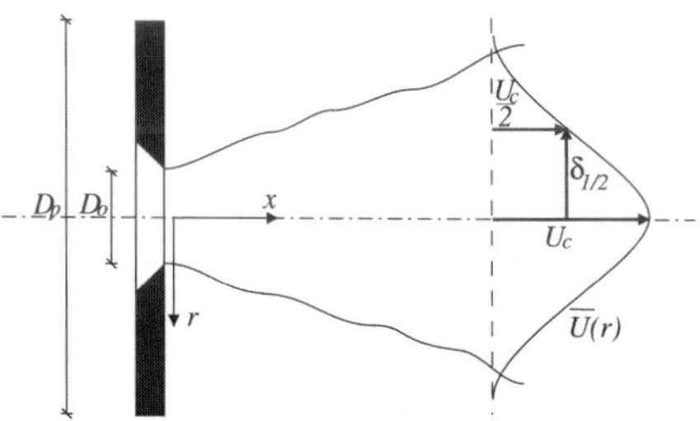

Fig. 7.19.   Geometry and parameters of round jet, orifice and coordinate system.

Details of the experimental apparatus and methodology for the PIV results presented here can be found in.[43]

7.5.2.   *Turbulence scales in turbulent jets: estimated Taylor and Kolmogorov micro-scales*

Turbulent jets contain motions with a broad range of length scales. The smallest of the turbulent length scales are much larger than molecular scales, which justifies the assumption that turbulence is a continuum phenomenon. This smallest turbulent length scale is known as the Kolmogorov micro-scale and is defined as

$$\eta = \left(\frac{\nu^3}{\epsilon}\right)^{\frac{1}{4}} \tag{7.58}$$

where ϵ is the dissipation of turbulent kinetic energy per unit mass. Equation 7.58 indicates that the smallest scales are set by the viscosity and the rate at which energy is supplied by the largest-scale eddies. The scale at which most of the energy resides is called l, the integral length scale of the turbulence. Because turbulence kinetic energy is extracted from the mean flow at these largest scales, they are often referred to as the energy-containing range. The integral length scale is a measure of the largest separation distance over which components of the eddy velocities at two distinct points are correlated. Thus, l characterizes the energy-containing range of eddy length scales. Intermediate between these two scales are the inertial subrange scales for which turbulence kinetic energy is neither generated nor destroyed but is transferred from larger to smaller scales. Smaller-scale eddies are generated from the larger eddies through the nonlinear process of vortex stretching. Typically, energy is transferred from the largest eddies to the smallest ones on a timescale of about one large-eddy turnover, characterised by l/u, where u is the characteristic root-mean-square of the fluctuating velocity. In between these two length scales there is also the Taylor micro-scale, λ, which is the length scale at which viscous dissipation begins to affect the eddies. Thus, it marks the transition from the inertial subrange to the dissipation range and characterizes the eddy scale for the inertial subrange eddies.

In the context of this chapter the spatial resolution is of importance and in particular the ability of instantaneous non-intrusive whole field velocimetry techniques like PIV or the version used here, the multigrid cross-correlation digital PIV (MCCDPIV), to resolve all relevant dynamic length scales. Therefore, it is of interest to estimate a priori the size of the Taylor micro-scale and Kolmogorov micro-scale for a free round turbulent jet. The estimation of these two length scales can be carried out using classical scaling arguments[44] (see Dimotakis[45]) and making a number of assumptions about the turbulence in the far field of a a jet (see Dowling & Dimotakis[46]) yielding an estimate for the Taylor micro-scale,

$$\frac{\lambda}{\delta_{\frac{1}{2}}(x)} \approx 2.3\, Re^{-\frac{1}{2}} \tag{7.59}$$

where $\delta_{\frac{1}{2}}(x)$ is the characteristic half-width or radius of the jet and x is the axial flow direction of the jet. The Kolmogorov micro-scale can be estimated as,[45]

$$\frac{\eta}{\delta_{\frac{1}{2}}(x)} \approx 0.95\, Re^{-\frac{3}{4}}. \tag{7.60}$$

A round jet is a constant Reynolds number flow where the width of the jet increases in the axial flow direction. Therefore, both the λ and η are expected to increase with x. In which case the most restrictive conditions regarding the spatial resolution would occur near the jet origin (Note that the derivations of the estimates λ and η assumed far field jet conditions and strictly speaking they do not apply near the jet origin) and thus for the present purpose $\delta_{\frac{1}{2}}(x) = d/2$. The estimated length scales are given in Table 7.5.2..

Re	Re_T	$\delta_{\frac{1}{2}}(0)$ (mm)	λ (mm)	η (mm)
2000	62.6	5	0.26 $(0.026 D_0)$	0.21 $(0.021 D_0)$
4000	88.5	5	0.18 $(0.018 D_0)$	0.16 $(0.016 D_0)$

7.5.3. *Issues with spatial resolution: velocity measurement*

Assume that the velocity sampling distance is given by Δ and that the PIV measurement volume is given by $IW^2 \times \Delta_{laser}$, where IW is the square interrogation window length and Δ_{laser} is the thickness of the laser sheet that is used to illuminate the tracer particles. For the subsequent analysis it will be assumed that $\Delta = IW$, *i.e.* the measurement integration is equivalent to velocity sampling. However, more typical is that $\Delta = IW/2$, *i.e.* two times oversampling. From Nyquist criterion of discrete sampling, the smallest resolved length scale is given by:

$$l_{resolved} \geq 2\,\Delta. \tag{7.61}$$

The largest scales in the turbulent flow will be of the size of the half width of the jet, $\delta_{1/2}(x)$ and therefore the the largest imaging area required will be $= 2 \times (3\,\delta_{1/2}(x))$, *i.e.* $(3\,\delta_{1/2}(x))$ on either side of the centre line of the round jet if no symmetry of the jet is assumed. Thus, the following ratio can be formed

$$\frac{\text{Largest Imaging Area}}{\text{Smallest Resolved Length Scale}} = \frac{3\,\delta_{1/2}(x)}{\Delta} \tag{7.62}$$

$$= \frac{3\left[\frac{D_0}{2} + 0.1\,x\right]}{\Delta}. \tag{7.63}$$

From the scaling arguments the following the ratios can be formed

$$\frac{\text{Largest Imaging Area}}{\eta} = 6.32\,Re^{3/4} \tag{7.64}$$

$$\frac{\text{Largest Imaging Area}}{\lambda} = 2.61\,Re^{1/2}. \tag{7.65}$$

$$\tag{7.66}$$

Defining the following non-dimensional quantities

$$\alpha \equiv \frac{\Delta}{D_0} \tag{7.67}$$

$$x' \equiv \frac{x}{D_0} \tag{7.68}$$

and equating corresponding ratios yields relationships that indicate for a given Δ/D_0, the downstream jet location where sufficient spatial resolution exists to resolve either of these microscales as a function of Reynolds number. The summary of these relationships is shown in Fig. 7.20. for a range of spatial resolution which are related to the CCD sensor characteristics, *i.e.* its physical size and the number of pixels.

Now we illustrate the application of Fig. 7.20. for the presented PIV measurements of the turbulent jet at $Re = 4000$. The CCD array used for these experiments with the optical magnification and the IW size used for the MCCDPIV analysis results in $Delta \approx 0.06D_0$. From the results of Fig. 7.20. this means that the largest scales and the and the Taylor micro-scale can only be resolved simultaneously for $x/D_0 \geq 10$, whereas the largest scales and the Kolmogorov viscous scale can only be resolved simultaneously for $x/D_0 \geq 600$.

The type of instantaneous 2C-2D velocity field measurement that is possible and that provides insight into the complex structure of turbulent flows (in this case the near jet region of turbulent jet flows) is shown in Fig. 7.21.. In this instantaneous measurements only every second velocity vector is shown for clarity and the non-dimensional velocity vectors are coloured with non-dimensional out-of-plane vorticity. The vorticity has been calculated using the technique proposed in.[37,39]

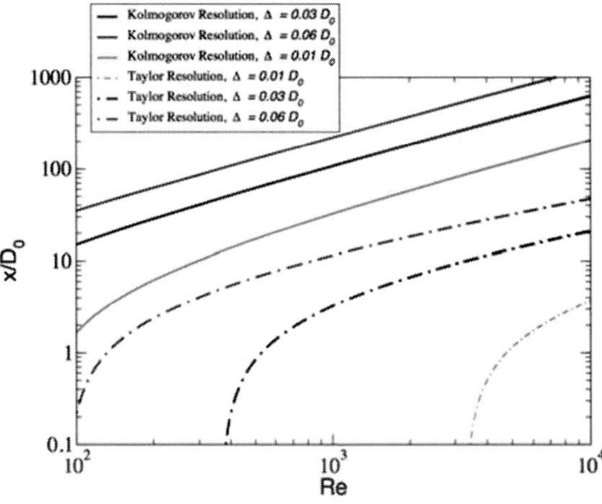

Fig. 7.20.   Relationship between turbulent jet Reynolds number and downstream position where the largest and smallest scales can be measured simultaneously for a given spatial resolution of a CCD sensor.

7.6.   Concluding remarks

This chapter provides an introduction to laser-based velocimetry and PIV. A detailed mathematical analysis of cross-correlation of single exposed image pairs is provided which results in a general equation for the cross-correlation function. This equation is then used to investigate one of the effects associated with spatial resolution, namely the effect of velocity gradients on the cross-correlation function of single exposed image pairs. Important conclusions regarding the permissible magnitude of the velocity gradient across the interrogation window are deduced from this analysis. Another effect associated with spatial resolution is concerned with the spatial differentiation of PIV measurements. This effect is also mathematically analysed and conditions are given that need to be satisfied to keep the bias error in spatial differentiation of PIV measurements, and thus quantities like vorticity and rate-of-strain measurements, within acceptable bounds. Finally a framework is presented to develop the necessary conditions for the simultaneous measurement of the large scales and the smallest scales in turbulent flow - this is done in the example of a turbulent round jet.

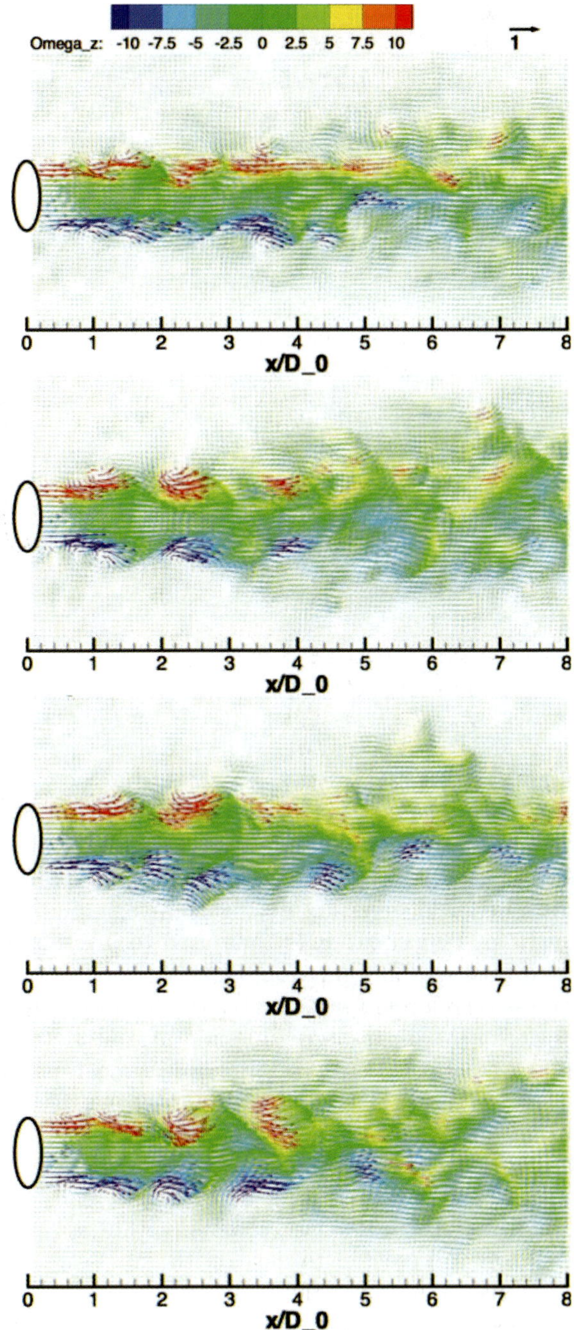

Fig. 7.21. Non-dimensional typical instantaneous 2C-2D jet velocity fields at $Re = 4000$ coloured by out-of-plane vorticity. Only every second vector is shown.

Acknowledgments

This work was supported by the Australian Research Council. I would like to thank my students and colleagues at LTRAC who have provided invaluable inspiration, support and comment to the work presented in this chapter and Mr Sachin Tomar for generating Figure 7.4..

Bibliography

1. F. Durst, A. Melling, and J.H. Whitelaw. *Principles and Practice of Laser Anemometry.* Academic Press Inc., 1976.

2. T. D. Dudderar, R. Meynart, and P. G. Simpkins. Full-field laser metrology for fluid velocity measurement. *Opt. Lasers Eng.*, 9:163 – 199, 1988.

3. J. M. Burch and J. M. J. Tokarski. Production of multiple beam fringes from photographic scatters. *Optics Acta*, 15:101–111, 1968.

4. D.B. Baker and M.E. Fourney. Measuring fluid velocities with speckle patterns. *Opt. Lett.*, 1:135 – 137, 1977.

5. P. G. Simpkins and T. D. Dudderar. Laser speckle measurement in transient bernard convection. *J. Fluid Mech.*, '89:665 – 671, 1978.

6. R. Meynart. Instantaneous velocity field measurements in unsteady gas flow speckle velocimetry. *Appl. Opt.*, 22:535–540, 1983.

7. R. J. Adrian. Scattering particle characteristics and their effect on pulsed laser measurements of fluid flow: speckle velocimetry vs. particle velocimetry. *Appl. Opt.*, 23:1690–1691, 1984.

8. J. Kompenhans and J. Reithmuth. Two-dimensional flow field measurements in wind tunnels by means of particle velocimetry. In *Int Congr Appl Lasers and Electro-Optics, San Diego, USA*, 1987.

9. L. Lourenco and A. Krothapalli. The role of photographic parameters in laser speckle or particle image displacement velocimetry. 1987.

10. A. Vogel and W. Lauterborn. Time resolved particle image velocimetry used in the investigation of cavitation bubble dynamics. *Appl. Opt.*, 27:1869–1876, 1988.

11. R.J. Adrian. Image shifting technique to resolve directional ambiguity in double-pulsed velocimetry. *Appl. Optics*, 25(1):3855 – 3858, 1986.

12. R. F. Shepherd, R. F. LaFontaine, L. W. Welch, J. Soria, and I. G. Pearson. Measurement of instantaneous flows using particle image velocimetry. In *Second World Conference on Experimental Heat Transfer, Fluid Mechanics and Thermodynamics*, Dubrovnik, Yugoslavia, 1991.

13. J. Wu, J. Soria, K. Hourigan, M. C. Welsh, and L. W. Welch. Flow-excited acoustic resonance in a duct: the feedback mechanism. In *Fifth Int. Conf. on Flow Induced Vibrations, Brighton, U.K*, 1991.

14. Z.-C. Liu, C.C. Landreth, R.J. Adrian, and T.J. Hanratty. High resolution measurement of turbulent structure in a channel with particle image velocimetry. *Exp. Fluids*, 10:301–312, 1991.

15. I. Grant, E. Owens, and Y.Y. Yan. Particle image velocimetry measurements of the separated flow behind a rearward facing step. *Exp. Fluids*, 12:238–244, 1992.

16. M.P. Arroyo and J.M. Saviron. Rayleigh-bernard convection in a small box: spatial features and thermal dependence of the velocity field. *J. Fluid Mech.*, 235:325–348, 1992.

17. C. Shinh, L. Lourenco, L. Vandommelen, and A. Krothapalli. Unsteady flow past an airfoil pitching at a constant rate. *AIAA Journal*, 30:1153 – 1161, 1992.

18. A. Shekarriz, T. C Fu, J. Katz, H. L. Liu, and T. T. Huang. Study of junction and tip vortices using particle image velocimetry. *AIAA J.*, 30:145–152, 1992.

19. R. Dong, S. Chu, and J. Katz. Quantitative visualisation of the flow within the volute of a centrifugal pump. part a: Technique. *Trans. ASME*, 114:390–395, 1992.

20. R. Dong, S. Chu, and J. Katz. Quantitative visualisation of the flow within the volute of a centrifugal pump. part b: Results and analysis. *Trans. ASME*, 114:396–403, 1992.

21. A. Shekarriz, T. C Fu, J. Katz, and T. T. Huang. Near-field behaviour of a tip vortex. *AHA J.*, 31:112–118, 1992.

22. W. Lauterborn and A. Vogel. Modern optical techniques in fluid mechanics. *Ann. Rev. Fluid Mech.*, 16:223–44, 1984.

23. L. Hesselink. Digital image processing in flow visualisation. *Ann. Rev. Fluid Mech.*, 20:421–485, 1988.

24. R.J. Adrian. Particle imaging techniques for experimental fluid mechanics. *Ann. Rev. Fluid Mech.*, 23:261–304, 1991.

25. I. Grant. Particle Image Velocimetry: A review. In *Proc. Instn. Mech. Engrs.*, volume 221, pages 55–76, London, U.K., 1997.

26. Richard B. Miles and Walter R. Lempert. Quantitative flow visualization in unseeded flows. *Ann. Rev. Fluid Mech.*, 29:285–326, 1997.

27. J. P. Bonnet, D. Gresillon, and P. Taran. Nonintrusive measurements for high-speed, supersonic, and hypersonic flows. *Ann. Rev. Fluid Mech.*, 30:231 – 273, 1998.

28. N.J. Lawson. Aerospace applications of laser-based flow measurements. *Journal of Aerospace Engineering*, 218:33–57, 2004.

29. R.J. Adrian. Twenty years of particle image velocimetry. *Exp. Fluids*, 39(2):159 – 169, 2005.

30. Y.C. Cho and H. Park. Instantaneous velocity field measurement of objects in coaxial rotation using digital image velocimetry. In *Proc. of SPIE Ultrahigh- and High-Speed Photography, Videography, Photonics, and Velocimetry '90*, volume 1346, San Diego, CA, USA, 1990.

31. C.E. Willert and M. Gharib. Digital particle image velocimetry. *Experiments in Fluids*, 10:181 – 193, 1991.

32. R. D. Keane and R. J. Adrian. Theory of cross-correlation analysis of PIV images. *Appl. Sci. Res.*, 49:191 – 215, 1992.

33. M. Raffel, C. Willert, and J. Kompenhans. *Particle Image Velocimetry*. Springer-Verlag, 1998.

34. R. D. Keane and R. J. Adrian. Optimisation of particle image velocimetry. Part i: Double pulsed systems. *Meas. Sci. Technol.*, 1:1202 – 1215, 1990.

35. R. D. Keane and R. J. Adrian. Optimisation of particle image velocimetry. Part ii: Multi-pulsed systems. *Meas. Sci. Technol.*, 2:963 – 974, 1991.

36. J. Soria. Multigrid approach to cross-correlation digital PIV and HPIV analysis. In *Proceedings of 13th Australasian Fluid Mechanics Conference*. Monash University, Melbourne, Australia, 1998.

37. A. Fouras and J. Soria. Accuracy of out-of-plane vorticity measurements using in-plane velocity vector field data. *Exp. Fluids*, 25:409 – 430, 1998.

38. J. Soria, R. Sondergaard, B.J. Cantwell, M.S. Chong, and A.E. Perry. A study of the fine-scale motions of incompressible time-developing mixing layers. *Phys. Fluids*, 6(2):871–884, February 1994.

39. J. Soria. An investigation of the near wake of a circular cylinder using a video-based digital cross-correlation particle image velocimetry technique. *Experimental Thermal and Fluid Science*, 12:221 – 233, 1996.

40. J. Soria. An adaptive cross-correlation digital PIV technique for unsteady flow investigations. In A.R. Masri and D.R. Honnery, editors, *Proc. 1st Australian Conference on Laser Diagnostics in Fluid Mechanics and Combustion*, pages 29 – 48, Sydney, NSW, Australia, Dec 1996. University of Sydney, University of Sydney.

41. J. Soria. Digital cross-correlation particle image velocimetry measurements in the near wake of a circular cylinder. In *International Colloquium on Jets, Wakes and Shear Layers*, Melbourne, Australia, 1994.

42. K. von Ellenrieder, J. Kostas, and J. Soria. Measurements of a wall-bounded turbulent, separated flow using HPIV. *Journal of Turbulence*, 2:1–15, 2001.

43. Julio Soria, Kamalluddien Parker, Rhys Cowling, and Virginia Palero. Experimental study of the near jet region of a pulsed free jet using mccdpiv. In *AIAA Paper 60964*, 36 AIAA Fluid Dynamics Conference, 2006.

44. H. Tennekes and J.L. Lumley. *A First Course in Turbulence*. The MIT Press, 1972.

45. P.E. Dimotakis. The mixing transition in turbulent flows. Caltech asci technical report 067, California Institute of Technology, 2000.

46. D. R. Dowling and P. E. Dimotakis. Similarity of the concentration of gas-phase turbulent jets. *J. Fluid Mech.*, 218:109–141, 1990.

Chapter 8

Vortex Flows in Optical Fields

Anton S. Desyatnikov

Nonlinear Physics Centre,
Research School of Physical Sciences and Engineering,
The Australian National University, Canberra ACT 0200, Australia
E-mail: asd124@rsphysse.anu.edu.au

We present an introduction into the basic physics and an overview of recent developments in Singular Optics, the novel discipline of modern optics studying electro-magnetic vortices. We focus on vortex solitons and related topics of nonlinear singular optics as well as discuss the future perspectives and applications of twisted light.

Contents

8.1. Introduction

Geometrical structures appear as the result of interference of light waves on different scales, ranging from caustics on the scale much larger than the length of wave, to phase dislocations and polarization singularities on the finest sub-wavelength scale.[1] The ultimately small spatial structure of light is an isolated point of zero intensity, where the field phase is unde-

349

fined, or *singular*.[2] Singular Optics is a novel and flourishing branch of modern optics[3] that studies the fine structure of light and the physics of phase dislocations in electromagnetic waves. Phase singularities are found throughout optical physics, as they arise naturally within diffraction patterns, speckle fields and various natural optical phenomena.[1] A physically intriguing feature is that the flow of electromagnetic energy in the vicinity of a singularity is curved, and this twisted flow of light is referred to as an *optical vortex*.[4]. The rotation of the electric field naturally leads to an optical orbital angular momentum,[5] similar to the angular momentum of a rotating fluid in the (micro-) whirlpool. While the ability of twisted light to carry robust eddies and finite angular momentum finds practical applications in various branches of modern physics,[6] the vortex phenomenon in general is a fundamental and unifying concept.

Phase singularities are recognized as important features common to all waves. Since the seminal work of Nye and Berry in 1974,[2] the optics of phase dislocations has become recognized for its significance and it is emerging now as a new discipline. During recent years, the number of publications devoted to the study of optical vortices has grown dramatically.[7,8] However, the earliest known scientific description of phase singularity was made in the 1830's by Whewell, as discussed by Berry.[9] While Whewell studied the ocean tides, he came to the extraordinary conclusion that rotary systems of tidal waves possess a singular point at which all cotidal lines meet and at which tide height vanishes. Vortices in waves possess a phase singularity and a rotational flow around the singular point, and they can be found in physical systems of different nature and scale, ranging from water whirlpools and atmospheric tornadoes to quantized vortices in superfluids and quantized lines of magnetic flux in superconductors.[10]

In a broad perspective, the study of optical vortices brings inspiring similarities between different and seemingly disparate fields of physics; the comparison of singularities of optical and other origins leads to theories that transcend the confines of specific fields. Vortices play an important role in many branches of physics, even those not directly related to wave propagation. An example is the Kosterlitz–Thouless phase transition[11] in solid-state physics models, characterized by creation of tightly bound pairs of point-like vortices that restore the quasi-long-range order of a two-dimensional model at low temperatures. Such vortex-induced phase transitions can be observed in superfluid helium films, thin superconducting films, and surfaces of solids, as well as in models of interest to particle physicists and cosmologists. The extensive study of vortices in Bose–Einstein condensate (BEC)[12–15] promises a deeper understanding of links between the physics of superfluidity, condensation, and nonlinear singular optics.

The study of optical vortices and associated localized objects is important from the viewpoint of both fundamental and applied physics. The unique nature of vortex fields is expected to lead to applications in many areas that include optical data storage, distribution, and processing. Optical vortices propagating, e.g., in air, have been suggested also for the establishment of optical interconnects between electronic chips and boards,[16] as well as free-space communication links,[17,18] based on the multidimensional alphabets afforded by the corresponding angular momentum states.[19] The optical angular momentum can be transferred to trapped particles, initiating their rotation (*optical spanners*), and the dark core of optical vortex, as well as the ability to use light vortices to create reconfigurable patterns of complex intensity,[20] became the physical instrument to create *optical tweezers* for manipulation of micro-[21,22] and nano-particles,[23] as well as guiding cold atoms.[24] Optical metrology,[25] interferometry,[26,27] fiber optics,[28] and even astronomy[29] represent well established fields where the use of optical vortices brings unexpected advances. The concepts of singular optics can be also applied to the wave physics in left-handed metamaterials,[30] and they can further be expanded to the physics of femtosecond pulses[31] as well as to cover new wavelength domains, e.g., to describe vortices in X-ray optics.[32]

In nonlinear optical media, self-focusing can arrest the natural spreading of the wavepackets (diffraction) and localize vortices in the form of "doughnut"-shaped *vortex solitons*.[33] Light can be guided by the light itself,[34] in particular by the waveguides created by optical vortices.[35,36] However, the modulational instability usually breaks the symmetry of the looped energy flow and vortices split to several spiraling beams. In dissipative optical systems, such as wide aperture optical resonators, the presence of energy sources and sinks provides an additional stabilization and supports the self-organization of optical fields, or pattern formation. Optical patterns exhibit phase transitions between different symmetries, including ordered states of vortices, as well as the transitions to the chaotic regimes and optical turbulence.

In this contribution, we describe the basic concepts of the nonlinear physics of optical vortices in the context of the propagation of singular beams in nonlinear media. In particular, we overview the recent advances in the study of optical vortices propagating in self-focusing nonlinear media, which leads to the azimuthal instability of a vortex-carrying beams.[34] We summarize different physical settings where such a nonlinearity can support novel types of stable or quasi-stable self-trapped beams carrying nonzero angular momentum, such as vortex solitons, soliton clusters, azimuthons, etc. We also describe the properties of vortex beams created by partially incoherent light and of discrete vortices in periodic photonic lattices. In addition, we present some of the experimental results demonstrating the propagation of singular optical beams in nonlinear media.

8.2.   Vortices in coherent and partially-coherent light waves

To introduce the basic notion of phase singularities we follow here the Ph.D. thesis by Mark Dennis.[37] Let us consider a complex scalar field $\psi(\boldsymbol{r};t) =$ Re $\psi + i$Im ψ, such as the envelope of an electric field of a fully coherent light wave varying smoothly in space and/or time. For simplicity, we fix the moment of time $t = t_0$ and closely study the distribution of the field in the cross-section $z = z_0$, so that $\Psi(x,y) = \psi(x,y,z_0;t_0)$. Complex function Ψ can be characterized by its special points, such as zeros and poles, however, we assume that the value of our physical field is finite, $|\Psi| < \infty$, and we look for field zeros $\Psi = 0$. The latter condition is equivalent to the system of two equations, Re $\Psi(x,y) = 0$ and Im $\Psi(x,y) = 0$, each defines a curve in the plane (x,y). The position of the field zero lies on the intersection of two curves, such as the ones sketched in Fig. 8.1.(a). Now we choose the origin of our coordinate system to lie at the intersection point and, close to the origin, we approximate functions Re $\Psi(x,y) \approx x$ and Im $\Psi(x,y) \approx y$ [see Fig. 8.1.(b)], so that the field is given by $\Psi(x,y) = x + iy$.

Next step is separation of variables in polar coordinates, $\rho = \sqrt{x^2 + y^2}$ and $\varphi = \tan^{-1} y/x$, see Fig. 8.1.(b). The field can be written as $\Psi(\rho,\varphi) \approx \rho\exp(i\varphi)$. Thus, the phase is a linear function of the polar angle, Arg$\Psi = \varphi$, plotted as the surface in Fig. 8.1.(c). It has a helicoidal shape structurally resembling a screw dislocation in a crystal lattice with the value undetermined, or *singular*, at the origin $\rho \to 0$, while the amplitude vanishes as $|\Psi| = \rho$. In general, the strength of phase dislocation (sometimes referred to as the vortex topological charge) can be defined in terms of the field amplitude $|\Psi|$ and phase ArgΨ as the circulation of the phase gradient around the singularity,

$$S = \frac{1}{2\pi} \oint \nabla \mathrm{Arg}\Psi \mathrm{dl}, \tag{8.1}$$

here dl is the element of an arbitrary counter-clockwise path closed around the dislocation. The result is an integer because the phase changes by a multiple of 2π. For example, in the example above the topological charge is $+1$. Under appropriate conditions, it also measures an orbital angular momentum of the vortex associated with the helical wave-front structure. The phase gradient direction swirls around the singular line much like fluid in a whirlpool.

If a light wave is characterized by an extra parameter, e.g., the wave polarization, its mathematical representation is no longer a scalar but a

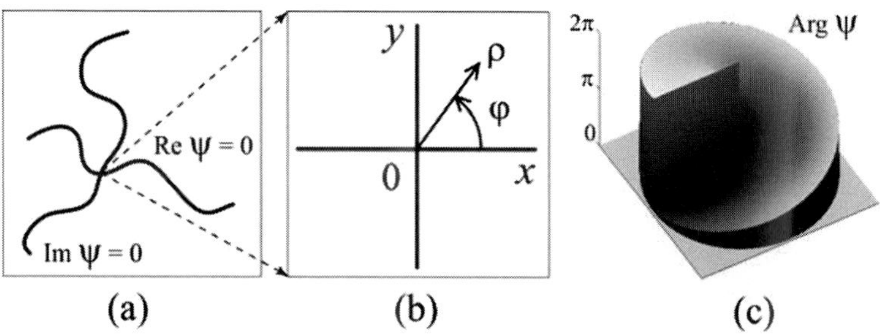

Fig. 8.1.   Sketch of the system of two equations $\Psi(x, y) = 0$ (a), its approximation close to the origin (b), and the phase of the field, $\mathrm{Arg}\Psi \approx \varphi$ (c), in the vicinity of dislocation at $\rho \to 0$. The circular aperture applied for better visibility.

vector field. In vector fields, several types of line singularity exist; for example, those analogous to disclinations in liquid crystals, which could be edge type, screw type, or mixed edge-screw type, that could move relative to background wave fronts and could interact in several different ways.[2,3] In *the linear theory of waves*, each wave dislocation could be understood as a simple consequence of destructive wave interference. In this contribution we mostly address screw phase singularities existing in scalar wave fields and thus we concentrate our analysis on the corresponding vortices. However, other types of singularities whose analysis falls beyond the scope of this review, such us polarization singularities,[38,39] do exist and exhibit fascinating properties.

When a light beam is polychromatic and/or partially incoherent, the phase front topology of the vortex-carrying beam is not well defined, and statistical description is required to quantify the beam structure.[40] For the spatially coherent *polychromatic* light beams the appearance of spectral anomalies near intensity zeros results in the coloring of phase singularities.[41–43] The white vortex beam can only be formed if the vortices associated with each constituent wavelength propagate coaxially.[44–46] This condition is in general difficult to satisfy.[47] Furthermore, in the *incoherent* limit neither the helical phase nor the characteristic zero intensity at the vortex core can be observed. Nevertheless, several recent theoretical and experimental studies have shed light on the question how phase singularities can be unveiled in partially coherent light fields.[48–53] In particular, Palacios *et al.*[52] used both experimental and numerical techniques to explore how a beam transmitted through a vortex phase mask changes as the transverse coherence length at the input of the mask varies. Assuming a quasi-monochromatic, statistically stationary light source and ignoring temporal coherence effects, they demonstrated that robust attributes of the vortex remain in the beam, most prominently in the form of a ring dislocation in the vortex cross-correlation function.

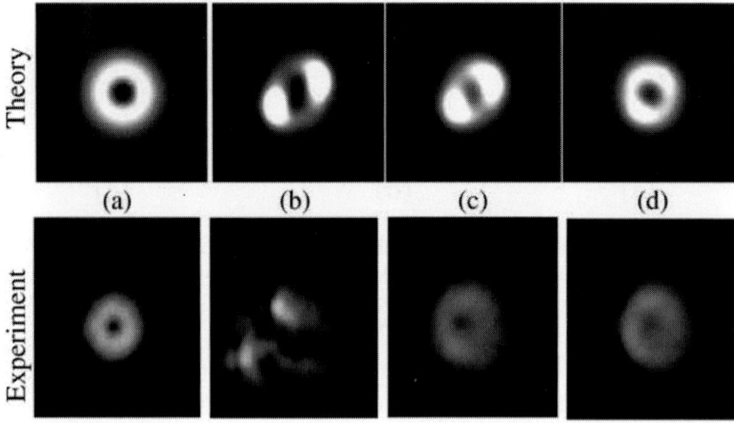

Fig. 8.2.   Numerical and experimental results showing the stabilization of the vortex with growing incoherence: (a) input intensity, (b) vortex after 9 mm of propagation for the coherent case, (c) less coherent vortex, and (d) least coherent vortex.

Vortex generated in a spatially partially-coherent beam can become invisible at the focal point for relatively weak incoherence of the input beam but its correlation properties can be altered by the beam focusing.[54]   In particular, an initially invisible vortex imprinted into a highly incoherent white-light beam may reappear again after the focus what implies that the information about the vortex can be effectively transferred from the cross-correlation function to its intensity distribution. Furthermore, the existence of a stable ring structures in the cross-correlation function appear to be very important to the vortices in nonlinear optical media.

Propagating in nonlinear coherent systems, optical vortices become highly unstable when the nonlinear medium is self-focusing. However, when spatial incoherence of light exceeds a certain threshold, the stable propagation of optical vortices in self-focusing nonlinear media is possible and has been recently demonstrated in experiments conducted with a biased photorefractive SBN crystal.[55,56]   The propagation of partially incoherent optical vortices in a photorefractive medium has been studied numerically[55] using the coherent density approach.[57,58]   The top row in Fig. 8.2.(a–d) shows numerical results for the propagation of an input Gaussian beam carrying a phase dislocation (a) after the total propagation in a nonlinear medium for the coherent light (b) and two different partially incoherent beams (c, d). The most obvious difference to the scenario of the propagation of a coherent vortex is that the vortex decay undergoes a visible delay when the degree of incoherence grows.

Therefore, despite the common notion of the ring vortex solitons as being unstable objects, Fig. 8.2. demonstrates one of the possibilities for experimental observation of self-trapped optical vortices. In the next section we discuss in detail their structure, stability properties, and generalizations.

8.3. Self-trapped vortex beams in nonlinear optical media

A laser beam with a phase singularity generally has a doughnut-like shape and diffracts when it propagates in free space. However, when the vortex-bearing beam propagates in *a nonlinear medium*, a variety of interesting effects can be observed. Nonlinear optical media are characterized by an electromagnetic response that depends on the strength of the propagating light. The polarization of such a medium can be described as $P = \chi^{(1)} E + \chi^{(2)} E^2 + \chi^{(3)} E^3$, where E is the amplitude of the light wave's electric field, and the coefficients characterize both the linear and the nonlinear response of the medium.[59–61] The $\chi^{(1)}$ coefficient describes the linear refractive index of the medium. When $\chi^{(2)}$ vanishes (as happens in the case of centro-symmetric media), the main nonlinear effect is produced by the third term that can be presented as an intensity-induced change of the refractive index proportional to $\chi^{(3)} E^3$. An important consequence of such intensity-dependent nonlinearity is the spontaneous focusing of a beam that is due to the lensing property of *a self-focusing medium* (i.e. when $\chi^{(3)} > 0$). This focusing action of a nonlinear medium can precisely balance the diffraction of a laser beam, resulting in the creation of optical solitons, which are self-trapped light beams that do not change shape during propagation.[62]

A stable bright spatial soliton is radially symmetric, and it has no nodes in its intensity profile. If, however, a beam with elaborate geometry carries a topological charge and propagates in a self-focusing nonlinear medium, it has a doughnut like structure. However, such a doughnut beam is unstable, and it decays into a number of more fundamental bright spatial solitons, such an example is shown in Fig. 8.3.. The resulting field distribution does not preserve the radial symmetry, and the vortex beam decays into several solitons that repel and twist around one another as they propagate. This rotation is due to the angular momentum of the vortex beam transferred to the splinters.

Remarkably, the behavior of a laser beam in *a self-defocusing nonlinear medium* (i.e. when $\chi^{(3)} < 0$) is distinctly different. Such a medium cannot produce a lensing effect and therefore cannot support bright solitons. Nevertheless, a negative change of the refractive index can compensate for spreading in light intensity of the dip, thus creating a dark soliton,[62] a self-trapped, localized low-intensity state (a dark hole) in a uniformly illuminated background. Vortices of single and multiple topological charges can be created in both linear and nonlinear media by use of, e.g., computer-generated holograms or spatial light modulators. Propagating through a nonlinear self-defocusing medium, such as a vortex-carrying beam, creates a self-trapped state, *a dark vortex soliton.* Dark vortex solitons have been observed experimentally in different materials with self-defocusing nonlinearity, such as slightly absorbent liquids, vapors of alkali metals, and photorefractive crystals.[63–65]

8.3.1. *Bright vortex solitons*

In an isotropic optical medium with a local nonlinear response the propagation of a paraxial light beam is described by the well-known generalized nonlinear Schrödinger equation[34] (NLS). In the dimensionless form, the generalized NLS equation has the form,

$$i\frac{\partial E}{\partial z} + \Delta_\perp E + F(I)E = 0, \tag{8.2}$$

where E is the complex envelope of the electric field, $\Delta_\perp = \partial^2/\partial x^2 + \partial^2/\partial y^2$ is the transverse Laplacian, and z is the propagation distance measured in units of the diffraction length L_D. Function $F(I)$ describes the nonlinear properties of an optical medium, and it is assumed to depend on the total beam intensity, $I \equiv |E|^2$. Examples of these Kerr-like materials include two-level model of resonant gases, $F = I$, or pure Kerr nonlinearity; the so-called saturable nonlinearity, $F = I(1 + \alpha I)^{-1}$, or its low-intensity expansion, the cubic-quintic model, $F = I - \alpha I^2$, etc. Here the parameter α defines the nonlinearity saturation.

In a self-focusing medium, i.e. $F(I) \geq 0$, the diffraction of a light beam can be compensated by the nonlinearity and the balance between these two counter-acting "forces" corresponds to the stationary state. *Spatial optical solitons* are stationary spatially localized solutions of the NLS equation (8.2) which do not change their intensity profile during propagation.[66–68] Such a definition covers many different types of stationary beams with a finite power and, in general, the spatial solitons can be found in a generic form,

$$E(x, y, z) = U(x, y) \exp\left[ikz + i\phi(x, y)\right], \tag{8.3}$$

where the real functions U and ϕ are the soliton amplitude and phase, respectively, and k is the soliton propagation constant. Substituting Eq. (8.3) into Eq. (8.2), we arrive at the system of coupled equations for the soliton amplitude and phase,

$$\Delta_\perp U - kU - (\nabla\phi)^2 U + F(U^2)U = 0, \qquad (8.4)$$

$$\Delta_\perp \phi + 2\nabla\phi\nabla \ln U = 0. \qquad (8.5)$$

We start with the solutions of the system (8.4), (8.5) with a constant phase, taking $\phi = 0$ without restriction of generality. In this case, it can be shown that the only type of a structure localized in both transverse dimensions should possess a radial symmetry, i.e. $U(x,y) = U(r)$, where $r = \sqrt{x^2 + y^2}$. Solutions of this type include *the fundamental (bell-shaped) soliton* (see Fig. 8.3.(b) for $m = 0$) and higher-order modes with several rings surrounding the central peak.[69–72] The number of radial nodes, defined by the index n, distinguishes the higher-order radially-symmetric spatial solitons. The main parameter characterizing the spatial soliton is its power

$$P = \int |E|^2 \, \mathrm{d}\mathbf{r} = \int U^2 \, \mathrm{d}\mathbf{r}, \qquad (8.6)$$

being the integral of motion associated with phase invariance of a solution to Eq. (8.2). Radial modes with n-rings in the intensity profile, U_n, belong to the different branches of the dependence $P(k)$, i.e. they form a discrete set of soliton families, $P_n(k)$. The generalization of each soliton family includes transversely moving solitons, obtained by applying the Galilean transformation, $\mathbf{r} \to \mathbf{r} - 2\mathbf{q}z$ and $\phi \to \phi + \mathbf{q}(\mathbf{r} - \mathbf{q}z)$, where $\mathbf{v} = 2\mathbf{q}$ is the soliton transverse velocity. Such moving solitons can be characterized by the soliton *linear momentum*

$$\mathbf{L} = \mathrm{Im} \int E^*\nabla E \, \mathrm{d}\mathbf{r} = \int \nabla\phi \, U^2 \, \mathrm{d}\mathbf{r}, \qquad (8.7)$$

which is defined for the fundamental soliton as $\mathbf{L} = \mathbf{q}P$.

A novel class of spatially localized beams in self-focusing nonlinear media, associated with the rotation of the field phase, was introduced by Kruglov.[73] The beam phase has a spiral structure with a singularity at the origin, as the one shown in Fig. 8.1.(c), representing a phase dislocation of the wave front in the form of an optical vortex.[74] The intensity of such a beam vanishes at the beam center, and, at the same time, the beam remains localized (i.e. its intensity decays at infinity) propagating in the form of a ring-like beam, see Fig. 8.3.(a, b). The existence of the ring-profile solitary waves can be explained intuitively. Indeed, quasi-one-dimensional solitary waves, i.e. the (1+1)-dimensional solitary waves embedded into a (2+1)-dimensional bulk medium, undergo *the transverse modulational instability*.[75] One of the possible ways to suppress this instability is to consider a ring-profile structure created from a (1+1)-dimensional soliton stripe wrapped around its tail.[76] It can be shown that such a soliton ring

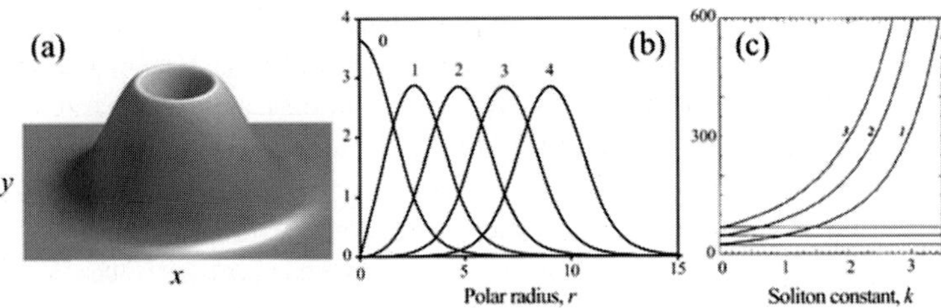

Fig. 8.3.   Examples of the stationary radially symmetric soliton solutions of Eq. (8.2) with saturable nonlinearity $F = I/(1 + \alpha I)$, characterized by the topological charge m indicated next to the curves, the case $m = 0$ corresponds to the fundamental soliton; (a) intensity distribution of a single-charged vortex, $m = 1$, (b) radial profiles for $\alpha = 0.5$ and $k = 1$, (c) soliton power vs. soliton parameter k for $\alpha = 0.2$. The absolute values for soliton power in (c) correspond to the diffraction coefficient $1/2$ and should be doubled for Eq. (8.2) with diffraction coefficient 1.

with the phase depending on the radial coordinate can display a stabilization effect, and even an initially expanding beam can shrink and eventually collapse.[76–78] The beams studied in Ref.[79] provide another example of nonstationary ring-profile solitary waves.

Introduced in[73] ring-profile *vortex solitons* represent the first example of a spatial soliton with the field depending on the azimuthal coordinate $\varphi = \tan^{-1}(y/x)$. They can be found as the solutions to Eqs. (8.4), (8.5) with a rotating spiral phase ϕ in the form of *a linear function of the polar angle* φ, i.e. $\phi = m\varphi$. Substituting this expression into Eqs. (8.4), (8.5) we obtain

$$\frac{\mathrm{d}^2 U}{\mathrm{d}r^2} + \frac{1}{r}\frac{\mathrm{d}U}{\mathrm{d}r} - \frac{m^2}{r^2}U - kU + F\left(U^2\right)U = 0, \qquad (8.8)$$

where the radially symmetric amplitude $U(r)$ vanishes at the center, $U \sim r^{|m|}$ at $r \to 0$. The rotation velocity should be quantized by the condition of the field univocacy Eq. (8.1), and therefore m is integer (see Fig. 8.3.(b, c)). The index m stands for a phase twist around the intensity ring, and it is usually called winding number, topological index or topological charge of the solitary wave. Such winding number distinguishes *azimuthal* higher-order stationary states, in addition to the radial modes, so that the full set of radially-symmetric spatial soliton families can be denoted by $P_{n,m}(k)$, with radial and azimuthal quantum numbers n and m. Figure 8.3.(b) illustrates the envelopes $U(r)$ for single ring ($n = 0$) vortex solitons with different topological charges m. Corresponding families are divided by the minimal threshold power, necessary for the formation of corresponding higher-order states, the dependencies $P_{0,m}(k)$ for a fixed nonzero saturation $\alpha = 0.2$ are shown[80] in Fig. 8.3.(c). The threshold power is defined in the limit of zero saturation $\alpha = 0$, i.e. in pure Kerr medium $F(I) = I$, where soliton constant k plays a role of a scaling parameter, see horizontal lines in Fig. 8.3.(c). Subsequently, vortex solitons were re-discovered in other studies[77,79,81] and for other types of nonlinear optical media, including quadratic nonlinear media.[80,82–84]

The important integral of motion associated with this type of the solitary waves is the beam *angular momentum*, $\left[M\mathbf{e}_z = \mathrm{Im} \int E^*(\mathbf{r} \times \nabla E) \, \mathbf{dr}, \right]$ which can be expressed through the soliton amplitude and phase,

$$M = \int \frac{\partial \phi}{\partial \varphi} \, U^2 \, \mathbf{dr}. \tag{8.9}$$

It is important to point out that the nonvanishing angular momentum is an overall property of the light beam, not necessarily directly connected to the quantum properties at the single photon level. Similarly, the angular momentum is not necessarily associated with optical singularities although in practice the two phenomena may occur together.[85–87] The angular momentum of a paraxial light beam can be separated into spin and orbital parts,[88,89] where *the spin momentum* is associated with the polarization structure of the light, and *the orbital momentum* is associated with the spatial structure of the beam, in particular, the beam carrying optical vortices. Therefore, the angular momentum of a scalar vortex soliton defined by Eq. (8.9) should be identified as an orbital angular momentum. However, as we describe below, the ring-profile vortex optical beams experience *the azimuthal instability* in nonlinear media, and they decay into a number of *moving* fundamental solitons. Because the input beams carry the overall angular momentum given by Eq. (8.9), the splinters fly off the ring along the tangential trajectories. Therefore, in the soliton community it has become customary to refer to the *soliton spin angular momentum*, and thus to *spinning solitons*, and to use *orbital angular momentum* in the case of several interacting solitons. This notion leads to the description of the break-up of a vortex due to modulational instability in terms of the transformation of the initial soliton spin angular momentum to the net orbital angular momentum of moving splinters.[80,84] The ratio of the soliton angular momentum to its power can be identified with *the soliton spin*, $S = M/P$, and for a vortex soliton the spin is equal to its nonzero topological charge $S = m$ (cf. Eq. (8.1)). We notice, however, that such a notation might be confusing when it is used in other areas where the concepts of spin angular momentum and orbital angular momentum are employed in the rigorous, proper sense.

8.3.2.　*Modulational instability*

As was shown in many numerical and analytical studies, the ring-like vortex beams in self-focusing nonlinear media are subject to the azimuthal symmetry-breaking modulational instability, a specific type of transverse modulational instability similar to one which is responsible for filamentation of the beams and generation of trains of optical solitons.[75] This effect should not be confused with the well-known collapse instability[90] which is eliminated in the nonlinear media with saturable and quadratic nonlinearity. As a result, stable fundamental solitons have been found in two and three spatial dimensions and the instability criterion was established by Vakhitov and Kolokolov.[91] It states that the principal mode of the nonlinear wave equation (8.2) is stable if its integrated intensity Eq. (8.6) has a positive slope, $\partial P/\partial k > 0$, and the latter is possible if the nonlinearity grows monotonically with intensity, i.e. for $\mathrm{d}F/\mathrm{d}I \geq 0$. Both these conditions are satisfied for the vortex solitons in, e.g., self-focusing media with saturation, see Fig. 8.3.(c). However, this property does not guarantee the vortex stability against azimuthal perturbations.

So far, no universal criterion, similar to the Vakhitov–Kolokolov stability criterion for the fundamental optical solitons, has been suggested for the stability of vortex solitons, and the vortex stability should be addressed separately for different models. Below, we present the stability analysis developed in Refs.[84,92,93]

We assume that there exists a stationary solution $E = E_0$ solving Eq. (8.2). The *linear stability* of this solution can be established by the behavior of an additional small perturbation $|p| \ll |E_0|$. Substituting the perturbed solution $E = E_0 + p$ into Eq. (8.2) and linearizing it with respect to a small perturbation p, we obtain

$$\mathrm{i}\frac{\partial p}{\partial z} + \Delta_\perp p + \left(F_0 + |E_0|^2 F_0'\right) p + E_0^2 F_0' p^* = 0, \qquad (8.10)$$

where $F_0 = F(|E_0|^2)$ and $F_0' = (\mathrm{d}F/\mathrm{d}I)|_{I=|E_0|^2}$. Equation (8.10) describes the evolution of initially small perturbation p corresponding to the solution E_0: if p does not grow with the beam propagation, the stationary solution E_0 is linearly stable.

The linear equation (8.10) can be solved by the separation of variables, depending on the geometry of underlying stationary point E_0. For the case of radially symmetric vortices determined by (8.8), i.e. $E_0 = U(r)\exp(\mathrm{i}m\varphi + \mathrm{i}kz)$, the perturbation p should posses an azimuthal periodicity and, therefore, it can be represented as a Fourier series

$$p(r,\varphi,z) = \sum_{n=-\infty}^{\infty} p_n(r,z)\exp(\mathrm{i}n\varphi). \qquad (8.11)$$

Substituting this decomposition into Eq. (8.10) and matching terms with equal angular dependence, we obtain the infinite set of equations for complex modal functions p_n. However, for any integer s, only two modes p_{m+s} and p_{m-s} are actually coupled and build a closed system:

$$\left\{ i\frac{\partial}{\partial z} + \hat{L}^{\pm} \right\} p_{m\pm s} + \exp(i2kz) A p^{*}_{m\mp s} = 0, \tag{8.12}$$

where $A \equiv U^2 F_0'$ and $\hat{L}^{\pm} \equiv d^2/dr^2 + r^{-1}d/dr - (m \pm s)^2 r^{-2} + F_0 + A$. Solution to these equations is given by $p_{m+s}(r,z) = u_s(r)\exp(ikz + \gamma_s z)$ and $p_{m-s}(r,z) = v_s^*(r)\exp(-ikz + \gamma_s^* z)$, with complex perturbation wave number γ_s and the modes u_s and v_s that solve an eigenvalue problem

$$i\gamma_s \begin{pmatrix} u_s \\ v_s \end{pmatrix} = \begin{bmatrix} k - \hat{L}^{+} & -A \\ A & -k + \hat{L}^{-} \end{bmatrix} \begin{pmatrix} u_s \\ v_s \end{pmatrix}. \tag{8.13}$$

In these notations, if the eigenvalue γ_s has a positive real part for some s, the perturbation modes $p_{m\pm s}$ grow exponentially with the *growth rate* Re $\gamma_s > 0$; and such modes are called *instability modes*. The equivalent representation of perturbation $p(r,\varphi,z) = \exp(im\varphi + ikz)\{u_s(r)\exp(is\varphi + \gamma_s z) + v_s^*(r)\exp(-is\varphi + \gamma_s^* z)\}$ guarantees, due to the completeness of the basis of azimuthal harmonics Eq. (8.11), that all possible perturbations are taken into account.

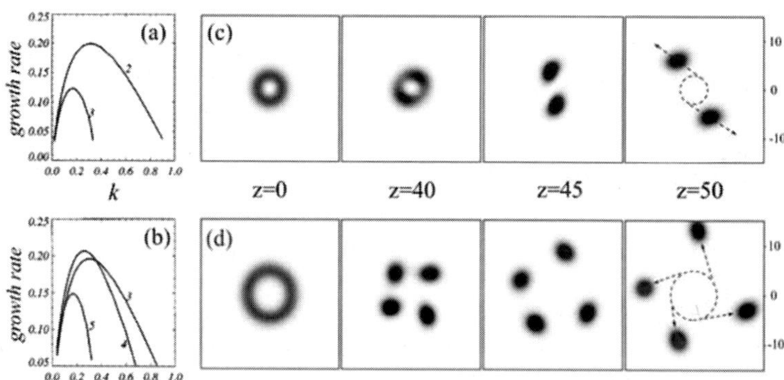

Fig. 8.4.   Growth rate, Re $\gamma_s(k)$, of the instability modes for (a) single- and (b) double-charged vortex solitons in saturable medium, $F(I) = I/(1+I)$. Corresponding values of perturbation index s are shown next to the curves. Break-up of the vortex solitons with the maximum growth rate and the charges (c) $m = 1$ and (d) $m = 2$. Dashed curves at $z = 50$ show the peak intensity of the initial rings and trajectories of the solitons flying away after the decay.

The instability growth rate depends on the vortex power,[80] as is shown in Figs. 8.4.(a, b) for a particular case of spatial vortex solitons with the topological charges $m = 1$ and $m = 2$ in a saturable medium with $\alpha = 1$. For any s, the growth rate vanishes in the linear limit $k \to 0$, and also in the opposite limit of infinite power, $k \to 1/\alpha$, but at least one mode has a nonzero positive value in the whole domain of the vortex soliton existence, $k \in [0, 1/\alpha]$. Thus, all vortex solitons are *linearly unstable* in saturable media. Similar results hold for the parametric interaction in a quadratic medium.

The index of the modes with the highest growth rate depends on the model and the mode topological charge. For example, a single-charged vortex *always* has the $s = 2$ mode growing faster than other modes in a saturable medium, while this can be the $s = 3$ mode in a quadratic medium. In general, the higher-charge vortices allow for competition between different modes with close values of the growth rate, such as the modes with $s = 3$ and $s = 4$ for $m = 2$, see Fig. 8.4.(b).

The symmetry-breaking instability of the ring-like vortex beams has been observed experimentally in saturable vapors,[94–96] photorefractive[97] and quadratic[98] nonlinear media. In all such cases, the generation of different numbers of fundamental solitons due to the ring instability was observed. Figure 8.4. shows a numerical example of the breakup scenario of the ring-like vortex solitons, when the initial stationary ring-profile structure decays, under the action of a numerical noise, into two (the case $m = 1$, see Fig. 8.4.(a)) or four (the case $m = 2$, see Fig. 8.4.(b)) fundamental solitons. The number of splinters coincides exactly with the topological index of the instability mode with highest growth rate, see Fig. 8.4.(a, b), thus the predictions of a linear stability analysis are in an excellent agreement with the numerical solution of the full system. The detailed stability analysis of solitary waves with central phase dislocation were reported in Ref.[80] for both saturable self-focusing and quadratic nonlinear media.[93] In the experiments, the excitation of the ring-profile vortex beams is conducted by pumping the nonlinear media by suitable approximations of Laguerre–Gaussian modes, and it was shown earlier that not only the spontaneous symmetry-breaking instability of the vortex solitons could be observed in that geometry but also that such excitation conditions offer additional possibilities by inducing suitable instabilities.[82,83]

An analytical approach to the study of the filament dynamics after the breakup, based on the conservation of the beam angular momentum and Hamiltonian, was developed in.[80] Given initial values of the conserved quantities, it is possible to predict the features of the filament trajectories and estimate their number. Two different analytical expressions for the velocity of the filaments in the transverse plane were derived, both formulas giving a reasonably good quantitative predictions for the velocities. The formula based on angular momentum being particularly simple and effective: the escape velocity can be estimated as $v \approx |m|/R$, where R is the initial radius of the vortex ring. The important conclusion, giving an insight into the underlying physics of the beam breakup, is that when filaments move out along tangents to the initial ring, they carry away its angular momentum. In our notation we can describe this effect as the transformation of initial spin angular momentum of the vortex soliton to the angular momentum of the splinters spiraling out.

Stabilization of coherent vortex solitons against the azimuthal instabilities remains a major challenge in the physics of spatial vortex solitons. Several theoretical models were suggested which support stable vortex solitons, including the formation of vortex solitons in the presence of competing nonlinearities, nonlocal nonlinear media,[99,100] or Bessel photonic lattices.[101] Very recently, observation of stable self-trapped vortex rings where reported for the nonlinear media with an infinite range of nonlocality.[102]

8.3.3. *Azimuthons*

In this subsection we discuss the self-trapped optical beams carrying angular momentum which do not necessarily correspond to the stationary states, and their angular momentum manifests itself in the complex interaction of simple spatial solitons and leads to their spiraling. Multi-soliton complexes with an imposed angular momentum, such as necklaces and soliton clusters, can also be regarded as multi-soliton spiraling beams.

Soliton spiraling. Spiraling of two spatial solitons was suggested theoretically in.[103] This should occur when two fundamental solitons collide with trajectories that are not lying in a single plane, so that they form a two-body system with nonzero orbital angular momentum. Then, if the mutual interaction is attractive, the centrifugal repulsive force can be balanced out, and two solitons orbit about each other in a double-helix structure. In quadratic media, soliton spiraling was predicted and studied in detail theoretically[104,105] and experimentally.[106] Depending on initial soliton states, such as soliton velocities, relative phase, and the impact parameter, soliton can reflect, spiral, and fuse.

The spiraling of *mutually incoherent* spatial solitons was observed experimentally[107] and studied theoretically[108,109] as a possible scenario for a dynamically stable two-soliton bound state. It is associated with large-amplitude oscillations of a dipole-mode vector state generated by the interaction of two initially mutually incoherent optical beams.[110] In anisotropic photorefractive medium, however, the mutually incoherent solitons demonstrate much complicated anomalous interaction.[111] Anomalous interaction results in complex trajectories which typically show partial mutual spiral-

 Anton S. Desyatnikov

ing, followed by damped oscillations and the fusion of solitons.[112,113] Nevertheless, the fascinating analogy between spiraling solitons and mechanical two-body system is applicable if one takes into account the anisotropic nature of spatial screening solitons interacting in photorefractive medium.[114]

Optical necklace beams. Since the decay of the ring-profile vortex solitons is associated with the growing azimuthal modulation of their intensity and the symmetry-breaking instability, one may try to stabilize the ring structure by imposing the initial intensity and phase modulation. The azimuthally modulated rings resemble "optical necklaces"; they are closely related to the higher-order guided modes, and also to suitable superpositions of Laguerre–Gaussian beams. In Kerr media, the first experimental results on the self-trapping of necklace-type beams were reported in.[115] Figure 8.5.(a) shows the experimental data for the case in which an input beam in the form of a higher-order Laguerre–Gaussian mode was launched at the input of a CS_2 cell.[116] Such self-trapped structures are remarkably stable and allow one to transport optical beams with powers several times the critical power at which a Gaussian beam would otherwise collapse because of self-focusing; they disintegrate for input intensities lower than the self-trapping intensity.

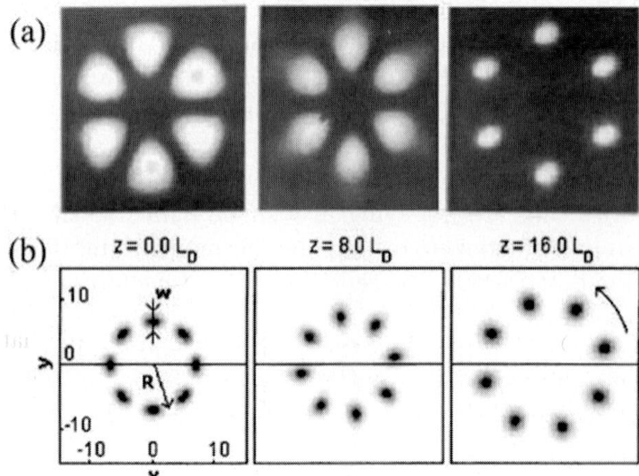

Fig. 8.5. (a) Experimental demonstration of the necklace beams. Propagation of a Laguerre–Gaussian beam inside a Kerr medium: (from left to right) input beam, diffracted beam at low intensity, and the self-trapped necklace beam at high intensities. (b) Rotating and expanding necklace with integer spin.

The concept of necklace beams was developed by numerical studies of the propagation of azimuthally modulated bright rings in Kerr medium.[117,118] In contrast to all previous studies in this model, the self-trapped beams localized in transverse plane can preserve their shape during the propagation and escape the collapse instability. It is possible if such beams are constructed as a "necklace" of the out-of-phase "pearls", each carrying the power less than critical power of catastrophic self-focusing. Each petal thus slowly diffracts, if it propagates alone, and this diffraction is greatly suppressed within the ring, because of collective self-trapping of the ring as a whole. Due to the repulsion between neighboring petals the ring expands self-similarly.

Necklaces with additional phase modulation, introducing a nonzero angular momentum, exhibit a series of phenomena typically associated with rotation of rigid bodies and centrifugal force effects.[119] The modulated ring-profile structure does not decay but, instead, it expands with small rotation. These ringlike structures were introduced[119] as the first example of optical beams with *fractional spin* and *rotating intensity*. Note, that the anticlockwise direction of the rotation is determined by the gradient of phase, which grows *anticlockwise* for $m > 0$, see Fig. 8.5.(b). The angular velocity vanishes as the ringlike structure expands, in analogy with Newtonian mechanics and "scatter on ice" effect.

Soliton clusters. In saturable media, the fundamental solitons are stable and demonstrate the particle-like robust interaction, e.g., spiraling out of the initial vortex ring after it breaks. The analogy with particles and forces between them applied to spatial solitons allows one to search for the bound states of several solitons corresponding to the balance between all acting forces, as in classical mechanics. Indeed, any ring-shaped array of solitons, resembling a necklace, will be radially unstable due to an effective tension induced by bending of the soliton array. Nevertheless, a simple physical mechanism will provide stabilization of the ringlike configuration of N solitons if we introduce an effective centrifugal force that can balance out the tension effect and stabilize the ringlike *soliton cluster*.[120] The requirement for stabilization is an additional phase on the scalar field that twists by $2\pi m$ along the soliton ring. This phase induces *a net angular momentum* and the soliton clusters rotate with an angular velocity which depends on the number of solitons N and phase charge m.

Analysis of many-soliton clusters shows that the radially stable configurations can be achieved if the phase is uniformly growing along the cluster, i.e., the relative phase between two neighboring solitons in the ring θ is given by

$$\theta = \frac{2\pi m}{N}, \tag{8.14}$$

here m is an integer topological charge. Applying the effective-particle approach[120] it was shown that an effective interaction energy for the soliton cluster can be classified in a simple way by its extremal points. The existence of a minimum point suggests that such a configuration describes a stable or long-lived ring-like cluster of a particular number of solitons. The general rule, generalizing a two-soliton phase sensitive interaction, predicts the existence of a bound state of N solitons if the nonzero phase step θ is equal or less $\pi/2$, i.e. the interaction is attractive and can be balanced out by the net centrifugal force for $0 < \theta \le \pi/2$. Therefore, soliton cluster with topological charge m can be quasi-stationary only for $N \ge 4m$, see Eq. (8.14). In addition, "excited" states are possible with the radius of the cluster oscillating near the minimum of interaction potential during the propagation. These prediction were verified by a series of numerical simulations for different N-soliton rings and their propagation in a local saturable medium.

The concept of soliton clusters was extended to the case of light propagation in quadratically nonlinear media in.[121] Due to the phase relation between the fundamental wave and the second harmonic beam, the topological charge in the second-harmonic field is double the fundamental wave charge. Therefore, the corresponding phase jump between neighboring solitons Eq. (8.14) is doubled. Metastable, robust propagation of clusters in media with competing quadratic and self-defocusing nonlinearities was reported in,[122] followed by similar findings in the cubic-quintic model.[123] Similar to the extension of the concept of two-dimensional vortex solitons into spatiotemporal domain,[124] the bound states of three-dimensional solitons, or light bullets, can be constructed using the approach described above for two-dimensional *cw* beams. Tree-dimensional, robust "soliton molecules" were introduced in[125,126] using this concept.

Azimuthons. Further studies revealed *an important missing link* between the radially symmetric vortices and rotating soliton clusters, namely a novel class of spatially localized self-trapped optical beams in nonlinear media, the so-called *azimuthons*.[127] In particular, there are soliton clusters whose visible rotation can be directed alongside or opposite to the direction of the energy flow, because the associated angular momentum has two different contributions. The first contribution is due to the internal energy flow; it comes from a nontrivial phase as in the case of the radially-symmetric vortex solitons; this reflects the wave nature of the self-trapped beams. The second contribution appears only when the rotating beam is modulated or fragmented, and it has a "particle" origin. While the latter contribution is important for the soliton clusters, the former one dominates for strongly overlapping beams when the "particle identity" in the beam structure is lost. Surprisingly, these two contributions can be of the opposite signs giving a birth to the nonrotating modulated singular beams described here as stationary azimuthons.

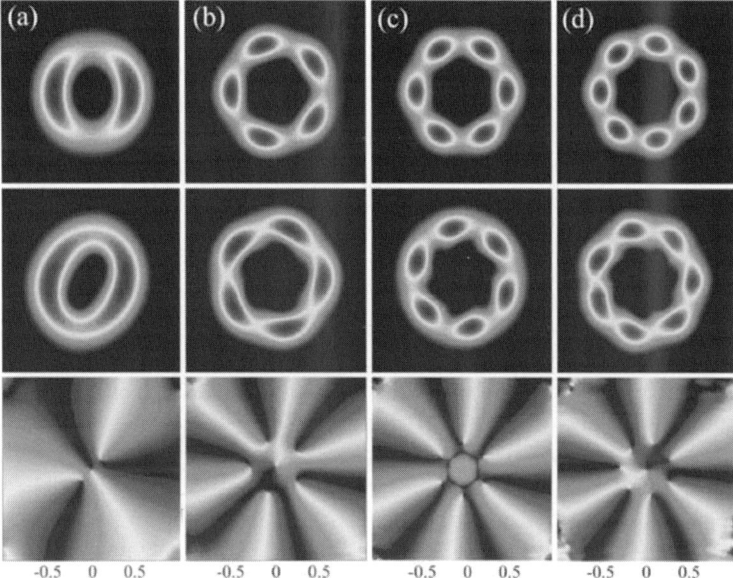

Fig. 8.6. (a) Propagation of the azimuthon with $m = 3$ and $N = 2$. Examples of higher-order azimuthons with $m = 6$ for the cases: (b) $N = 5$, (c) $N = 6$, and (d) $N = 7$. Top: variational solutions; middle: after propagation of (a) $z = 10$ and (b-d) $z = 2$; bottom: corresponding phases.

However, in all types of local nonlinear media described above, soliton clusters and azimuthons appear only as the metastable objects. On the other side, the stabilization of spatially localized vortex solitons against symmetry breaking azimuthal instability was reported in the nonlocal media.[99,100,102] It has been already shown that the spatially nonlocal nonlinear response may bring interesting novel physics also introducing new effects such as strong modification of modulational instability,[128] suppression of beam collapse,[129] dramatic change of the soliton interaction,[130,131] as well as formation of multi-soliton bound states.[132] Many of the predicted and demonstrated properties of nonlocal nonlinear systems suggest that in non-linear optical media we should expect stabilization of different multihump nonlinear structures such as azimuthons. It is further supported by a number of experimental observations of self-trapping effects and spatial solitons in different types of nonlocal nonlinear media such as atomic vapors,[133] nematic liquid crystals,[134] and optical media with thermal self-action such as lead glasses.[102]

The simplest example of multihump solitons predicted in nonlinear optics is the dipole-like structures composed of two interacting fundamental beams with opposite phases that undergo angular rotation during propagation.[135] Nonlocality can provide an effective physical mechanism for stabilizing these dipole solitons which are known to be unstable in all types of realistic nonlinear media with a local response. Because the dipole soliton can be viewed as a special type of azimuthon,[127] the same stabilization mechanism applies to other types of multihump solitons and higher-order azimuthons.[136] Furthermore, in a sharp contrast to local media,[127] nonlocal nonlinearity allows to increase the family with novel types of azimuthons. Their general condition for existence, $N \geq 2$, becomes independent on the beam topological charge. The topological structure of azimuthons with $N < 2m$ is presented in Fig. 8.6. by a circular array of N single-charge vortices with one additional dislocation at the origin with the charge given by the rule $m - N$, so that the total topological charge of azimuthon m is calculated as an algebraic sum of charges of all vortices.

8.4. Discrete vortices

The optical vortices discussed above propagate in homogeneous nonlinear media. When refractive index is periodically modulated, it modifies the wave diffraction properties, and it can affect strongly both nonlinear propagation and localization of light.[34,137] As a result, periodic photonic structures and photonic crystals recently attracted a lot of interest due to the unique ways they offer for controlling light propagation. In particular, many nonlinear effects, including formation of lattice solitons, have been demonstrated experimentally for one- and two-dimensional optically-induced photonic lattices.[138–140] The concept of optically-induced lattices arises from the possibility to modify the refractive index of a nonlinear medium with periodic optical patterns, and use a weaker probe beam to study scattering of light from the resulting periodic photonic structure. Current experiments employ photorefractive crystals with strong electro-optic anisotropy to create a *linear* optically-induced lattice with a polarization orthogonal to that of a probe beam, which also eliminates the nonlinear interaction between the beam and the lattice.

A vortex beam propagating in an optical lattice can be stabilized by the effective lattice discreteness in a self-focusing nonlinear media creating a two-dimensional *discrete vortex soliton*. This has been shown in several theoretical studies of the discrete[141–146] and continuous models with an external periodic potential.[147–149]

To provide evidence for the existence of the discrete vortex solitons two separate groups performed independent experimental investigations.[150,151] Both relied on optical induction to create a nonlinear lattice in a pho-

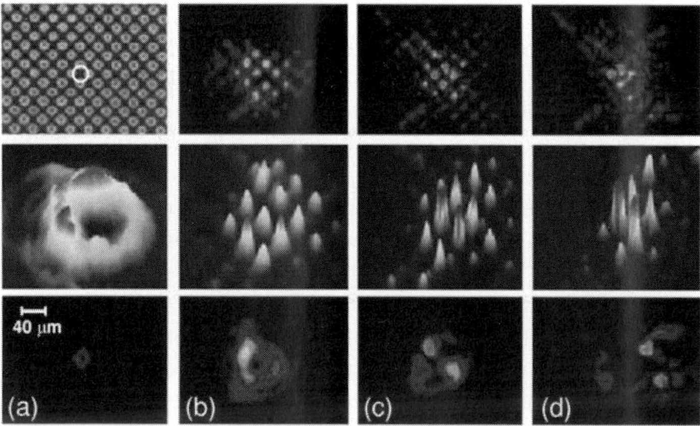

Fig. 8.7. Experimental observation of an optical vortex propagating with (top panel) and without (bottom panel) an optically-induced lattice. Middle panel – three dimensional representation. (a) input intensity pattern of the lattice (top) and input (middle) and linear output vortex beam (bottom). (b–d) Intensity patterns of the vortex beam at crystal output with a bias field of 600, 1200, and 3000 V/cm, respectively. The bright ring in the lattice pattern (a) indicates the location of the input vortex.

tosensitive (photorefractive) material.[152] In this technique the ordinarily polarized light is periodically modulated (by interference or by imaging a mask) to induce a 2D array of waveguides in an anisotropic photorefractive crystal. A separate probe beam of extraordinary polarization acquires a vortex structure by passing through a phase mask and is then launched into the array. The degree of nonlinearity is controlled by applying a voltage across the c-axis of the crystal (photorefractive screening nonlinearity) and controlling the intensity of the probe beam.

Typical experimental results are summarized in Fig. 8.7.. A two-dimensional square lattice was first created, with its principal axes oriented in the diagonal directions shown in the top panel of Fig. 8.7.(a). The resulting periodic structure acts as a square array of optically induced waveguides for the probe beam. The vortex beam, shown in the bottom panel of Fig. 8.7.(a), was then launched straight into the middle of the lattice "cell" of four waveguides, as indicated by a bright ring in the lattice pattern. Due to the coupling between closely spaced waveguides of the lattice, the vortex beam exhibits discrete diffraction when the nonlinearity is low (Fig. 8.7.(b)), whereas it forms a discrete vortex soliton at an appropriate level of higher nonlinearity (see Fig. 8.7.(c, d), the top-middle panel). As predicted, the observed discrete diffraction and discrete self-trapping of the vortex beam in the photonic lattice is remarkably different from that in a homogeneous medium (see Fig. 8.7.(bottom row)).

In a number of recent publications, authors develop the concept of higher-order lattice solitons, such as dipole solitons,[153] "quasivortices",[145] higher-band vortex solitons,[154] or even periodic soliton trains.[155,156] These multi-humped states can be constructed as the bound states of individual lowest order lattice solitons.[157] Though the simplest stationary solutions with in-phase solitons also exist, usually the stable configuration requires the edge-type π-phase dislocations between neighboring sites.[158] The approach developed in[159] includes into this rich family the higher-order solitons with nested screw phase dislocations and nonzero angular momentum.

In analogy with the periodic lattices, the photonic structures with discrete rotational symmetries of a finite order received a special attention.[160,161] Examples include the Bessel-type lattices[162–166] and photonic crystal fibers.[167,168] It was shown that while such structures do not allow for conservation of the angular momentum, they can be characterized by the so-called "angular Bloch momentum", conserved during the evolution of spatially localized modes.[160] At the same time, these structures introduce essential limitations to the symmetry of the supported localized modes, e.g., the topological index of a vortex m can not exceed the cut-off value $|m| \leq n/2$,[161] determined by the symmetry order n of the photonic structure. The selection rules then apply not only to the existence, but also to the stability properties of the discrete vortices.[165] The interplay between the symmetries of the angularly-modulated modes and corresponding waveguiding nonlinear structures leads to possibilities of the so-called vortex transmutations, where the initial individual vortex is mapped into another with a different topological charge.[169]

Stable discrete vortex solitons carry an angular momentum in a structural environment which does not support the conservation of the angular momentum because the medium periodicity breaks the rotational symmetry. As a result, novel types of the vortex dynamics such as the vortex charge flipping[159,170] and vortex transmutations[169] should be observed. We note that inversion of the topological charge of the free-propagating singular beam was predicted[171] and observed[172] for a single-charge "non-canonical" (i.e., elliptically deformed) vortex, as well as for higher-order singular beams focused by an astigmatic lens.[173] In these experiments, the intensity distribution of a vortex undergoes drastic deformation in the focus, where the charge-flipping occurs. In sharp contrast, in periodic media the vortex intensity distribution may practically remain the same, as it is seen in Fig. 8.7., the feature which was commonly believed to guarantee the stability of phase topology as well.[150,151] Recent experiments[174] revealed that in nonlinear periodic media the vortex beams demonstrate remarkable features of *topological transformations*: while the energy of the beam remains localized by the lattice, its phase structure changes substantially. This topological reaction results in a reverse of the vortex transverse energy flow, which is associated with flipping of the sign of the vortex angular momentum. Most probably, the topological transformations can be accurately described and explained by the presence and development of oscillatory instability (internal) modes of the discrete vortex solitons, similar to the breather solutions in discrete systems.[175]

8.5. Vortices in lasers and optical turbulence

Above, we described the vortex solitons and related phenomena in optical systems with the help of the conservative NLS-type equations. The NLS equation can be linked to a more general dissipative Ginzburg–Landau (GL) model as its conservative limit. The theoretical approach was introduced[176] to describe the phenomena of superconductivity,[177] further developed for the type II superconductors,[178] more common in nature, where the flux penetrates the superconductor in the form of a regular array of flux tubes or vortices (Abrikosov vortices). In two dimensions, the *complex* GL equation (CGL) admits extensively studied quantized vortices,[10,179] the stationary limit of the latter equation is also known as the stationary Gross–Pitaevskii equation. A large number of publications is devoted to the study of vortices in boson condensates (such as superconductors and superfluids), described by the CGL equation (see, e.g., Refs.,[180–183] and references therein). In the context of nonlinear optics, the latter model corresponds to the stationary NLS with self-defocusing type nonlinearity. It is interesting to note that related *integrable* complex sine-Gordon model admits the continuous families of the nonradially symmetric dark vortex solitons,[184] while their existence remains an open question for the nonintegrable CGL equation.[185]

The analogy between superfluids and laser optics was recognized as early as 1970.[186] In particular, the vortex solutions to laser equations were found in[187,188] and intensively studied latter, both theoretically and experimentally.[189] The concept of vortices in lasers is connected to the *dissipative optical solitons* (DOS), or auto-solitons. This kind of solitons was initially predicted for wide-aperture nonlinear interferometers excited by external radiation[190] and for laser systems with saturable absorbers.[191] If the material relaxation time is much smaller than that for the field in optical resonator (the so-called class-A laser), the master Maxwell–Bloch equations can be reduced to a CGL equation[192]

$$\frac{\partial E}{\partial \zeta} = (\delta + \mathrm{i})\Delta_d E + f\left(|E|^2\right) E + E_i. \tag{8.15}$$

Here the evolution variable ζ stands for time, in the resonator schemes, or the propagation coordinate z, in a bulk medium, δ is the effective diffusion coefficient, and the diffraction operator Δ_d is acting in $d = 1,\ 2,\ 3$ "transverse" coordinates. The parameter E_i represents an external driving plane-wave field, and the nonlinearity $f\left(|E|^2\right)$ is a complex function, so that the conservative limit Eq. (8.2) is given by $\delta = E_i = 0$ and $f\left(|E|^2\right) = iF\left(|E|^2\right)$.

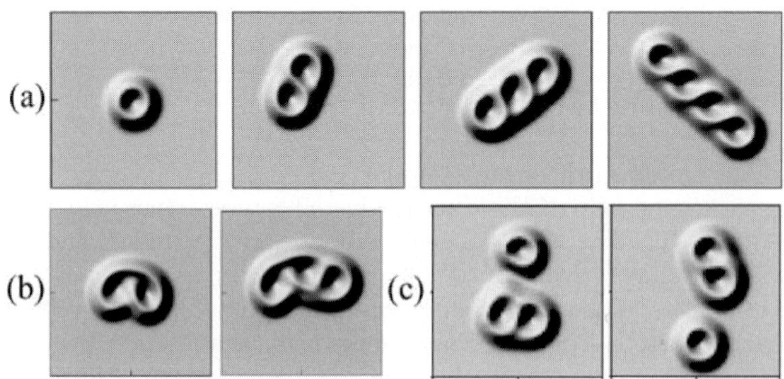

Fig. 8.8.　(a, b) The instantaneous transverse distributions of intensity of laser radiation for the regime of rotating chains with (a) different number of single-charged dislocations with $m = +1$, and (b) the opposite topological charges of the links $m = -1, +1$ (on the left) and $m = -1, +1, +1$ (right). (c) Interaction between laser vortices. The instantaneous transverse intensity distributions of laser radiation for the regime of three solitons with topological charges $m = -1$ during the evolution at dimensionless time $t = 0$ and $t = 1250$.

The wide-aperture interferometers, filled with passive or active nonlinear media (*optical cavities*) and excited by external radiation E_i, exhibit optical bistability.[193] Due to the *spatial hysteresis*[194] there possible domain walls connecting two stable and otherwise spatially homogeneous transmitted waves. Domain walls are usually identified with switching waves, and they represent building blocks for DOS's – the localized solutions emerging as bound states of switching waves. This implies the discreetness of the DOS spectrum in contrast to the continuous "families" of conservative solitons. The external radiation E_i determines the frequency and the phase of DOS's, or "cavity solitons"; they exist on the nonzero background and have oscillating tails. Several DOSs may interact[195–199] and form bound states or clusters of cavity solitons.[200] In contrast to the conservative model, the clusters may appear due to the oscillating effective interaction potential[201,202] with a discrete set of the equilibrium distances between cavity solitons. A large number of publications is devoted to the study of cavity solitons and their link to spontaneous pattern formation, as also discussed in the recent review papers,[203–209] and reflected in the comprehensive list of references prepared by.[210]

Interesting application of vortices as the pump beams in externally driven cavities was suggested by.[194,211] For degenerate optical parametric oscillator it was demonstrated,[212] that the stable domain walls appear as being trapped in the beam. In the vertical-cavity surface-emitting lasers the cavity solitons perform a uniform rotary motion along the crater of a doughnut-shaped holding beam.[213] Micro-cavities offer novel possibilities for the cavity soliton generation and control,[214–217] and the examples include the spontaneous generation of the "optical vortex crystals".[16]

In lasers, the phase of the field is free ($E_i = 0$ in Eq. (8.15)) and the topological solitons are possible.[207] The comprehensive overview of the theoretical studies and the experimental generation of laser vortices can be found in.[189,218] In lasers with saturable absorbers, i.e. when nonlinearity in Eq. (8.15) is given by $f\left(|E|^2\right) = -1 + g_0/(1 + |E|^2) - a_0/(1 + b|E|^2)$, novel types of solitons may appear, such as transversely asymmetric and rotating structures without phase dislocations and radially symmetric vortices with higher-order topological charges.[219–222] Furthermore, bright dissipative vortex solitons can form strongly coupled "vortex clusters", as shown in Fig. 8.8.(a,b), as well as weakly coupled bound states.[223,224] The latter exhibit spontaneous rotation during the evolution, see Fig. 8.8.(c).

Stabilization of dissipative vortex solitons in the cubic-quintic CGL and new types of radially symmetric solitons, such as erupting, flat-top, and composite vortices, were reported recently in.[225,226] In the same model, the existence of stable clusters of dissipative solitons rotating around a central vortex core was predicted in.[227] Another example of the dissipative optical systems supporting topological spatial solitons is given by the nonlinear interferometer formed by a liquid crystal light valve with a feedback. The so-called "triangular solitons" with the rich structure of phase singularities were found in this system.[228,229]

Optical turbulence. Solitons appear as an optical analog of the small spatial scales observed in fluids experiment and they are related to the spatiotemporal instabilities in nonlinear optics, as was shown by J. V. Moloney[230] for the optical bistable ring resonator. Different routes to the optical turbulence in passive all-optical resonators were predicted[231] and observed experimentally.[232] It was also shown that transverse effects play a prominent role in the development of temporal instabilities (pulsations) in lasers,[233] where the competition between longitudinal and transverse modes takes place. Competition of transverse cavity modes can lead to the spontaneous formation of stable patterns, including "vortex crystals", see Refs.[234,235] and references therein. Furthermore, mode competition can lead to the dynamical periodic or quasiperiodic states with correlated move-

 Anton S. Desyatnikov

ment of localized dark spots (vortices), including their nucleations.[236,237] An important issue is the control of the dynamics and stabilization of transverse patterns which can be achieved by spatial/spectral filtering.[238] Some of the related results were reviewed in Ref.[239]

8.6. Conclusions and Outlook

We have presented an overview of exciting research in the field of *nonlinear singular optics* that studies the propagation of optical vortices and optical beams carrying an angular momentum in nonlinear media. Understanding and controlling the properties of optical vortices could lead to applications in the near future, ranging from optical communications and data storage to the trapping, control, and manipulation of particles and cold atoms. Indeed, optical vortices provide an efficient way to control light by creating reconfigurable waveguides in bulk media. The study of phase singularities in optical parametric processes not only suggests novel directions of fundamental research in optics but also provides links to other branches of physics.

Acknowledgments

This work was supported by the Australian Research Council and the Australian National University. Author acknowledges essential help and critical reading of this contribution by Yuri S. Kivshar.

Bibliography

1. J. F. Nye, *Natural focusing and fine structure of light*, (IOP Publishing, Bristol, 1999).
2. J. F. Nye and M. V. Berry, Dislocations in wave trains, *Proc. R. Soc. London Ser. A-Math. Phys. Eng. Sci.* **336**, 165–190 (1974).
3. M. S. Soskin and M. V. Vasnetsov, Singular optics, Ed. E. Wolf, *Progress in Optics*, vol. 42 (North-Holland, Amsterdam, 2001), pp. 219–276.
4. M. Vasnetsov and K. Staliunas, Eds., *Optical Vortices*. vol. 228, *Horizons in World Physics*, (Nova Sciences Publishers, New York, 1999).
5. L. Allen, M. Padgett, and M. Babiker, The orbital angular momentum of light, Ed. E. Wolf, *Progress in Optics*, vol. XXXIX (North-Holland, Amsterdam, 1999), pp. 291–372.
6. L. Allen, S. M. Barnet, and M. J. Padgett, Eds., *Optical Angular Momentum* (IOP Publishing, Bristol, 2003).

7. See the extensive list of references on optical vortices: http://www.u.arizona.edu/~grovers/SO/so.html.

8. Publications initiated by a series of international conferences on Singular Optics: *Proc. SPIE* **3487** (1998); *Proc. SPIE* **4403** (2001); the special issues of the journals: *J. Opt. B* **4** (2002), and *J. Opt. A* **6** (2004).

9. M. Berry, Making waves in physics - three wave singularities from the miraculous 1830s, *Nature* **403**, 6765 (2000).

10. L. M. Pismen, *Vortices in Nonlinear Fields*, (Clarendon Press, Oxford, 1999).

11. J. M. Kosterlitz and D. J. Thouless, Ordering, metastability and phase-transitions in 2 dimensional systems, *J. Phys. C* **6**, 1181–1203 (1973).

12. J. E. Williams and M. J. Holland, Preparing topological states of a Bose-Einstein condensate, *Nature* **401**, 568–572 (1999).

13. M. R. Matthews, B. P. Anderson, P. C. Haljan, D. S. Hall, C. E. Wieman, and E. A. Cornell, Vortices in a Bose-Einstein condensate, *Phys. Rev. Lett.* **83**, 2498–2501 (1999).

14. K. W. Madison, F. Chevy, W. Wohlleben, and J. Dalibard, Vortex formation in a stirred Bose-Einstein condensate, *Phys. Rev. Lett.* **84**, 806–809 (2000).

15. C. Raman, J. R. Abo-Shaeer, J. M. Vogels, K. Xu, and W. Ketterle, Vortex nucleation in a stirred Bose-Einstein condensate, *Phys. Rev. Lett.* **87**, 210402–4 (2001).

16. J. Scheuer and M. Orenstein, Optical vortices crystals: Spontaneous generation in nonlinear semiconductor microcavities, *Science* **285**, 230–233 (1999).

17. G. Gibson, J. Courtial, M. J. Padgett, M. Vasnetsov, V. Pas'ko, S. M. Barnett, and S. Franke-Arnold, Free-space information transfer using light beams carrying orbital angular momentum, *Opt. Express* **12**, 5448–5456 (2004).

18. Z. Bouchal and R. Celechovsky, Mixed vortex states of light as information carriers, *New J. Physics* **6**, 131–622 (2004).

19. G. Molina-Terriza, J. P. Torres, and L. Torner, Management of the angular momentum of light: Preparation of photons in multidimensional vector states of angular momentum, *Phys. Rev. Lett.* **88**, 013601–4 (2002).

20. K. T. Gahagan and G. A. Swartzlander, Simultaneous trapping of low-index and high-index microparticles observed with an optical-vortex trap, *J. Opt. Soc. Am. B* **16**, 533–537 (1999).

21. N. R. Heckenberg, M. E. J. Freise, T. A. Nieminen, and H. Rubinsztein-Dunlop. In eds. M. Vasnetsov and K. Staliunas, *Optical Vortices*, vol. 228 (Nova Sciences Pub., New York, 1999).

22. D. G. Grier, A revolution in optical manipulation, *Nature* **424**, 810–816 (2003).

23. R. Agarwal, K. Ladavac, Y. Roichman, G. Yu, C. M. Lieber, and D. G. Grier, *Opt. Express* **13**, 8906 (2005).

24. N. Chattrapiban, E. A. Rogers, I. A. Arakelyan, R. Roy, and W. T. Hill, *J. Opt. Soc. Am. B* **23**, 94 (2006).

25. E. Hasman, G. Biener, A. Niv, and V. Kleiner, Ed. E. Wolf, *Progress in Optics*, vol. 47 (North-Holland, Amsterdam, 2005), p. 215.

26. J. Masajada, *Opt. Lett.* **234**, 373 (2004).

27. S. Fürhapter, A. Jesacher, S. Bernet, and M. Ritsch-Marte, *Opt. Lett.* **30**, 1953 (2005).

28. A. V. Volyar, Fiber singular optics, *Ukr. J. Phys. Opt.* **3**, 69 (2004).

29. G. Foo, D. M. Palacios, and G. A. S. Jr., *Opt. Lett.* **30**, 3308 (2005).

30. I. V. Shadrivov, A. A. Sukhorukov, and Y. S. Kivshar, Beam shaping by a periodic structure with negative refraction, *Appl. Phys. Lett.* **82**, 3820–3822 (2003).

31. K. Bezuhanov, A. Dreischuh, G. G. Paulus, M. G. Schatzel, and H. Walther, Vortices in femtosecond laser fields, *Opt. Lett.* **29**, 1942–1944 (2004).

32. A. G. Peele, P. J. McMahon, D. Paterson, C. Q. Tran, A. P. Mancuso, K. A. Nugent, J. P. Hayes, E. Harvey, B. Lai, and I. McNulty, Observation of an x-ray vortex, *Opt. Lett.* **27**, 1752–1754 (2002).

33. A. S. Desyatnikov, Y. S. Kivshar, and L. Torner, Optical vortices and vortex solitons, Ed. E. Wolf, *Progress in Optics*, vol. 47 (North-Holland, Amsterdam, 2005), pp. 291–391.

34. Y. S. Kivshar and G. P. Agrawal, *Optical Solitons: From Fibers to Photonic Crystals* (Academic Press, San Diego, 2003).

35. A. G. Truscott, M. E. J. Friese, N. R. Heckenberg, and H. Rubinsztein-Dunlop, Optically written waveguide in an atomic vapor, *Phys. Rev. Lett.* **82**, 1438–1441 (1999).

36. C. T. Law, X. Zhang, and G. A. Swartzlander, Waveguiding properties of optical vortex solitons, *Opt. Lett.* **25**, 55–57 (2000).

37. M. R. Dennis, PhD thesis (University of Bristol, UK, 2001).

38. I. Freund, Emergent polarization singularities, *Opt. Lett.* **29**, 539–541 (2004).

39. I. Freund, Polarization singularities in optical lattices, *Opt. Lett.* **29**, 875–877 (2004).

40. M. Born and E. Wolf, *Principles of Optics*, 7th (expanded) ed. (Cambridge U. Press, Cambridge, 1999).

41. G. Gbur, T. D. Visser, and E. Wolf, Anomalous behaviour of spectra near phase singularities of focused waves, *Phys. Rev. Lett.* **88**, 013901 (2002).

42. M. Berry, Coloured phase singularities, *New J. Phys.* **4**, 66 (2002).

43. M. Berry, Exploring the colours of dark light, *New J. Phys.* **4**, 74 (2002).

44. J. Leach and M. J. Padgett, Observation of chromatic effects near a while-light vortex, *New J. Phys.* **5**, 154 (2003).

45. A. V. Volyar, Y. A. Egorov, A. F. Rubass, and T. A. Fadeeva, Fine structure of white optical vortices in crystals, *Tech. Phys. Lett.* **30**, 701 (2004).

46. M. S. Soskin, P. V. Polyanskii, and O. O. Arkhelyuk, Computer-synthesized hologram-based rainbow optical vortices, *New J. Phys.* **6**, 196 (2004).

47. O. V. Angelsky, S. G. Hanson, A. P. Maksimyak, and P. P. Maksimyak, On the feasibility for determining the amplitude zeroes in polychromatic fields, *Opt. Express.* **13**, 4396 (2005).

48. S. A. Ponomarenko, A class of partially coherent vortex beams carrying optical vortices, *J. Opt. Soc. Am. A.* **18**, 150 (2001).

49. G. Gbur and T. D. Visser, Coherence vortices in partially coherent beams,

      Opt. Commun. **222**, 117–125 (2003).

50. G. V. Bogatyryova, C. V. Fel'de, P. V. Polyanskii, S. A. Ponomarenko, M. S. Soskin, and E. Wolf, Partially coherent vortex beams with a separable phase, *Opt. Lett.* **28**, 878–880 (2003).

51. H. F. Schouten, G. Gbur, T. D. Visser, and E. Wolf, Phase singularities of the coherence functions in young's interference pattern, *Opt. Lett.* **28**, 968–970 (2003).

52. D. M. Palacios, I. D. Maleev, A. S. Marathay, and G. A. Swartzlander, Spatial correlation singularity of a vortex field, *Phys. Rev. Lett.* **92**, 143905–4 (2004).

53. I. D. Maleev, D. M. Palacios, A. S. Marathay, and G. A. Swartzlander, Spatial correlation vortices in partially coherent light: theory, *J. Opt. Soc. Am. B.* **21**, 1895–1900 (2004).

54. V. Shvedov, W. Krolikowski, A. Volyar, D. N. Neshev, A. S. Desyatnikov, and Y. S. Kivshar, Focusing and correlation properties of white-light optical vortices, *Opt. Express* **13**, 7393 (2005).

55. C. C. Jeng, M. F. Shih, K. Motzek, and Y. Kivshar, Partially incoherent optical vortices in self-focusing nonlinear media, *Phys. Rev. Lett.* **92**, 043904–4 (2004).

56. K. Motzek, Y. S. Kivshar, M.-F. Shih, and G. A. Swartzlander, Spatial coherence singularities and incoherent vortex solitons, *J. Opt. Soc. Am. B.* **22**, 1437 (2005).

57. D. N. Christodoulides, T. H. Coskun, M. Mitchell, and M. Segev, Theory of incoherent self-focusing in biased photorefractive media, *Phys. Rev. Lett.* **78**, 646–649 (1997).

58. C. Anastassiou, M. Soljacic, M. Segev, E. D. Eugenieva, D. N. Christodoulides, D. Kip, Z. H. Musslimani, and J. P. Torres, Eliminating the transverse instabilities of Kerr solitons, *Phys. Rev. Lett.* **85**, 4888–4891 (2000).

59. Y. R. Shen, *The Principles of Nonlinear Optics* (John Wiley and Sons, New York, 1984).

60. P. N. Butcher and D. Cotter, *The elements of Nonlinear Optics* (Cambridge, UK, 1992).

61. R. W. Boyd, *Nonlinear Optics* (Academic Press, San Diego, 1992).

62. Y. S. Kivshar and B. Luther-Davies, Dark optical solitons: physics and applications, *Phys. Rep.* **298**, 81–197 (1998).

63. G. A. Swartzlander and C. T. Law, Optical vortex solitons observed in Kerr nonlinear media, *Phys. Rev. Lett.* **69**, 2503–2506 (1992).

64. A. V. Mamaev, M. Saffman, and A. A. Zozulya, Vortex evolution and bound pair formation in anisotropic nonlinear optical media, *Phys. Rev. Lett.* **77**, 4544–4547 (1996).

65. Z. Chen, M. Segev, D. W. Wilson, R. E. Muller, and P. D. Maker, Self-trapping of an optical vortex by use of the bulk photovoltaic effect, *Phys. Rev. Lett.* **78**, 2948–2951 (1997).

66. M. Segev, Optical spatial solitons, *Opt. Quantum Electron.* **30**, 503–533 (1998).

67. G. I. Stegeman and M. Segev, Optical spatial solitons and their interactions: Universality and diversity, *Science* **286**, 1518–1523 (1999).

68. Y. S. Kivshar and G. I. Stegeman, Spatial optical solitons, *Opt. Photon. News* **13**,59–63 (2002).

69. H. A. Haus, *Appl. Phys. Lett.* **8**, 128 (1966).

70. Z. K. Yanauskas, *Sov. Radiopys.* **9**, 261 (1966).

71. J. M. Soto-Crespo, D. R. Heatley, E. M. Wright, and N. N. Akhmediev, Stability of the higher-bound states in a saturable self-focusing medium, *Phys. Rev. A.* **44**, 636–644 (1991).

72. D. E. Edmundson, Unstable higher modes of a three-dimensional nonlinear Schrödinger equation, *Phys. Rev. E.* **55**), 7636–7644 (1997).

73. V. I. Kruglov and R. A. Vlasov, Spiral self-trapping propagation of optical beams in media with cubic nonlinearity, *Phys. Lett. A.* **111**, 401–404 (1985).

74. Y. S. Kivshar and E. A. Ostrovskaya, Optical vortices: folding and twisting waves of light, *Opt. Photon. News* **12**, 26–31 (2001).

75. Y. S. Kivshar and D. E. Pelinovsky, Self-focusing and transverse instabilities of solitary waves, *Phys. Rep.* **331**, 118–195 (2000).

76. P. S. Lomdahl, O. H. Olsen, and P. L. Christiansen, Return and collapse of solutions to the non-linear Schrödinger-equation in cylindrical symmetry, *Phys. Lett. A.* **78**, 125–128 (1980).

77. V. V. Afanasjev, Rotating ring-shaped bright solitons, *Phys. Rev. E.* **52**, 3153–3158 (1995).

78. C. Anastassiou, C. Pigier, M. Segev, D. Kip, E. D. Eugenieva, and D. N. Christodoulides, Self-trapping of bright rings, *Opt. Lett.* **26**, 911–913 (2001).

79. V. I. Kruglov, Y. A. Logvin, and V. M. Volkov, The theory of spiral laser-beams in nonlinear media, *J. Mod. Opt.* **39**, 2277–2291 (1992).

80. D. V. Skryabin and W. J. Firth, Dynamics of self-trapped beams with phase dislocation in saturable Kerr and quadratic nonlinear media, *Phys. Rev. E* **58**, 3916–3930 (1998).

81. J. Atai, Y. J. Chen, and J. M. Sotocrespo, Stability of 3-dimensional self-trapped beams with a dark spot surrounded by bright rings of varying intensity, *Phys. Rev. A* **49**, R3170–R3173 (1994).

82. L. Torner and D. V. Petrov, Azimuthal instabilities and self-breaking of beams into sets of solitons in bulk second-harmonic generation, *Electron. Lett.* **33**, 608–610 (1997).

83. L. Torner and D. V. Petrov, Splitting of light beams with spiral phase dislocations into solitons in bulk quadratic nonlinear media, *J. Opt. Soc. Am. B* **14**, 2017–2023 (1997).

84. W. J. Firth and D. V. Skryabin, Optical solitons carrying orbital angular momentum, *Phys. Rev. Lett.* **79**, 2450–2453 (1997).

85. L. Allen, Introduction to the atoms and angular momentum of light special issue, *J. Opt. B: Quantum Semicl. Opt.* **4**, S1–S6 (2002).

86. M. Berry, M. Dennis, and M. Soskin, The plurality of optical singularities, *J. Opt. A-Pure Appl. Opt.* **6**, S155–S156 (2004).

87. M. Padgett, J. Courtial, and L. Allen, Light's orbital angular momentum, *Phys. Today* **57**, 35–40 (2004).

88. C. Cohen-Tannoudji, J. Dupont-Roc, and G. Grynberg, *Photons and Atoms: Introduction to Quantum Electrodynamics* (Wiley, New York, 1989).

89. S. M. Barnett, Optical angular-momentum flux, *J. Opt. B: Quantum Semicl. Opt.* **4**, S7–S16 (2002).

90. L. Berge, Wave collapse in physics: principles and applications to light and plasma waves, *Phys. Rep.* **303**, 260–370 (1998).

91. N. G. Vakhitov and A. A. Kolokolov, Stationary solutions of the wave equation in the medium with nonlinearity saturation, *Izv. Vyssh. Uchebn. Zaved. Radiofiz.* **16**, 1020–1028 (1973).

92. J. M. Soto-Crespo, E. M. Wright, and N. N. Akhmediev, Recurrence and azimuthal-symmetry breaking of a cylindrical gaussian beam in a saturable self-focusing medium, *Phys. Rev. A.* **45**, 3168–3175 (1992).

93. J. P. Torres, J. M. Soto-Crespo, L. Torner, and D. V. Petrov, Solitary-wave vortices in quadratic nonlinear media, *J. Opt. Soc. Am. B.* **15**, 625–627 (1998).

94. V. Tikhonenko, J. Christou, and B. Luther-Davies, Spiraling bright spatial solitons formed by the breakup of an optical vortex in a saturable self-focusing medium, *J. Opt. Soc. Am. B* **12**, 2046–2052 (1995).

95. V. Tikhonenko, J. Christou, and B. Luther-Davies, Three dimensional bright spatial soliton collision and fusion in a saturable nonlinear medium, *Phys. Rev. Lett.* **76**, 2698–2701 (1996).

96. M. S. Bigelow, P. Zerom, and R. W. Boyd, Breakup of ring beams carrying orbital angular momentum in sodium vapor, *Phys. Rev. Lett.* **92**, 083902–4 (2004).

97. Z. Chen, M.-F. Shih, M. Segev, D. W. Wilson, R. E. Muller, and P. D. Maker, Steady-state vortex-screening solitons formed in biased photorefractive media, *Opt. Lett.* **22**, 1751–1753 (1997).

98. D. V. Petrov, L. Torner, J. Martorell, R. Vilaseca, J. P. Torres, and C. Cojocaru, Observation of azimuthal modulational instability and formation of patterns of optical solitons in a quadratic nonlinear crystal, *Opt. Lett.* **23**, 1444–1446 (1998).

99. A. I. Yakimenko, Y. A. Zaliznyak, and Y. S. Kivshar, Stable vortex solitons in nonlocal self-focusing nonlinear media, *Phys. Rev. E* **71**, 065603 (2005).

100. D. Briedis, D. E. Petersen, D. Edmundson, W. Krolikowski, and O. Bang, Ring vortex solitons in nonlocal nonlinear media, *Opt. Express* **13**, 435 (2005).

101. Y. V. Kartashov, V. A. Vysloukh, and L. Torner, Stable ring-profile vortex solitons in Bessel optical lattices, *Phys. Rev. Lett.* **94**, 043902–4 (2005).

102. C. Rotschild, O. Cohen, O. Manela, M. Segev, and T. Carmon, Solitons in nonlinear media with an infinite range of nonlocality: First observation of coherent elliptic solitons and of vortex-ring solitons, *Phys. Rev. Lett.* **95**, 213904 (2005).

103. L. Poladian, A. W. Snyder, and D. J. Mitchell, Spiraling spatial solitons, *Opt. Commun.* **85**, 59–62 (1991).

104. V. V. Steblina, Y. S. Kivshar, and A. V. Buryak, Scattering and spiraling of solitons in a bulk quadratic medium, *Opt. Lett.* **23** , 156–158 (1998).

105. A. V. Buryak and V. V. Steblina, Soliton collisions in bulk quadratic media: Comprehensive analytical and numerical study, *J. Opt. Soc. Am. B* **16**, 245–255 (1999).

106. C. Simos, V. Coudrec, and A. Barthelemy, Expriences sur les interactions de type particulaire de solitons multicolores: fusion, rpulsion, enroulement en spirale, *J. Phys. IV.* **12**, 211–212 (2002).

107. M.-F. Shih, M. Segev, and G. Salamo, Three-dimensional spiraling of interacting spatial solitons, *Phys. Rev. Lett.* **78**, 2551–2554 (1997).

108. A. S. Desyatnikov and A. I. Maimistov, Interaction of two spatially separated light beams in a nonlinear Kerr medium, *JETP* **86**, 1101–1106 (1998).

109. J. Schjodt-Eriksen, M. R. Schmidt, J. J. Rasmussen, P. L. Christiansen, Y. B. Gaididei, and L. Berge, Two-beam interaction in saturable media, *Phys. Lett. A* **246**, 423–428 (1998).

110. D. V. Skryabin, J. M. McSloy, and W. J. Firth, Stability of spiralling solitary waves in hamiltonian systems, *Phys. Rev. E.* **66**, 055602–4 (2002).

111. W. Krolikowski, M. Saffman, B. Luther-Davies, and C. Denz, Anomalous interaction of spatial solitons in photorefractive media, *Phys. Rev. Lett.* **80**, 3240–3243 (1998).

112. W. Krolikowski, B. Luther-Davies, C. Denz, J. Petter, C. Weilnau, A. Stepken, and M. Belic, Interaction of two-dimensional spatial incoherent solitons in photorefractive medium, *Appl. Phys. B.* **68**, 975–982 (1999).

113. C. Denz, W. Krolikowski, J. Petter, C. Weilnau, T. Tschudi, M. R. Belic, F. Kaiser, and A. Stepken, Dynamics of formation and interaction of photorefractive screening solitons, *Phys. Rev. E.* **60**, 6222–6225 (1999).

114. M. R. Belic, A. Stepken, and F. Kaiser, Spatial screening solitons as particles, *Phys. Rev. Lett.* **84**, 83–86 (2000).

115. A. Barthelemy, C. Froehly, M. Shalaby, P. Donnat, J. Paye, and A. Migus, *Ultrafast Phenomena VIII* (Springer, Berlin, 1993), pp. 299–305.

116. A. Barthelemy, C. Froehly, and M. Shalaby, *Proc. SPIE Int. Soc. Opt. Eng.* **2041**, 104 (1994).

117. M. Soljacic, S. Sears, and M. Segev, Self-trapping of "necklace" beams in self-focusing Kerr media, *Phys. Rev. Lett.* **81**, 4851–4854 (1998).

118. M. Soljacic and M. Segev, Self-trapping of "necklace-ring" beams in self-focusing Kerr media, *Phys. Rev. E* **62**, 2810–2820 (2000).

119. M. Soljacic and M. Segev, Integer and fractional angular momentum borne on self-trapped necklace-ring beams, *Phys. Rev. Lett.* **86**, 420–423 (2001).

120. A. S. Desyatnikov and Y. S. Kivshar, Rotating optical soliton clusters, *Phys. Rev. Lett.* **88**, 053901–4 (2002).

121. Y. V. Kartashov, G. Molina-Terriza, and L. Torner, Multicolor soliton clusters, *J. Opt. Soc. Am. B* **19**, 2682–2691 (2002).

122. Y. V. Kartashov, L. C. Crasovan, D. Mihalache, and L. Torner, Robust propagation of two-color soliton clusters supported by competing nonlinearities, *Phys. Rev. Lett.* **89**, 273902–4 (2002).

123. D. Mihalache, D. Mazilu, L. C. Crasovan, B. A. Malomed, F. Lederer, and L. Torner, Robust soliton clusters in media with competing cubic and quintic nonlinearities, *Phys. Rev. E* **68**, 046612–8 (2003).

124. A. Desyatnikov, A. Maimistov, and B. Malomed, Three-dimensional spinning solitons in dispersive media with the cubic-quintic nonlinearity, *Phys. Rev. E.* **61**, 3107–3113 (2000).

125. L. C. Crasovan, Y. V. Kartashov, D. Mihalache, L. Torner, Y. S. Kivshar, and V. M. Perez-Garcia, Soliton "molecules": Robust clusters of spatiotemporal optical solitons, *Phys. Rev. E* **67**, 046610–5 (2003).

126. D. Mihalache, D. Mazilu, L. C. Crasovan, A. Malomed, F. Lederer, and L. Torner, Soliton clusters in three-dimensional media with competing cubic and quintic nonlinearities, *J. Opt. B: Quantum Semicl. Opt.* **6**, S333–S340 (2004).

127. A. S. Desyatnikov, A. A. Sukhorukov, and Y. S. Kivshar, Azimuthons: Spatially modulated vortex solitons, *Phys. Rev. Lett.* **95**, 203904 (2005).

128. W. Krolikowski, O. Bang, N. I. Nikolov, D. Neshev, J. Wyller, J. Rasmussen, and D. Edmundson, Modulational instability, solitons, and beam propagation in spatially nonlocal nonlinear media, *J. Opt. B: Quantum Semicl. Opt.* **6**, S288 (2004).

129. O. Bang, W. Krolikowski, J. Wyller, and J. J. Rasmussen, Collapse arrest and soliton stabilization in nonlocal nonlinear media, *Phys. Rev. E* **66**, 046619 (2002).

130. M. Peccianti, K. A. Brzdakiewicz, and G. Assanto, Nonlocal spatial soliton interactions in nematic liquid crystals, *Opt. Lett.* **27**, 1460 (2002).

131. N. I. Nikolov, D. Neshev, W. Krolikowski, O. Bang, J. J. Rasmussen, and P. L. Christiansen, Attraction of nonlocal dark optical solitons, *Opt. Lett.* **29**, 286 (2004).

132. Z. Xu, Y. V. Kartashov, and L. Torner, Upper threshold for stability of multipole-mode solitons in nonlocal nonlinear media, *Opt. Lett.* **30**, 3171 (2005).

133. D. Suter and T. Blasberg, Stabilization of transverse solitary waves by a nonlocal response of the nonlinear medium, *Phys. Rev. A* **48**, 4583 (1993).

134. C. Conti, M. Peccianti, and G. Assanto, Route to nonlocality and observation of accessible solitons, *Phys. Rev. Lett.* **91**, 073901 (2003).

135. S. Lopez-Aguayo, A. S. Desyatnikov, Y. S. Kivshar, S. Skupin, W. Krolikowski, and O. Bang, Stable rotating dipole solitons in nonlocal optical media, *Opt. Lett.* **31**, 1100 (2006).

136. S. Lopez-Aguayo, A. S. Desyatnikov, and Y. S. Kivshar, Azimuthons in nonlocal nonlinear media, *Opt. Express* **14**, in press (2006).

137. D. N. Christodoulides, F. Lederer, and Y. Silberberg, Discretizing light behaviour in linear and nonlinear waveguide lattices, *Nature* **424**, 817–823 (2003).

138. J. W. Fleischer, T. Carmon, M. Segev, N. K. Efremidis, and D. N. Christodoulides, Observation of discrete solitons in optically induced real time waveguide arrays, *Phys. Rev. Lett.* **90**, 023902–4 (2003).

139. D. Neshev, E. Ostrovskaya, Y. Kivshar, and W. Krolikowski, Spatial solitons in optically induced gratings, *Opt. Lett.* **28**, 710–712 (2003).

140. J. W. Fleischer, M. Segev, N. K. Efremidis, and D. N. Christodoulides, Observation of two-dimensional discrete solitons in optically induced nonlinear

photonic lattices, *Nature* **422**, 147–150 (2003).

141. M. Johansson, S. Aubry, Y. B. Gaididei, P. L. Christiansen, and K. O. Rasmussen, Dynamics of breathers in discrete nonlinear Schrödinger models, *Physica D.* **119**, 115–124 (1998).

142. B. A. Malomed and P. G. Kevrekidis, Discrete vortex solitons, *Phys. Rev. E.* **64**, 026601–6 (2001).

143. P. G. Kevrekidis, B. A. Malomed, A. R. Bishop, and D. J. Frantzeskakis, Localized vortices with a semi-integer charge in nonlinear dynamical lattices, *Phys. Rev. E* **65**, 016605–7 (2002).

144. P. G. Kevrekidis, B. A. Malomed, and Y. B. Gaididei, Solitons in triangular and honeycomb dynamical lattices with the cubic nonlinearity, *Phys. Rev. E* **66**, 016609–10 (2002).

145. P. G. Kevrekidis, B. A. Malomed, Z. G. Chen, and D. J. Frantzeskakis, Stable higher-order vortices and quasivortices in the discrete nonlinear Schrödinger equation, *Phys. Rev. E* **70**, 056612–9 (2004).

146. P. G. Kevrekidis, B. A. Malomed, D. J. Frantzeskakis, and R. Carretero-Gonzalez, Three-dimensional solitary waves and vortices in a discrete nonlinear Schrödinger lattice, *Phys. Rev. Lett.* **93**, 080403–4 (2004).

147. J. Yang and Z. H. Musslimani, Fundamental and vortex solitons in a two-dimensional optical lattice, *Opt. Lett.* **28**, 2094–2096 (2003).

148. B. B. Baizakov, B. A. Malomed, and M. Salerno, Multidimensional solitons in periodic potentials, *Europhys. Lett.* **63**, 642–648 (2003).

149. J. Yang, Stability of vortex solitons in a photorefractive optical lattice, *New J. Phys.* **6**, 47 (2004).

150. D. N. Neshev, T. J. Alexander, E. A. Ostrovskaya, Y. S. Kivshar, H. Martin, I. Makasyuk, and Z. G. Chen, Observation of discrete vortex solitons in optically induced photonic lattices, *Phys. Rev. Lett.* **92**, 123903–4 (2004).

151. J. W. Fleischer, G. Bartal, O. Cohen, O. Manela, M. Segev, J. Hudock, and D. N. Christodoulides, Observation of vortex-ring "discrete" solitons in 2D photonic lattices, *Phys. Rev. Lett.* **92**), 123904–4 (2004).

152. N. K. Efremidis, S. Sears, D. N. Christodoulides, J. W. Fleischer, and M. Segev, Discrete solitons in photorefractive optically induced photonic lattices, *Phys. Rev. E* **66**, 046602–5 (2002).

153. J. K. Yang, I. Makasyuk, A. Bezryadina, and Z. Chen, Dipole solitons in optically induced two-dimensional photonic lattices, *Opt. Lett.* **29**, 1662–1664 (2004).

154. O. Manela, O. Cohen, G. Bartal, J. W. Fleischer, and M. Segev, Two-dimensional higher-band vortex lattice solitons, *Opt. Lett.* **29**, 2049–2051 (2004).

155. Z. G. Chen, H. Martin, E. D. Eugenieva, J. J. Xu, and A. Bezryadina, Anisotropic enhancement of discrete diffraction and formation of two-dimensional discrete-soliton trains, *Phys. Rev. Lett.* **92**, 143902–4 (2004).

156. Y. V. Kartashov, V. A. Vysloukh, and L. Torner, Soliton trains in photonic lattices, *Opt. Express.* **12**, 2831–2837 (2004).

157. P. G. Kevrekidis, B. A. Malomed, and A. R. Bishop, Bound states of two-dimensional solitons in the discrete nonlinear Schrödinger equation, *J. Phys.*

     A. **34**, 9615–9629 (2001).

158. Y. V. Kartashov, A. A. Egorov, L. Torner, and D. N. Christodoulides, Stable soliton complexes in two-dimensional photonic lattices, *Opt. Lett.* **29**, 1918–1920 (2004).

159. T. J. Alexander, A. A. Sukhorukov, and Y. S. Kivshar, Asymmetric vortex solitons in nonlinear periodic lattices, *Phys. Rev. Lett.* **93**, 063901–4 (2004).

160. A. Ferrando, Discrete-symmetry vortices as angular Bloch modes, *Phys. Rev. E* **72**, 036612 (2005).

161. A. Ferrando, M. Zacarés, and M.-A. García-March, Vorticity cutoff in nonlinear photonic crystals, *Phys. Rev. Lett.* **95**, 043901 (2005).

162. Y. V. Kartashov, V. A. Vysloukh, and L. Torner, Rotary solitons in Bessel optical lattices, *Phys. Rev. Lett.* **93**, 093904–4 (2004).

163. Y. V. Kartashov, A. A. Egorov, V. A. Vysloukh, and L. Torner, Rotary dipole-mode solitons in Bessel optical lattices, *J. Opt. B: Quantum Semicl. Opt.* **6**, 444–447 (2004).

164. Y. V. Kartashov, A. A. Egorov, V. A. Vysloukh, and L. Torner, Stable soliton complexes and azimuthal switching in modulated Bessel optical lattices, *Phys. Rev. E.* **70**, 065602–4 (2004).

165. Y. V. Kartashov, A. Ferrando, A. A. Egorov, and L. Torner, Soliton topology versus discrete symmetry in optical lattices, *Phys. Rev. Lett.* **95**, 123902 (2005).

166. R. Fischer, D. N. Neshev, S. Lopez-Aguayo, A. S. Desyatnikov, A. A. Sukhorukov, W. Krolikowski, and Y. S. Kivshar, Observation of light localization in modulated Bessel optical lattices, *Opt. Express* **14**, 2825 (2006).

167. A. Ferrando, M. Zacarés, P. F. de Córdoba, D. Binosi, and J. Monsoriu, Vortex solitons in photonic crystal fibers, *Opt. Express* **12**, 817 (2004).

168. A. Ferrando, M. Zacarés, P. Andreés, P. F. de Córdoba, and J. Monsoriu, Nodal solitons and the nonlinear breaking of discrete symmetry, *Opt. Express* **13**, 1072 (2005).

169. A. Ferrando, M. Zacarés, M.-A. García-March, J. A. Monsoriu, and P. F. de Córdoba, Vortex transmutation, *Phys. Rev. Lett.* **95**, 123901 (2005).

170. A. Bezryadina, E. Eugenieva, and Z. Chen, Self-trapping and flipping of double-charged vortices in optically induced photonic lattices, *Opt. Lett.* **31**, 2458 (2006).

171. G. Molina-Terriza, E. M. Wright, and L. Torner, Propagation and control of noncanonical optical vortices, *Opt. Lett.* **26**, 163 (2001).

172. G. Molina-Terriza, J. Recolons, J. P. Torres, L. Torner, and E. M. Wright, Observation of the dynamical inversion of the topological chage of an optical vortex, *Phys. Rev. Lett.* **87**, 023902 (2001).

173. A. Y. Bekshaev, M. S. Soskin, and M. V. Vasnetsov, Transformation of higher-order optical vortices upon focusing by an astigmatic lens, *Opt. Commun.* **241**, 237 (2001).

174. A. Bezryadina, D. N. Neshev, A. S. Desyatnikov, J. Young, Z. Chen, and Y. S. Kivshar, Observation of topological transformations of optical vortices in two-dimensional photonic lattices, *Opt. Express* **14**, in press (2006).

175. M. Öster and M. Johansson, Stable stationary and quasi-periodic discrete

vortex-breathers with topological charge s = 2, *Phys. Rev. E* **73**, 066608 (2006).

176. V. L. Ginzburg and L. D. Landau, *Zh. Éksp. Teor. Fiz.* **20**, 1064 (1950).

177. M. Cyrot, Ginzburg-Landau theory for superconductors, *Rep. Prog. Phys.* **36**, 103–158 (1973).

178. A. A. Abrikosov, *JETP* **5**, 1174 (1957).

179. M. C. Cross and P. C. Hohenberg, Pattern formation outside of equilibrium, *Rev. Mod. Phys.* **65**, 851–1112 (1993).

180. Y. N. Ovchinnikov and I. M. Sigal, Ginzburg-Landau equation in static vortices, *Partial Differential Equations and their Applications* **12**, 199–220 (1997).

181. Y. N. Ovchinnikov and I. M. Sigal, The Ginzburg-Landau equation III. vortex dynamics, *Nonlinearity* **11**, 1277–1294 (1998).

182. Y. N. Ovchinnikov and I. M. Sigal, Long-time behaviour of Ginzburg-Landau vortices, *Nonlinearity* **11**, 1295–1309 (1998).

183. Y. N. Ovchinnikov and I. M. Sigal, The energy of Ginzburg-Landau vortices, *Eur. J. Appl. Math.* **13**, 153–178 (2002).

184. I. V. Barashenkov, V. S. Shchesnovich, and R. M. Adams, Noncoaxial multivortices in the complex Sine-Gordon theory on the plane, *Nonlinearity* **15**, 2121–2146 (2002).

185. Y. N. Ovchinnikov and I. M. Sigal, Non-radially symmetric solutions to the Ginzburg-Landau equation, *preprint* (2000).

186. R. Graham and H. Haken, Laserlight - first example of a second-order transition far away from thermal equilibrium, *Z Phys* **237**, 31 (1970).

187. P. Coullet, L. Gil, and F. Rocca, Optical vortices, *Opt. Commun.* **73**, 403–408 (1989).

188. C. Tamm and C. O. Weiss, Bistability and optical switching of spatial patterns in a laser, *J. Opt. Soc. Am. B.* **7**, 1034–1038 (1990).

189. C. O. Weiss, M. Vaupel, K. Staliunas, G. Slekys, and V. B. Taranenko, Solitons and vortices in lasers, *Appl. Phys. B* **68**, 151–168 (1999).

190. N. N. Rozanov and G. V. Khodova, Switching of a bistable interferometer on localized inhomogeneities, *Opt. Spektrosk.* **65**, 1399–1401 (1988).

191. N. N. Rozanov and S. V. Fedorov, Diffraction waves of switching and autosolitons in a sturable-absorber laser, *Opt. Spektrosk.* **72**, 1394–1399 (1992).

192. P. Mandel, *Theoretical Problems in Cavity Nonlinear Optics* (Cambridge Uni. Press, 1997).

193. H. M. Gibbs, *Optical Bistability: Controlling Light with Light* (Academic Press, New York, 1985).

194. N. N. Rosanov, *Spatial Hysteresis and Optical Patterns* (Springer, Berlin, 2002).

195. V. V. Afanasjev, B. A. Malomed, and P. L. Chu, Stability of bound states of pulses in the Ginzburg-Landau equations, *Phys. Rev. E* **56** , 6020–6025 (1997).

196. P. L. Ramazza, E. Benkler, U. Bortolozzo, S. Boccaletti, S. Ducci, and F. T. Arecchi, Tailoring the profile and interactions of optical localized structures, *Phys. Rev. E* **65**, 066204–4 (2002).

197. B. Schapers, M. Feldmann, T. Ackemann, and W. Lange, Interaction of localized structures in an optical pattern-forming system, *Phys. Rev. Lett.* **85**, 748–751 (2000).

198. B. Schapers, T. Ackemann, and W. Lange, Properties of feedback solitons in a single-mirror experiment, *IEEE J. Quantum Electron.* **39** , 227–237 (2003).

199. M. Tlidi, A. G. Vladimirov, and P. Mandel, Interaction and stability of periodic and localized structures in optical bistable systems, *IEEE J. Quantum Electron.* **39**, 216–226 (2003).

200. A. G. Vladimirov, J. M. McSloy, D. V. Skryabin, and W. J. Firth, Two-dimensional clusters of solitary structures in driven optical cavities, *Phys. Rev. E* **65**, 046606–11 (2002).

201. B. A. Malomed, Bound solitons in the nonlinear Schrödinger-Ginzburg-Landau equation, *Phys. Rev. A* **44**, 6954–6957 (1991).

202. B. A. Malomed, Potential of interaction between two- and three-dimensional solitons, *Phys. Rev. E* **58**, 7928–7933 (1998).

203. F. Prati, M. Brambilla, and L. A. Lugiato, Pattern formation in lasers, *Rivista del Nuovo Cimento* **17**, 1–85 (1994).

204. L. A. Lugiato, M. Brambilla, and A. Gatti, Optical pattern formation, *Advances in atomic, molecular, and optical physics* vol. 40, (Academic Press, 1999), pp. 229–306.

205. F. T. Arecchi, S. Boccaletti, and P. Ramazza, Pattern formation and competition in nonlinear optics, *Phys. Rep.* **318**, 1–83 (1999).

206. N. N. Rozanov, Dissipative optical solitons, *Usp. Fiz. Nauk* **170**, 462–465 (2000).

207. W. J. Firth and C. O. Weiss, Cavity and feedback solitons, *Opt Photonics News* **13**, 54–58 (2002).

208. U. Peschel, D. Michaelis, and C. O. Weiss, Spatial solitons in optical cavities, *IEEE J. Quantum Electron.* **39**, 51–64 (2003).

209. L. A. Lugiato, Introduction to the feature section on cavity solitons: An overview, *IEEE J. Quantum Electron.* **39**, 193–196 (2003).

210. P. Mandel and M. Tlidi, Transverse dynamics in cavity nonlinear optics (2000-2003), *J. Opt. B: Quantum Semicl. Opt.* **6**, R60–R75 (2004).

211. N. N. Rosanov, Mechanics of diffraction autosolitons: Interaction with wavefront dislocations and formation of multiparticle configurations, *Opt Spectrosc Engl Trans* **73**, 187–193 (1992).

212. G. L. Oppo, A. J. Scroggie, and W. J. Firth, Characterization, dynamics and stabilization of diffractive domain walls and dark ring cavity solitons in parametric oscillators, *Phys. Rev. E* **63**, 066209–16 (2001).

213. S. Barland, M. Branbilla, L. Columbo, L. Furfaro, M. Giudici, X. Hachair, R. Kheradmand, L. A. Lugiato, T. Maggipinto, G. Tissoni, and J. Tredicce, Cavity solitons in a vcsel: reconfigurable micropixel arrays, *Europhys. News* **34** (2003).

214. S. Barland, J. R. Tredicce, M. Branbilla, L. A. Lugiato, S. Balles, M. Giudici, T. Maggipinto, L. Spinelli, G. Tissoni, T. Knödl, M. Miller, and R. Jäger, Cavity solitons as pixels in semiconductor microcavities, *Nature* **419**, 699–

702 (2002).

215. P. Debernardi, G. P. Bava, F. M. di Sopra, and M. B. Willemsen, Features of vectorial modes in phase-coupled vcsel arrays: Experiments and theory, *IEEE J. Quantum Electron.* **39**, 109–119 (2003).

216. T. Maggipinto, M. Brambilla, and W. J. Firth, Characterization of stationary patterns and their link with cavity solitons in semiconductor microresonators, *IEEE J. Quantum Electron.* **39**, 206–215 (2003).

217. K. J. Vahala, Optical microcavities, *Nature* **424**, 839–846 (2003).

218. C. O. Weiss, K. Staliunas, M. Vaupel, V. B. Taranenko, G. Slekys, and Y. Larionova, Vortices and spatial solitons in optical resonators, and the relations to other fields of physics, In: *World Sc. Lectures in Compl. Syst.*, eds. R. Ball and N. Akhmediev, (World Scientific, Singapore, 2003), pp. 315–368.

219. N. N. Rozanov, A. V. Fedorov, S. V. Fedorov, and G. V. Khodova, New types of localized structures in laser radiation, *Opt. Spektrosk.* **79**, 868–870 (1995).

220. S. V. Fedorov, N. N. Rosanov, A. N. Shatsev, N. A. Veretenov, and A. G. Vladimirov, Topologically multicharged and multihumped rotating solitons in wide-aperture lasers with a saturable absorber, *IEEE J. Quantum Electron.* **39**, 197–205 (2003).

221. N. N. Rozanov, S. V. Fedorov, and A. N. Shatsev, Nonstationary multivortex and fissionable soliton-like structures of laser radiation, *Opt Spectrosc Engl Trans* **95**, 843–848 (2003).

222. N. N. Rozanov, S. V. Fedorov, A. N. Shatsev, and N. A. Loiko, A three-soliton structure with strong and weak coupling in a wide-aperture laser with saturable absorption, *Opt Spectrosc Engl Trans* **97**, 88–90 (2004).

223. N. N. Rozanov, S. V. Fedorov, and A. N. Shatsev, Energy-flow patterns and bifurcations of two-dimensional cavity solitons, *JETP* **98**, 427–437 (2004).

224. N. N. Rosanov, S. V. Fedorov, and A. N. Shatsev, Curvilinear motion of multivortex laser-soliton complexes with strong and weak coupling, *Phys. Rev. Lett.* **95**, 053903 (2005).

225. L. C. Crasovan, B. A. Malomed, and D. Mihalache, Erupting, flat-top, and composite spiral solitons in the two-dimensional Ginzburg-Landau equation, *Phys. Lett. A* **289**, 59–65 (2001).

226. L. C. Crasovan, B. A. Malomed, and D. Mihalache, Stable vortex solitons in the two-dimensional Ginzburg-Landau equation, *Phys. Rev. E* **63**, 016605–6 (2001).

227. D. V. Skryabin and A. G. Vladimirov, Vortex induced rotation of clusters of localized states in the complex Ginzburg-Landau equation, *Phys. Rev. Lett.* **89**, 044101–4 (2002).

228. P. L. Ramazza, U. Bortolozzo, and L. Pastur, Phase singularities in triangular dissipative solitons, *J. Opt. A-Pure Appl. Opt.* **6**, S266–S270 (2004).

229. U. Bortolozzo, L. Pastur, P. L. Ramazza, M. Tlidi, and G. Kozyreff, Bistability between different localized structures in nonlinear optics, *Phys. Rev. Lett.* **93**, 253901–4 (2004).

230. J. V. Moloney, Self-focusing optical turbulence, *Phys. Rev. Lett.* **53**, 556

(1984).

231. J. V. Moloney, Many-parameter routes to optical turbulence, *Phys. Rev. A* **33**, 4061 (1986).

232. W. J. Firth, R. G. Harrison, and I. A. Al-Saidi, Instabilities and routes to chaos in passive all-optical resonators containing a molecular gas, *Phys. Rev. A* **33**, 2449 (1986).

233. L. A. Lugiato, F. Prati, L. M. Narducci, P. Ru, J. R. Tredicce, and D. K. Bandy, Role of transverse effects in laser instabilities, *Phys. Rev. A* **37**, 3847 (1988).

234. M. Brambilla, F. Battipede, L. A. Lugiato, V. Penna, F. Prati, C. Tamm, and C. O. Weiss, Transverse laser patterns. I. Phase singularity crystals, *Phys. Rev. A* **43**, 5090 (1991).

235. M. Brambilla, L. A. Lugiato, V. Penna, F. Prati, C. Tamm, and C. O. Weiss, Transverse laser patterns. II. Variational principle for pattern selection, spatial multistability, and laser hydrodynamics, *Phys. Rev. A* **43**, 5114 (1991).

236. M. Brambilla, M. Cattaneo, L. A. Lugiato, R. Pirovano, F. Prati, A. J. Kent, G.-L. Oppo, A. B. Coates, C. O. Weiss, C. Green, E. J. D'Angelo, and J. R. Tredicce, Dynamical transverse laser patterns. I. Theory, *Phys. Rev. A* **49**, 1427 (1994).

237. A. B. Coates, C. O. Weiss, C. Green, E. J. D'Angelo, and J. R. Tredicce, M. Brambilla, M. Cattaneo, L. A. Lugiato, R. Pirovano, F. Prati, A. J. Kent, and G.-L. Oppo, Dynamical transverse laser patterns. I. Experiments, *Phys. Rev. A* **49**, 1452 (1994).

238. D. Hochheiser and J. V. Moloney and J. Lega, Controlling optical turbulence, *Phys. Rev. A* **55**, R4011 (1997).

239. J. V. Moloney, Spontaneous generation of patterns and their control in nonlinear optical systems, *J. Opt. B: Quantum Semiclass. Opt.* **1**, 183 (1999).